S0-DUV-108

The Future of Nonfuel Minerals in the U.S. and World Economy

The Future of Nonfuel Minerals in the U.S. and World Economy

Input-Output Projections, 1980–2030

Wassily Leontief
James C.M. Koo
Sylvia Nasar
Ira Sohn
Institute for Economic Analysis
New York University

LexingtonBooks
D.C. Heath and Company
Lexington, Massachusetts
Toronto

Library of Congress Cataloging in Publication Data
Main entry under title:

The future of nonfuel minerals in the U.S. and world economy.

Includes index.
1. Mineral industries—United States. 2. Mineral industries. 3. Input-output analysis—United States. 4. Input-output analysis. I. Leontief, Wassily W., 1906-
HD9506.U62F87 1983 338.2'0973 82-48956
ISBN 0-669-06377-0

 This book was prepared with the support of NSF Grant AER-77-14602. However, any opinions, findings, conclusions, or recommendations herein are those of the authors and do not necessarily reflect the views of the National Science Foundation.

Published simultaneously in Canada

Printed in the United States of America

International Standard Book Number: 0-669-06377-0

Library of Congress Catalog Card Number: 82-48956

Contents

Figures

Tables

Preface and Acknowledgments

This book is the product of research that began several years ago. Although its gestation period was comparatively long, at least for a research project in economics, it must be considered analogous to the construction of a large infrastructure project such as an interstate highway system or a large hydroelectric power plant in the back country. Like any such major undertaking, the project was delayed by design errors, construction mishaps, above-normal lags in the delivery of construction materials, labor-management disputes, and, as a result of all of these problems, egregious cost overruns. Although the primary methodological technique used in the project—input-output analysis—is a well-proven tool of economic analysis, it, too—not unlike the tools and construction materials used in remote development sites—had to be adapted to the untested, and often inclement, local environment. Today, thirty years after construction began, no fair-minded citizen doubts the need for and usefulness of the interstate highway system, even though there is a definite lack of consensus on the way that the network of roads is maintained and improved. We can only hope that our work will share a similar fate.

Our research efforts were double-pronged: the first aim was methodologically or conceptually oriented; the second was empirical or practical. The former poses the intellectual challenge of improving on the description, and incorporation into a single, internally consistent framework, of physical data that too often remain in the domain of the physical sciences. In fact, these materials are the stuff of which our standard of living is made. Our empirical effort attempts to provide some guidance to those who will design and implement national and even global policies concerning future economic growth and development into the first thirty years of the twenty-first century.

We are grateful to Professor Michael B. Bever of the Massachusetts Institute of Technology who provided estimates of the prospective changes in materials use, sector by sector and mineral by mineral, for the U.S. economy to the year 2000. We extend our thanks and appreciation to Ernest Battifarano, who, for the U.S. model, assisted in the mathematical formulation, prepared all the programs, and executed all the computations—all a number of times. We wish to thank John Kambhu for his important contribution; he replaced information in the North American region of the World Model with disaggregated information for Canada and the United States.

We are indebted to Jorge Mariscal for his assistance in developing the methodology used to incorporate the twenty new nonfuel minerals into the World Model framework and for seeing the computations of the global projections of minerals through to their completion. We would like to acknowledge Richard Kleinberg's work in preparing the projected vectors of final demand used in the U.S. model and developing methodology used in the analysis of the projections. We are grateful to the other members of the computer staff, in particular Eva Chan and Bryan Woodley, both of whom worked under the general supervision of Daniel Szyld.

We also wish to acknowledge the role of other graduate and undergraduate assistants who were employed on the project at various times and in various capacities: Raymond Antes, Paloma Avila, Daniel Berg, Debesh Chakraborty, Anne Johnson, David Ly, Thijs ten Raa, and Anthony Small. Special thanks are extended to Yvette Ballard who organized the bibliography and helped in the preparation of the final draft of the study.

We acknowledge those, both in the private and public sector, who provided technical information and insights into the workings of the minerals sectors in the economy: the commodity experts of the Divisions of Ferrous, Nonferrous and Nonmetallic Minerals at the U.S. Bureau of Mines; Norman Saunders and Arthur Andreassen of the Office of Economic Growth in the Bureau of Labor Statistics; Harry Postner of the Economic Council of Canada; Statistics Canada and The Canadian Department of Mines; ASARCO, Inc.; Federated Metals Co., Inc. (a subsidiary of ASARCO); Revere Brass and Copper Co., Inc.; Phelps-Dodge, Inc.; and INCO, Inc.

In the early stages of the project, we sought and received technical advice on the treatment of minerals in the national economy from the members of our Advisory Committee: Anne Carter of Brandeis University; Leonard Fischman, then at Resources for the Future; Gary Kingston of the U.S. Bureau of Mines; and Hans Thalheim of ASARCO, Inc. To all of them, we are grateful.

Major support for this research was provided by a grant from the National Science Foundation (AER 77-14602). Supplementary funding was provided under a contract from the U.S. Bureau of Mines (Contract number J0188147). We would also like to thank individuals at the National Science Foundation and the U.S. Bureau of Mines for their patience and encouragement throughout the tenure of the study. They include Lynn Pollnow, Program Director at the National Science Foundation; and John Connelly, Contracts Specialist; William Mo and K.L. Wang of the Branch of Economic Analysis, all of the U.S. Bureau of Mines. We would also like to acknowledge the Control Data Corporation for its generous support of our research activities over the past two years.

Last and, of course, not least, we are grateful to Holly Lammers and Mimi Forrest, who, under the supervision of Mary Parker, prepared the final manuscript for publication.

Needless to say, the views and conclusions contained in this book are those of the authors and should not be interpreted as necessarily representing the official policies, views, or recommendations of the National Science Foundation, the Interior Department's Bureau of Mines, or the U.S. government.

Part I
Nonfuel Minerals in the U.S. Economy

1 Introduction

Seven years ago, members of the National Commission on Supplies and Shortages (among whom were Donald Rice, president of The Rand Corporation; Alan Greenspan, chairman of the Council of Economic Advisers; James Lynn, director of the Office of Management and Budget; and Professor Hendrik Houthakker of Harvard University) issued *Government and the Nation's Resources* [32]. This report considered possible institutional adjustments that would improve the government's ability to detect and anticipate significant problems with the supply of materials. Leaving energy aside, the report concentrated on shifts in demand and supply that might be large and abrupt enough to exceed the immediate adjustment capabilities of the materials-producing and materials-using industries.

In its recommendations the Commission emphasized the need to improve the understanding of the role of materials in our economy, especially its effect on sectors and industry, and issues raised by growing international interdependence. The lack of adequate data, in its opinion, poses much less of a problem than does inadequate analysis of existing data. Moreover, questions requiring detailed understanding of the increasing interdependence of the world economy will more and more dominate high-level policy discussion.

This study was designed to address this problem and to advance the understanding of shifts in the supply and demand for principal mineral resources under conditions of economic growth and technological advance. The conventional economic approach to the study of an individual commodity market involves derivation of statistical supply and demand curves, one showing the amount of the commodity in question offered for sale, the other indicating the quantity that would be demanded—at any given price. Although this partial approach might suffice for an elementary theoretical explanation of the operation of an ideal competitive market, it would not be satisfactory for the description and systematic analysis of all the factors and the many complex interrelationships that actually affect the levels of production, consumption, and price. Even a simple list of these factors and interrelationships would include long arrays of items, such as volume and quality of specific reserves and resources, alternative methods of extraction, and the technical characteristics of both industries directly involved in processing or using a particular material and all sectors of the economy involved directly or indirectly in providing the inputs used by these industries or depending either directly or indirectly on their outputs. The shapes and the shifts in the statistically derived supply and demand curves could in principle reflect all these factors. In practice, however, it would prove impossible to do.

In this study, the position of each resource is determined not through a step-by-step pursuit of all the twists and turns of each one of these strands, but by means of placing it within the framework of a detailed input-output model of the American, and whenever necessary, of the entire world economy.

The dilemma of having to choose between the micro- and the macroeconomic approach—that is, between inspecting individual trees and looking at the general outline of the forest—is solved by a modeling approach that describes the entire forest in terms of all the trees of which it is comprised.

Objectives of the Study

Once the requisite data are collected, appropriately adjusted to ensure compatability with the model and inserted into their proper places in the framework, various questions about the present and future production and consumption of each type of the actually or potentially important mineral resources can be posed and answered systematically. For instance:

> What are the magnitudes of the structural changes brought about by alternative U.S. mineral policies regarding self-sufficiency or import-dependence for minerals?
>
> How are the mineral sectors affected by a determined policy of increased (or decreased) recycling?
>
> What effects did the expected changes in materials consumption—that is, technological changes in various sectors—have on the minerals industry specifically and all other sectors of the economy in general?

One of the two principal objectives of this project is to use the data bank and analytical approach to answer these and other questions. More generally, we hope to provide specific answers to current policy questions.

This objective is accomplished by preparing a set of projections of production, processing, and final use of a selected group of mineral resources for three periods: from 1972 to 1980, from 1990 to 2000, and through 2030. Such detailed, comprehensive, and internally consistent projections are practically indispensable for the formulation of long-run policies aimed at securing an adequate supply of mineral resources.

Nearly all aspects of changes in economic activity—from technological change and materials substitution in the production process to the evolving tastes and preferences of consumers that determine, in part, the bundle of goods and services produced—are addressed in this study. In addition, worldwide demographic, developmental, environmental, and political issues also affect the future position of the nonfuel minerals sectors in the United States and the world economy. Twenty-six nonfuel minerals (both metallic and nonmetallic) were included in the study:

Iron and Ferrous Metals

Iron
Nickel
Manganese
Chromium
Silicon
Tungsten
Molybdenum
Vanadium

Nonferrous Metals

Aluminum
Copper
Lead
Gold
Silver
Platinum
Titanium
Zinc
Magnesium
Tin
Mercury

Fertilizer-Related Minerals

Potash
Phosphate rock

Miscellaneous Chemicals

Sulphur
Fluorine
Chlorine
Soda ash
Boron

The list is self-explanatory. The first group, which corresponds to the steel industry, includes iron and the major elements used for the production of ferroalloys. The second group contains the major nonferrous metals; the third group includes minerals necessary to produce potassium- and phosphor-based fertilizers. The miscellaneous chemicals category comprises elements used for producing some of the more common bases and inorganic acids widely used in the mineral and metallurgical industries.

The eight ferrous metals, eleven nonferrous metals, and seven nonmetallic elements were selected using *Mineral Facts and Problems,* published by the U.S. Bureau of Mines, as a guideline [45]. This work projects demand (both low and high estimates) for the year 2000 for over 50 elements and compounds throughout the United States and the world. The low estimate of the projected U.S. and world demand for the selected elements was at least $200 million (1970 dollars). Level of demand is a useful criterion to indicate the more important elements.

In addition, several materials were excluded from the nonmetallic minerals category:

1. All construction or construction-related materials, such as sand, gravel, broken stone, gypsum, asbestos, and calcium
2. Gases, such as nitrogen and argon
3. Synthetic industrial diamonds.

The resulting list of twenty-six nonfuel minerals is a manageable number and contains over 90 percent of the dollar value of the entire mining industry.

Organization of the Study

The book is organized in two parts. Part I, containing chapters 1 to 6, concentrates on a detailed description of the current uses and sources of supply of the twenty-six nonfuel minerals in the U.S. economy and projects future U.S. consumption and production of these nonfuel minerals under a host of alternative assumptions through 2000.

More specifically, chapter 2 is composed of three sections: a description of the role of nonfuel minerals in the U.S. economy; a review of past and current U.S. policy with respect to nonfuel minerals; and a brief review of the literature of past studies on long-term projections of the production and consumption of nonfuel minerals in the U.S. and the world economy. Chapter 3 describes the methodology used in the projections and the technical approach used to integrate the detailed list of nonfuel minerals into the general analytical framework.

Chapter 4 presents the different sets of assumptions that affect the levels of production and consumption of minerals—such as recycling rates, import dependency, technological change, and economic growth. In chapter 5, the results of the computations generated by the alternative sets of assumptions are presented from various vantage points—from the more general market-share concept to the more detailed consuming-sector approach—as well as a comparison of the effects on total consumption or production of each mineral resulting from a change in one or more of the various assumptions. Chapter 6 describes the use of minerals from yet another point of view—use through GDP component, by levels, percentages, and intensities.

Part II, including chapters 7 and 8, describes the use of each of the twenty-six minerals in a global context. It also projects the levels needed to satisfy future global requirements by 2030 under alternative rates of growth in world economic activity.

That is, chapter 7 integrates the U.S. economy into the wider framework of the world economy and incorporates the current description and projections of the future levels of global consumption and production, region by region, for the twenty nonfuel minerals not previously included in the World Input-Output Model [22]. Chapter 8 is a summary of both U.S. and world projections for 2000 and 2030 under low and high world economic-growth assumptions. Chapter 8 also discusses problems of potential imbalances in resources—that is, the world's currently estimated stock of resources versus cumulative consumption on a global scale.

Also included are eight technical appendixes. The appendixes include the information from which the model's data base was constructed and from which the assumptions used in the computations were formulated.

2 Resources and the Economy

The Role of Resources in the U.S. Economy

In 1978, the nonfuel minerals mining sectors in the United States contributed less than 1 percent to national income, employed less than 1 percent of the total labor force, and accounted for less than 0.5 percent of the total energy consumed in the economy [44]. These conventional macroeconomic aggregates suggest that the mining sectors have received attention far in excess of their apparent importance in the national economy. Yet these data fail to capture the critical role that nonfuel minerals play in the national economy. Their pivotal position can be demonstrated only through a detailed and fundamental understanding of their use. Only after the nonfuel minerals extraction sectors and their output, metallic and nonmetallic minerals, are placed in their respective positions in the national and world economies do these statistics assume different proportions.

In the wake of unprecedented increases in world petroleum prices in 1973 and 1978, we have been compelled to consider whether there are sufficient quantities of other basic materials to meet national and global requirements to the year 2000 and beyond. For example, at current prices, is the world economy capable of meeting the rising demand for food in light of current and expected future growth rates in population and income? On the other hand, can we be reasonably sure of the availability of nonfuel minerals, the essential building blocks of a modern economy, to satisfy the expected world demand?

Since the publication of the Club of Rome Report [29] in 1972, minerals specialists have been divided into two opposing camps: the catastrophists and the cornucopians. The first group, whose position is reminiscent of the doomsday arguments first articulated by the eighteenth-century moral philosopher Thomas Malthus, centers its argument on the allegedly immutable laws of nature. That is, there are not enough primary resources to meet the ever-increasing demand made by an expanding population over the long term.

The cornucopians, on the other hand:

> argue that most of the essential raw materials are in infinite supply: that is, when society exhausts one raw material, it will turn to lower-grade, inex-

haustible substitutes. Eventually, society will subsist on renewable resources and on elements such as iron and aluminum that are practically inexhaustible. According to this view, society will settle into a steady state of substitution and recycling [14].

The answer probably lies somewhere in between (although closer to the cornucopian camp), if past history serves as a guide to the present and future use of minerals. In the past, advance in knowledge and its implementation through innovation has consistently enabled mankind to repeal or at least delay the Malthusian laws. With this metaphysical argument behind us, this work, along with other recent minerals studies [8, 22, 28, 38], has more modest goals—that is, of projecting future minerals production and consumption levels for the U.S. and world economy for the remainder of this century and the first thirty years of the next. In acknowledgment of the fact that the future is uncertain, we consider several possible futures or scenarios, each based on a set of reasonably likely assumptions.

A Review of Recent U.S. Minerals Policy

A brief summary of past U.S. minerals policy and some studies of future U.S. minerals requirements (and in some cases world requirements) serve as a prologue for the description of the methodological approach, the data base, the alternative scenarios, and the results of our computations.

After World War Two and the Strategic and Critical Materials Stock Piling Act of 1946, the Paley Commission prepared its historic report, *Resources for Freedom,* in which it urged that:

1. Statistics on the relation of the supply of materials to the demand should be carefully maintained and closely watched
2. Resource conservation and development should be pressed, with emphasis on new materials technology in substituting materials
3. A non-governmental institution should be established to monitor the state of minerals supply and demand
4. A new governmental institution close to the President should be established to formulate and direct national materials policy [36].

The Paley Commission report established for the first time the pivotal role played by basic materials in the U.S. economy. After the Korean conflict, materials surpluses reappeared and were absorbed as part of the national stockpile. Throughout the remainder of the 1950s and much of the 1960s, U.S. minerals policy consisted of managing the strategic stockpile. Most, if not all, materials research was directed to the development of military hardware or for the space program, including aerospace, electronics, and missile systems.

Only in the late 1960s, when environmental quality emerged as a national issue, were the minerals sectors judged in terms other than for their military importance. "It became evident to more analysts," as Franklin Huddle [18:20] reports, "that all forms of environmental pollution (except for such transients as thermal or noise pollution) were the consequence of the management of materials."

The National Commission of Materials Policy (NCMP), established in 1971, was formed on the recommendation of the ad hoc committee organized in 1969 by Senator J.C. Boggs. The Boggs report, as summarized by Huddle called attention to

> the need for an adequate supply of industrial materials for peace or war, the need to practice frugality in the light of growing world demands for these minerals, the need to satisfy ever-more-demanding requirements for improved properties of materials, and the need to manage materials so as to preserve a benign and unpolluted environment [18:20].

NCMP issued a report in 1973, stressing that the nation should "rely on market forces as a prime determinant of the mix of imports and domestic production in the field of materials but at the same time decrease and prevent wherever necessary a dangerous or costly dependence on imports." The commission argued for the need to "conserve our natural resources and environment by treating waste materials as resources and returning them either to use or in a harmless condition to the ecosystem" [18:21]. The report also recommended that "we institute coordinated resource policy planning which recognizes the interrelationships among materials, energy and environment" [18:21].

Since the 1973 NCMP report, numerous government projects have been planned or undertaken by such agencies as the Bureau of Mines, Geological Survey, Central Intelligence Agency, Department of Defense, General Services Administration, Department of Energy, the General Accounting Office, the Office of Technology Assessment, and the Congressional Research Service. These activities indicate the general acceptance of the NCMP recommendations.

In December 1976, the Temporary National Commission on Supplies and Shortages recommended the design of a tool or mechanism that would help to mitigate or prevent any future perceived crisis [32]. The issue seemed no longer to be a decision between a free-market and a planned-economy approach in forging a minerals policy. The new objective was rather to design a coordinated economic information system that could collect, process, and analyze information. In turn, policy makers would use this information to define possible solutions for the anticipated problems before they arose.

With concerns mounting in Congress about import dependence on such strategic minerals as cobalt, manganese, chromium, platinum, and bauxite, the National Materials and Minerals Policy, Research and Development Act of 1980 was passed. This act

> reiterates the intent of a law passed ten years earlier, which basically tells the government to organize a materials policy and encourage private enterprise. . . . Now, moves are afoot to replenish the country's strategic materials stockpile, revitalize mining and inject minerals consideration into every conceivably relevant aspect of domestic and foreign policy [17:305, 307].

After more than a quarter of a century of evolution towards the still-unrealized goal of a comprehensive and internally consistent approach to tracking the production and consumption of minerals for the national economy, the Paley Commission's 1952 report seems prophetic: "There must be, somewhere, a mechanism for looking at the problem as a whole, for keeping track of changing situations and the intervention of policies and programs" [36]. We hope and expect that this study brings us closer to the realization of that goal.

Resource Studies: A Review of the Literature

The first comprehensive, long-term overview of U.S. resource requirements and availabilities was completed over twenty years ago. *Resources in America's Future* by Landsberg, Fischman, and Fisher, not only considered alternative projections of the nation's vital statistics and the component parts of the GNP accounts to 2000, but also implicitly incorporated specific assumptions about technological change in the food, clothing, construction, transportation, packaging, paper products, and energy sectors. These assumptions were the bases of projections of land, lumber, water, chemical, minerals (fuel and nonfuel), and labor requirements that would satisfy the alternative projected GNP levels assumed in the study. The study also compared the projected resource requirements with the 1960 U.S. resource base and, where necessary, the world resource base of specific minerals in 1960.

The wide range spanned by the high- and low-consumption estimates can be explained by varying rates of economic growth and technological change between 1960 and 2000. The authors' twenty-year projections have much less variation between high and low estimates. Despite the high rates of economic growth registered through the 1960s and most of the 1970s, the U.S. economy has become more efficient in its use of the major primary metals. Almost without exception, the low estimates for 1980 projected by Landsberg and others were closest to the actual 1979 levels of consumption. Their projections and the actual figures for 1979 are reproduced in table 2–1.

A little more than a decade ago, Aurelio Peccei, founder of the Club of Rome, went on record as being "perplexed and worried by the orderless, torrential character of (the) precipitous human progress" during the twentieth century. Because the world's problems were global and therefore interconnected, a study was commissioned to quantify the interrelated current and future problems of mankind—problems of rapid population growth, resource depletion, pollution, and hunger. Whether or not the conclusions of the various reports [9, 22, 29, 37] issued in response are accurate, the work by Forrester [9] and Meadows [29] provided a quantum leap forward in the conception of the domain and timeframe of systems models.

For the first time, these key macrovariables—population, capital, resources, pollution, and food—were simultaneously modeled on a global

Table 2–1
U.S. Consumption of Principal Primary Metals for 1960 and 1979 and Projections for 1980 and 2000
(million metric tons)

Metal	*1960*		*1980*	*2000*	*1979 Actual*
		L	76	89	
Iron and steel	72.0	M	121	194	88.0
		H	176	378	
		L	3.4	6.6	
Aluminum	1.6	M	5.6	14.7	4.8
		H	10.1	31.1	
		L	2.2	2.8	
Copper	1.8	M	3.6	6.8	1.8
		H	5.2	14.0	
		L	1.2	1.7	
Zinc	0.9	M	1.8	3.4	0.9
		H	2.7	7.2	
		L	1.0	1.1	
Lead	1.0	M	1.5	2.3	0.7
		H	2.2	4.9	

Sources: Appendix tables A-16-9, 32, 44, 52, and 61 in Landsberg and others, *Resources in America's Future; Mineral Facts & Problems* (1981), pp. 22, 234, 466, 503, 1034.
Note: L = low projection, M = medium, and H = high.

scale, over decades, in an internally consistent way. Although subsequent models have taken issue with the functional relationships connecting the variables as well as the initial assumptions, these early studies seem to have achieved one of Peccei's main objectives—"to promote and disseminate a more secure in-depth understanding of mankind's predicament" [33:73]. A plethora of world models have been constructed in the wake of the Club of Rome Report. However, as far as the projections of future resource requirements are concerned, the model's aggregative treatment of minerals does not lend itself to analysis of such issues as resource depletion, substitution, import dependence, or recycling.

Other world models, such as the Mesarovic-Pestel [30], MOIRA [27], and the Latin American World Model [16], were constructed in light of the Club of Rome Report. None of these models has a detailed description or projections of nonfuel mineral resources as such. The *Global 2000 Report to the President* [6] provides a summary of the objectives, structure, and results of these models.

The World Input-Output Model [22], published in 1977, addresses the question of global resource requirements and the availability of food, mineral, and energy resources to meet the demands of accelerated world economic development to 2000. In addition to projecting the standard macroeconomic variables, such as gross domestic product (GDP), consumption, investment, and balance of payments, the model describes and projects from the 1970 base-year to 1980, 1990, and the year 2000 for the fifteen regions into which the world's countries were aggregated, a large number of sectoral outputs.[1] It includes thirty manufacturing and service sectors, four agricultural sectors, three energy resource outputs, and six nonfuel mineral outputs: iron, nickel, copper, lead, zinc, and aluminum. In addition to output levels, the model tracks imports and exports in minerals, as well as all other traded goods. For nonrenewable fuel and nonfuel minerals, the model traces the cumulative resources produced in each region after each decade. Given current reserve estimates and future demand projections, the study concludes that "only two of the metallic minerals, lead and zinc, are expected to 'run out' by the turn of the century" [22:5]. "However, other investigators expressed concern about the adequacy of the supply of other minerals such as asbestos, fluorine, gold, mercury, phosphorus, silver, tin, and tungsten" [22:6].

The most consistent and comprehensive U.S. government data-gathering, analysis, and information system of minerals availability and use, both current and projected, is published by the U.S. Bureau of Mines. Its quinquennial publication *Mineral Facts and Problems* supplements the annual *Minerals Yearbook* and *Minerals Commodity Profiles,* written by bureau commodity specialists. *Mineral Facts and Problems* [46], issued in 1976, offers alternative projections and forecasts of minerals consumption for

1985 and 2000. The projections are mostly based on GNP or an index of industrial production, and the forecasts implicitly incorporate expected technological and institutional changes.

Many studies have appeared in the last decade concerning short- and long-term projections of U.S. and world mineral requirements [8, 28, 38]. Some concentrate on the supply of and demand for one particular mineral; others discuss the problems of mineral supply and depletion and the still-unresolved controversy of increased mineral scarcity reflected in price increases attributed to mining inferior ore grades or development of remote mining sites. Increased energy and capital costs for extraction and processing facilities (influenced by higher interest rates, environmental factors, and higher royalties imposed by host governments) have also contributed to higher costs for minerals [39, 42]. H.J. Barnett and C. Morse [3] argue that because of new discoveries and technological change, relative prices of minerals in the national economy have been declining, and that there is strong reason to believe this trend will continue into the future.

The Bureau of Mines, with technical assistance from the U.S. Geological Survey, is undertaking a long-term project called Minerals Availability System [57]. It is designed to establish an inventory of mineral reserves and resources data, initially in the United States and eventually throughout the world. This profile of the world's resource base would be sharpened by taking into account physical, geographic, environmental, as well as economic determinants, in assessing the current and future levels of U.S. and world reserves.

The most often-quoted study on current and future minerals consumption, both for the U.S. economy and the rest of the world, is Malenbaum's *World Demand for Raw Materials in 1985 and 2000* [28]. The demand for minerals is derived from two sources: the level of gross domestic output adjusted for population and the intensity of use (defined as the ratio of consumption of materials to GDP in the same time period). Underlying the intensity-of-use concept is, of course, the composition of GDP—not only its disposition among the component parts that constitute final output but also the allocation of consumption, investment, government expenditures, and import and export levels among agricultural, manufacturing goods (durable and nondurable), and services, as well as the technological profile of a country which would produce these alternative compositions of final demand. According to Malenbaum,

> The intensity-of-use measure was, of course, obtained by dividing apparent consumption by the GDP figures discussed above. It is this measure that receives major attention in this study. As already indicated, the very concept has a technological ring. It is apparent that the values of these measures differ significantly among the world's regions, as they do in each region over time. Presumably, the measure for each mineral depends upon

> industrialization, economic sophistication, resource endowment and other factors for which the world does offer a wide range of patterns. And these factors change over time, notably with industrial and economic development. Use of material inputs must be associated with the specific products—the output of automobiles, or the miles people travel in a year, or the extent of electrification, as examples. And every measure of materials use should be accounted for fully in such itemized outputs. To do this would certainly be a major research undertaking even for the relatively few countries where statistics on use detail are available. Any projected use-intensity would require many specific assumptions about future product use. Hence, reliance needs to be put on more general attributes of the demand for materials, derived from past relationships and from the theory that helps explain past use of materials [28:16].

Malenbaum's intensity-of-use thesis implies that, in addition to the introduction of efficiencies in the use of metals in the production process, developed countries also introduce efficiencies in the use of metals in the consumption process by virtue of the "relatively intensive" use of services (figure 2-1). This implication poses yet another problem: are services per

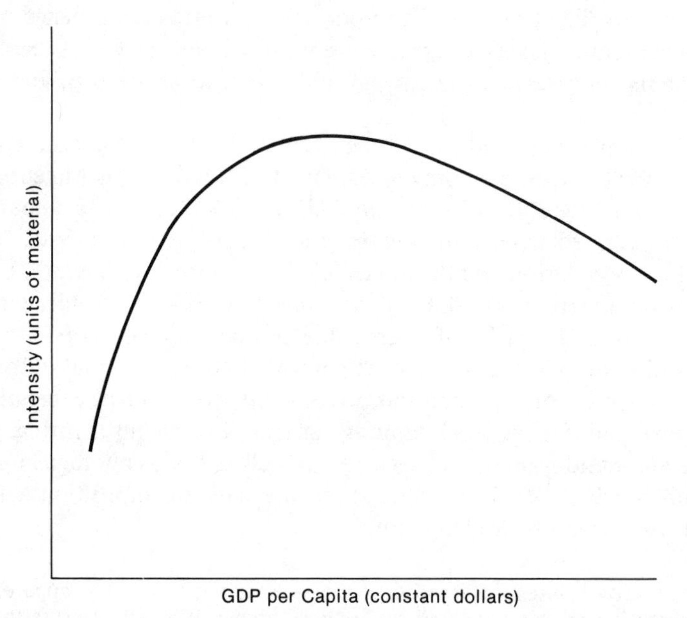

Note: Reproduced with permission from Malenbaum, *World Demand for Raw Materials in 1985 and 2000.* New York: McGraw Hill, 1978, figure 2-2, p. 18.

Figure 2-1. Intensity-of-Use and GDP per Capita: An Impressionistic Representation

dollar of output (directly or indirectly) more or less minerals-intensive than manufacturing or agricultural products? Vogely [56] and Landsberg [19] suggest that services may, in fact, be more metals-intensive, directly and indirectly, per dollar of output than manufacturing. The analysis in this study should provide some insight toward the resolution of this longstanding controversy.

To Choose A Future by Ridker and Watson [38] discusses long-term resource availability of nonfuel minerals in one chapter. It projects resource requirements and availabilities for seventeen "less-abundant ferrous, nonferrous, and non-metallic minerals" (all of which, with the exception of cobalt, are included in this study) to the year 2025 for the U.S. and world economies. Their projections trace the cumulative use of each mineral and estimate the year of exhaustion of its reserve and prospective resource base.

Although this study uses a similar method of projecting the baseline scenario as Ridker and Watson, the results of our computations systematically combine the various factors that determine the level of domestic production. These factors—such as the level of final demand, import dependence, recycling rates, and materials substitution—can be compared and analyzed directly. In a sense, this study extends Ridker and Watson's alternative future projections of the U.S. demand for minerals. However, this work differs in the numbers and types of scenarios and in the presentation of the underlying data and results for the United States.

The major difference, however, concerns the treatment of the rest of the world. Ridker and Watson implement the intensity-of-use concept and base their projections for regions other than the United States on projected growth of per-capita income. In contrast, we use a full-scale model of the world economy as a basis for two alternative projections of minerals demand and supply to the year 2030.

Another recent contribution to the expanding literature on long-term projections of the demand for nonfuel minerals on a global basis was made under contract with the Office of Science and Technology in the Executive Office of the President by Resources for the Future under the direction of Leonard L. Fischman. *World Minerals Trends and U.S. Supply Problems* [8] considers the historical patterns of consumption for seven nonfuel minerals—cobalt, chromium, manganese, aluminum, copper, lead, and zinc—from 1970 to 1977. Its projections of future patterns of consumption are driven for the most part by demographic and macroeconomic projections—that is, growth in GDP or GDP per capita. Because the study focuses on U.S. supply problems, trade in various forms of minerals ore, metal (smelted and refined), and scrap is presented. It also considers potential long-term supply problems as well as some of the possible short-term measures for alleviating shortages.

Several tables in chapter 7 compare the projections of some of the works discussed in this section with our estimates of U.S. and world minerals requirements in the year 2000 and beyond.

Note

1. The model's time frame was recently extended to 2030. Sixteen regions—North America was disaggregated into the United States and Canada—are now represented.

3 The Methodological Approach

This chapter presents the methodology used to incorporate the detailed description of the production and consumption of nonfuel minerals into the U.S. economy in the base-year and to project future levels of minerals production and consumption. The first section contains a brief general survey of the input-output technique of economic analysis. Next, we present the modifications required to make the general-purpose input-output table of the U.S. economy (published by the Bureau of Economic Analysis [43]) useful for a detailed study on nonfuel minerals. This section also describes the data sets incorporated in the model as well as their mathematical and computational characteristics. A description of the World Model system and its methodology is deferred until chapter 7, which discusses the integration of the U.S. economy along with the twenty-six nonfuel minerals into the world economy.

Input-Output Analysis

The American economy provides a relatively high annual standard of living for approximately 225 million people. This standard of living is shaped by three forces: a culture, which fixes what economists refer to as the "tastes and preferences" of consumers; a technology, which prescribes, in accordance with the existing state of knowledge, efficient means for producing the goods and services; and the endowments of primary factors of production, labor, capital, and natural resources, the stock of which limits what is produced domestically and what goods are imported. In addition to the physical constraints of factor endowments and the technological constraints imposed on production, various social, institutional, and environmental measures are designed and implemented by the government to influence or conform with consumer tastes and preferences.

Any comprehensive analysis of our increasingly complex economy must be based on a systematic description of the structure of production, households, and factor endowments. It must also include the political, demographic, social, environmental, and institutional phenomena around which the accommodating and often competing economic relationships are woven.

Our understanding about the way in which an economy hangs together has become both wider and deeper. Our methodological approach to economic analysis—what Schumpeter referred to as the ''economist's analytical tool kit''—now incorporates the input-output technique. This tool facilitates a structural description of the economy based on an empirical investigation of its component parts.

The input-output technique of economic analysis, first introduced some forty-five years ago, is specifically designed as a tool for the systematic analysis of the mutual interdependencies between the different parts of an economy. It describes the economy as a system of interdependent activities—both in a direct and in an indirect way. Hence, it is particularly well suited to both the preparation of internally consistent multisectoral projections of prevailing economic trends and to a detailed quantitative assessment of both the direct and indirect secondary effects of any single policy or combination of policies.

An input-output table provides a systematic picture of the flow of goods and services among all the producing and consuming sectors of a given economy—that is, among all the various branches of business, households, and government. It also registers the flow of goods and services out of a given region and the flow of goods and services received from the outside.

The input structure of each producing sector is explained in terms of its technology—that is, a set of technical coefficients specifying the amount of goods and services, including labor, required by a sector to produce a unit of its own output. Similarly, a separate set of capital coefficients would describe the stocks of buildings and equipment, as well as all kinds of working inventories that each producing sector has to maintain to transform the proper combination of its inputs into its final output of goods and services.

The inputs of primary natural resources, such as agricultural land, water, and minerals, required by all producing sectors of the economy as well as households—that is, consumption patterns—can also be depicted and analyzed in modern input-output analysis along with the production and consumption of ordinary goods. Prices are determined in an open input-output system from a set of equations that states that the price which each productive sector of the economy receives for its output must equal the total payments made by it, per unit of its product, for inputs purchased from itself and from other industries, plus value-added, which essentially represents payments for labor, capital, and taxes.

An input-output table shows the amounts of goods and services individual industries buy and sell to each other in a particular year. Table 3-1 is an aggregated version of the more detailed 1972 input-output table constructed by the Bureau of Economic Analysis (BEA). The BEA table was the basis for the minerals model described later.

Table 3–1
1972 U.S. Input-Output Flow Table

1. 1972 U.S. Interindustry Flows ($ Millions)

	Agri-culture	*Minerals*	*Construc-tion*	*Durables*	*Nondur-ables*	*Transpor-tation*	*Trade-margins*	*Finance*	*Services*	*Dummies*
1. Agriculture	26,435.6	1.1	329.2	1,919.5	38,917.8	212.9	1,566.5	740.5	765.7	198.6
2. Minerals	200.8	1,663.1	1,480.4	5,933.3	16,496.4	6,195.7	65.8	27.5	91.7	332.3
3. Construction	583.1	857.8	47.0	1,541.3	1,703.2	5,012.1	1,040.5	11,084.1	2,453.9	2,672.3
4. Durables	1,254.4	2,025.7	50,208.8	147,358.4	15,281.8	3,708.5	1,049.3	396.4	14,303.2	308.5
5. Nondurables	10,649.5	928.5	9,408.8	23,121.9	107,015.7	6,137.6	22,187.5	3,978.1	12,475.2	1,007.9
6. Transportation	2,138.9	950.1	5,324.6	16,885.8	17,107.2	21,578.1	10,934.7	4,554.8	8,973.9	2,916.5
7. Trade Margins	3,099.2	567.3	13,427.4	17,347.5	13,351.5	3,107.8	8,374.2	2,579.3	6,871.8	297.5
8. Finance	5,882.0	3,624.8	2,086.3	6,498.3	6,765.2	5,292.7	14,501.9	33,092.5	14,891.1	718.1
9. Services	1,372.0	733.1	7,200.9	11,567.3	12,779.2	7,312.3	15,994.5	8,465.7	17,609.1	764.9
10. Dummies	177.4	152.3	377.7	2,340.4	3,867.9	3,934.6	2,167.6	2,493.7	2,443.9	704.9

2. Value Added

	Wages	*Interest*	*Taxes*	*Total*
1. Agriculture	6,053.5	24,446.2	1,622.1	32,121.8
2. Minerals	6,273.9	11,206.1	1,401.7	18,881.7
3. Construction	60,155.1	14,733.7	1,217.7	76,106.5
4. Durables	126,080.0	41,784.7	2,661.4	170,526.1
5. Nondurables	76,721.8	31,905.7	14,339.8	122,867.3
6. Transportation	51,344.0	35,509.7	11,659.0	98,512.7
7. Trade Margins	107,595.8	39,397.7	40,057.6	187,051.1
8. Finance	39,537.0	111,429.1	34,088.3	185,054.4
9. Services	94,296.4	42,179.9	3,842.4	140,318.7
10. Dummies	149,605.0	1,629.0	51.1	151,285.1

Table 3-1 continued

	3. Final Demand					*4. Output*
	Consumption	*Investment*	*Government*	*Net Exports*	*Total*	*Total*
1. Agriculture	8,536.3	2,540.6	−1,125.7	2,916.1	12,867.3	83,954.7
2. Minerals	638.7	463.9	110.9	−3,314.9	−2,101.4	30,385.6
3. Construction	.0	99,086.5	39,900.0	15.8	139,002.3	165,997.6
4. Durables	66,370.2	76,540.0	32,380.8	−6,145.8	169,145.2	405,040.2
5. Nondurables	149,396.5	4,822.3	12,071.5	−7,074.9	159,215.4	356,126.1
6. Transportation	52,142.8	3,707.6	9,625.7	4,164.2	69,640.3	161,004.9
7. Trade Margins	176,494.0	11,109.0	1,276.3	7,030.6	195,909.9	264,933.4
8. Finance	147,573.1	4,426.1	4,989.7	2,125.2	159,114.1	252,467.0
9. Services	119,527.3	35.3	16,738.1	1,098.6	137,399.3	221,198.3
10. Dummies	17,392.5	−7,450.7	136,851.6	−4,427.1	142,546.3	161,206.7

The rows of table 3–1, for instance, trace the amount of a particular commodity that each industry delivered to itself, to other industries (part 1), and to final demand (part 3). The amount of value-added paid out by each sector as wages, interest, and taxes is shown in the lower left. Reading down a column of the intersectoral flows, the amount of each input purchased by an individual sector can be seen, and reading across the row of the value-added matrix the cost of primary inputs of labor and capital of that same sector can be seen. All transactions in the table are measured in millions of 1972 dollars valued in producers' prices.

Input coefficients are derived by dividing the column entries by the respective sectoral output (part 4). The coefficients show the amount of each input required to produce one dollar's worth of a sector's output. Thus, the columns of a coefficient matrix give a detailed, quantitative description of the inputs used by the individual sectors. Because each industry has its own column, the matrix is a structural description of the entire economy for a particular year. Input-output tables are prepared by the BEA every five years corresponding to the years in which a national census of manufactures is conducted.

Mathematically, the structure of an open static input-output model is simple [21]. Consider

$$X - AX = Y$$

where X is a vector of outputs of industries 1, . . . , n, and Y is a vector of deliveries by the n industries to final demand (personal consumption, investment, government, and foreign trade). A is a square matrix of input-output coefficients. Each element, a_{ij}, in the structural matrix A represents the amount of sector i's output purchased by sector j, per unit of sector j's output. Thus, each column of the A matrix describes the structure of a particular industry and serves as an implicit representation of that industry's technology. The matrix equation simply states that sectoral outputs X must be such that they satisfy both intermediate demand (as inputs into other sectors) AX and final demand Y.

The matrix of technical coefficients, A, is usually derived from published input-output tables, augmented, as in this study, by additional information. Y is actual and projected GDP, specified exogenously. The model is solved for X through the following equation:

$$X = (I - A)^{-1}Y$$

The elements of the inverse matrix $(I - A)^{-1}$ also have a special interpretation: each element, $\bar{a}_{ij}$, of the inverse matrix, represents the direct plus indirect requirements of sector i to deliver one unit of the output of sector j to final demand.

Modeling Nonfuel Minerals: IEA/USMIN

The United States Minerals Model (IEA/USMIN) developed at the Institute for Economic Analysis describes the production and consumption of twenty-six nonfuel metallic and nonmetallic minerals listed in chapter 1 within a detailed picture of the U.S. economy in the year 1972. The point of departure for the description of the flow of these materials into all other sectors of the economy is provided by the official BEA 1972 U.S. input-output table [43].

In the official BEA input-output table, the production of nonfuel minerals is described in terms of five producing sectors. The output of these sectors is traced through the minerals-processing sectors (BEA 37 and 38) to the minerals-consuming sectors.[1] All flows (and coefficients derived from them) are measured in millions of 1972 dollars.

In the enlarged minerals model in this study, the five BEA nonfuel minerals-producing sectors are disaggregated into twenty-six nonfuel minerals. All flows of minerals (and coefficients derived from them) in IEA/USMIN are described in physical amounts of minerals content. To achieve internal consistency with the 1972 base year, consumption of minerals by final demand components was also calculated. For nonfuel minerals, the final demand categories were, almost without exception, exports, imports, and changes in inventories.

The supply of minerals has a number of alternative sources: mine output, by-product output resulting from the production of other minerals, imports, decreases of private inventories, and, if applicable, releases of government stockpiles. All of these component parts of supply were calculated for the twenty-six nonfuel minerals to balance total supply (production plus imports plus decreases in business inventories and government stockpiles) and total demand (interindustry consumption plus exports plus increases in business inventories and government stockpiles).

Often, an additional source of supply for metallic minerals is available—secondary production, which is considered alongside the supply of primary metal. Accordingly, the model describes the production and consumption of recycled metals separately. However, it ensures at all times that the balance between production and consumption of minerals (now including secondary production and consumption) is not violated.

Structure of IEA/USMIN

The IEA/USMIN model depicts the structure of the United States economy in 1972 and the hypothetical structure of the economy in 1980, 1990, and 2000. The model consists of 321 equations containing 328 variables. (See appendix A for the names of the row and column variables.) Figures 3–1 and 3–2 describe schematically the model's structure.

In the algebraic formulation of the model, capital letters denote matrices or vectors and subscripts c and f denote coefficient and flow matrices or vectors, respectively. Individual elements of matrices or vectors are given in lower case with superscript (c) or (f) denoting a coefficient or flow. Subscripts for individual elements denote row and column place. References to dollar amounts are given in the model in billions of 1972 dollars. Physical units include metric units, labor hours, and Btu's.

For each time t, the following system is solved:

$$R_c \cdot Q_f = P_f \tag{3.1}$$

$$Q_f = R_c^{-1} \cdot P_f \tag{3.2}$$

where

$$R_c = \begin{bmatrix} (I - A_c) & 0 & I & 0 & 0 \\ -B_c & (I + C_c) & 0 & G_c & 0 \\ L_c & 0 & -I & 0 & 0 \\ M_c & N_c & 0 & H_c & 0 \\ D_c & 0 & 0 & 0 & -I \end{bmatrix}$$

A_c = a commodity-by-commodity input-output coefficients matrix (106 × 106) whose elements $a_{ij}^{(c)}$ give dollar amounts of input i required to produce one dollar's worth of output j (valued in base-year prices).

0 = a null matrix or vector.

I = an identity matrix.

$-B_c$ = an input-output coefficients matrix (36 × 106) whose elements $b_{ij}^{(c)}$ give the physical amount of mineral (primary or scrap) input i required to produce one dollar's worth of output j. Only minerals produced in the United States are included in this submatrix.

C_c = a diagonal by-product coefficients matrix (36 × 36) whose elements $c_{ij}^{(c)}$, give the physical amount of each mineral (primary or scrap) produced as a by-product per physical unit of its own-industry output.

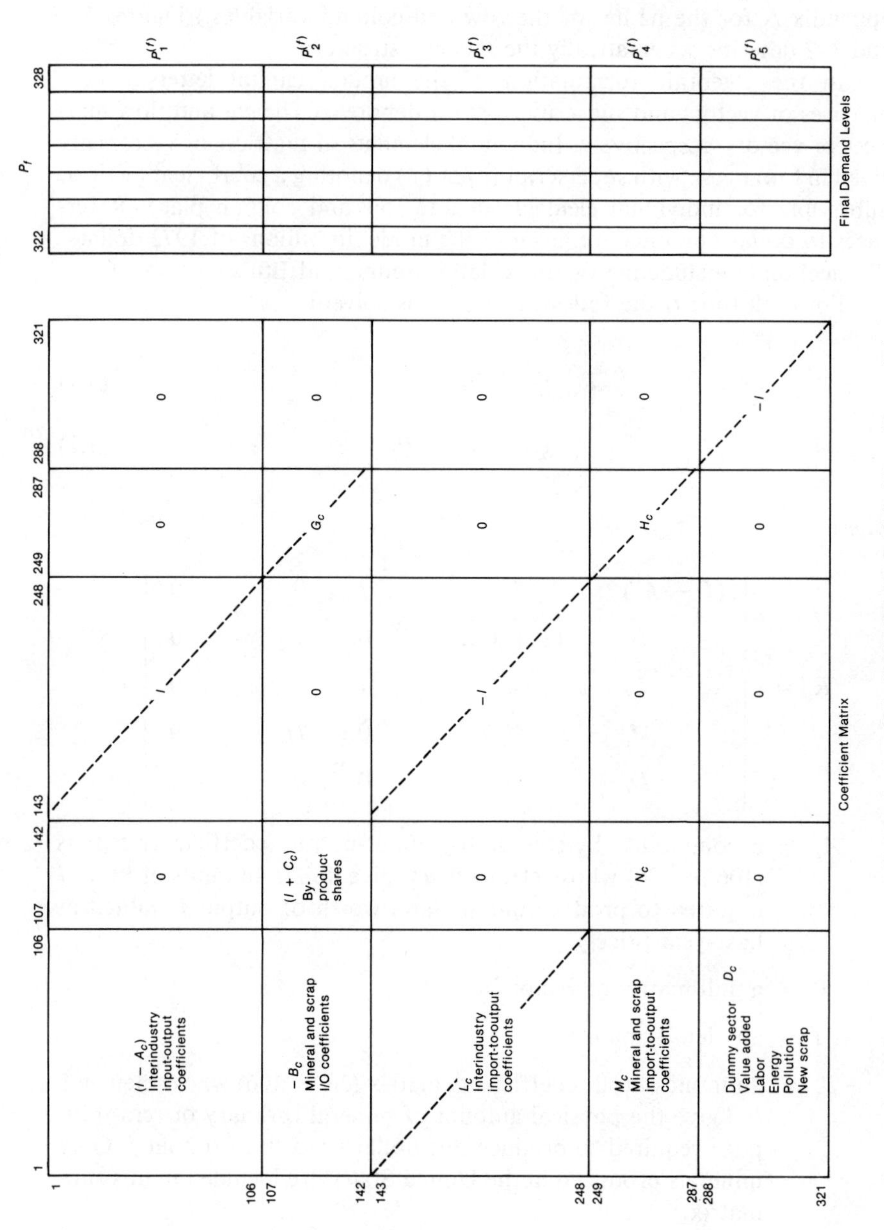

Figure 3–1. IEA/USMIN Model (One Decade)

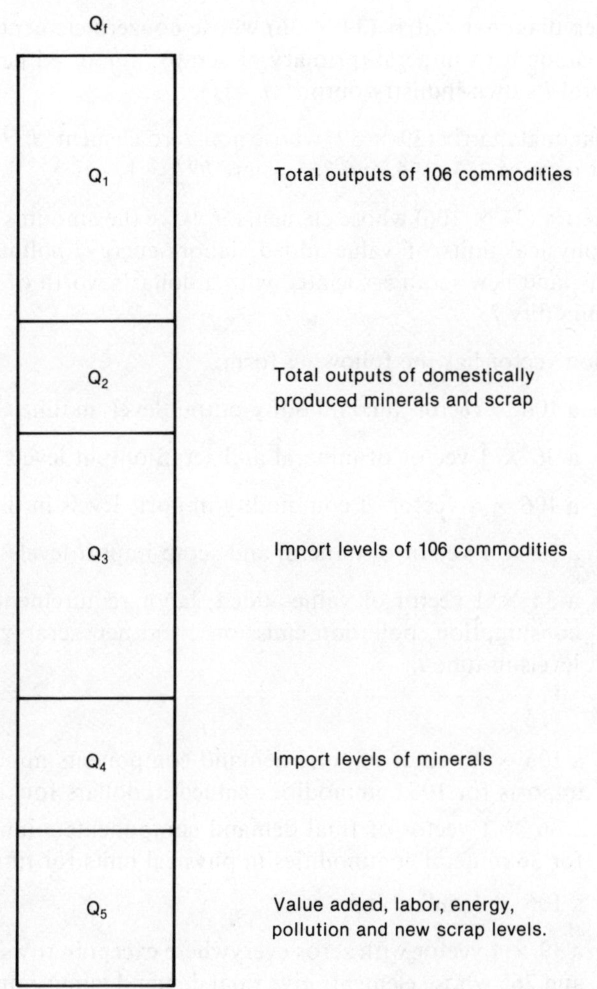

Figure 3-2. IEA/USMIN (One Decade) Solution Vector

G_c = a step-diagonal matrix (36 × 40) whose nonzero elements $g_{ij}^{(c)} = 1$.

L_c = a diagonal import coefficient matrix (106 × 106) whose elements $l_{ij}^{(c)}$ give the dollar amount of imports per dollar's worth of j ($i = j$).

M_c = a matrix (39 × 106) whose only nonzero elements appear in IEA/USMIN rows 253, 254, and 265. The elements of these three rows $m_{ij}^{(c)}$give the physical amount of noncompetitive import i per dollar's worth of output j ($i = j$).[2]

N_c = a step diagonal matrix (39 × 36) whose nonzero elements $n_{ij}^{(c)}$ give the physical amount of mineral (primary or scrap) i imported per physical unit of mineral j's own-industry output ($i = j$).

H_c = a diagonal matrix (39 × 39) whose non-zero elements $h_{ij}^{(c)} = -1$ except in rows 253, 254, and 265 where $h_{ij}^{(c)} = 1$.

D_c = a matrix (34 × 106) whose elements $d_{ij}^{(c)}$ give the amounts in dollars or physical units of value-added, labor, energy, pollution emissions, and new scrap associated with a dollar's worth of output of commodity j.

A solution vector has the following form:

Q_f =

- $q_1^{(f)}$ = a 106 × vector of commodity-output levels in time t.
- $q_2^{(f)}$ = a 36 × 1 vector of mineral and scrap-output levels in time t.
- $q_3^{(f)}$ = a 106 × 1 vector of commodity-import levels in time t.
- $q_4^{(f)}$ = a 39 × 1 vector of mineral and scrap-import levels in time t.
- $q_5^{(f)}$ = a 34 × 1 vector of value-added, labor requirements, energy consumption, pollution emissions, and new scrap-generation levels in time t.

and:

P_f =

- $p_1^{(f)}$ = a 106 × 1 vector of final demand components minus imports for 106 commodities valued in dollars for time t.
- $p_2^{(f)}$ = a 36 × 1 vector of final demand components minus imports for 36 mineral commodities in physical units for time t.
- $p_3^{(f)}$ = a 106 × 1 null vector.
- $p_4^{(f)}$ = a 39 × 1 vector with zeros everywhere except in rows 253, 254, and 265 whose elements give final demand minus import levels for noncompetitive imports in time t.
- $p_5^{(f)}$ = a 34 × 1 null vector.

The system specified in equation 3.1 is equivalent to the following set of equations:

$$(I - A_c) \cdot q_1^{(f)} + I \cdot q_3^{(f)} = p_1^{(f)} \quad (3.3)$$

$$-B_c \cdot q_1^{(f)} + (I + C_c) \cdot q_2^{(f)} + G_c \cdot q_4^{(f)} = p_2^{(f)} \quad (3.4)$$

$$L_c \cdot q_1^{(f)} - I \cdot q_3^{(f)} = p_3^{(f)} \quad (3.5)$$

$$M_c \cdot q_1^{(f)} + N_c \cdot q_2^{(f)} + H_c \cdot q_4^{(f)} = p_4^{(f)} \quad (3.6)$$

$$D_c \cdot q_1^{(f)} - I \cdot q_5^{(f)} = p_5^{(f)} \quad (3.7)$$

Equation 3.3 states that the gross domestic output of each commodity plus imports minus intermediate consumption must satisfy final demand. Similarly, equation 3.4 states that the domestic output of minerals (own-industry plus by-product) plus competitive imports minus intermediate consumption must equal final demand for minerals. Equation 3.5 states that the level of imports for each commodity is equal to a specified fraction of domestic (own-industry) output. Equation 3.6 states the same proposition for noncompetitive minerals (primary and scrap). Equation 3.7 states that the sum of each industry's value-added, labor inputs, and emmissions output equals the respective total for the economy as a whole.

Input-Output Data

The data forming the core of the model are of different vintages, which required varying degrees of adjustment, and include estimates as well as numbers taken directly from official statistics.

The IEA/USMIN model contains several thousand figures. Taken together, these data provide a complete, internally consistent description of the input structure of all individual producing and consuming sectors; the composition of GDP; import requirements; requirements for primary factors such as labor, energy and other minerals; and emissions of air and water pollutants by industries.

The core of the model is the interindustry matrix of input-output coefficients. Mining and primary-metals industries, as well as sectors that are either major end-users of metals or that produce important substitutes, are defined at the BEA 496-order level of industrial classification. The remaining industries are aggregated to the BEA 85-order level.

The extraction of iron, copper, aluminum, lead, zinc, and nickel is represented in several forms: copper mining has its own sector; iron and nickel mining are part of iron and ferroalloy ores mining; and aluminum, lead, and zinc are part of other nonferrous ores mining. Basic steel production is a single sector, as are primary copper, primary aluminum, primary zinc, and primary lead. Nickel alloys are partly included in the electrometallurgical sector and also in the sector other nonferrous primary metals. Secondary copper, aluminum, lead, zinc, and nickel are aggregated in the secondary nonferrous metals sector. Finally, all scrap is part of a single sector, scrap, used, and secondhand goods. Consumption is treated on a first-transactions basis rather than on the basis of end use, and all coefficients are expressed in 1972 dollars.

Another set of coefficients represents physical input-output relations controlling the production and consumption of nonfuel minerals. These coefficients are based on end use—that is, the last statistically identifiable sector in which a mineral substance is consumed, either in its original form (as ore or concentrate) or embodied in a finished product (lead in automobile batteries or copper in wire). This treatment of mineral inputs was necessary for several reasons. First, some of the metals are aggregated with other metals in some instances even at the most detailed, 496-order level of industrial classification. Second, the interindustry coefficients are expressed in dollars, not physical units. Third, modeling based on end-use consumption avoids distortions caused by differences in the level of aggregation.

There are separate coefficients for primary and secondary metal inputs for each mineral. The coefficients express, per billion dollars of each sector's output, the physical amount of contained minerals (in primary and secondary form) that is required. Each row, when multiplied by the solution vector, gives the complete supply-demand balance for the total amount of metal extracted from ore (or total metal recovered from scrap) used in all sectors of the economy. No distinctions are made among different commodities in which the metal is embodied, such as chemicals, metals, or refractories. The minerals coefficients were derived from physical supply-demand balance flows for 1972 taken from Bureau of Mines statistical reports [49, 50, 51, 52]. When consumption data provided by the Bureau of Mines were more aggregated than the IEA/USMIN level of industrial classification or when some portion of consumption was ascribed to other uses, specific additional assumptions had to be made to make end-use distributions. (See appendix B for the derivation of the base year input-output coefficients.)

Import-to-output ratios were derived for 106 commodities from the BEA input-output table and for minerals from import data reported by the Bureau of Mines [46, 49]. Sectoral requirements for labor were taken from a Bureau of Labor Statistics report [54] and energy requirements were taken from a study published by the Center for Advanced Computation at the University of Illinois at Urbana-Champaign [34]. Pollution-emissions data used to derive coefficients for 106 sectors came from two studies carried out at the National Bureau of Economic Research and Resources for the Future [12, 13].

GDP for 1972, 1980, 1990, and 2000 is represented in the model by seven column vectors that show the sectoral composition of personal consumption, inventory change, investment, exports, imports, and defense and nondefense government spending. For 1972, GDP data were taken from the BEA table [43]. For 1980, 1990, and 2000, official projections by the Bureau of Labor Statistics [53] were adapted for use in IEA/USMIN.

The difficult task of updating interindustry coefficients to reflect expected technological change was simplified by the use of updated matrices prepared for the World Model study [22]. The main effort was devoted to arriving at a reasonable appraisal of the future changes in mineral input coefficients—by surveying special studies of future trends in materials use, making extrapolations based on past trends, and then qualifying these crude estimates by discussions with experts on materials use. Labor and energy coefficients were also updated similarly.

Having reviewed the way nonfuel minerals were incorporated into the input-output table and a description of the structure of IEA/USMIN, we now can identify the variables that determine the amounts of minerals needed to satisfy future U.S. requirements to 2000. As mentioned in chapter 1, because the future can take any one of a number of different paths, by exploiting the flexibility of the input-output approach we can combine alternative sets of assumptions regarding the variables that enter into the determination of U.S. minerals requirements. Before presenting the alternative projections of future U.S. production and consumption of minerals, which are described in chapter 5, a detailed overview of the assumptions that were used in the design of the scenarios is presented in chapter 4.

Notes

1. BEA sectors 37 and 38 include the primary iron and steel manufacturing and primary nonferrous metals manufacturing sectors.

2. Noncompetitive imports or noncomparable imports are goods used to satisfy intermediate or final demand for which there is no corresponding domestic producing sector. Examples in the U.S. economy include coffee, bananas, and chromium.

4 Designing the Scenarios: The Assumptions

The input-output framework can be used to project the future levels of minerals and all other sectoral outputs needed to satisfy any given bill of final demand. Because the future is uncertain, it can take several, alternative paths, each based on one of several alternative sets of assumptions. In projecting the future levels of production and consumption of nonfuels minerals, it is necessary first to identify the determinants of the demand for minerals. The results of the computations presented in chapter 5 reflect the influence of prospective changes in the level of GDP, materials substitution resulting from technological change, changes in recycling rates, and changes in the degree of import dependence. The different sets of projections are based on changes in one, two, and, in some cases, simultaneous changes in all of the determinants of the demand for minerals. Before incorporating these projections in specific scenarios, we describe alternative assumptions about the changes introduced in GDP, materials substitution, recycling rates, and trade dependence.

GDP Projections

The size of the U.S. economy, measured by gross domestic product (GDP), was projected by the Bureau of Labor Statistics to grow in real terms from 1972 to 2000 at an average annual rate of 3.1 percent. Changes are also expected in the composition of the different goods and services entering into the different GDP components.[1] The GDP accounts and their actual 1972 and projected 2000 levels are shown in tables 4–1 and 4–2, respectively. In these two tables, different goods and services were grouped into ten broadly defined sectors, ranging from agriculture to mining, to dummy and special industries.[2] The total GDP was split into seven components: personal consumption, private capital investment, inventory change, imports, exports, government defense, and government nondefense expenditures. All quantities are measured in constant 1972 dollars.

Figures 4–1 and 4–2 complement these tables. Figure 4–1 shows—as a percentage of total GDP—the relative importance of each sector in the GDP accounts both for the base year, 1972, and in the projected target

Table 4–1
Composition of Final Demand in 1972
(billions of 1972 dollars)

Type of Economic Activity	*PCE*	*CAP*	*INV*	*EX*	*IM*	*NDG*	*DEF*	*Subtotal*
Agriculture, forestry, and fisheries	7.00	0.00	2.51	5.00	−2.04	−1.19	0.01	11.28
Mining	0.13	0.25	0.21	0.80	−4.07	0.03	0.00	−2.64
Construction	0.00	99.10	0.00	0.02	0.00	37.33	2.53	138.98
Manufacturing	214.42	72.76	13.65	37.58	−50.78	18.22	23.81	329.66
Transportation, communication, and utilities	55.10	3.38	0.53	5.76	−1.56	6.26	3.42	72.89
Wholesale and retail trade	140.00	10.20	1.00	4.10	2.99	1.94	0.77	161.00
Finance, insurance, and real estate	148.00	4.43	0.00	2.28	−0.16	4.81	0.16	159.52
Services	158.74	0.19	−0.16	1.34	−0.05	12.92	4.08	177.05
Government enterprises	4.16	0.00	0.00	0.14	0.00	0.81	0.22	5.34
Dummy and special industries[a]	10.53	−5.33	−7.38	15.80	−20.45	−0.20	0.00	−7.03
Subtotal	738.08	184.98	10.36	72.81	−76.13	80.93	35.01	
							GDP = 1,046.03[b]	

Notes:
PCE = Personal consumption expenditures
CAP = Gross private domestic fixed investment
INV = Change in business inventories
EX = Exports
IM = Imports
NDG = Government purchases, nondefense
DEF = Government purchases, defense

[a]To compare with the projected GDP for the year 2000, significantly large entries (government industries into NDG and DEF) are purposely deleted.
[b]Figure does not include government industries into NDF (35.6) and DEF (96.4).

Table 4–2
Composition of Final Demand in 2000
(billions of 1972 dollars)

Type of Economic Activity	*PCE*	*CAP*	*INV*	*EX*	*IM*	*NDG*	*DEF*	*Subtotal*
Agriculture, forestry, and fisheries	14.65	0.00	0.00	13.02	−3.23	−0.95	0.12	23.61
Mining	0.43	0.00	0.00	2.49	−16.38	0.31	0.07	−13.08
Construction	0.00	167.20	0.00	0.08	0.00	38.01	4.62	209.91
Manufacturing	564.55	193.24	0.00	128.80	−135.18	27.89	37.54	816.85
Transportation, communication, and utilities	155.00	8.56	0.00	19.47	−2.16	10.39	4.11	195.37
Wholesale and retail trade	321.00	27.57	0.00	13.40	9.02	3.55	1.85	376.39
Finance, insurance, and real estate	405.00	8.01	0.00	2.96	1.19	10.33	0.06	427.54
Services	400.90	0.00	0.00	3.26	0.00	25.80	5.18	435.15
Government enterprises	6.08	0.00	0.00	0.11	0.00	2.02	0.29	8.51
Dummy and special industries[a]	0.00	0.00	0.00	0.00	0.00	0.00	0.00	0.00
Subtotal	1,867.61	404.58	0.00	183.61	−146.74	117.36	53.84	
							GDP = 2,480.25[b]	

Notes:
PCE = Personal consumption expenditures
CAP = Gross private domestic fixed investment
INV = Change in business inventories
EX = Exports
IM = Imports
NDG = Government purchases, nondefense
DEF = Government purchases, defense

[a]Not available.

[b]Figure does not include dummy and special industries.

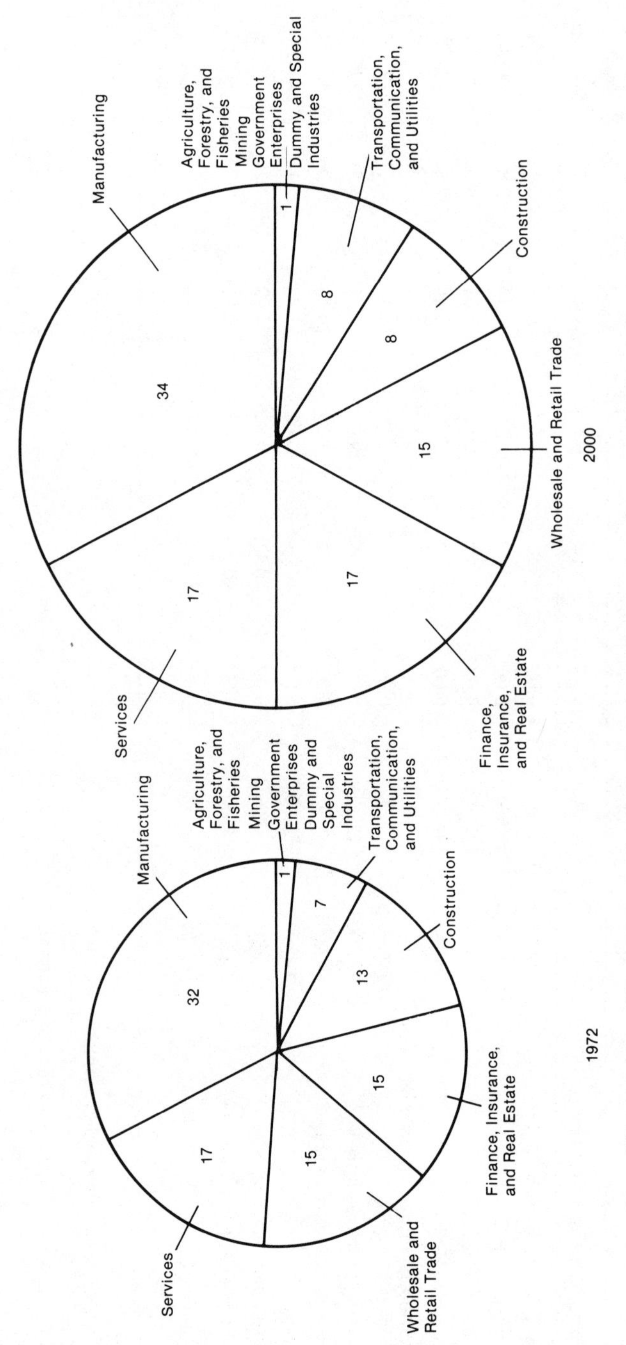

Figure 4–1. Final Demand by Type of Economic Activity in 1972 and 2000 (Percentage)

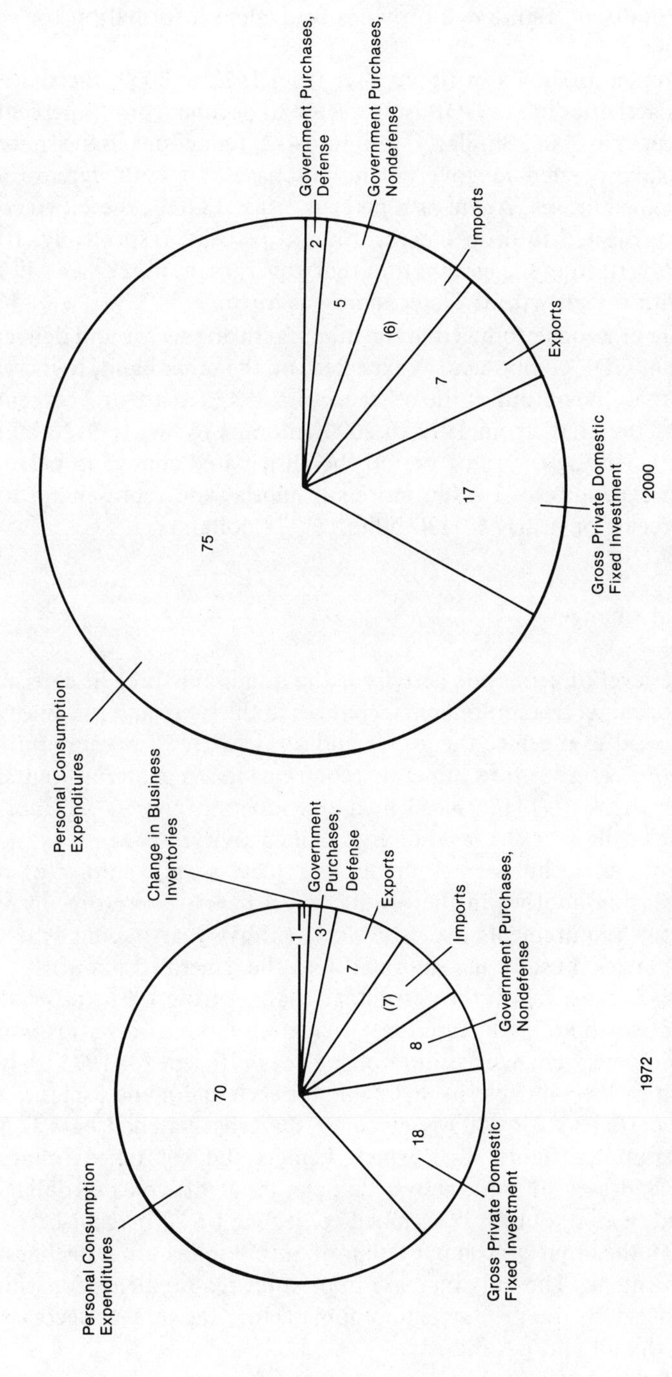

Figure 4–2. Final Demand by GDP Component in 1972 and 2000 (Percentage)

year, 2000. Similarly, figure 4–2 provides equivalent information for each GDP component.[3]

For example, as shown in figure 4–1, from 1972 to 2000, the share of construction activities in the GDP is projected to decline from 13 percent in 1972 to 8 percent in 2000. Similarly, in figure 4–2, reductions in the share of GDP are also projected in government purchases, in both defense and nondefense expenditures. Again as a percent of total GDP, these two components are projected to drop 1 percent and 3 percent, respectively, from 1972 to 2000, reflecting the assumption that government purchases will not keep pace with the growth of the economy as a whole.

The value of goods produced in the manufacturing sector and delivered to the different GDP components is expected, on the other hand, to increase much faster than the output of the other sectors. An increase of 2 percent in their share of the GDP from 1972 t0 2000 amounts to nearly $120 billion (1972 dollars). During the same period the anticipated change in personal consumption expenditures is even more substantial and represents a total gain of 5 percent, or nearly $1,130 billion (1972 dollars).

Technological Change

Although the level of economic activity is the dominant force in determining the level of minerals requirements, changes in the types and the amounts of materials used to produce the goods and services are of paramount importance in projecting future minerals requirements. A material that, for whatever reason, is no longer used in the production process will not be demanded, regardless of the level of economic activity.

The history of technological change is replete with examples of one material displacing another in the production process. Therefore, in projecting minerals requirements over a period of thirty years, some expected qualitative changes must be incorporated into the general framework.

Figures 4–3 to 4–28 show the percentage changes from 1972 to 2000 that are projected for the mineral input coefficients for those scenarios which employ technological-change assumptions. The coefficient for 1972 is given as 100 percent by the white bar in each table for each major end use; the corresponding coefficient for 2000 is given by the cross-hatched bars to the right. For example, figure 4–3, which depicts the coefficient changes assumed for end-uses of iron, shows that the input of iron per dollar of plumbing and heating output is assumed to decline by 20 percent between 1972 and 2000; the input of iron per dollar of automobile output declines by the same percentage. The only increase in per-unit requirements for iron is assumed to occur in the primary aluminum sector; the largest decrease is assumed for the oil and gas industry.

Although not shown in the tables, the same changes applied to primary

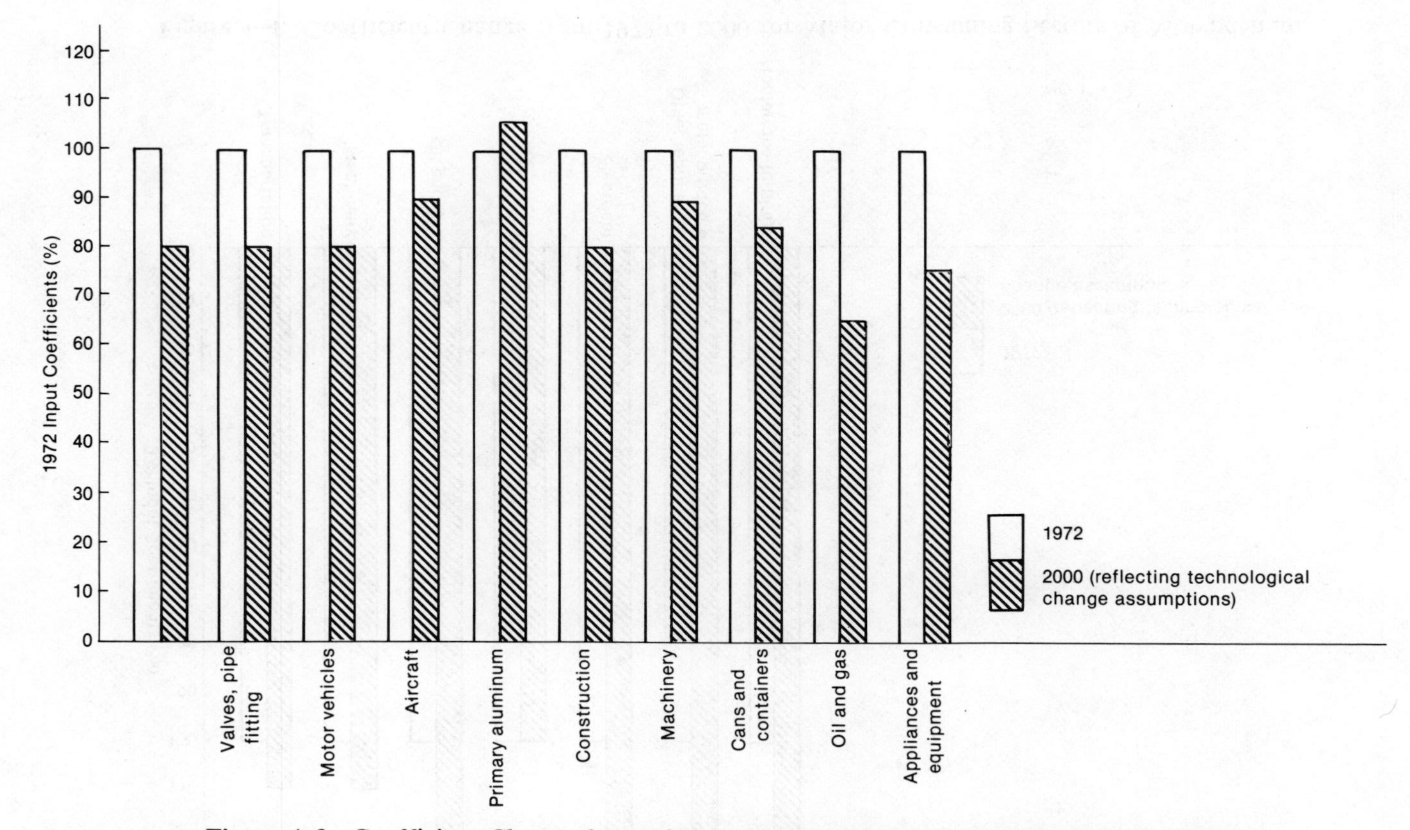

Figure 4–3. Coefficient Change from 1972 to 2000 for Major Consuming Sectors of Iron

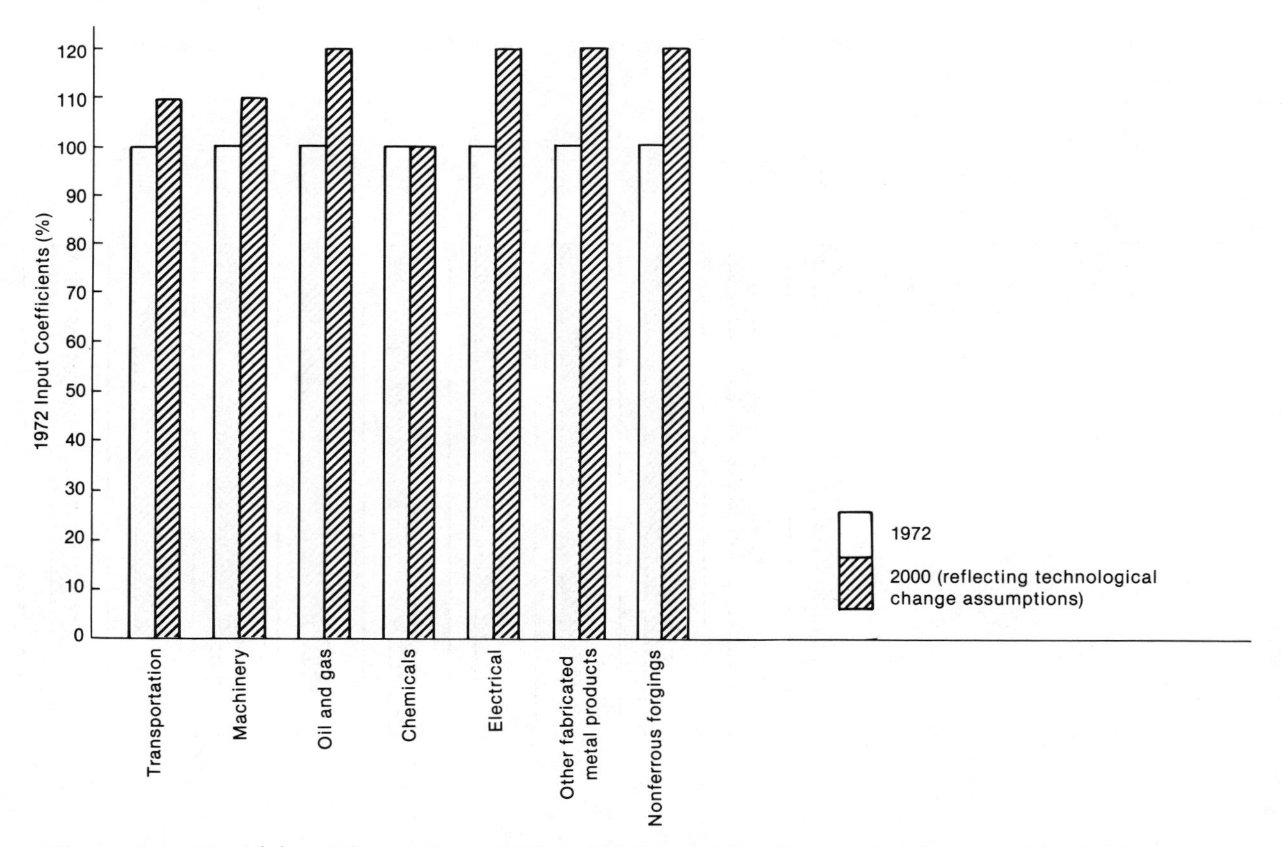

Figure 4–4. Coefficient Change from 1972 to 2000 for Major Consuming Sectors of Molybdenum

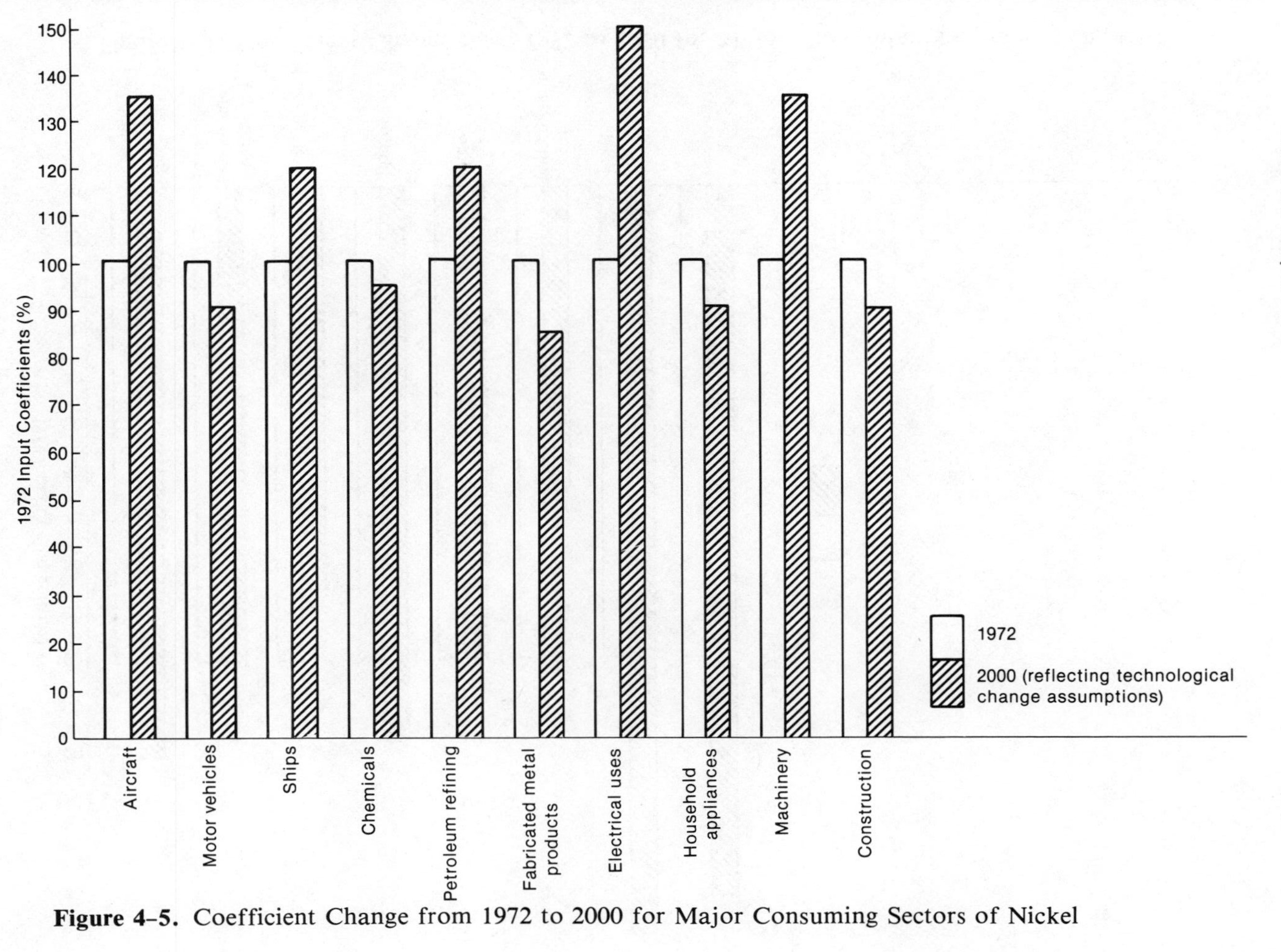

Figure 4–5. Coefficient Change from 1972 to 2000 for Major Consuming Sectors of Nickel

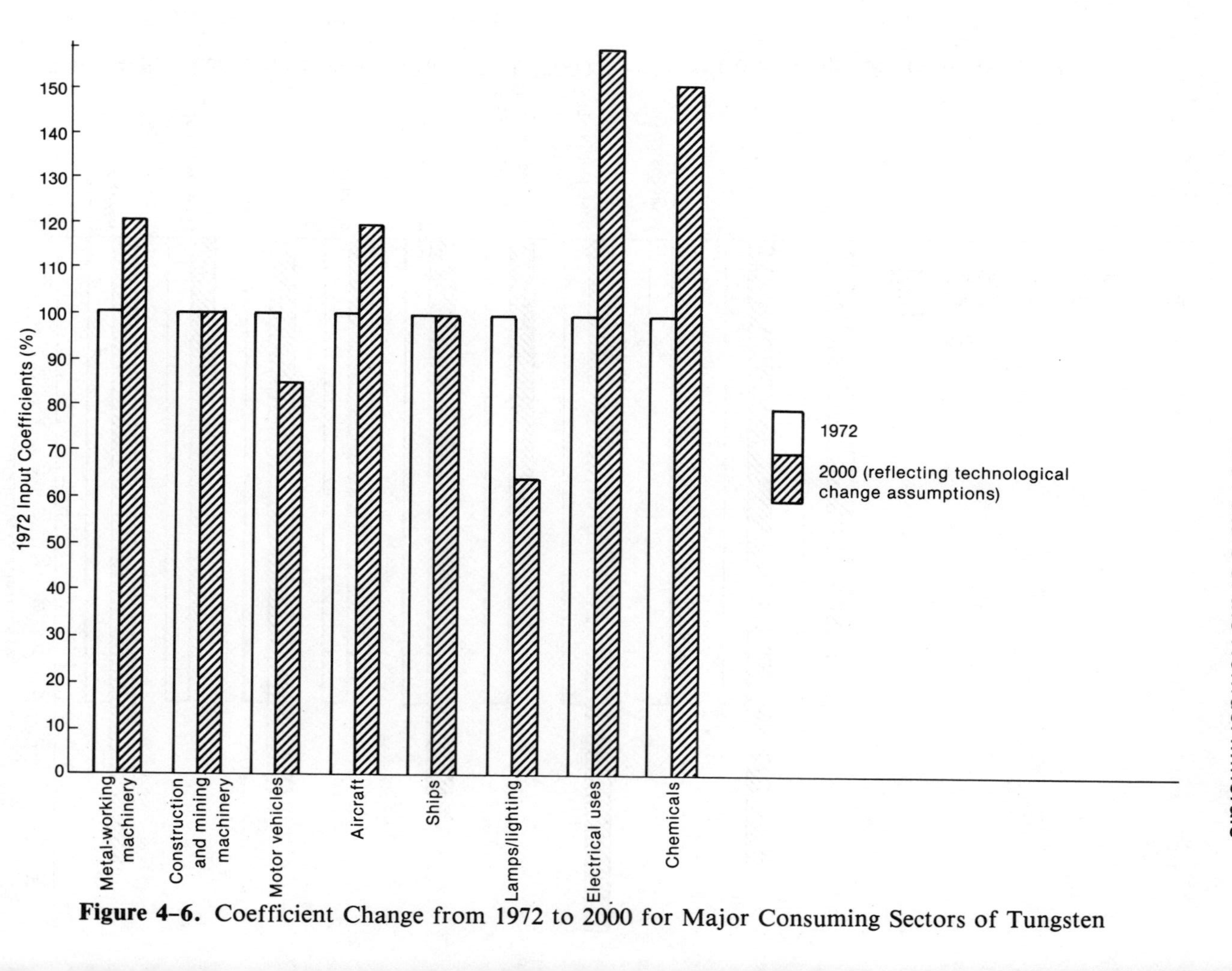

Figure 4–6. Coefficient Change from 1972 to 2000 for Major Consuming Sectors of Tungsten

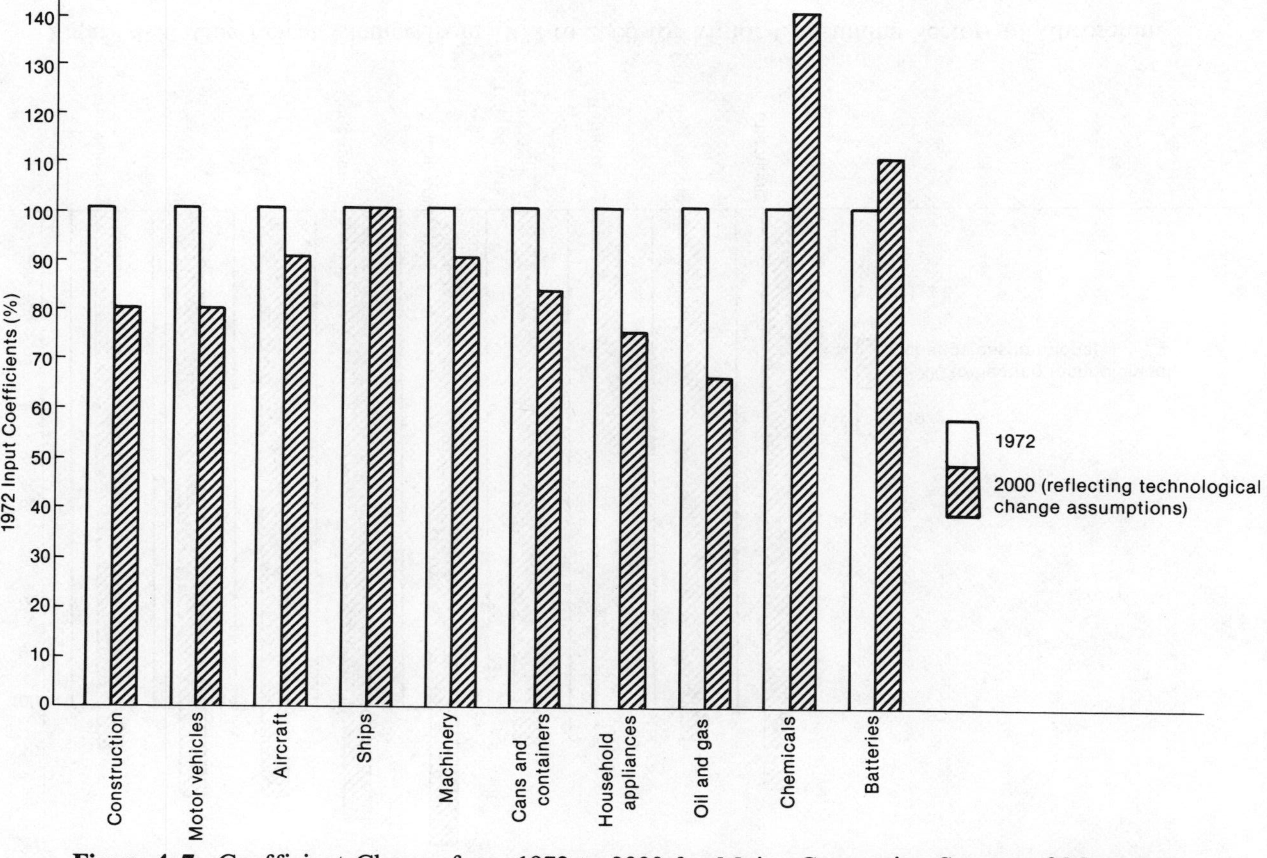

Figure 4–7. Coefficient Change from 1972 to 2000 for Major Consuming Sectors of Manganese

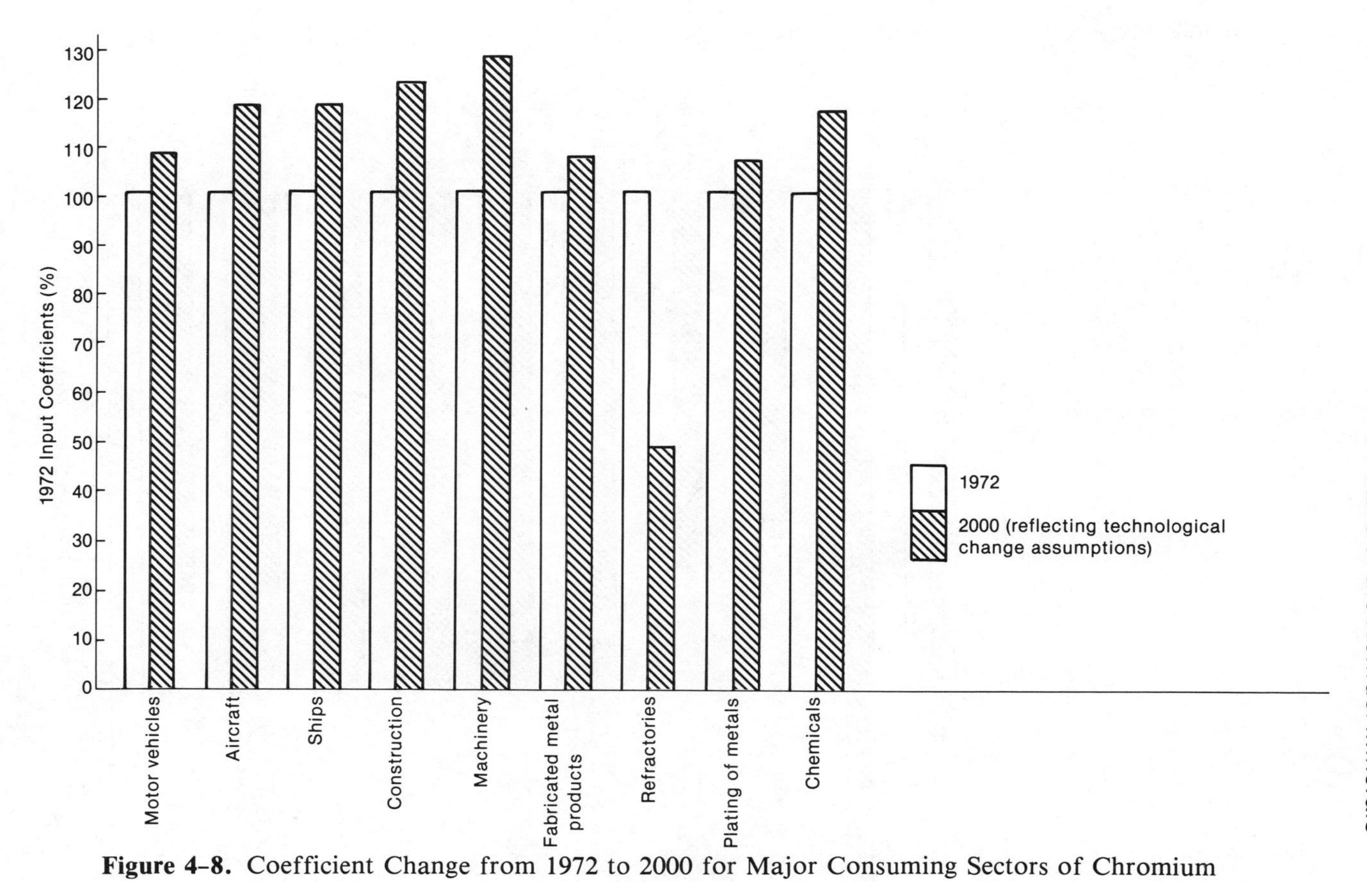

Figure 4-8. Coefficient Change from 1972 to 2000 for Major Consuming Sectors of Chromium

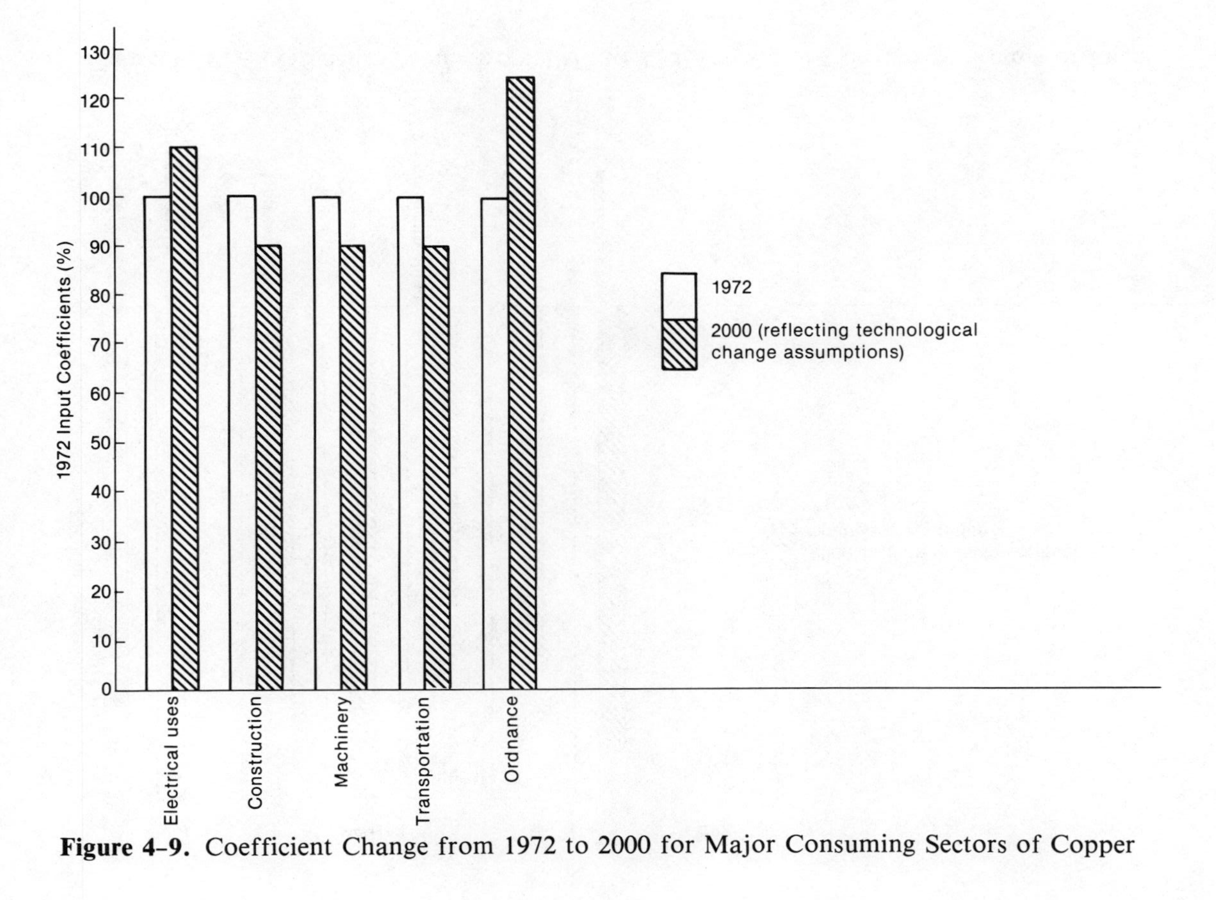

Figure 4–9. Coefficient Change from 1972 to 2000 for Major Consuming Sectors of Copper

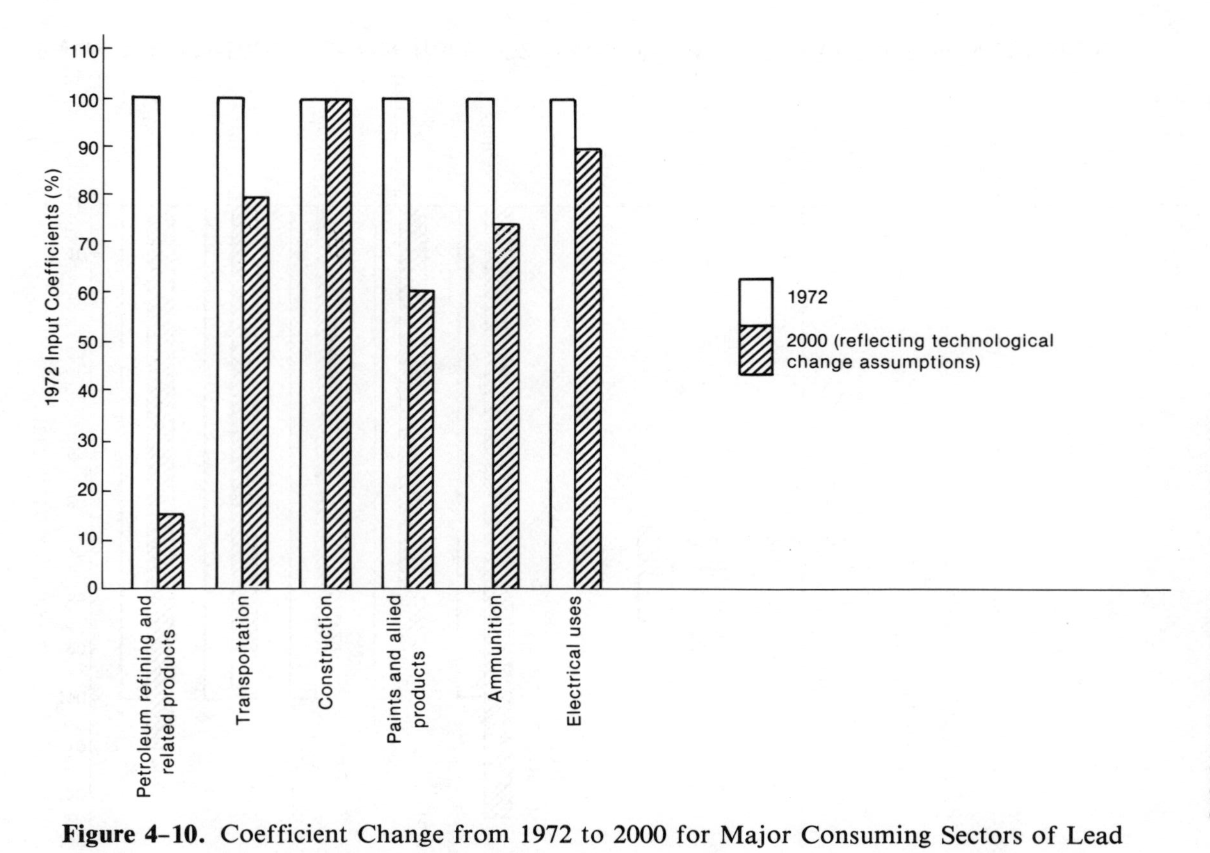

Figure 4–10. Coefficient Change from 1972 to 2000 for Major Consuming Sectors of Lead

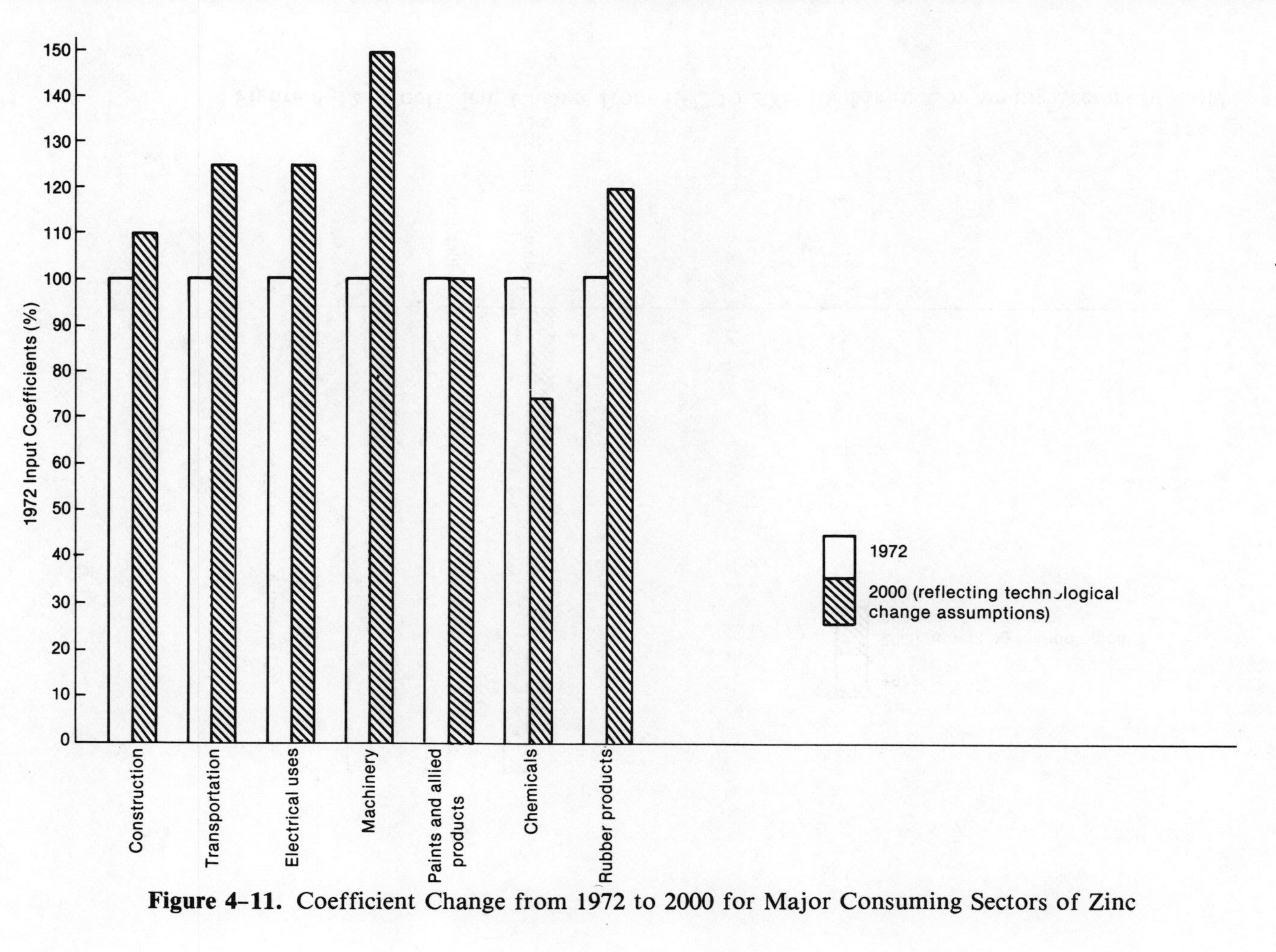

Figure 4–11. Coefficient Change from 1972 to 2000 for Major Consuming Sectors of Zinc

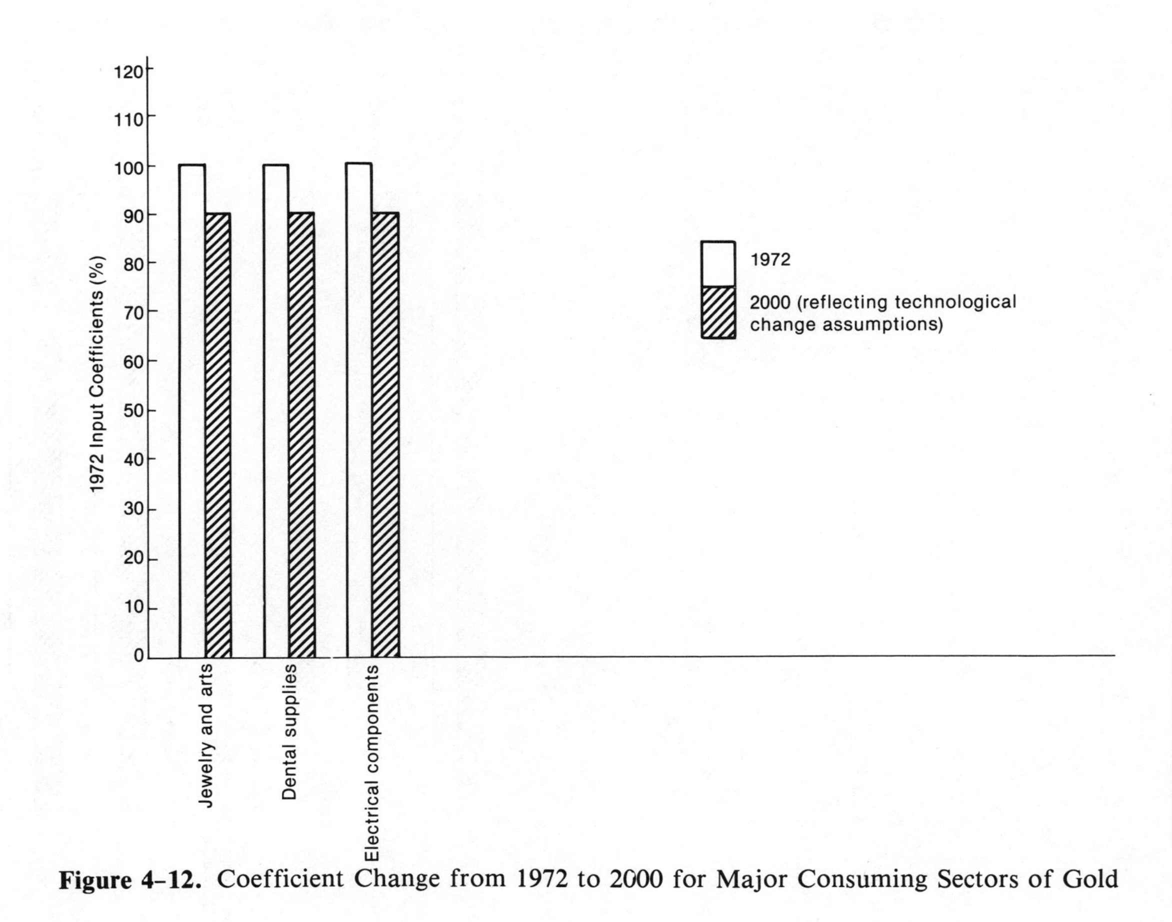

Figure 4–12. Coefficient Change from 1972 to 2000 for Major Consuming Sectors of Gold

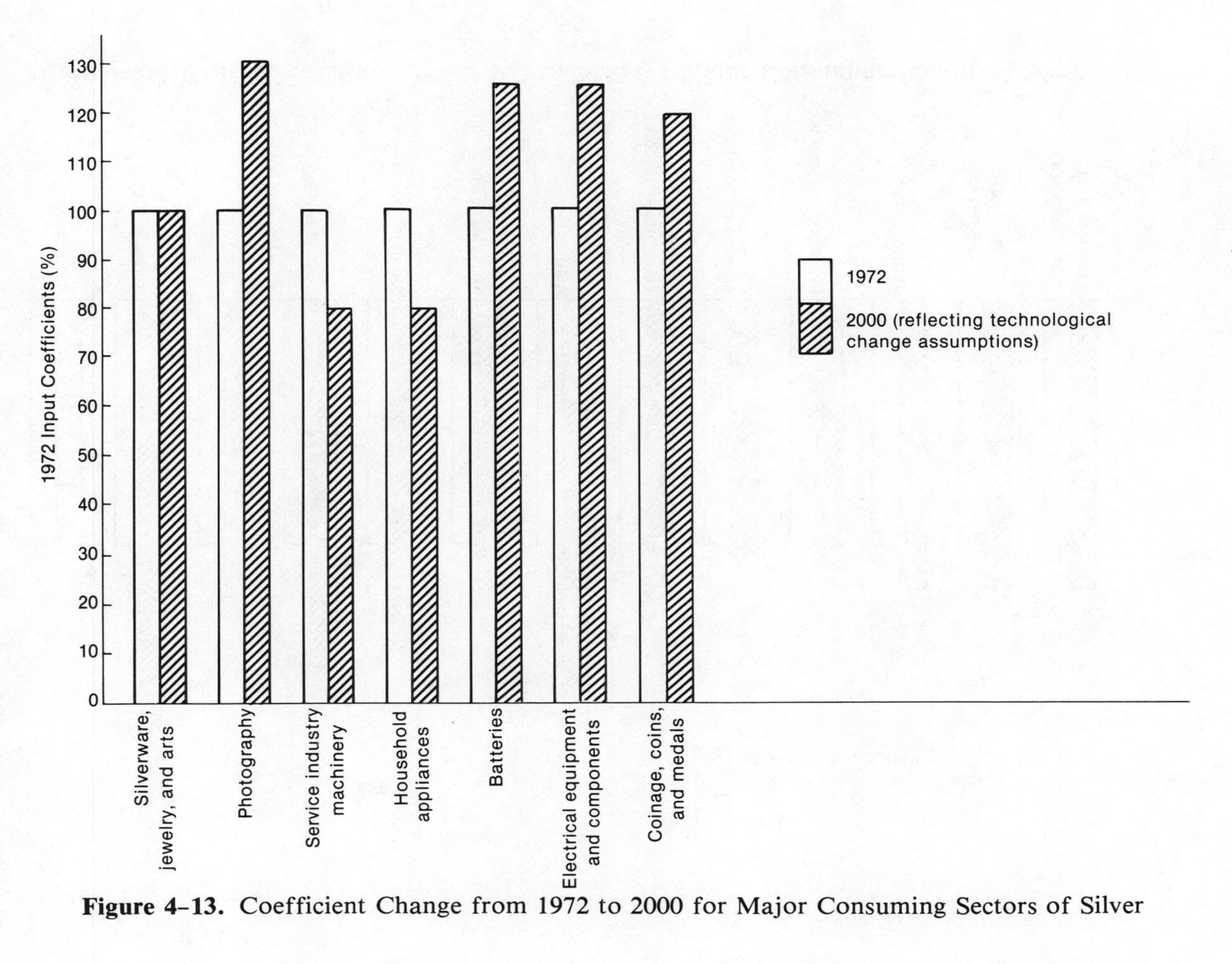

Figure 4–13. Coefficient Change from 1972 to 2000 for Major Consuming Sectors of Silver

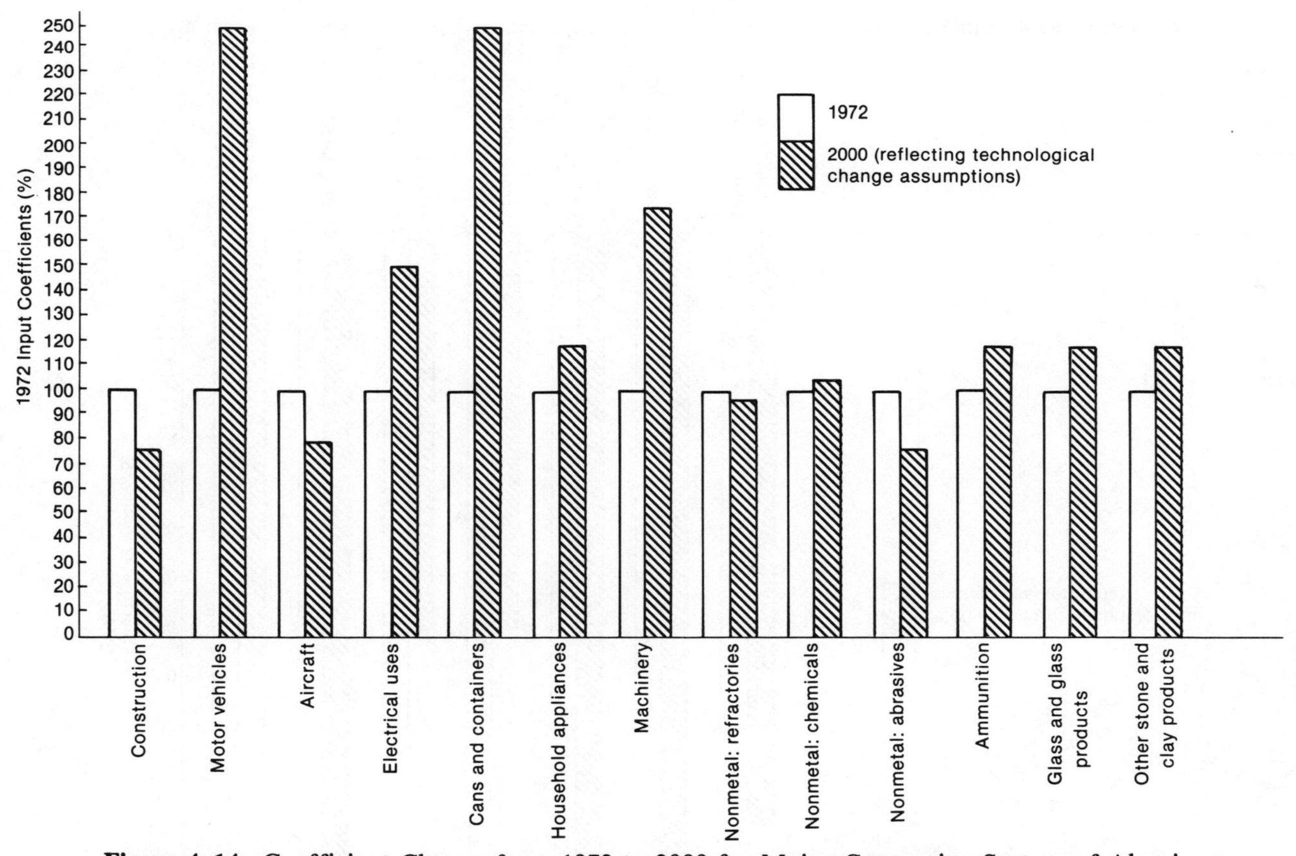

Figure 4–14. Coefficient Change from 1972 to 2000 for Major Consuming Sectors of Aluminum

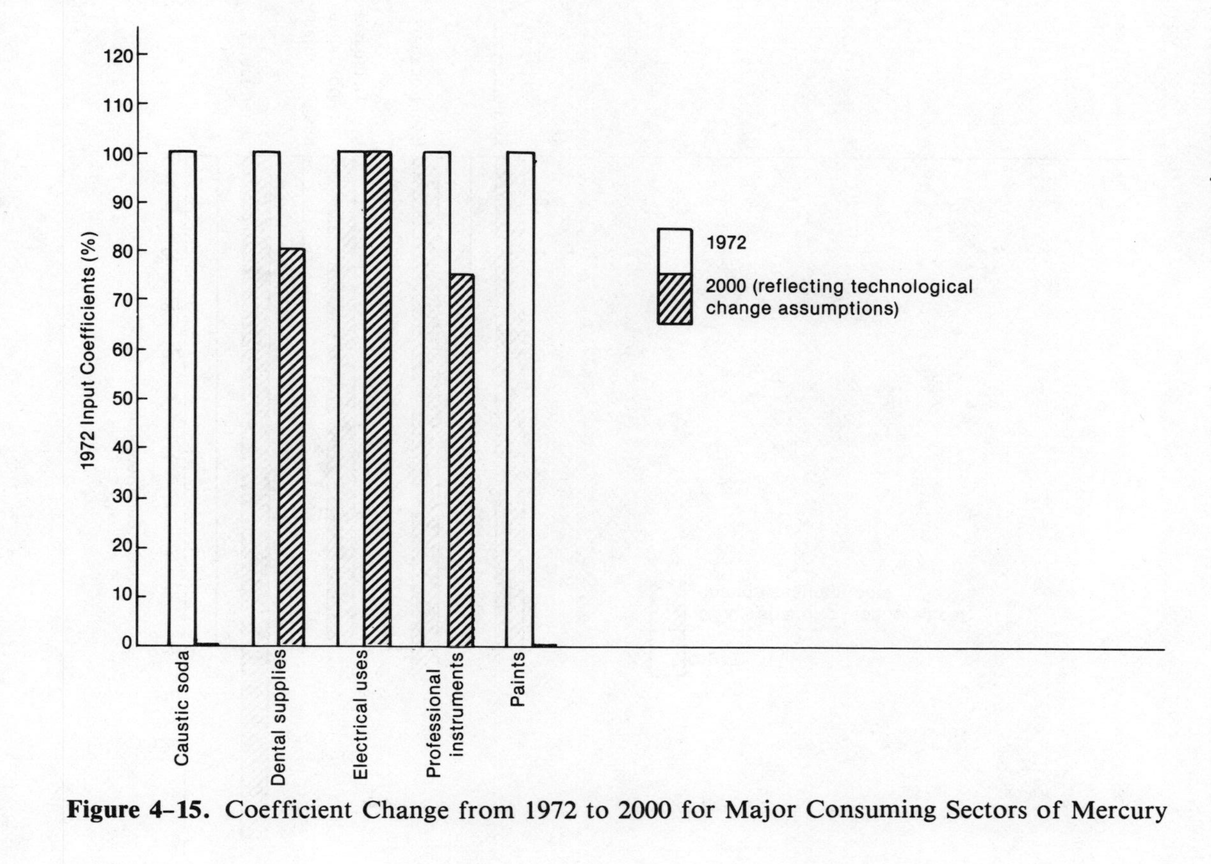

Figure 4–15. Coefficient Change from 1972 to 2000 for Major Consuming Sectors of Mercury

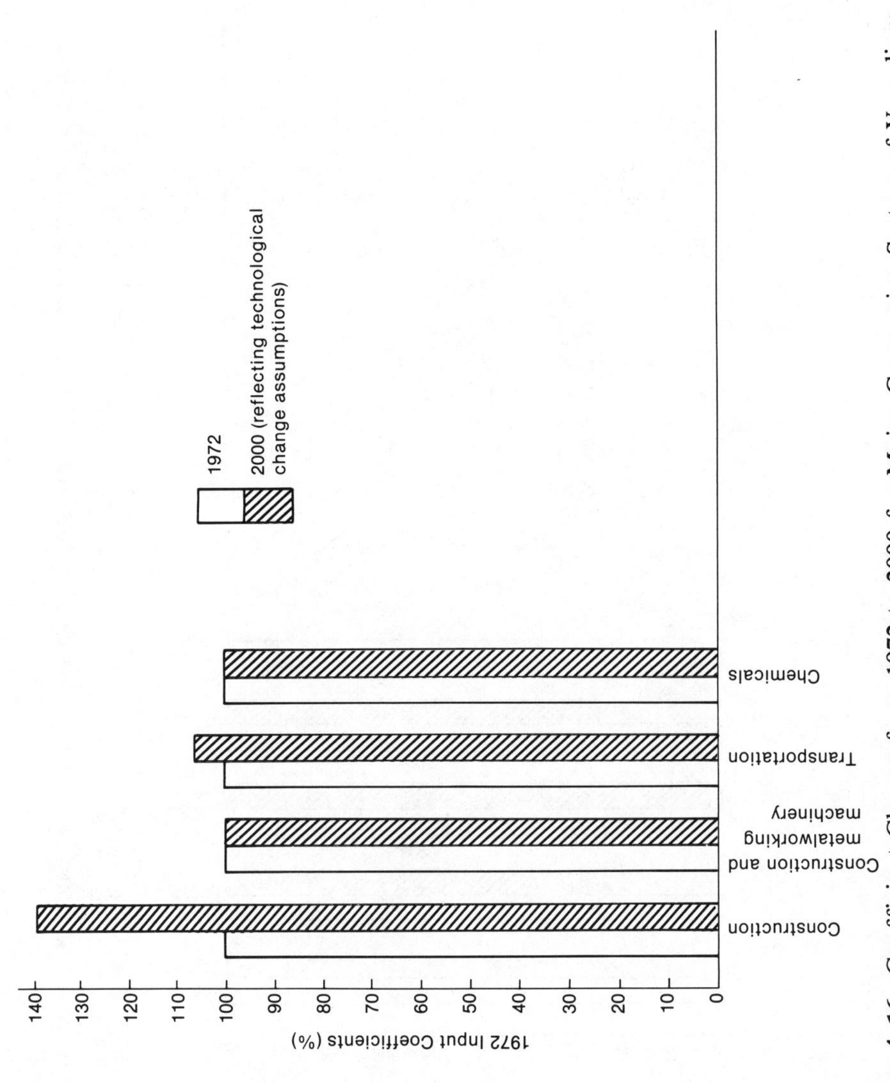

Figure 4–16. Coefficient Change from 1972 to 2000 for Major Consuming Sectors of Vanadium

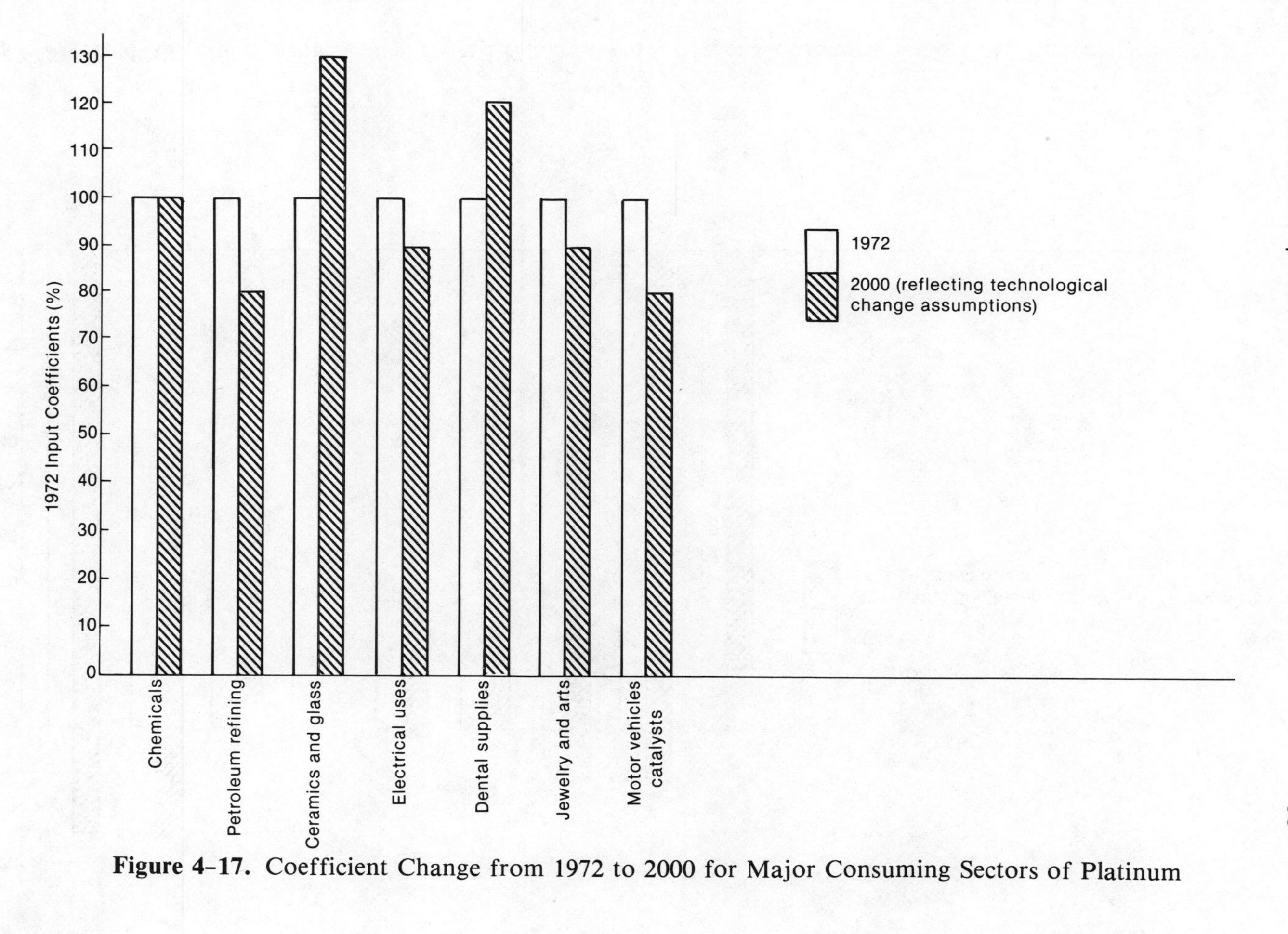

Figure 4-17. Coefficient Change from 1972 to 2000 for Major Consuming Sectors of Platinum

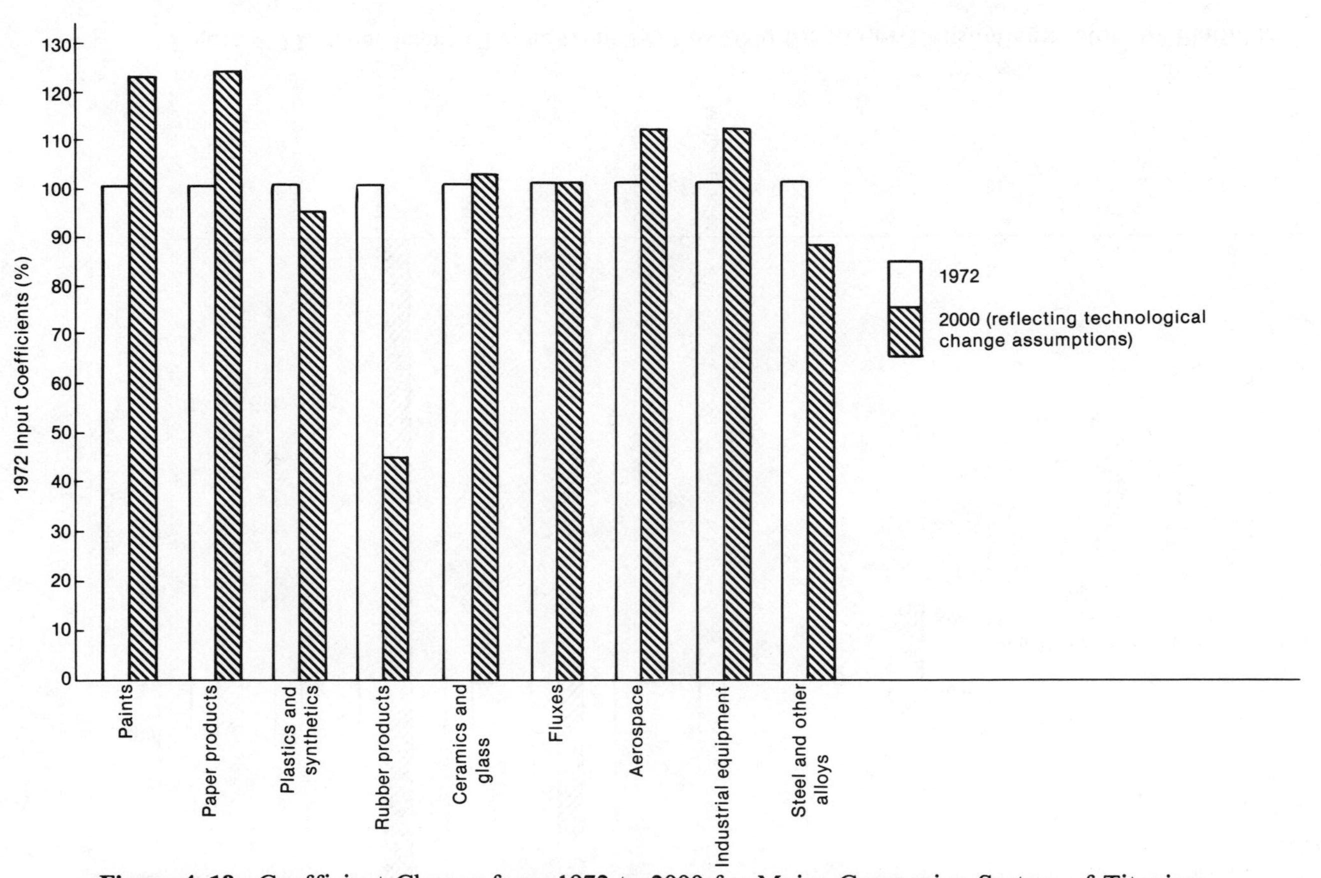

Figure 4–18. Coefficient Change from 1972 to 2000 for Major Consuming Sectors of Titanium

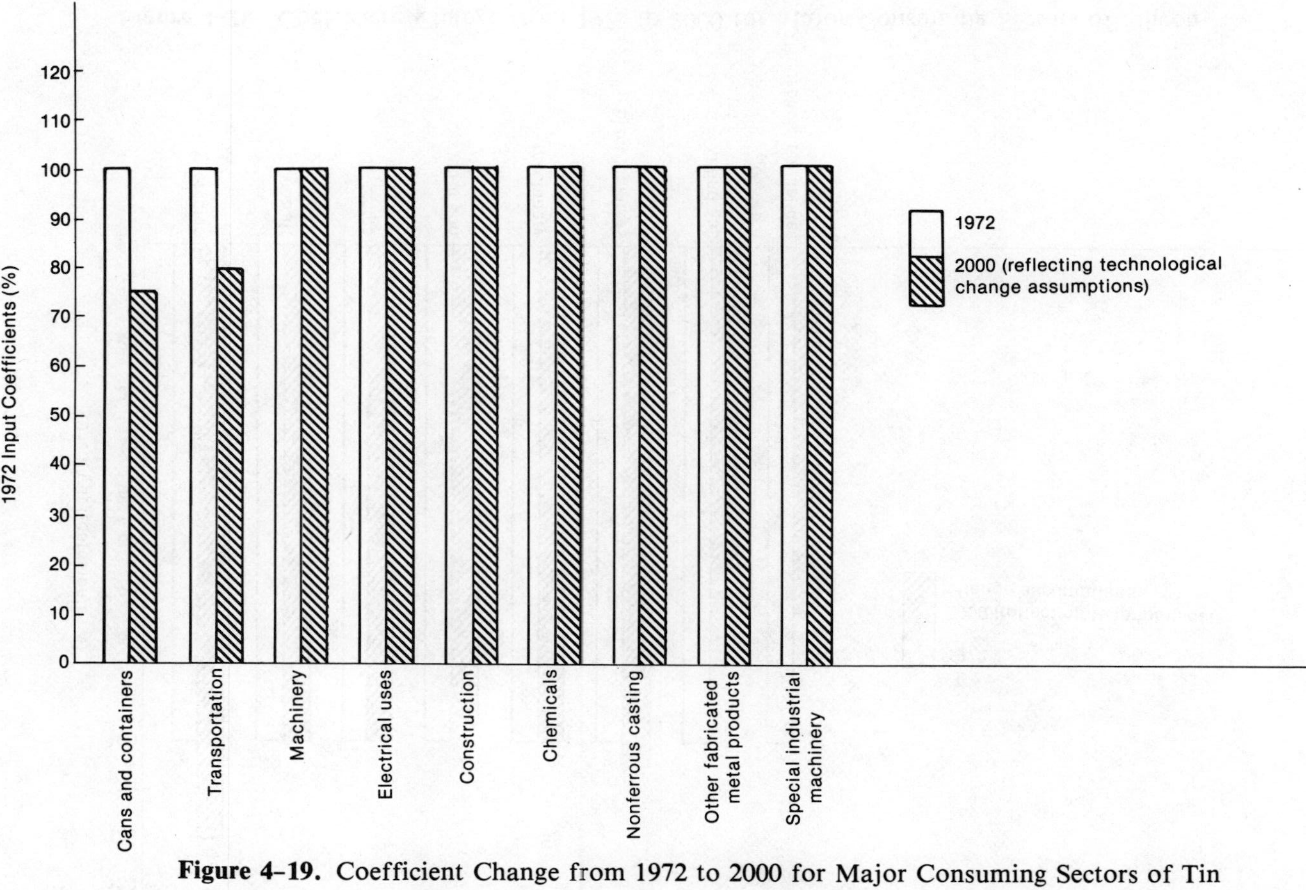

Figure 4–19. Coefficient Change from 1972 to 2000 for Major Consuming Sectors of Tin

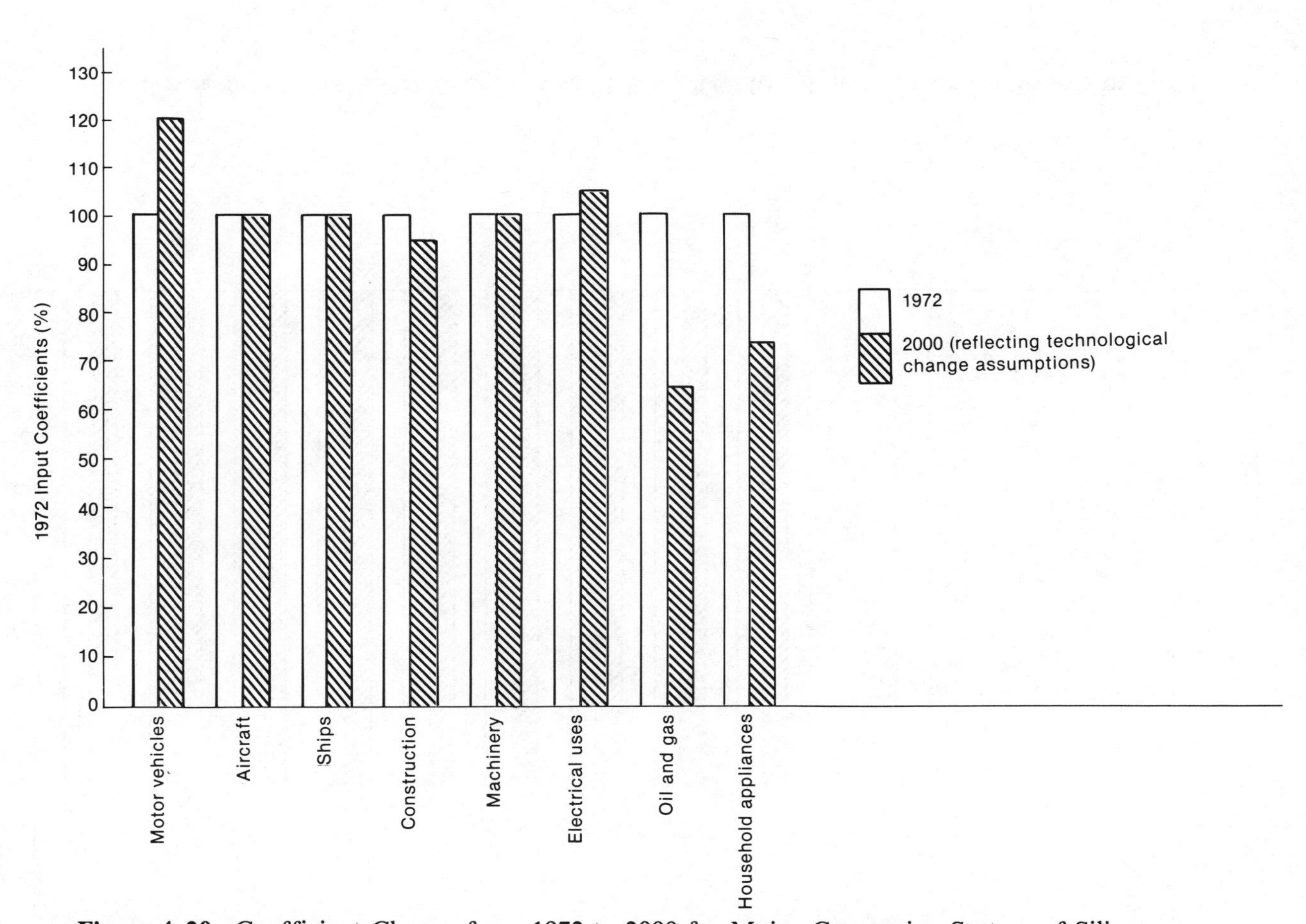

Figure 4–20. Coefficient Change from 1972 to 2000 for Major Consuming Sectors of Silicon

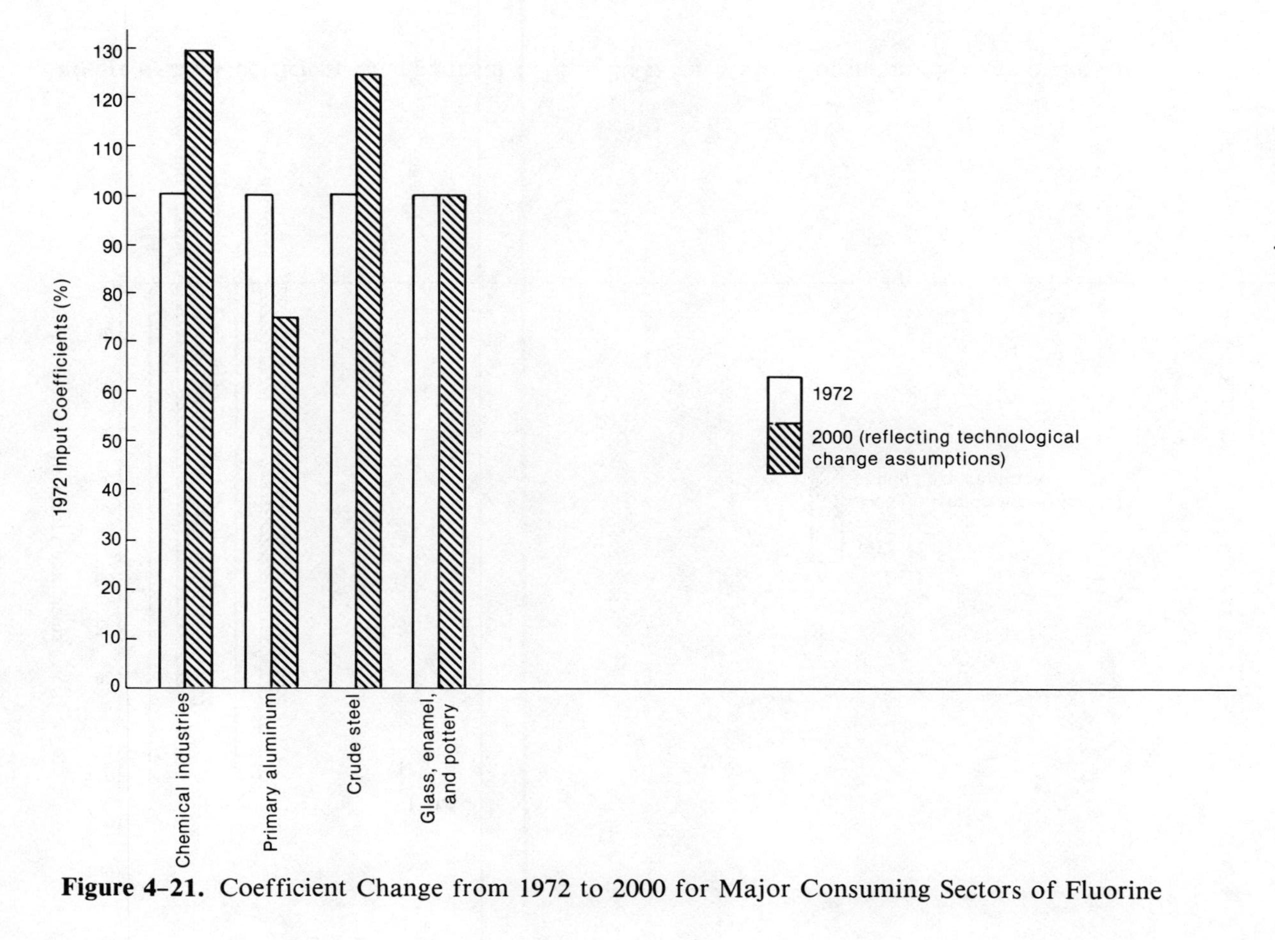

Figure 4–21. Coefficient Change from 1972 to 2000 for Major Consuming Sectors of Fluorine

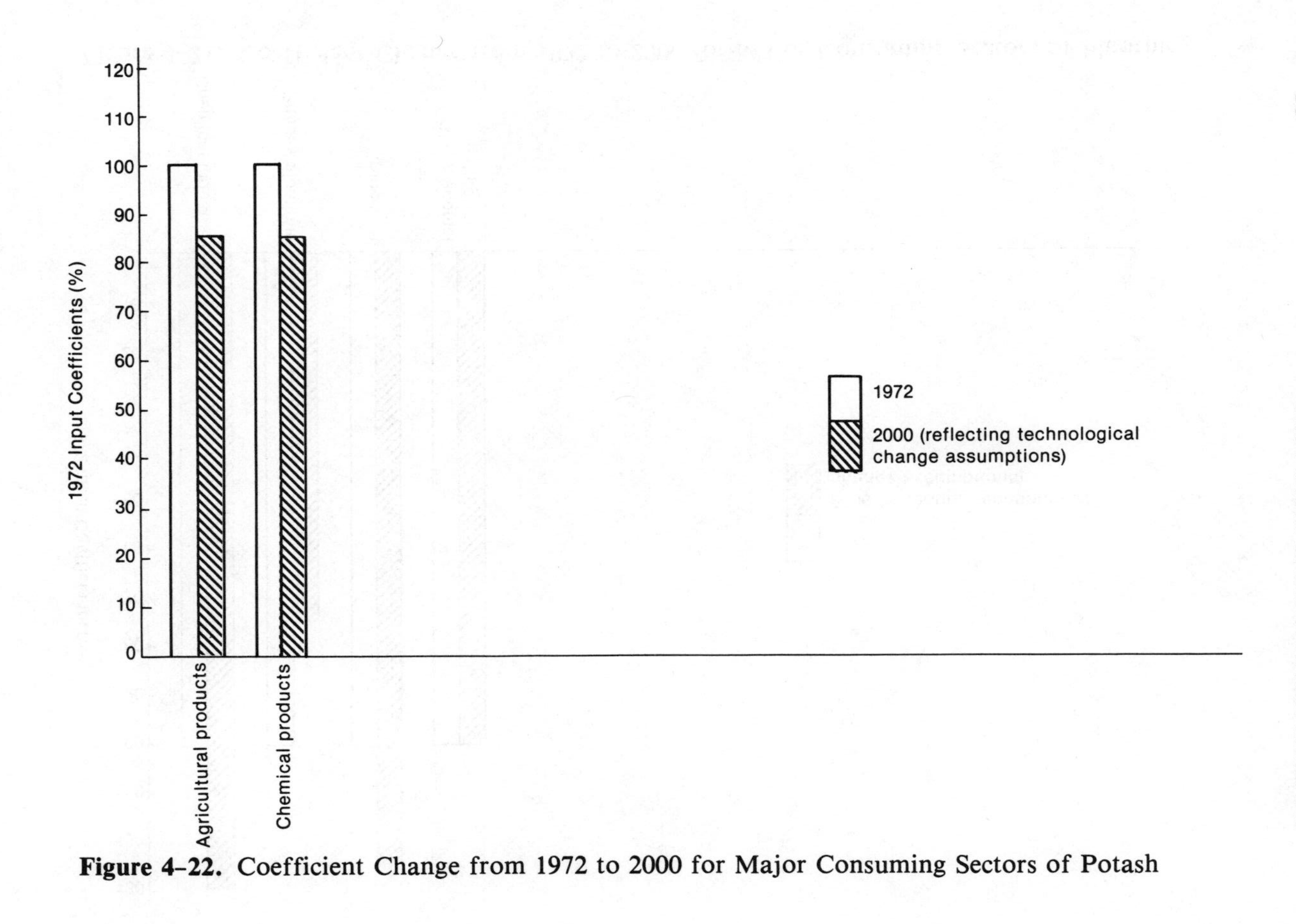

Figure 4–22. Coefficient Change from 1972 to 2000 for Major Consuming Sectors of Potash

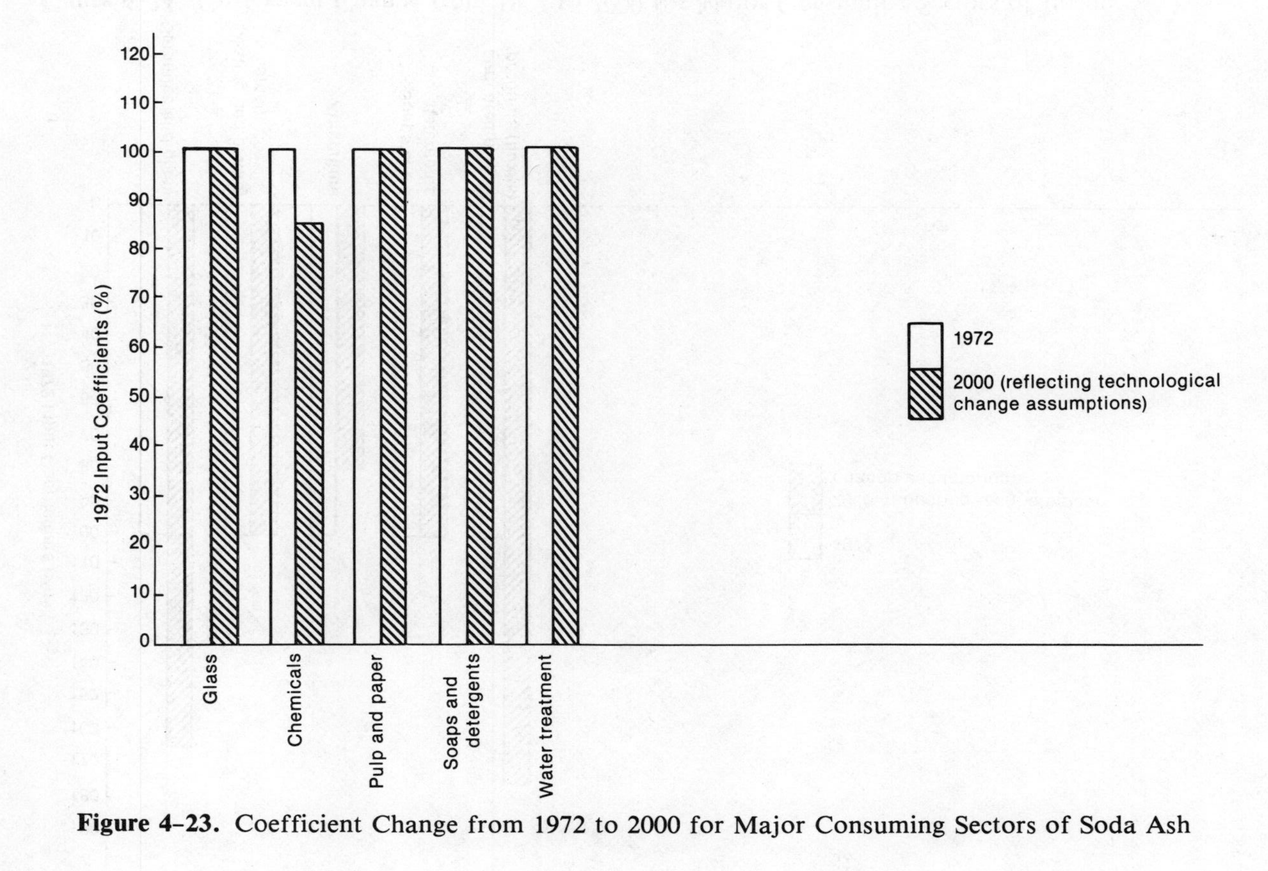

Figure 4–23. Coefficient Change from 1972 to 2000 for Major Consuming Sectors of Soda Ash

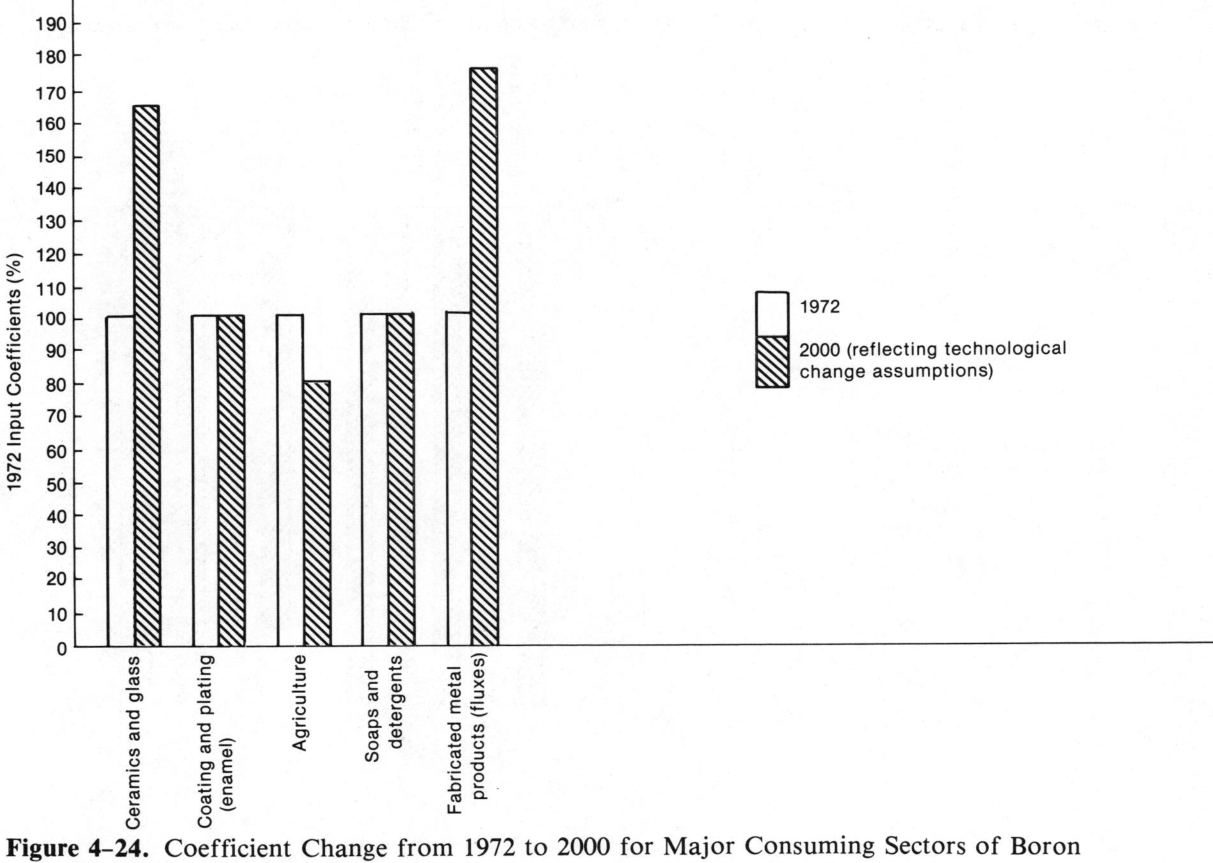

Figure 4–24. Coefficient Change from 1972 to 2000 for Major Consuming Sectors of Boron

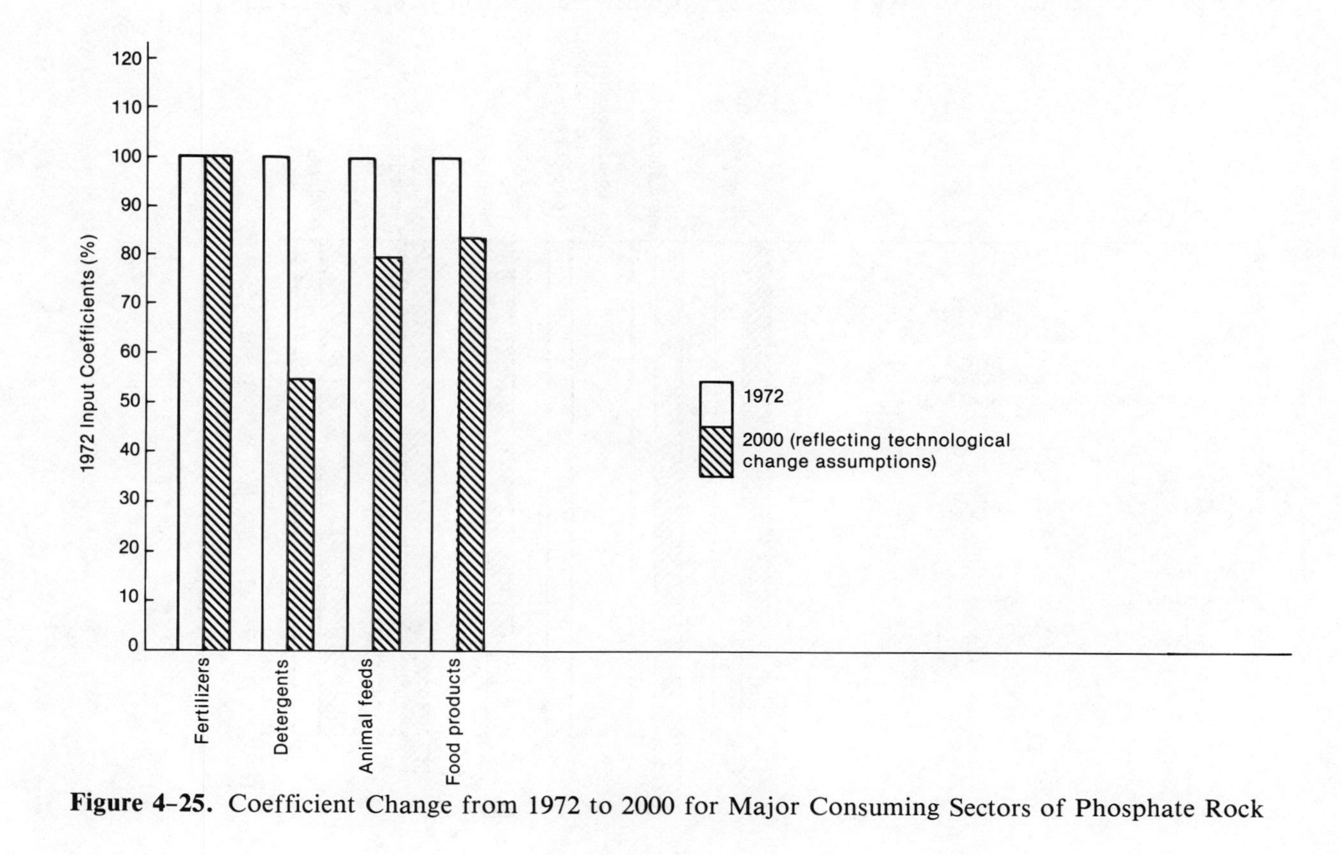

Figure 4–25. Coefficient Change from 1972 to 2000 for Major Consuming Sectors of Phosphate Rock

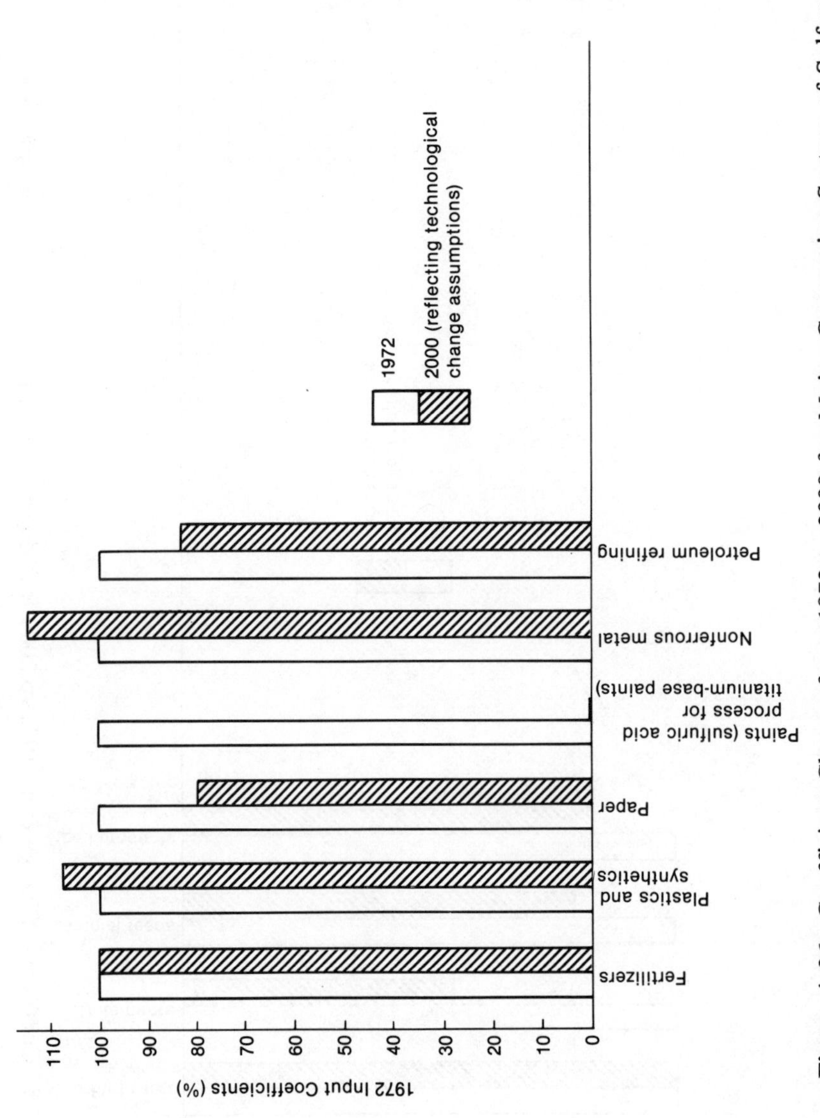

Figure 4–26. Coefficient Change from 1972 to 2000 for Major Consuming Sectors of Sulfur

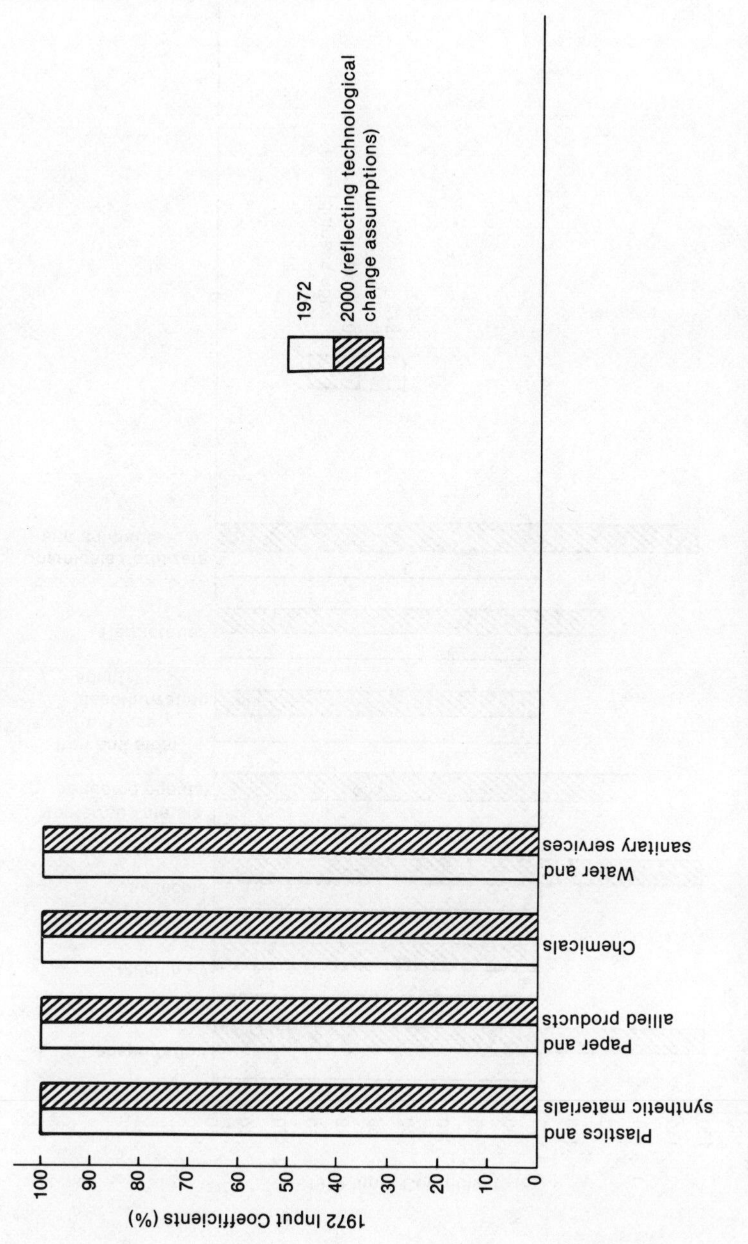

Figure 4–27. Coefficient Change from 1972 to 2000 for Major Consuming Sectors of Chlorine

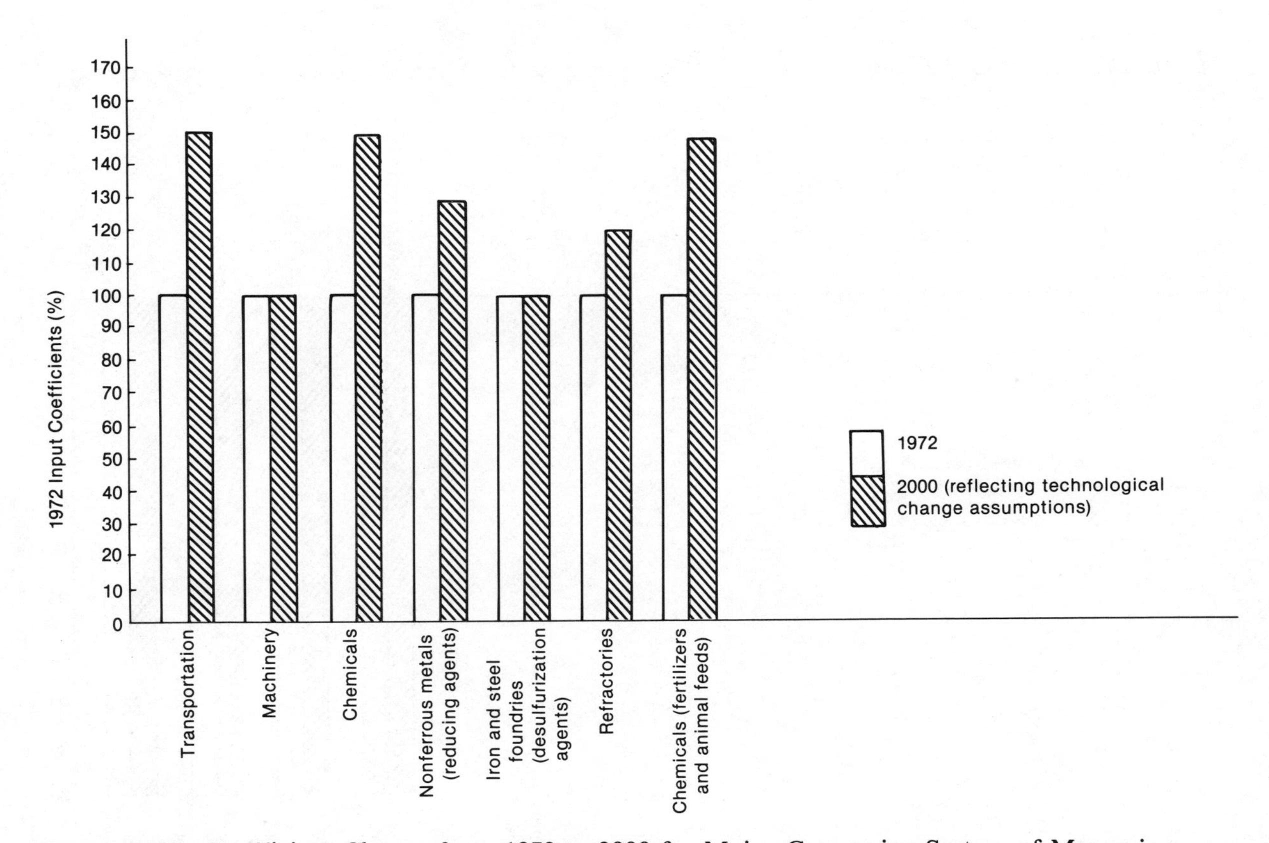

Figure 4-28. Coefficient Change from 1972 to 2000 for Major Consuming Sectors of Magnesium

inputs of metal are also applied to scrap inputs because judgments made about future use concerned total metal inputs regardless of source. (For a detailed description of the sources used to introduce changes in the minerals coefficients, see appendix C.)

Changes in Recycling Rates

Another important contribution to the supply of nonfuel minerals—in particular, to the supply of metals—is secondary production or recycled metals. Over 40 percent of total requirements for some metals are today supplied by recycled metals. Consequently, hypothetical changes in the rate of recycling are introduced systematically into the model.

Three different assumptions were selected as the basis for alternative projections of future requirements of primary and secondary metals: base, zero, and high recycling assumptions. The recycling rates introduced under the two latter assumptions are the upper and lower limits of the potential contribution of recycling. The recycling rates assumed for the base and high recycling cases are given in figure 4-29; under the zero recycling assumption, all metal inputs are assumed to come from primary sources and, as the name implies, the recycling rate is assumed to be zero.

Recycling rates are given indirectly in the primary and scrap input coefficients for each metal. Scrap coefficients refer to combined purchased scrap (prompt industrial plus old scrap). The primary and scrap coefficients show total metal inputs required per billion dollars of sectoral output. The recycling rate is the ratio of the scrap coefficient to the sum of the scrap and primary coefficients.

Under the base recycling assumption, the primary and scrap coefficients for 1972 are also used for 1980, 1990, and 2000. As a result, the recycling rate is a constant value for the entire period. This conservative view of future recycling rates resembles that of the Bureau of Mines projections, which are based on expectations of only modest increases in recycling rates for most metals over the next two decades.

Under the zero recycling assumption, scrap coefficients for all four years are reduced to zero, whereas primary coefficients are stepped up accordingly to equal total metal requirements per billion dollars of sectoral output. Under the high recycling assumption on the other hand, the values of scrap and primary coefficients for 2000 are changed to reflect estimates of the technically maximum recycling rate. Coefficients for 1980 are the same as for 1972 and coefficients for 1990 are taken as halfway between the 1980 and 2000 values.

For example, if the potential recovery of iron is assumed to be twice the current rate—so that 60 rather than 30 percent of total metal units are sup-

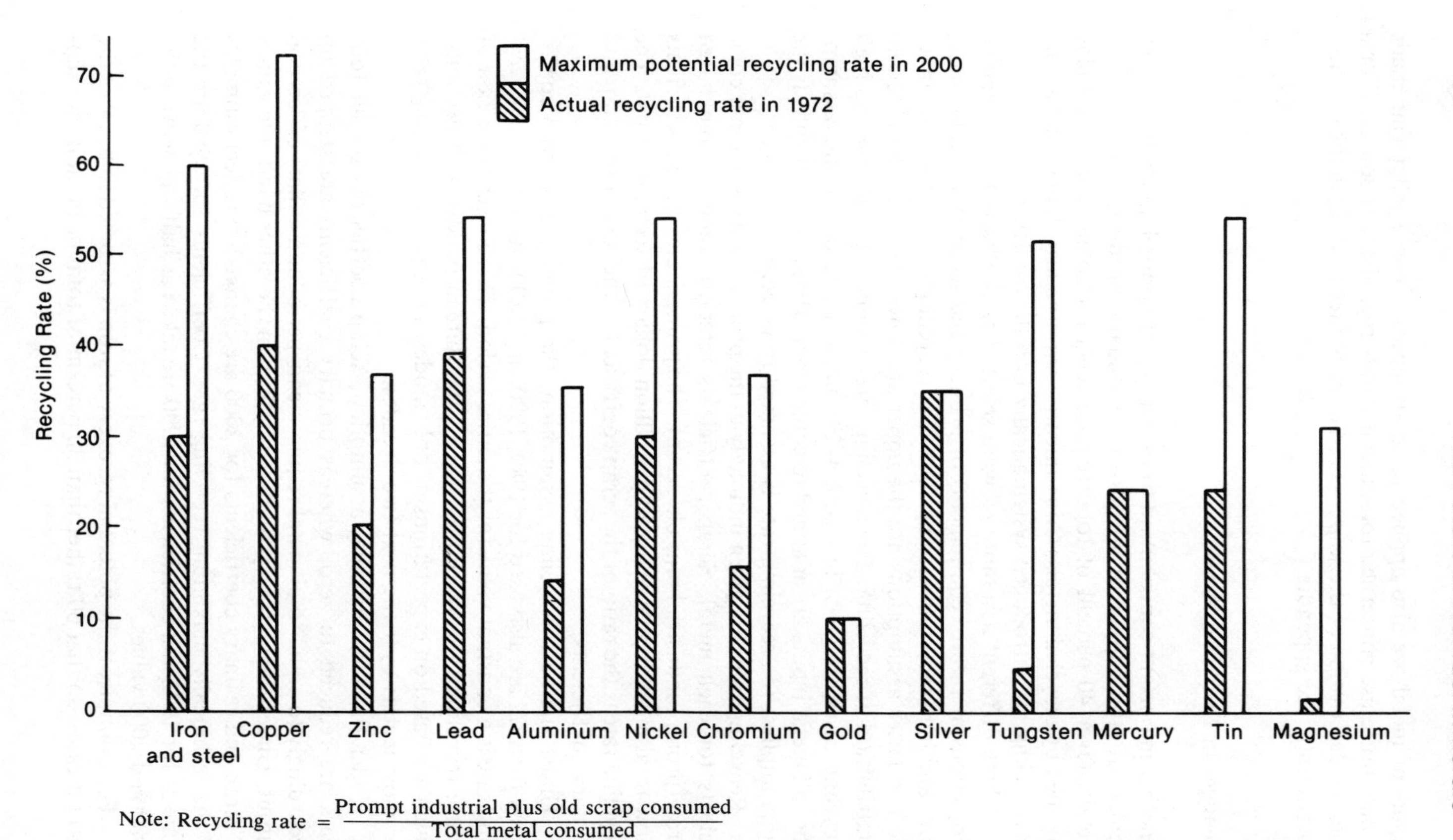

Note: Recycling rate = $\frac{\text{Prompt industrial plus old scrap consumed}}{\text{Total metal consumed}}$

Figure 4–29. Recycling Rates for Metals: Actual Rates in 1972 and Maximum Potential Rates in 2000

plied by scrap—base-year scrap coefficients are multiplied by a factor of two for 2000 and by a factor of 1.5 for 1990. Primary coefficients are set equal to the difference between total metal requirements minus the new (that is, stepped-up) scrap coefficients. The lead time required to implement innovative recovery techniques and changes in product and process design that would permit greater recovery from scrap is approximately ten years. Accordingly, the first appreciable changes in recycling rates are assumed under the high recycling assumption to occur in 1990.[4]

Changes in Import Requirements

For many nonfuel minerals, the major contribution to total supply for the United States is foreign imports. Even though the United States is endowed with substantial amounts of mineral resources, in many cases the indigenous deposits are not currently economical to mine or process. As a result, these minerals are being purchased abroad more cheaply.[5] Figure 4–30, reprinted from the Bureau of Mines [47], is an overview of the import dependence of the United States economy in 1979 and the various foreign sources of supply.

In some of the following computations, the level of domestic output of minerals was calculated on the basis of introducing alternative assumptions regarding import dependency on foreign sources of supply. The Bureau of Mines [46] implicitly projects these ratios—that is, a high and low level of imports to primary demand. These projected ratios were reconstituted into import coefficients, the ratio of imports to domestic output, so that they are compatible with the structure of the model. Figure 4–31 presents a comparison of the projected high and low values along with those observed in the 1972 base-year and used in all the other scenarios unless otherwise specified.

The Scenarios

Each row of the columns in figure 4–32 provides the particular assumption with respect to each of the variables fitted on the left that was used in each computation. For example, the first column represents a business-as-usual scenario—that is, with the exception of the higher level of gross domestic product, all other variables that affect the level of domestic mine output (including by-product output) or minerals consumption—such as technology, import coefficients, and recycling rates—are left unchanged from their 1972 base-year value. Scenarios 2 and 3 explore the effects on domestic mine output of alternative assumptions regarding the minimum and maximum degrees of import dependence on particular nonfuel minerals. (See the

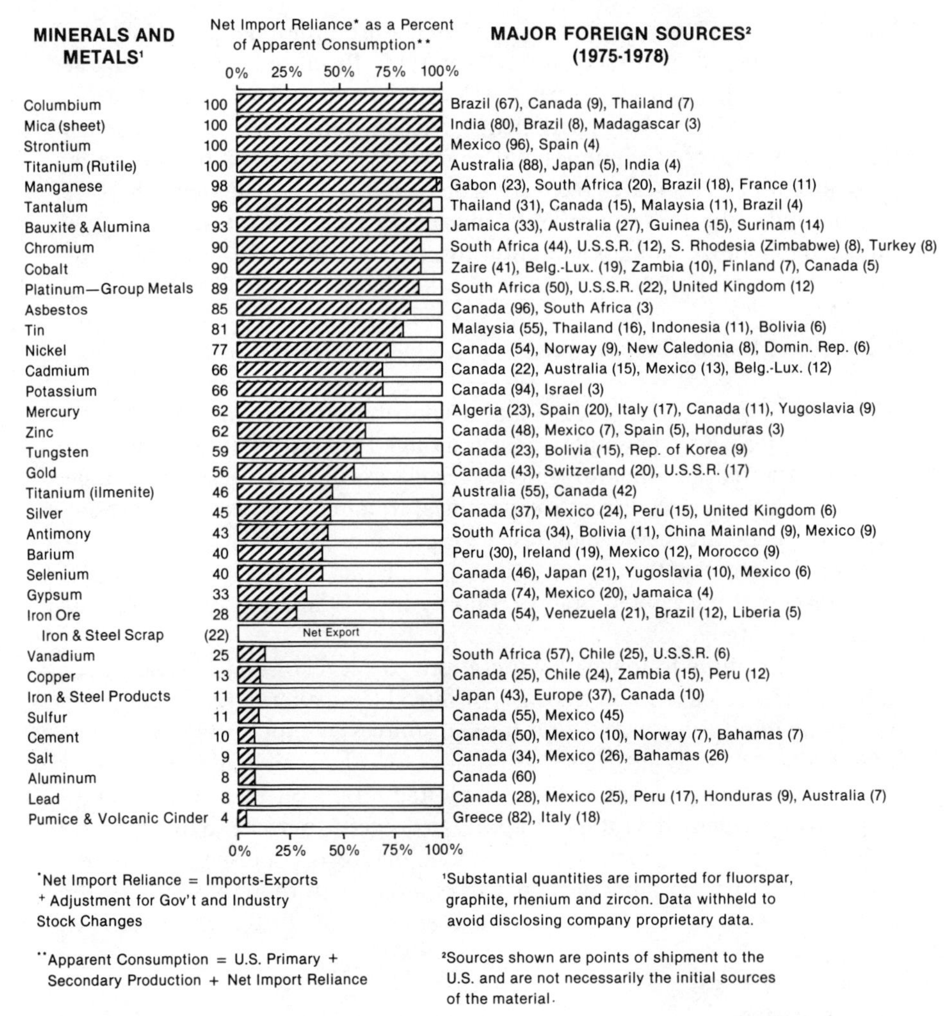

Source: April 1980–Bureau of Mines, US Department of Interior (Import-export data from Bureau of the Census).

Figure 4–30. U.S. Net Import Reliance on Selected Minerals and Metals as a Percentage of Consumption in 1979

previous section and appendix F for a detailed description of the methodology used to project these degrees of import dependency.)

In column 4 (scenario 4) of figure 4–32, the assumptions present the potential contribution that increased recycling can make to conserving

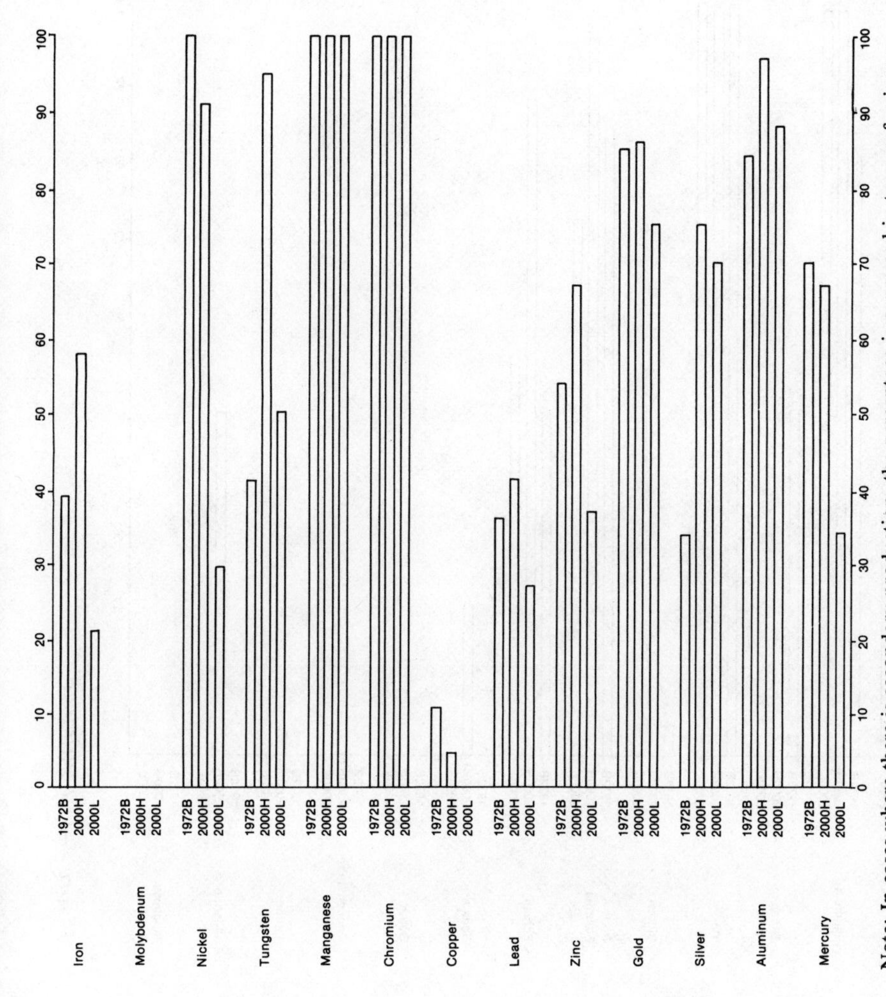

Note: In cases where there is secondary production, the percentage is expressed in terms of primary consumption only.

Figure 4–31. Imports as a Percentage of Domestic Consumption

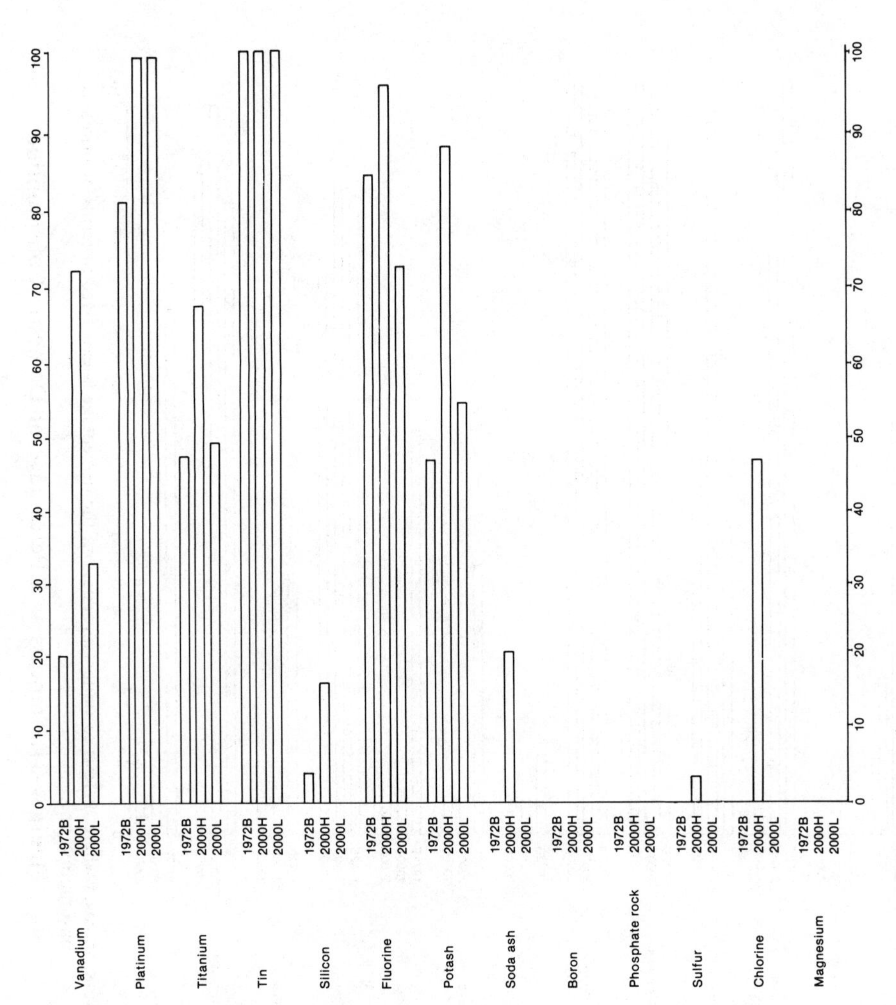

Figure 4–31 continued

Assumptions \ Scenarios	1 Base	2 Base Technology High Imports Base Recycling	3 Base Technology Low Imports Base Recycling	4 Base Technology Base Trade High Recycling	5 Base Technology Base Trade Zero Recycling	6 Technological Change High Imports Base Recycling	7 Technological Change Base Trade High Recycling	8 Technological Change Base Trade Base Recycling	9 Technological Change High Imports High Recycling	10 Technological Change Low Imports High Recycling	11 Technological Change Low Imports Base Recycling
GNP	Level and composition set by official BLS projections (Assumed average annual rate of growth from 1972 to 2000 equal to 3.1%)										
Technological Change	None					Technological change introduced in 1980, 1990, and 2000					
Imports to Output Ratios for Minerals	Set at 1972 Ratios	Increased Import Dependency to 2000	Decreased Import Dependency to 2000	Set at 1972 Ratios		Same as Scenario 2	Same as Scenario 1	Same as Scenario 1	Same as Scenario 2	Same as Scenario 3	Same as Scenario 3
Recycling Rates for Metals	Set at 1972 Recycling Rates			Increased Recycling Rates	Zero Recycling Rates	Same as Scenario 1	Same as Scenario 4	Same as Scenario 1	Same as Scenario 4	Same as Scenario 4	Same as Scenario 1
Domestic Mine Output	Endogenous										

Figure 4–32. Alternative Scenarios

domestic resources and reducing the level of foreign imports needed to satisfy the same bill of final demand used in scenarios 1, 2, and 3, in addition to maintaining the same structural matrix of input coefficients.

Scenario 5, on the other hand, describes the effects on the levels of imports and domestic reserves of a deliberate policy of zero recycling. (See the third section of this chapter and appendix F, for a detailed discussion of the methodology and assumptions used in these two scenarios.)

The one characteristic that scenarios 6 through 11 share—the bill of final demand notwithstanding—is the introduction of technological change and materials substitution into the matrix of technological coefficients. Since the end of World War Two, the mix of materials used to produce consumer durables and the nation's stock of plant and equipment have changed greatly. The introduction of aluminum in place of steel and tin and alloyed steel in place of carbon steel has affected greatly all the domestic nonfuel mineral sectors. The substitution of one mineral for another or for other materials—such as plastic for metals—is expected to continue in the future. Scenarios 6 through 11 were designed to capture the effects of these changes in the structural matrix. (See the second and third sections of this chapter and appendix C for a detailed presentation of the methodology and assumptions used to update the matrix of technological coefficients.)

In particular, scenario 8 isolates the effects on materials production and consumption by introducing technological change and materials substitution into the technological coefficients. Scenarios 6, 7, 9, 10, and 11 combine one or more changes of the other determinants of minerals production and consumption presented in scenarios 1 through 5. In this way, it is possible to see the individual contributions of each of the changes and, in addition, to see the reinforcing or offsetting effects of changes in two or more factors.

Notes

1. Sources of data used in this section are listed in appendix E.

2. Entries in the dummy and special industries sector included direct payments for noncompetitive imports, secondhand goods, direct payments for labor in government, compensation for household workers, net flow of foreign wages and other income, and adjustment for inventory valuation.

3. No projections were available for the distribution of the "output" from the dummy (and special) industries to the different GDP components for 1980, 1990, and 2000. In this particular sector, the 1972 direct payments to labor accounted for 94 percent of the output of the whole sector; conse-

quently this dominant entry was purposely excluded in 1972 to compare the base and the target years.

4. See appendix F for data sources.
5. See appendix D for current estimates of U.S. reserves and resources.

5 Alternative Projections to 2000

Chapter 3 presented the way in which the twenty-six nonfuel minerals were introduced into the 1972 input-output table of the U.S. economy. Chapter 4 described several sets of assumptions about income growth, technological change, recycling, and import dependency in nonfuel minerals under which the computations were carried out. This chapter presents the results of the alternative projections and, where instructive, a comparison between them.

To facilitate the analysis, the base-year data and projected results for 2000 are summarized in four sets of graphs and tables. The first set traces through the base-year and future sectoral mineral input requirements of forty-two selected sectors. These sectors collectively account for over 80 percent of the total U.S. consumption of each of the twenty-six nonfuel minerals. The second set of results is displayed in charts that depict the base-year distribution (in percentage terms) and the projected changes for the twenty-six nonfuel minerals in the major consuming sectors.

The effects on the projected levels of production and consumption resulting from changes in the assumptions underlying the alternative scenarios, such as the introduction of increased recycling and increased import dependency, in conjunction with income growth to the year 2000 for each of the twenty-six minerals and for the thirteen metals for which secondary production exists, are described in the last set of tables. The final section of this chapter summarizes the long-run effects of the alternative projections on U.S. cumulative production over twenty years compared to the 1980 level of estimated U.S. reserves for each of the twenty-six minerals.

Minerals Requirements by Consuming Sectors: 1972 and 2000

Figures 5-1 through 5-42 present a comparison of the mineral input requirements for the forty-two major minerals-consuming sectors for the 1972 base-year and 2000. For the year 2000, two different sets of minerals requirements for each of these consuming sectors are shown: the business-as-usual scenario (scenario 1); and the scenario embodying the projected technological changes in minerals use (scenario 8).

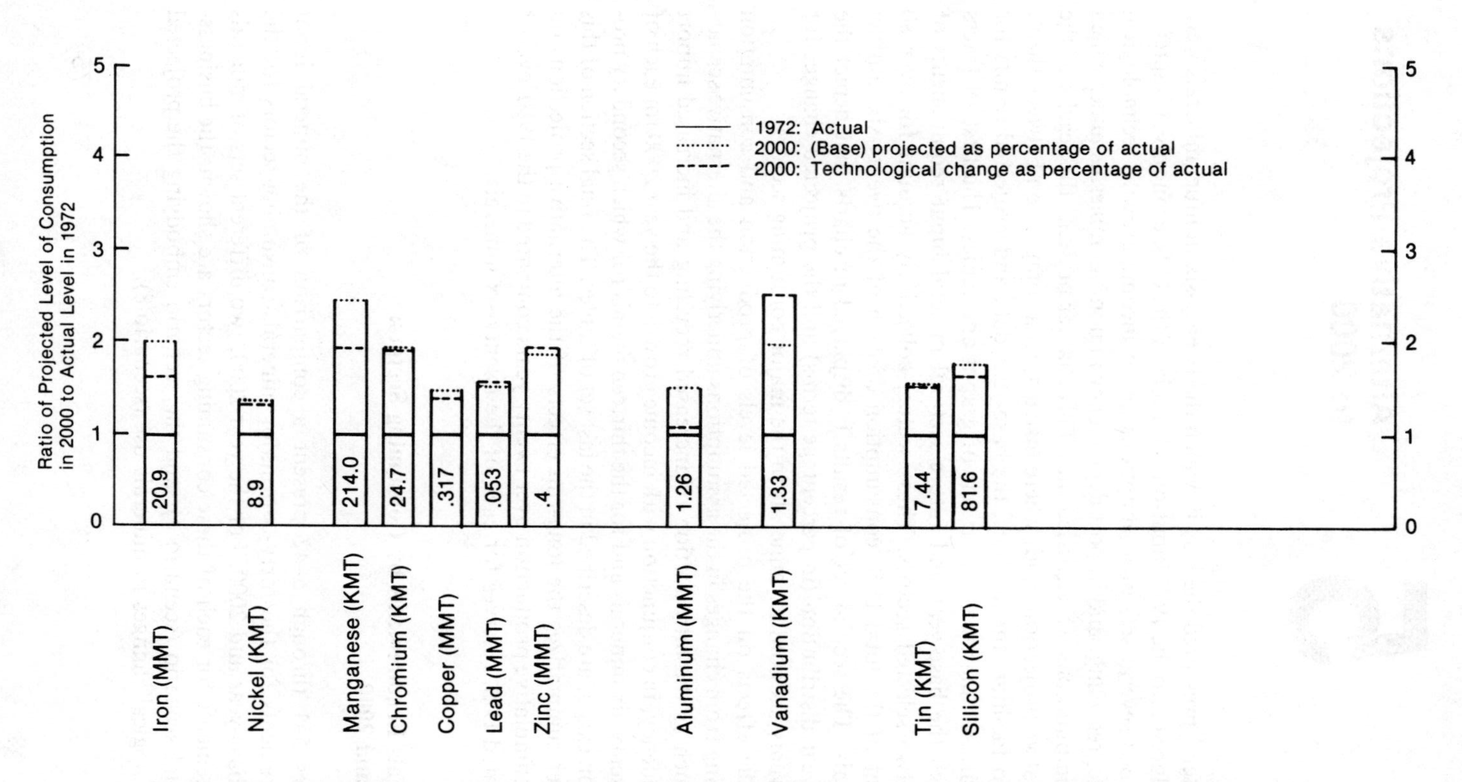

Figure 5–1. Mineral Input Requirements of New Construction (CON 1) under Alternative Scenarios

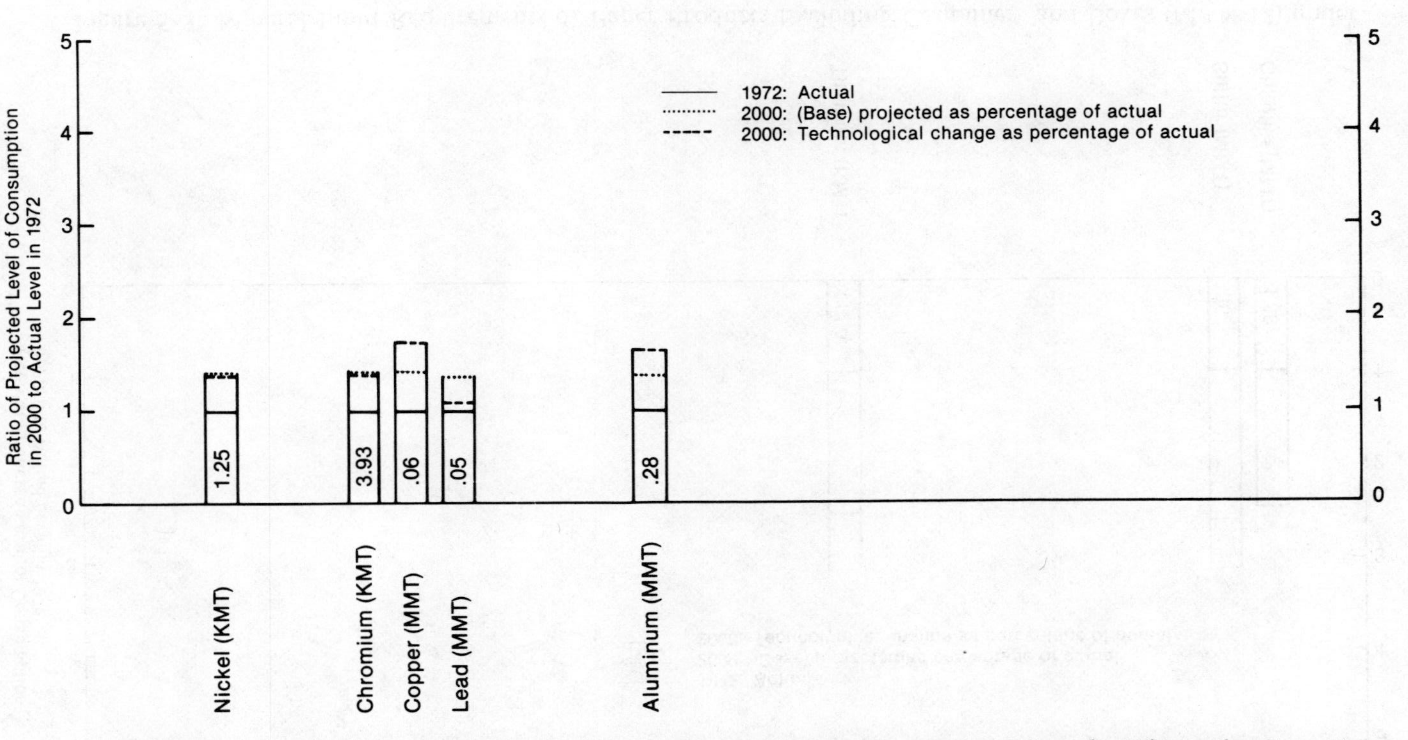

Figure 5–2. Mineral Input Requirements of Arms and Ammunitions (MAN 1) under Alternative Scenarios

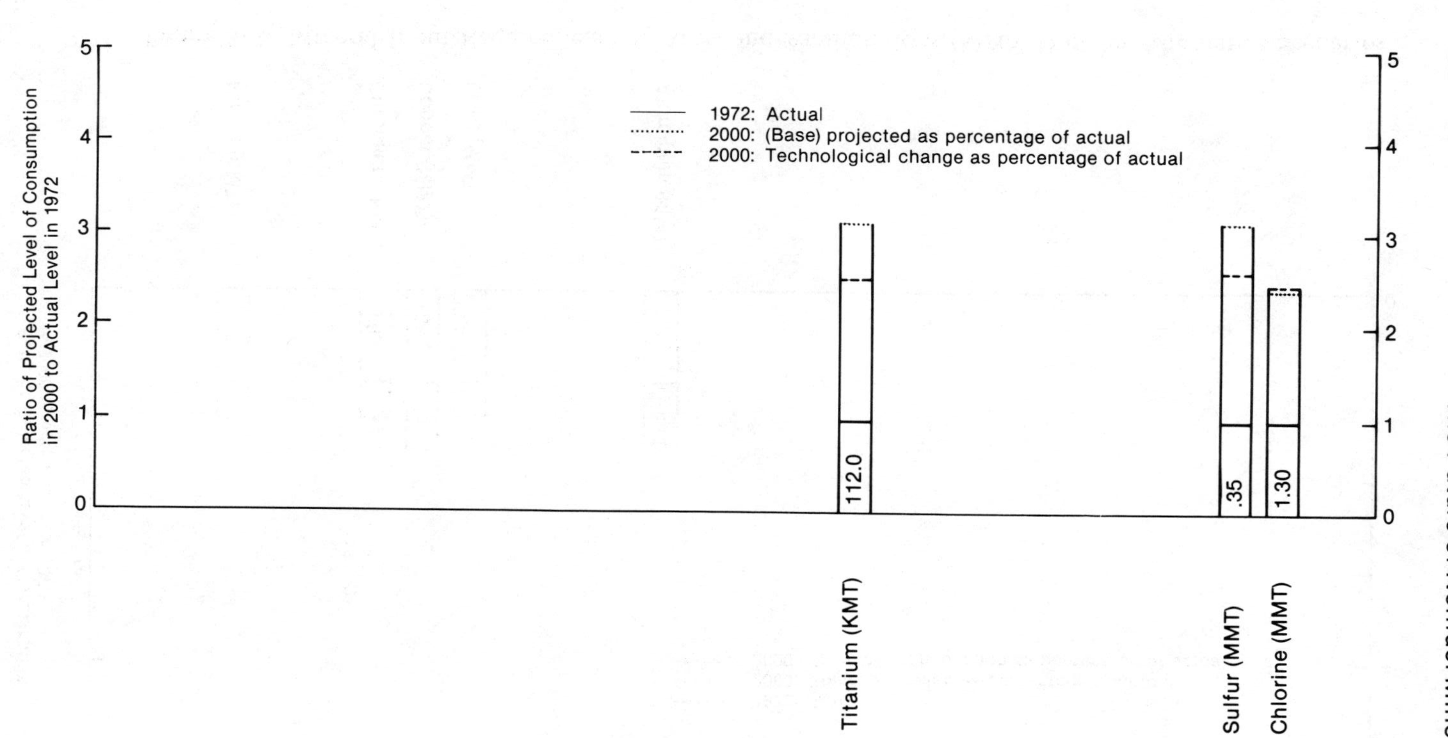

Figure 5–3. Mineral Input Requirements of Paper Products Excluding Containers and Boxes (MAN 12) under Alternative Scenarios

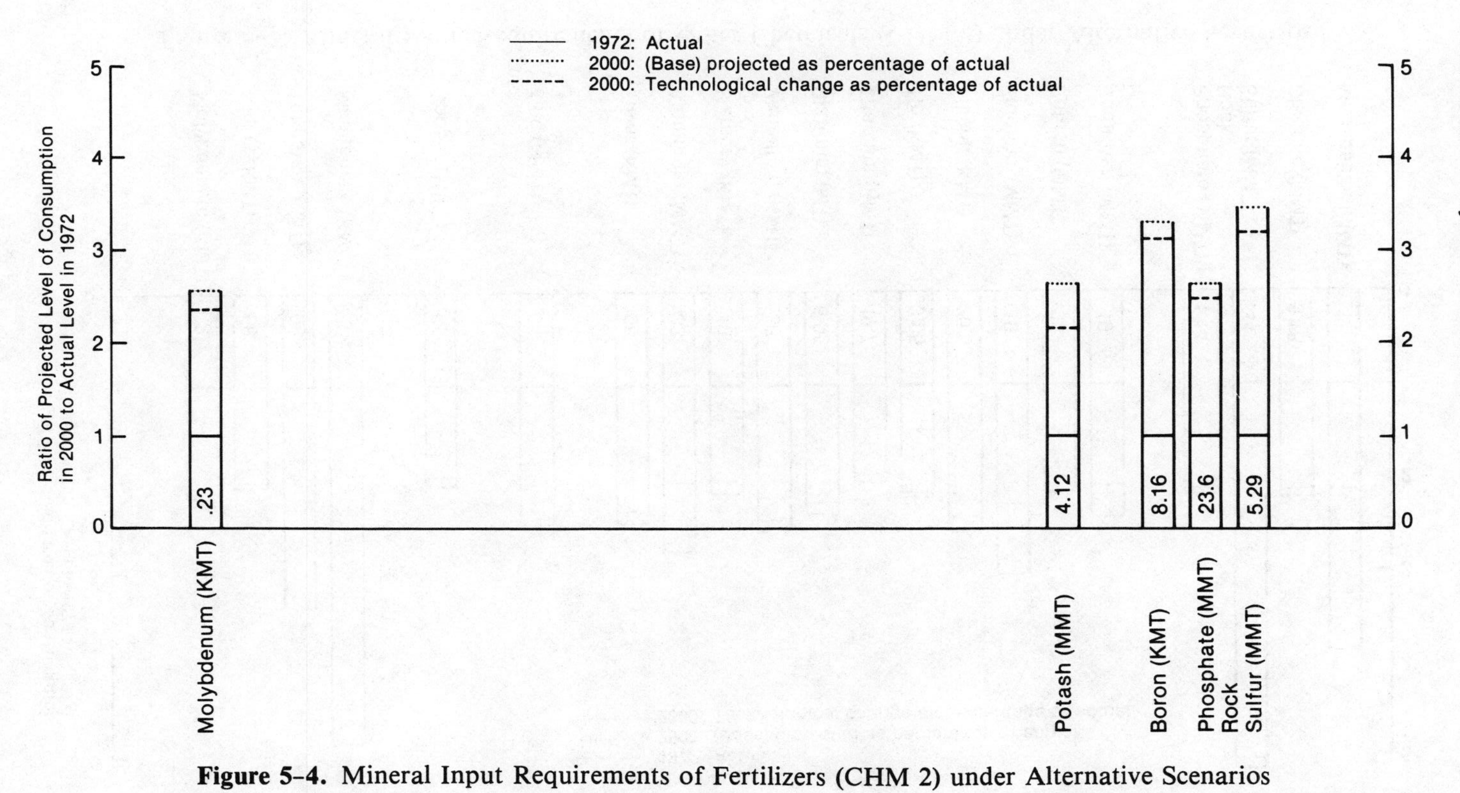

Figure 5-4. Mineral Input Requirements of Fertilizers (CHM 2) under Alternative Scenarios

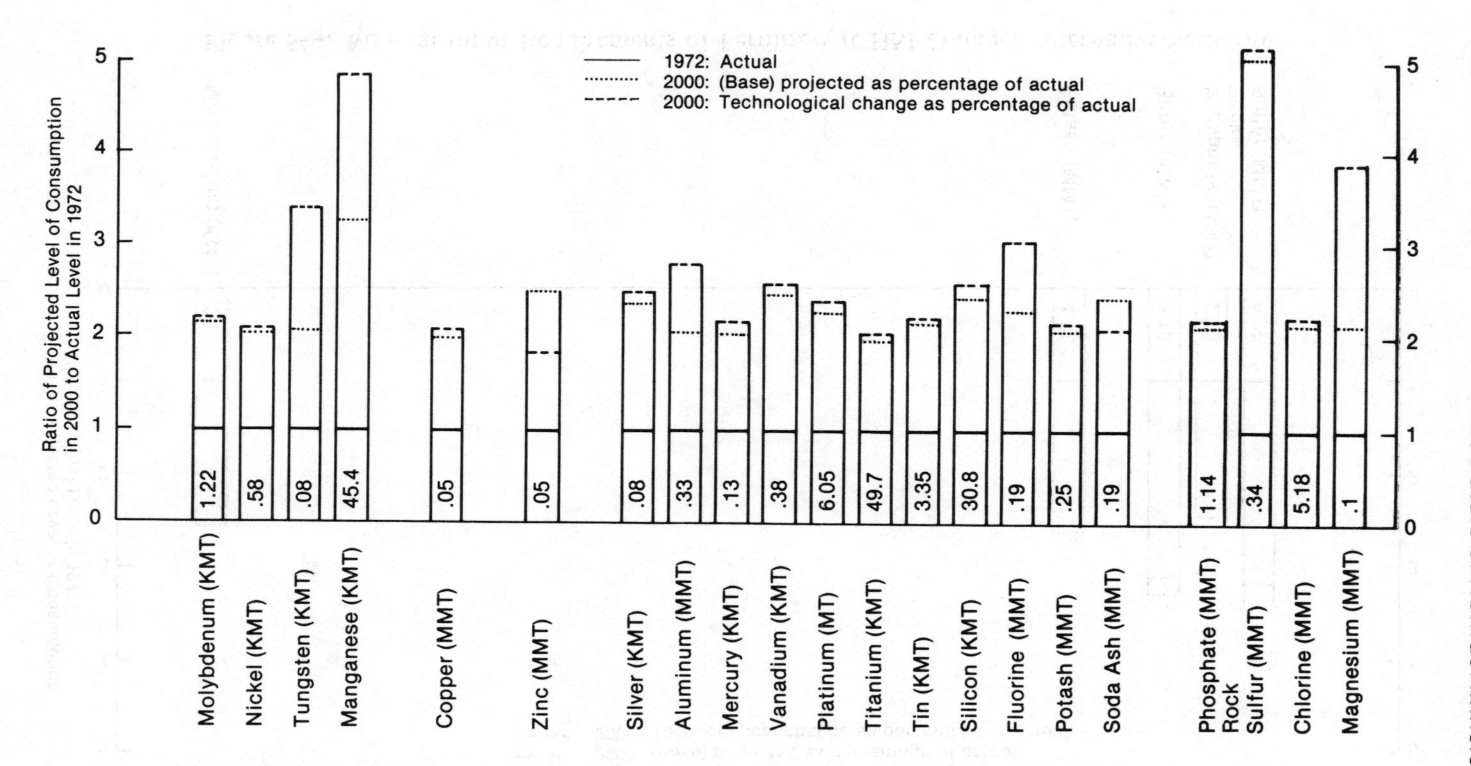

Figure 5–5. Mineral Input Requirements of Other Chemicals (CHM 3) under Alternative Scenarios

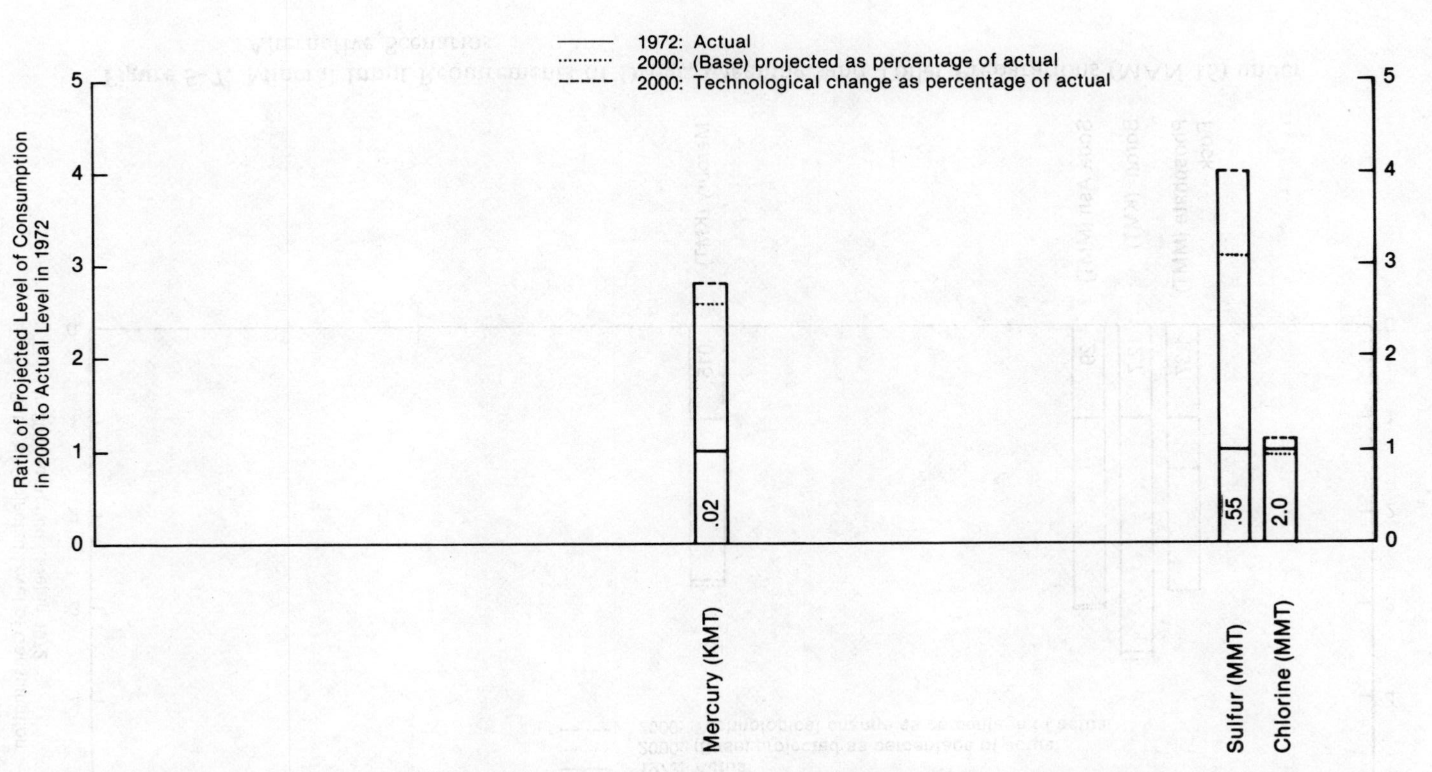

Figure 5–6. Mineral Input Requirements of Plastics and Synthetic Materials (MAN 15) under Alternative Scenarios

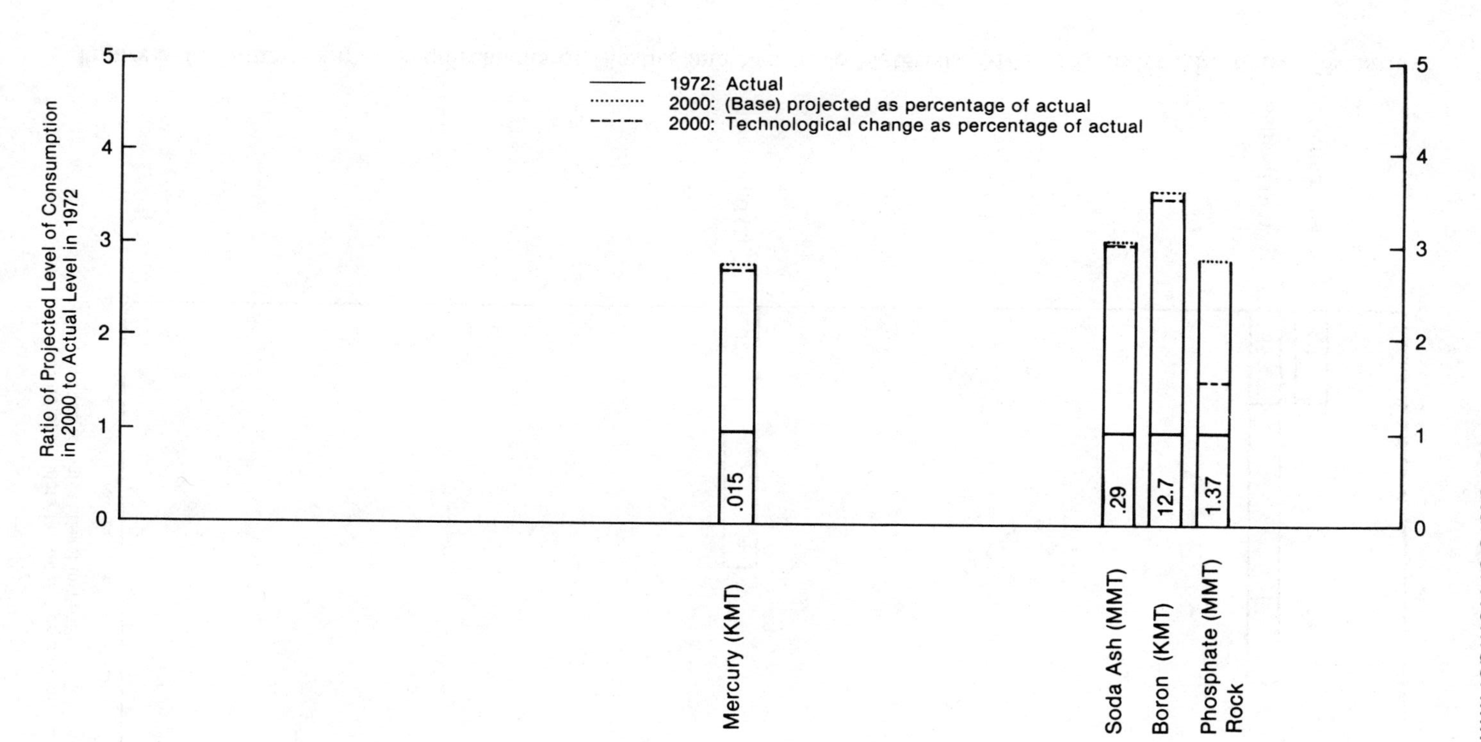

Figure 5–7. Mineral Input Requirements of Drugs, Cleaning, and Toilet Preparations (MAN 16) under Alternative Scenarios

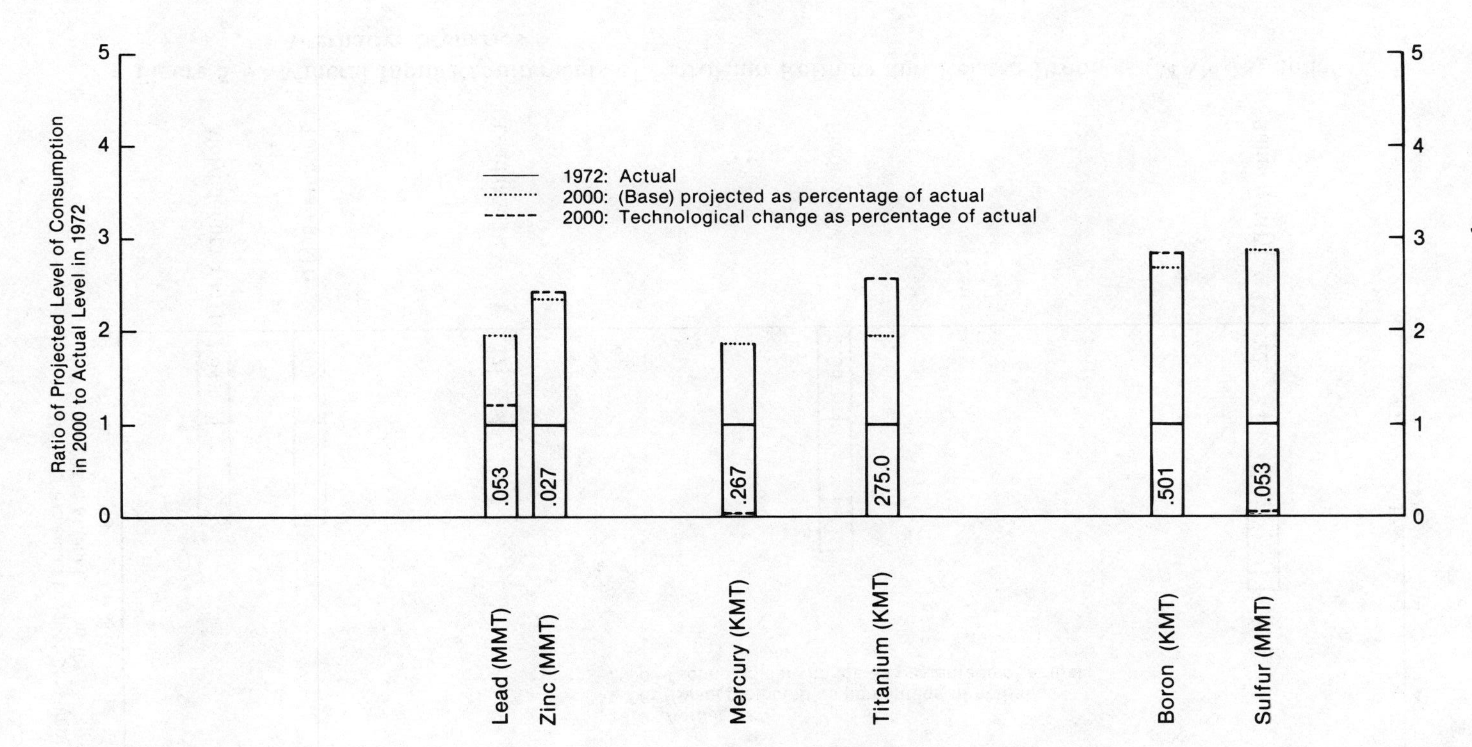

Figure 5-8. Mineral Input Requirements of Paints and Allied Products (MAN 17) under Alternative Scenarios

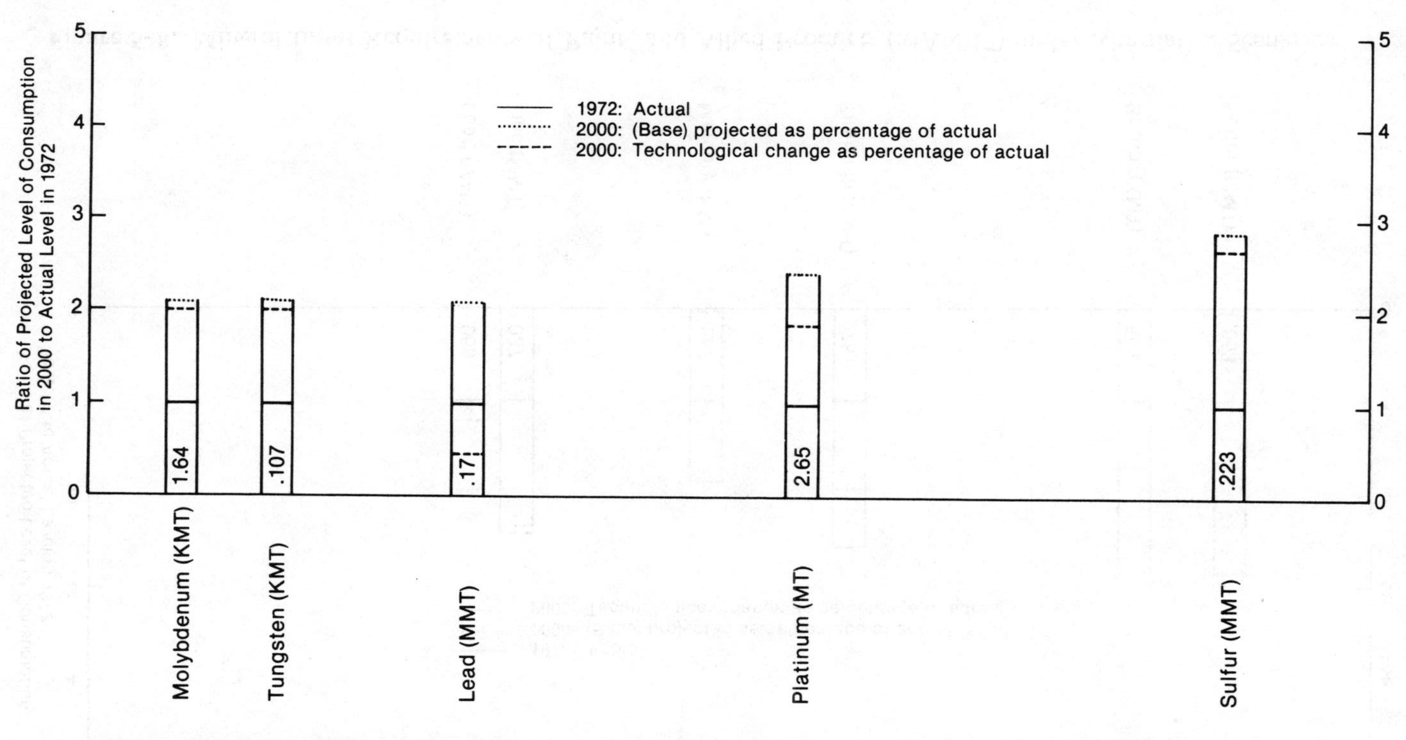

Figure 5–9. Mineral Input Requirements of Petroleum Refining and Related Products (MAN 18) under Alternative Scenarios

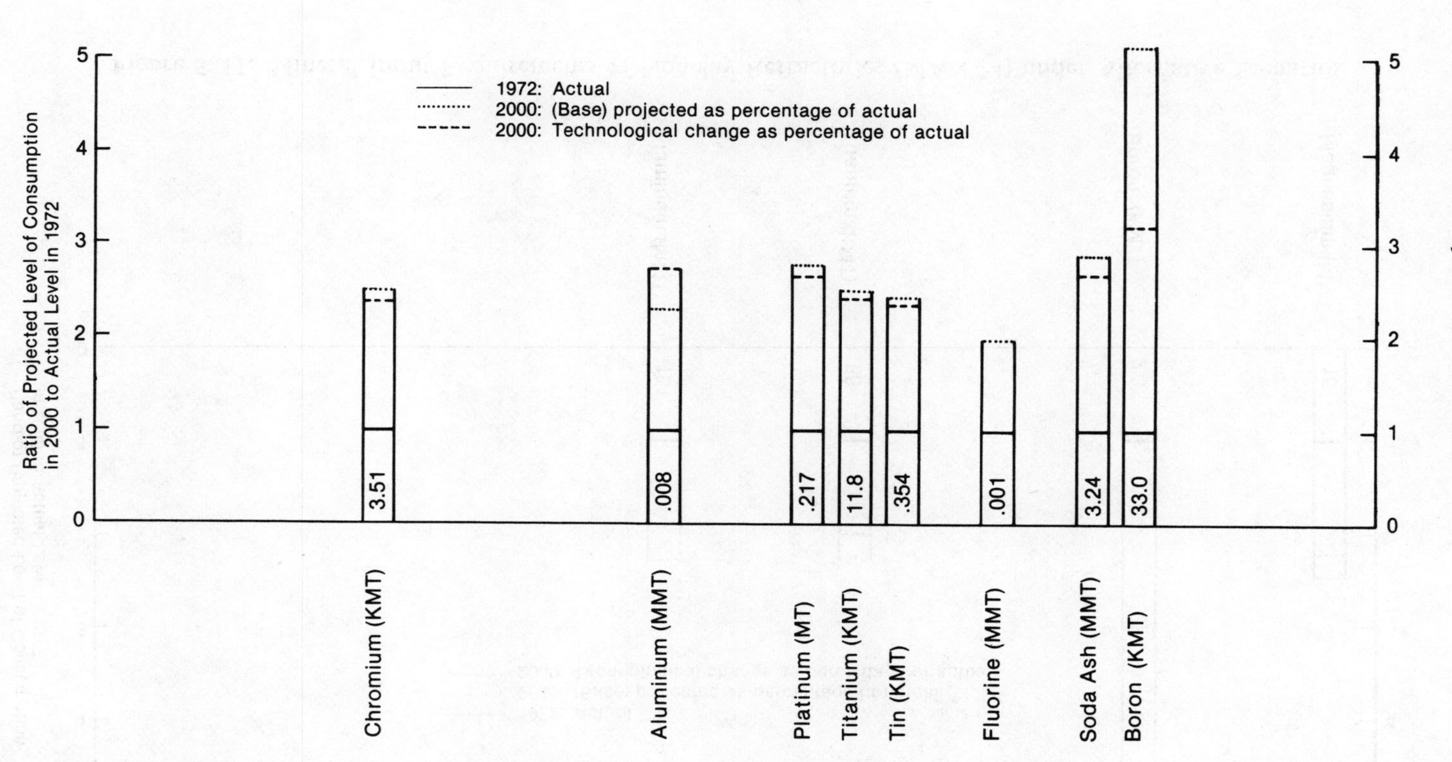

Figure 5–10. Mineral Input Requirements of Glass and Glass Products (MAN 22) under Alternative Scenarios

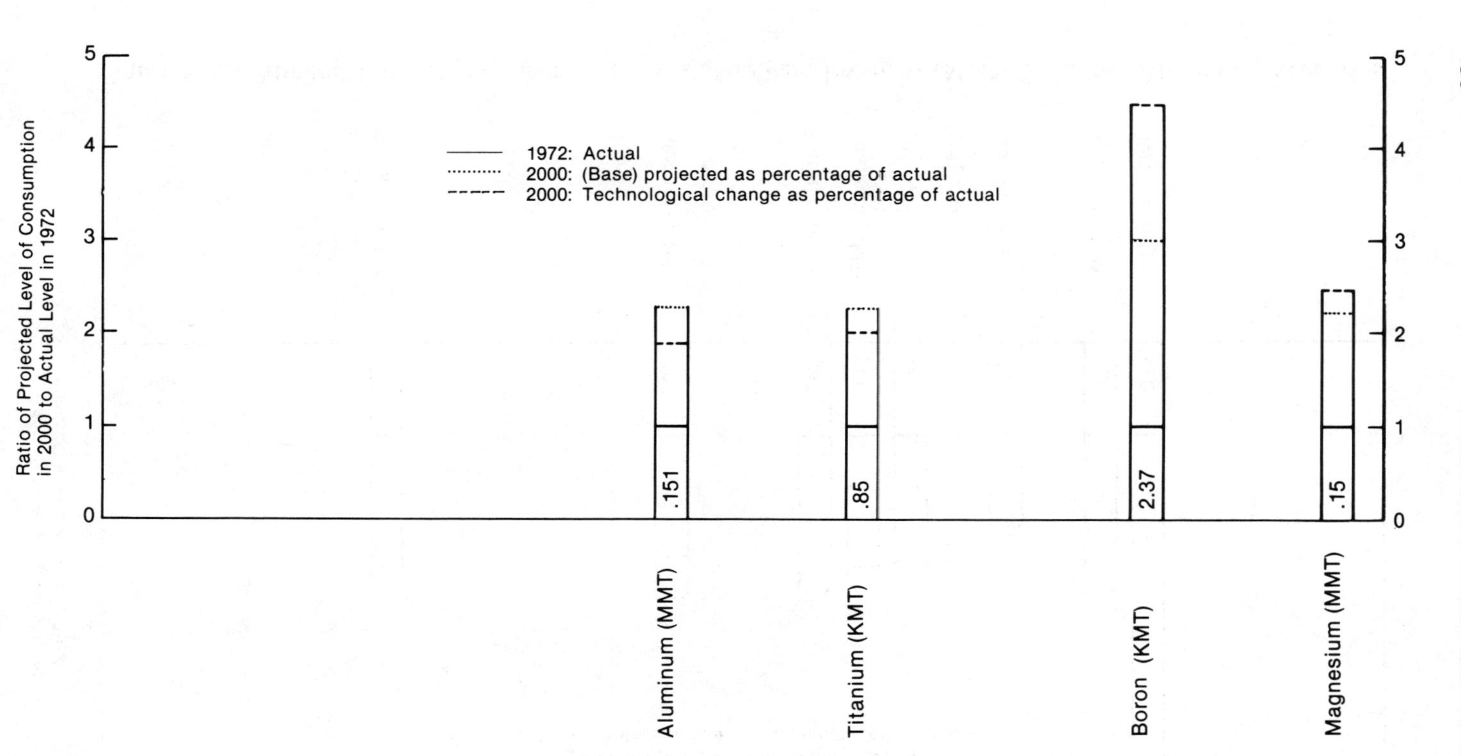

Figure 5–11. Mineral Input Requirements of Nonclay Refractories (MAN 24) under Alternative Scenarios

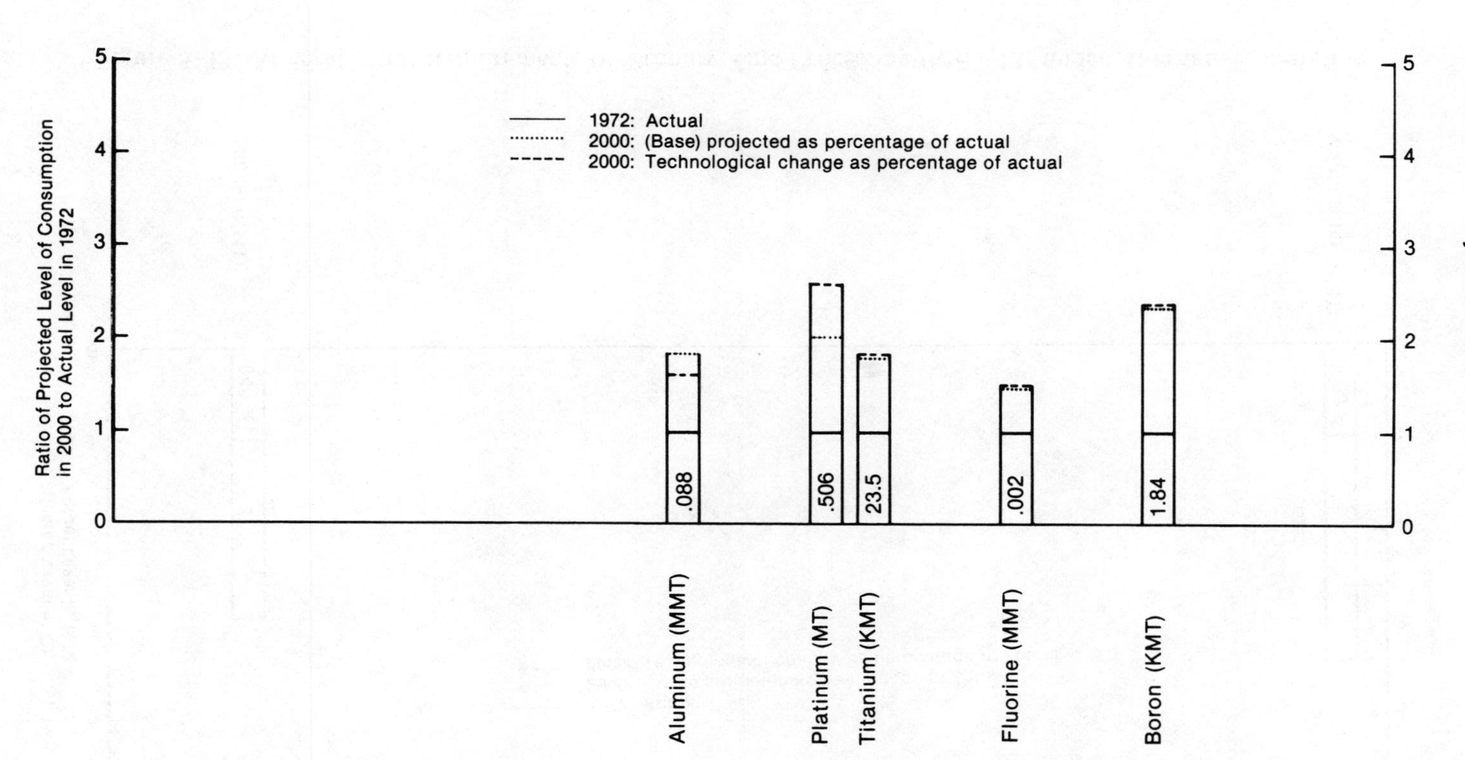

Figure 5-12. Mineral Input Requirements of Other Stone and Clay Products (MAN 25) under Alternative Scenarios

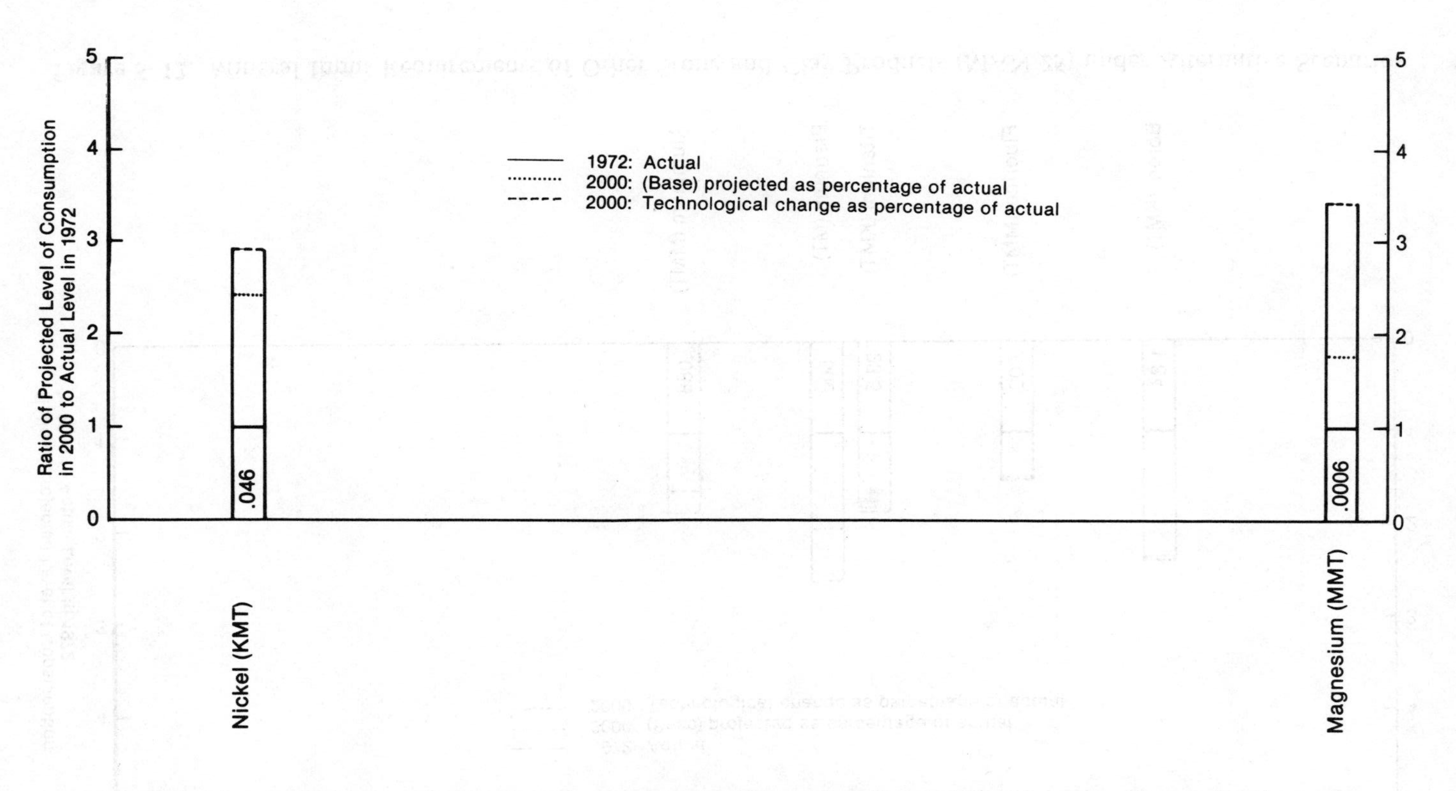

Figure 5-13. Mineral Input Requirements of Primary Zinc Processing (MET 12) under Alternative Scenarios

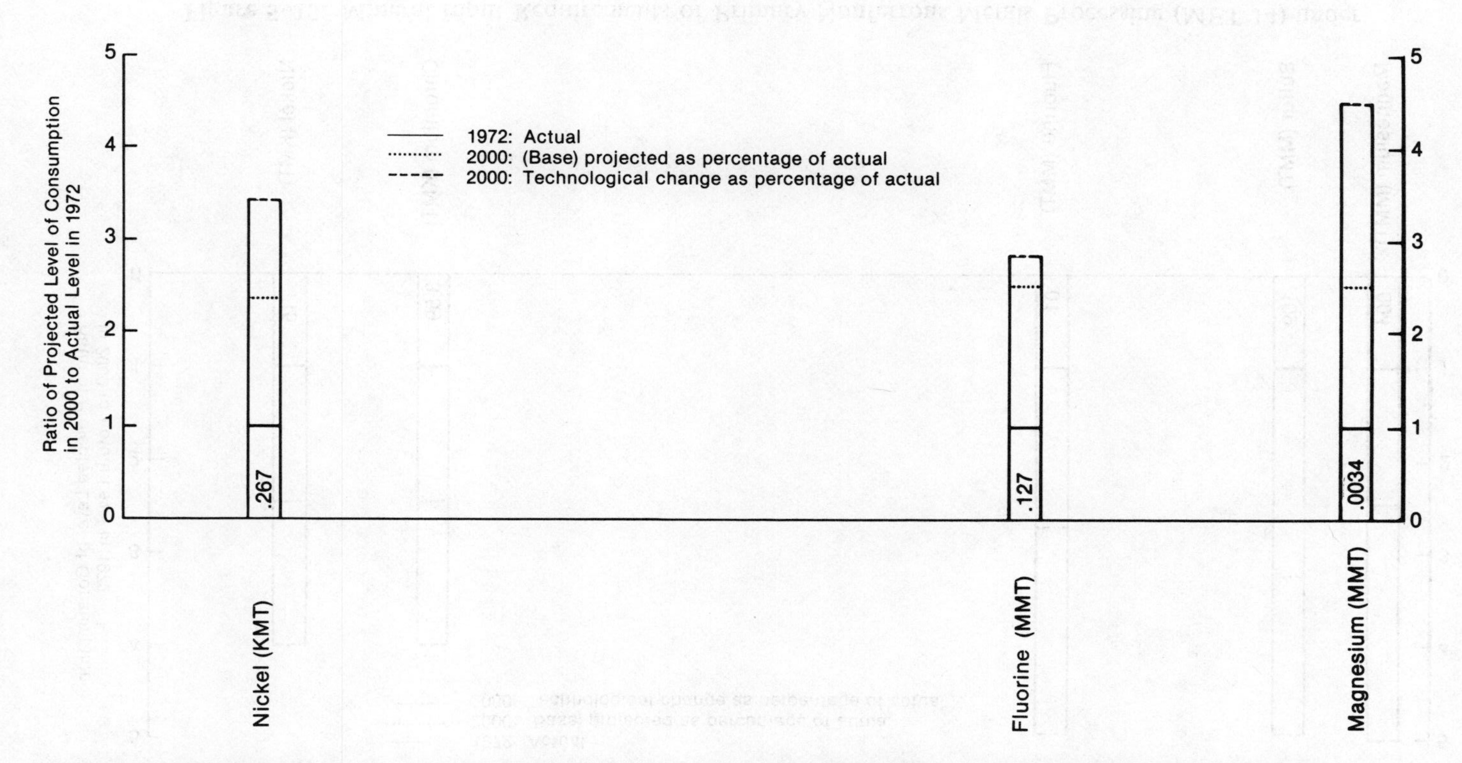

Figure 5-14. Mineral Input Requirements of Primary Aluminum Processing (MET 13) under Alternative Scenarios

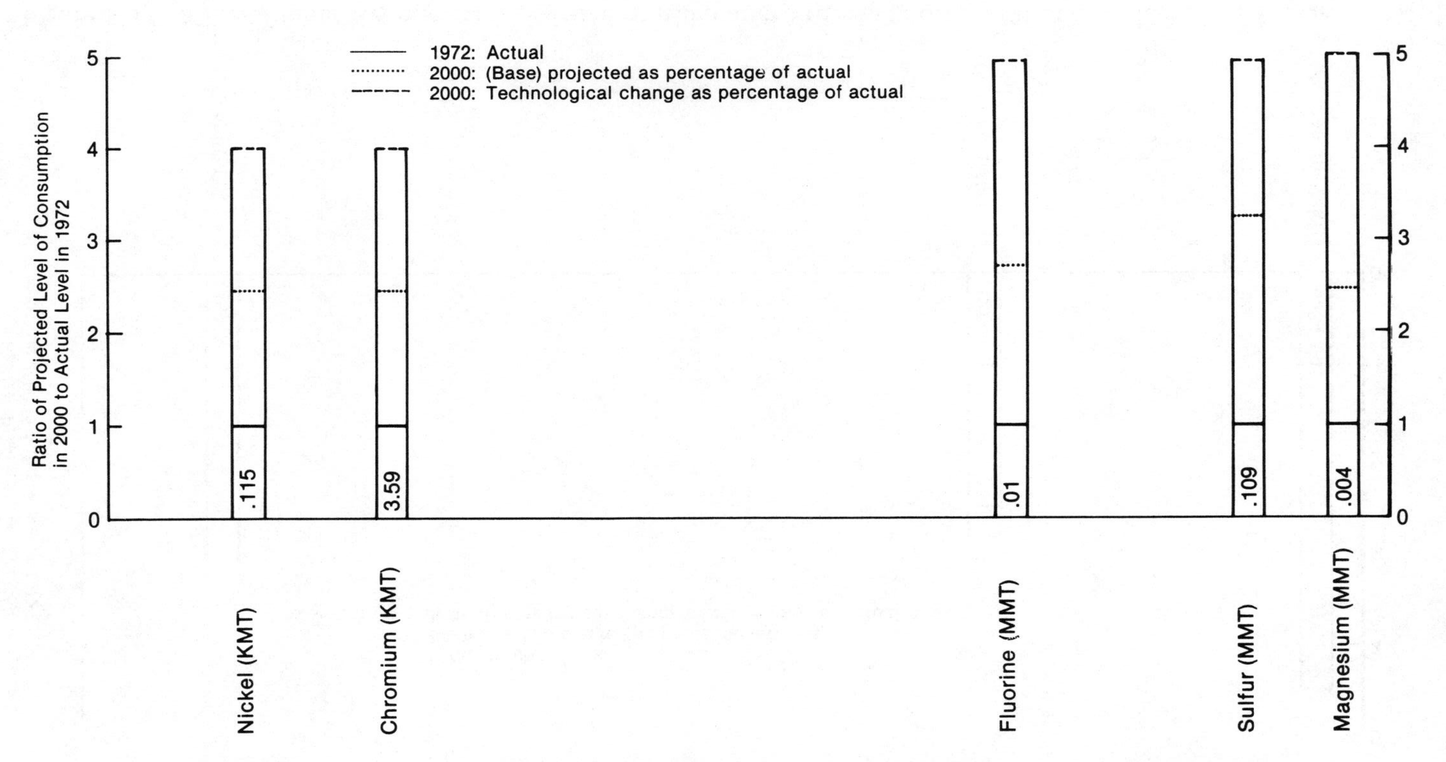

Figure 5–15. Mineral Input Requirements of Primary Nonferrous Metals Processing (MET 14) under Alternative Scenarios

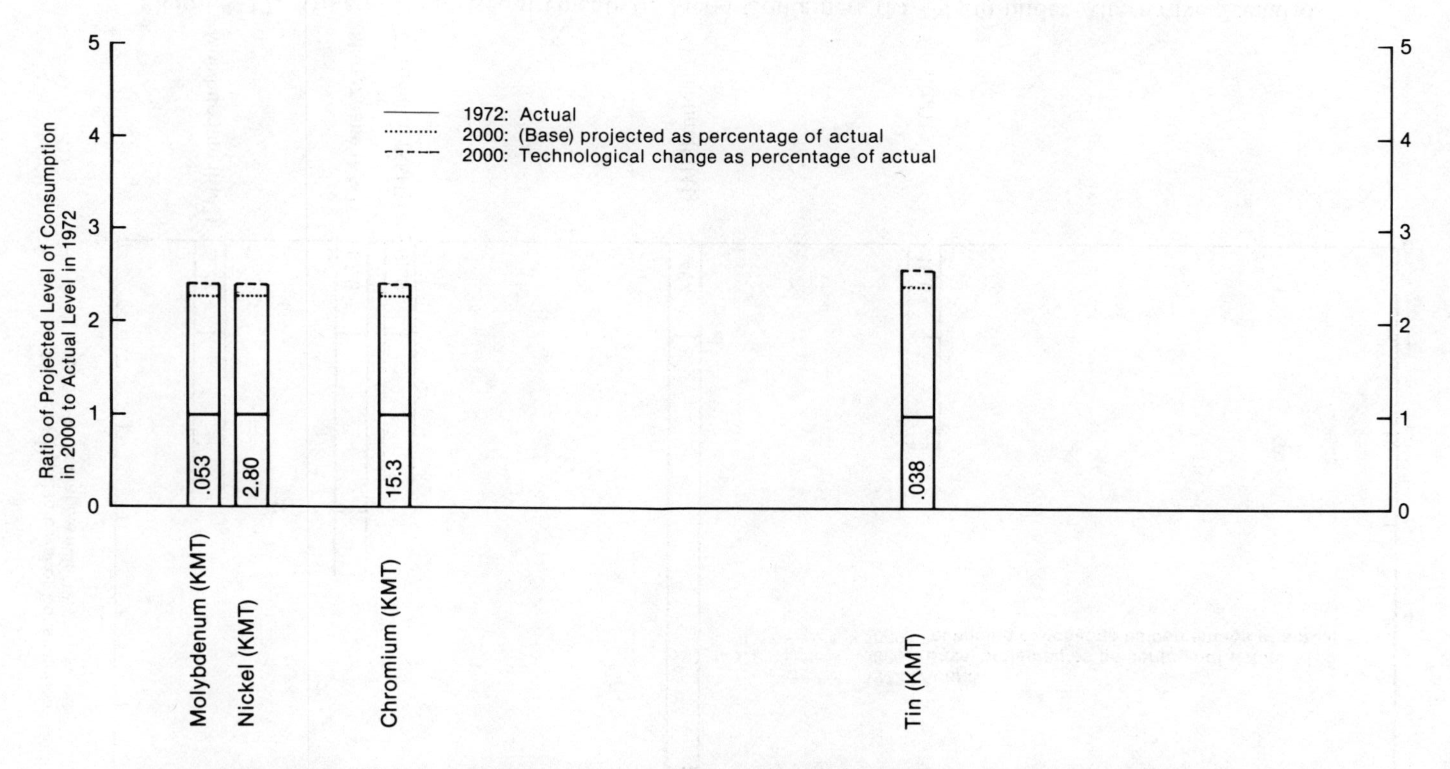

Figure 5–16. Mineral Input Requirements of Nonferrous Casting (MET 22) under Alternative Scenarios

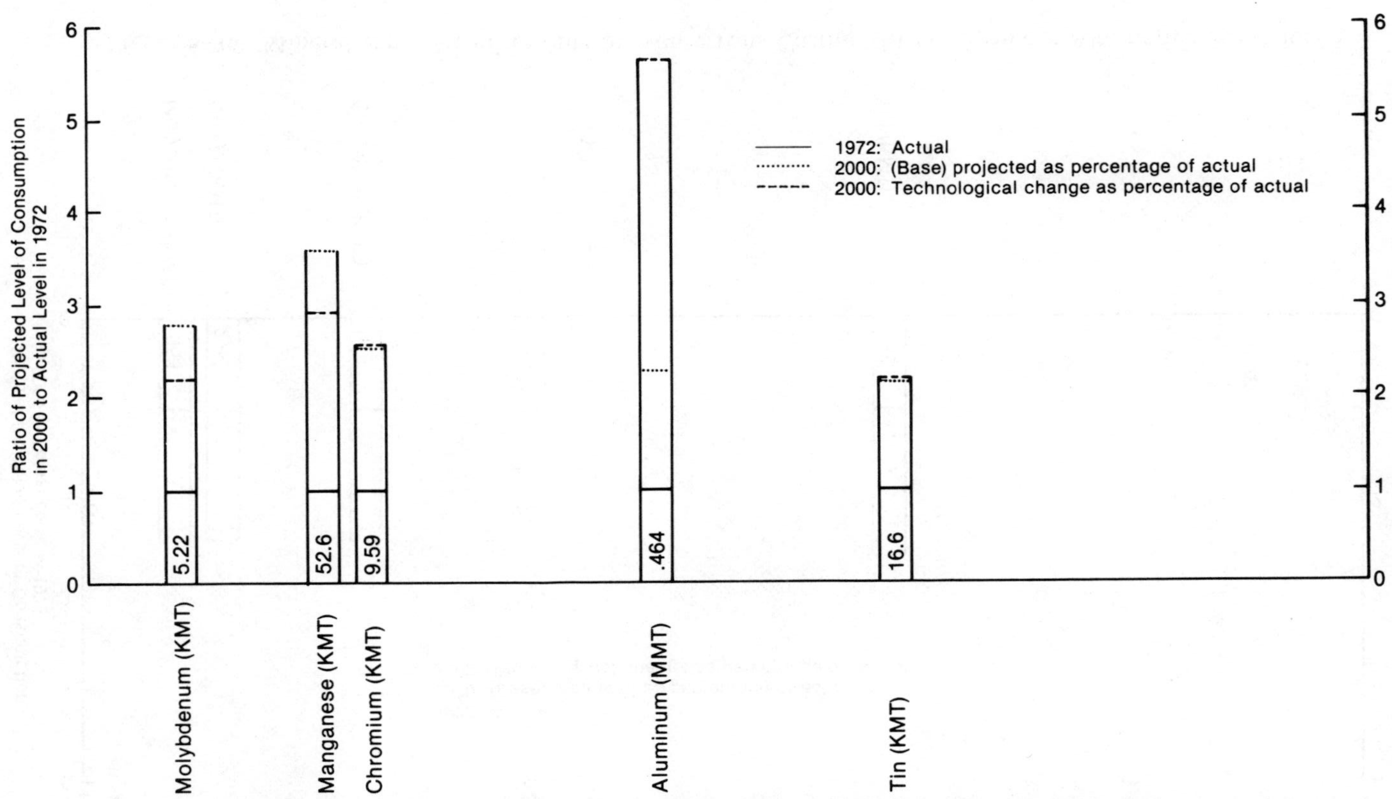

Figure 5–17. Mineral Input Requirements of Metal Containers (MAN 26) under Alternative Scenarios

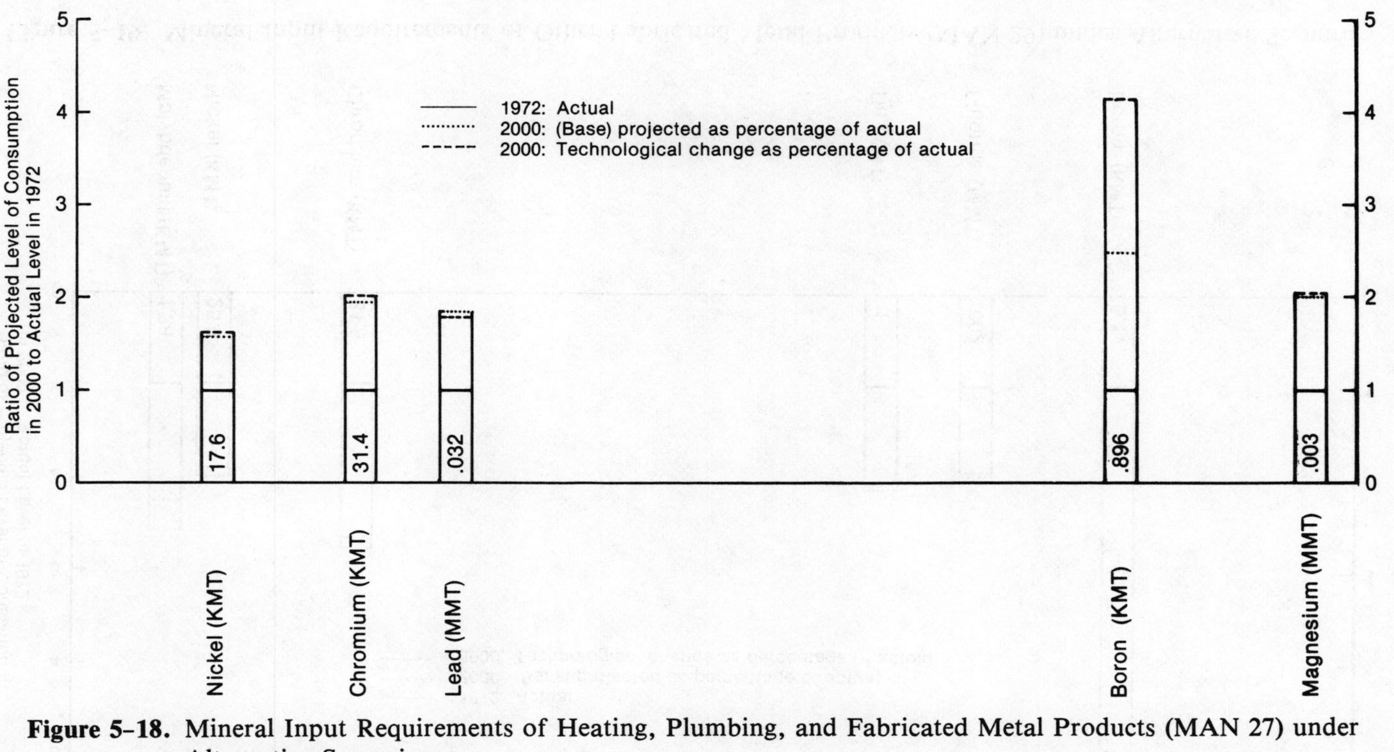

Figure 5–18. Mineral Input Requirements of Heating, Plumbing, and Fabricated Metal Products (MAN 27) under Alternative Scenarios

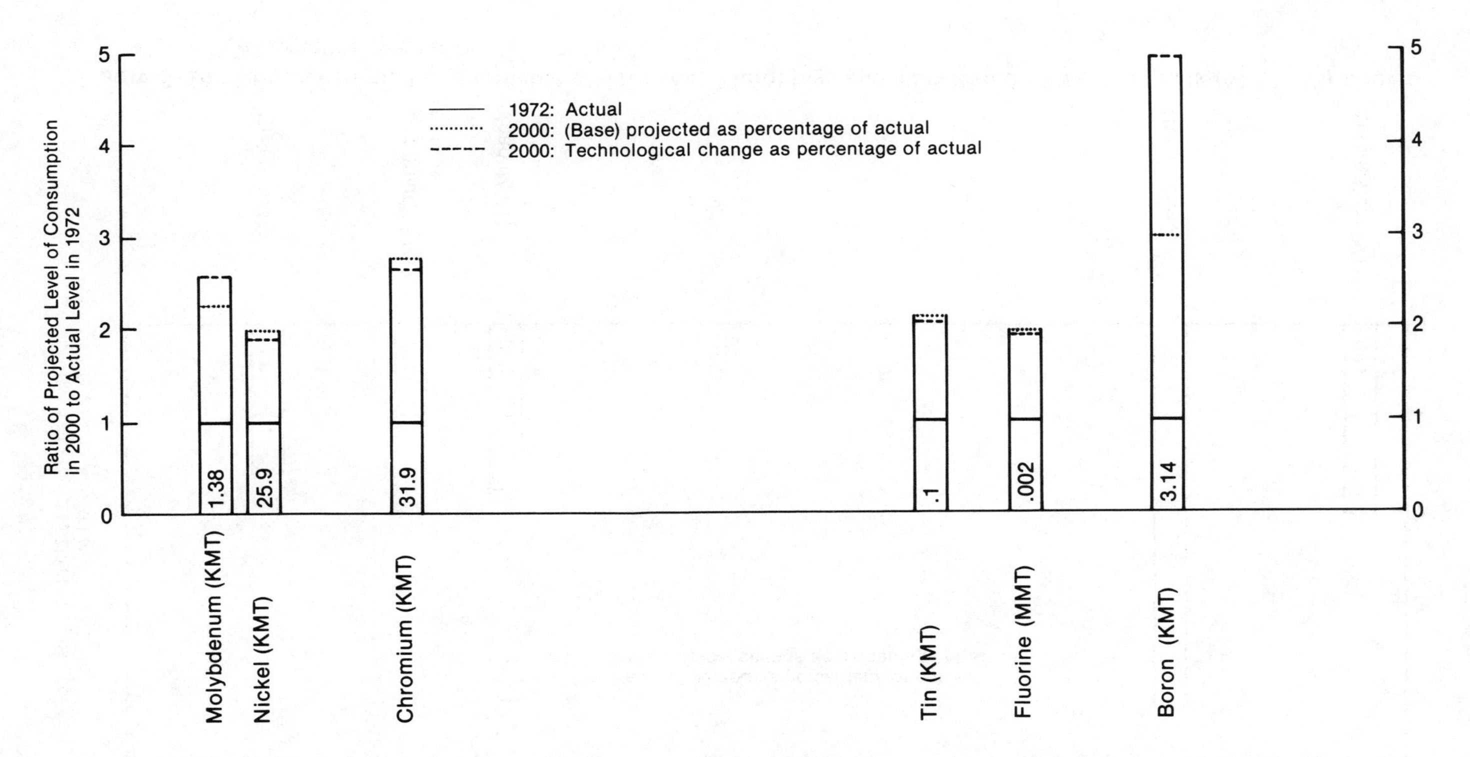

Figure 5–19. Mineral Input Requirements of Other Fabricated Metal Products (MAN 29) under Alternative Scenarios

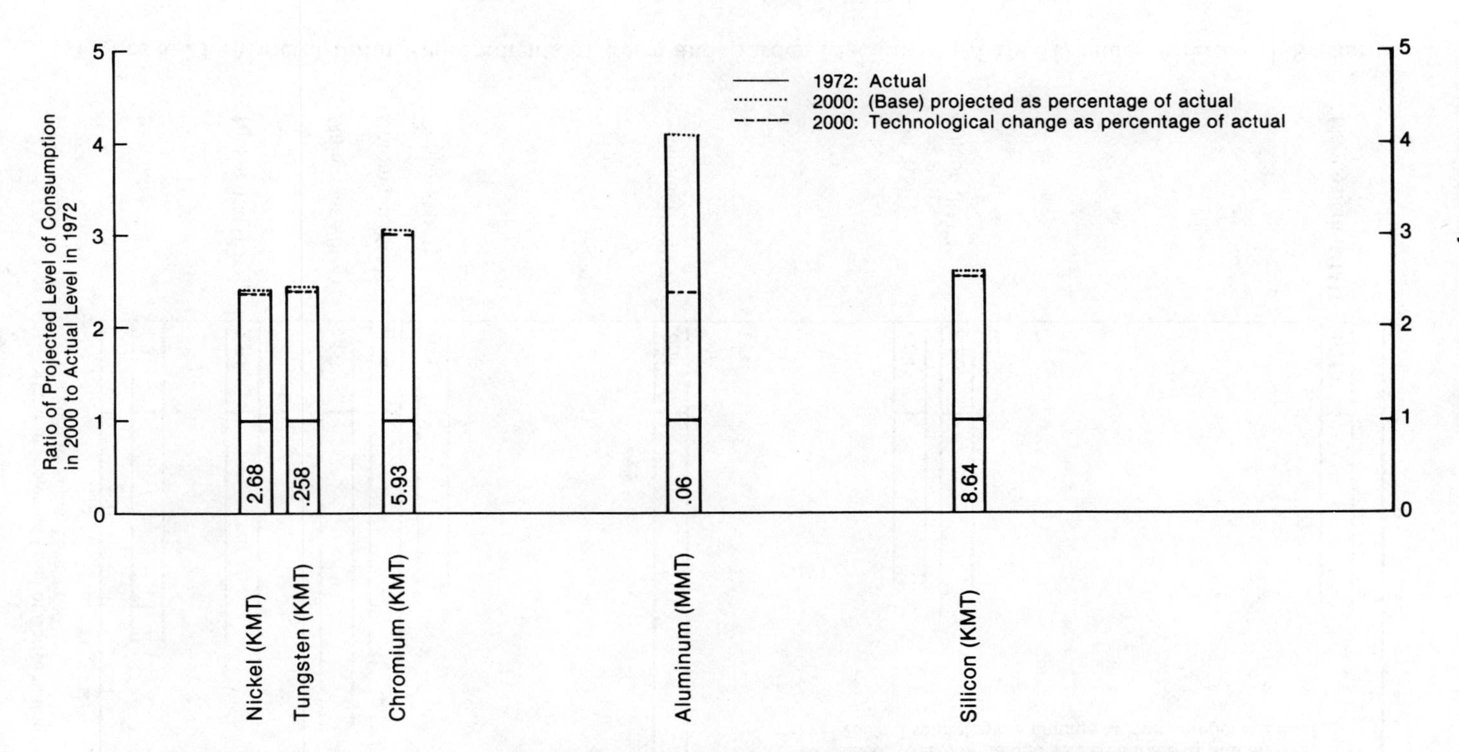

Figure 5-20. Mineral Input Requirements of Engines and Turbines (MAN 30) under Alternative Scenarios

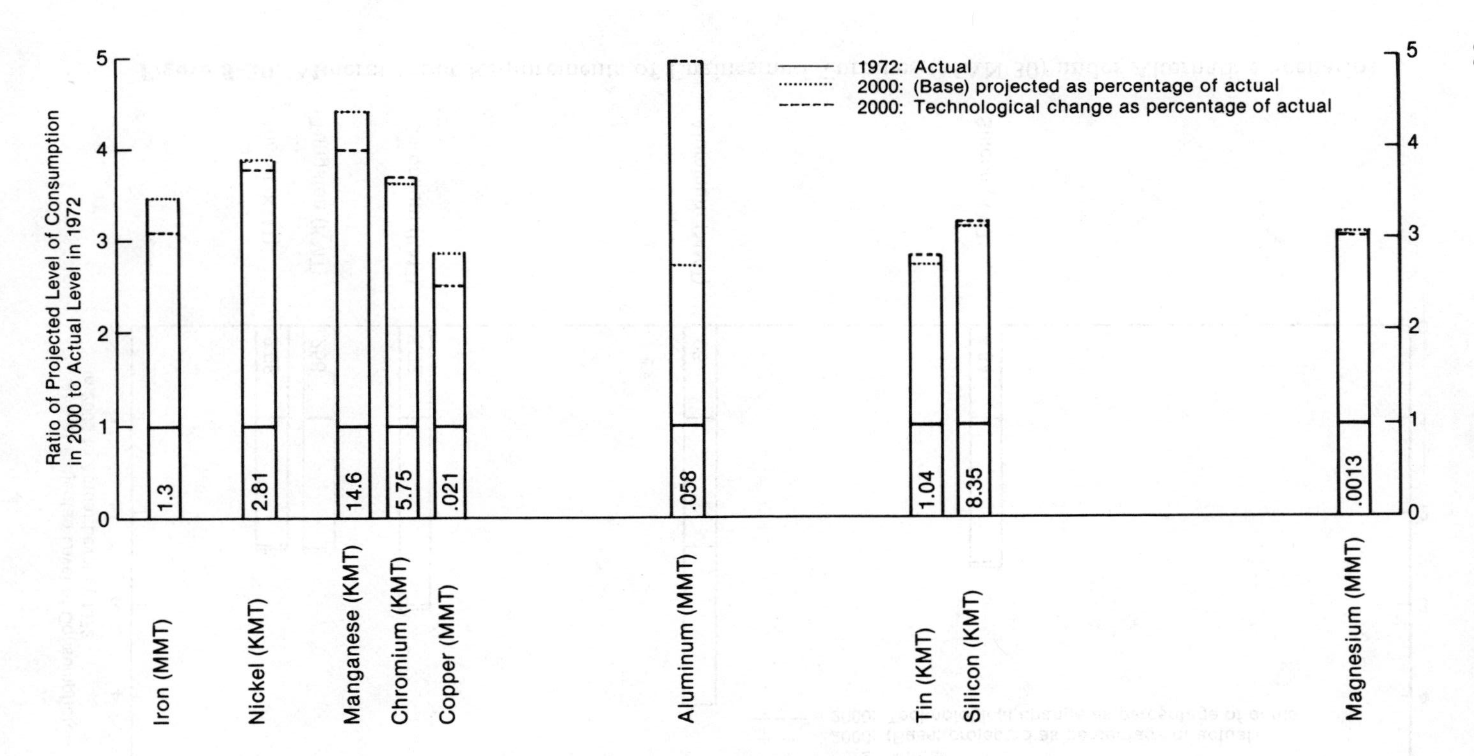

Figure 5–21. Mineral Input Requirements of Farm and Garden Machinery (MAN 31) under Alternative Scenarios

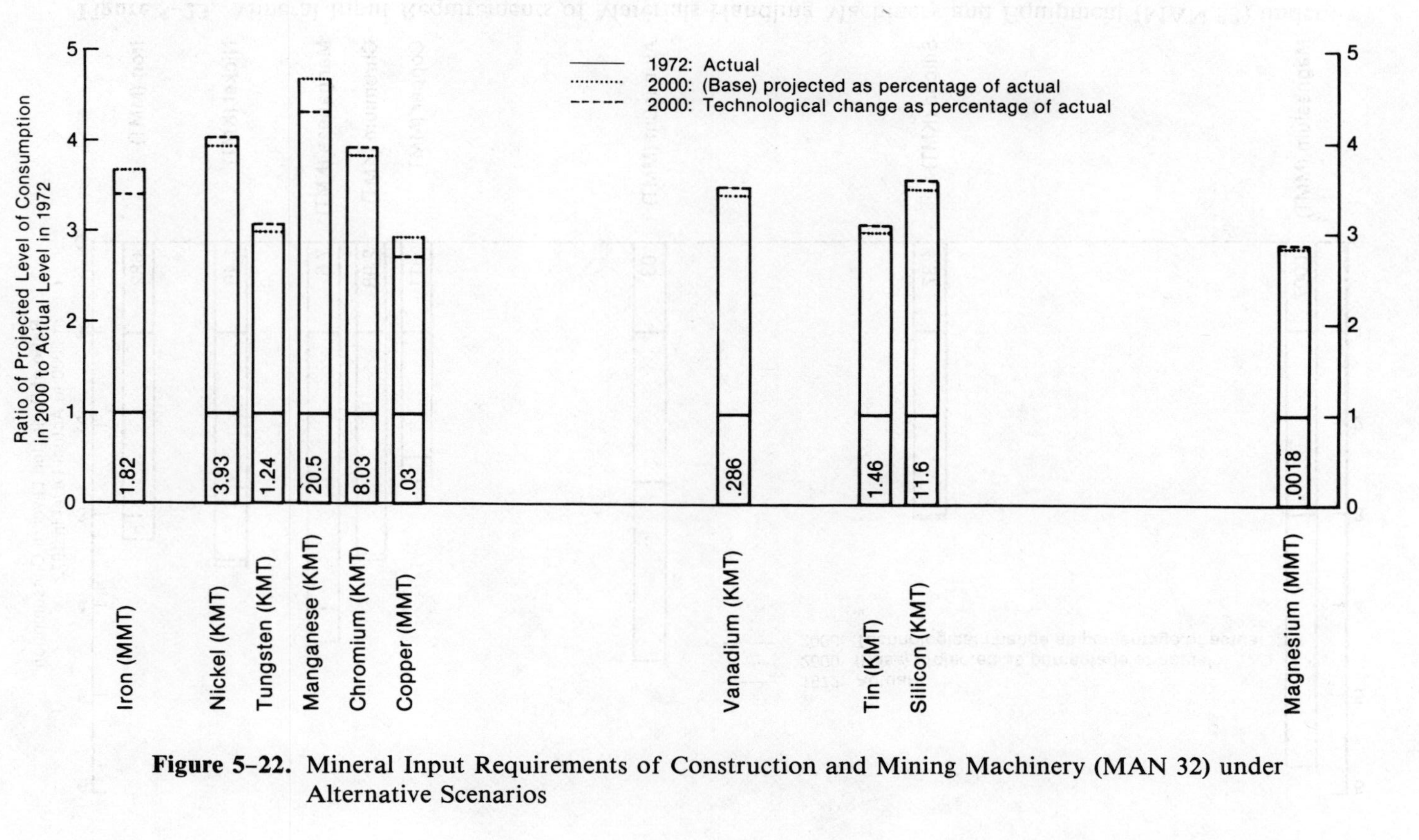

Figure 5–22. Mineral Input Requirements of Construction and Mining Machinery (MAN 32) under Alternative Scenarios

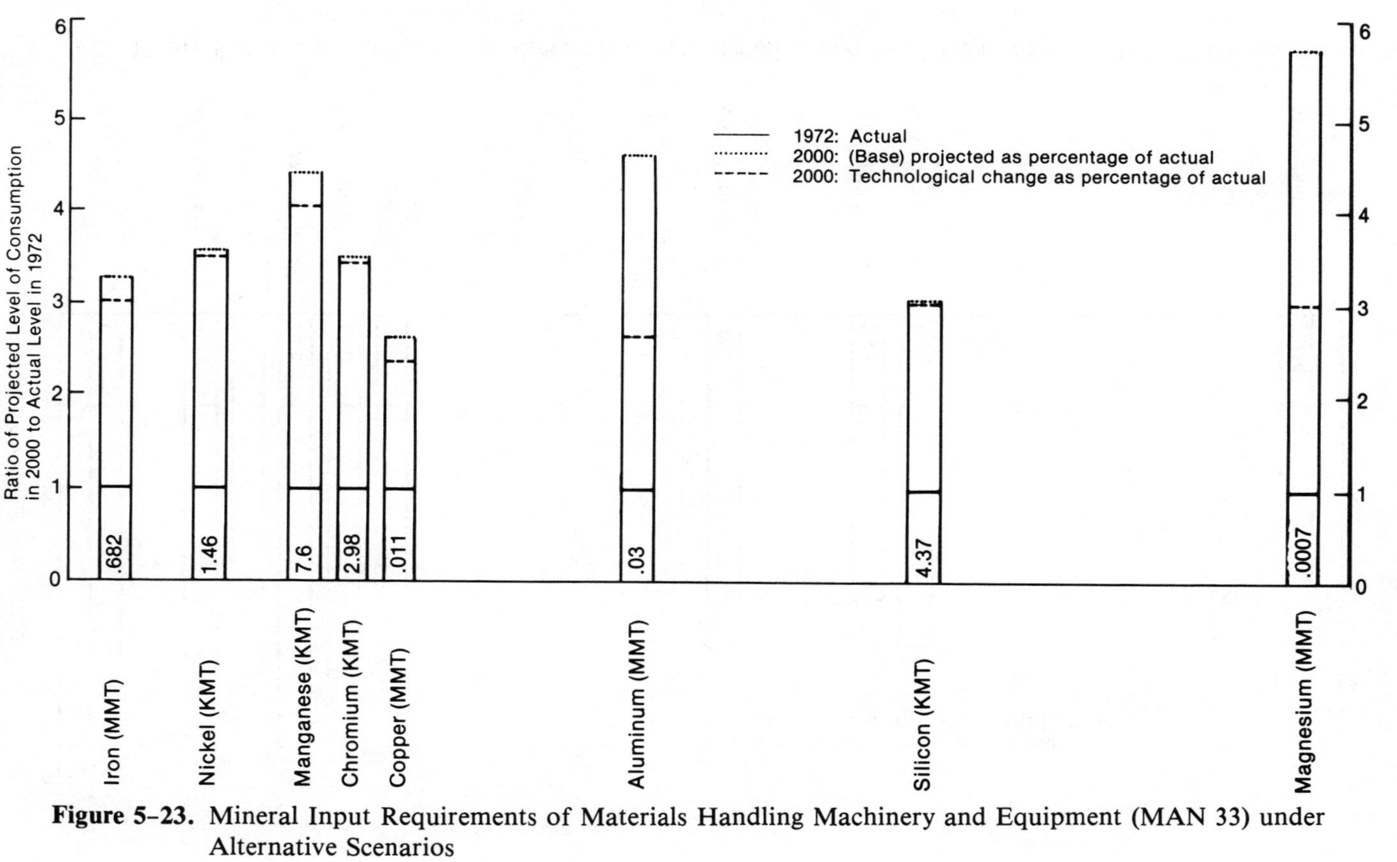

Figure 5–23. Mineral Input Requirements of Materials Handling Machinery and Equipment (MAN 33) under Alternative Scenarios

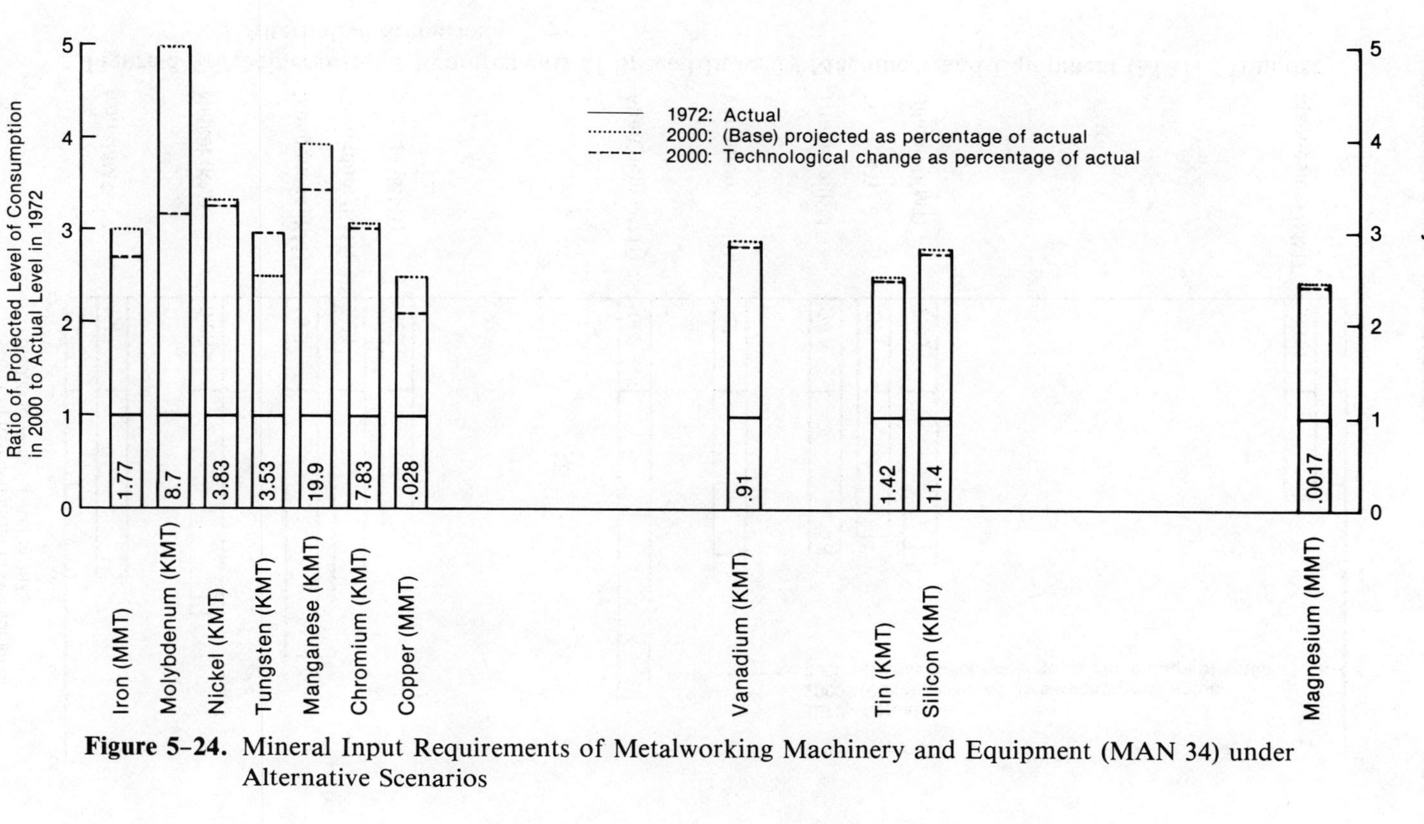

Figure 5–24. Mineral Input Requirements of Metalworking Machinery and Equipment (MAN 34) under Alternative Scenarios

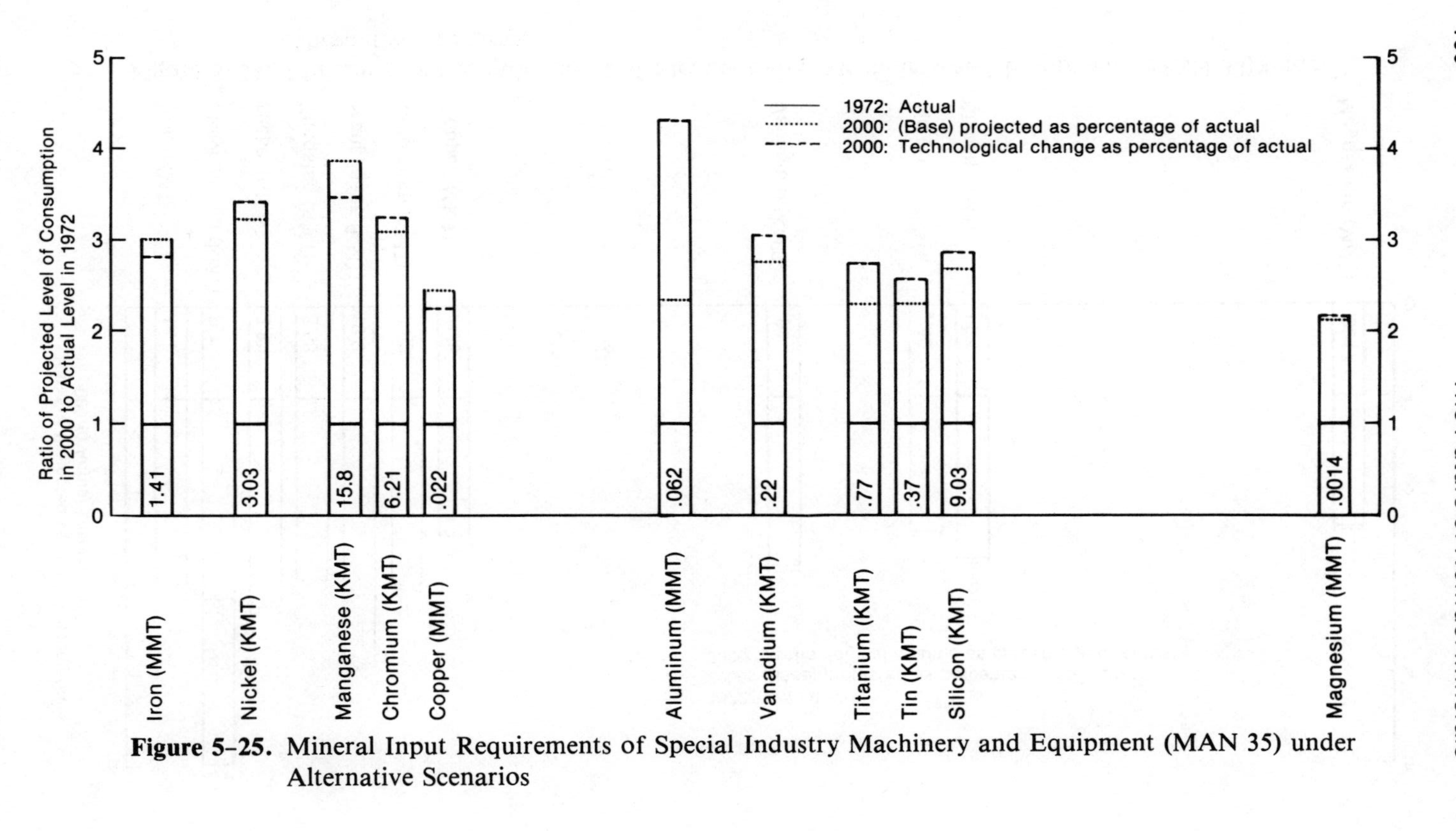

Figure 5–25. Mineral Input Requirements of Special Industry Machinery and Equipment (MAN 35) under Alternative Scenarios

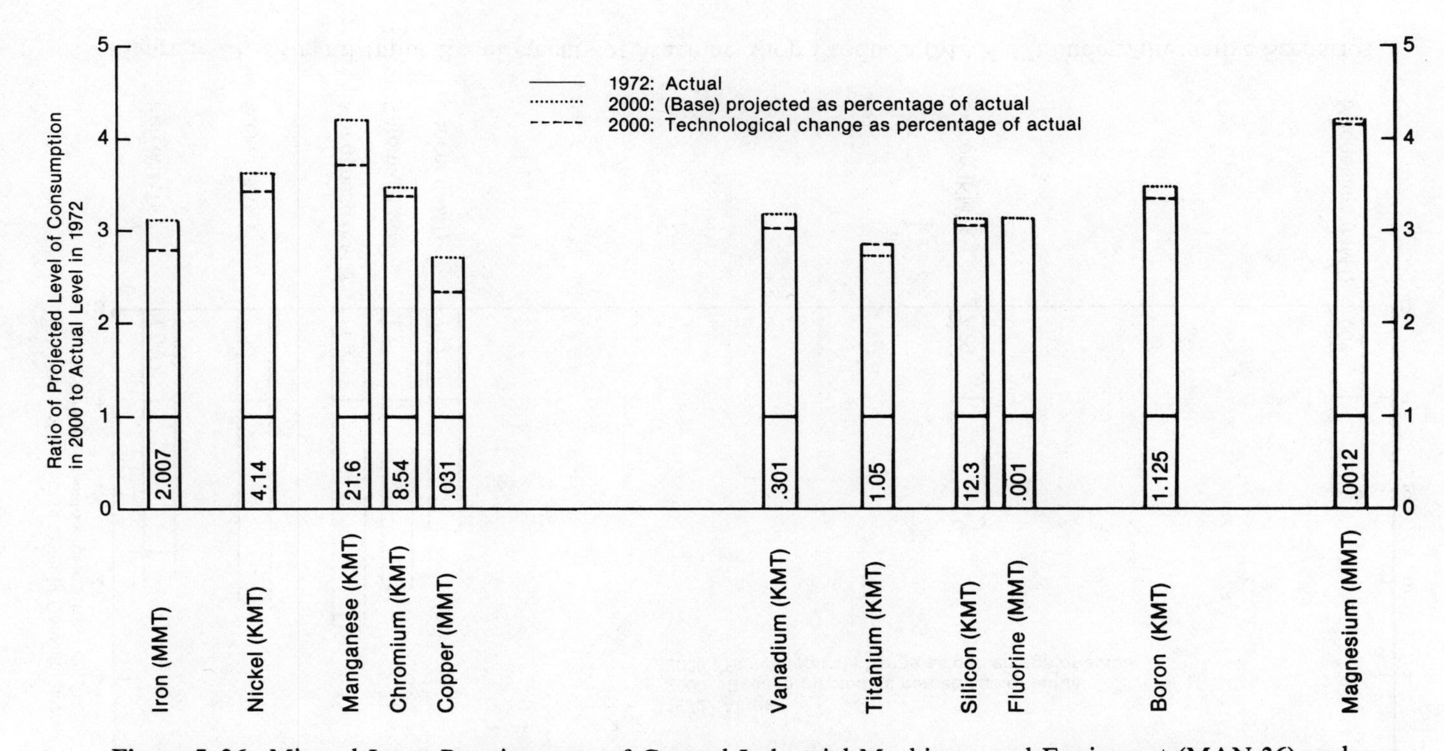

Figure 5–26. Mineral Input Requirements of General Industrial Machinery and Equipment (MAN 36) under Alternative Scenarios

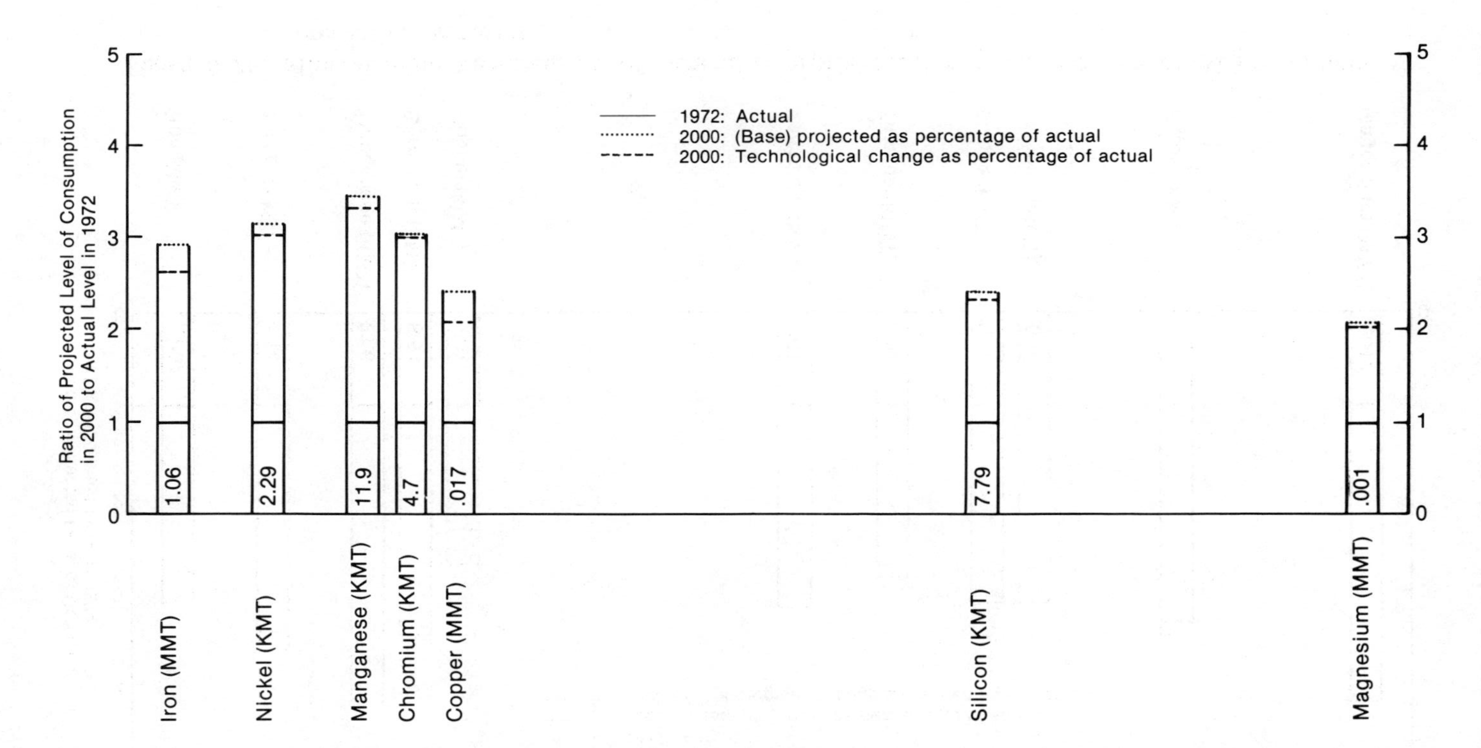

Figure 5–27. Mineral Input Requirements of Machine Shop Products (MAN 37) under Alternative Scenarios

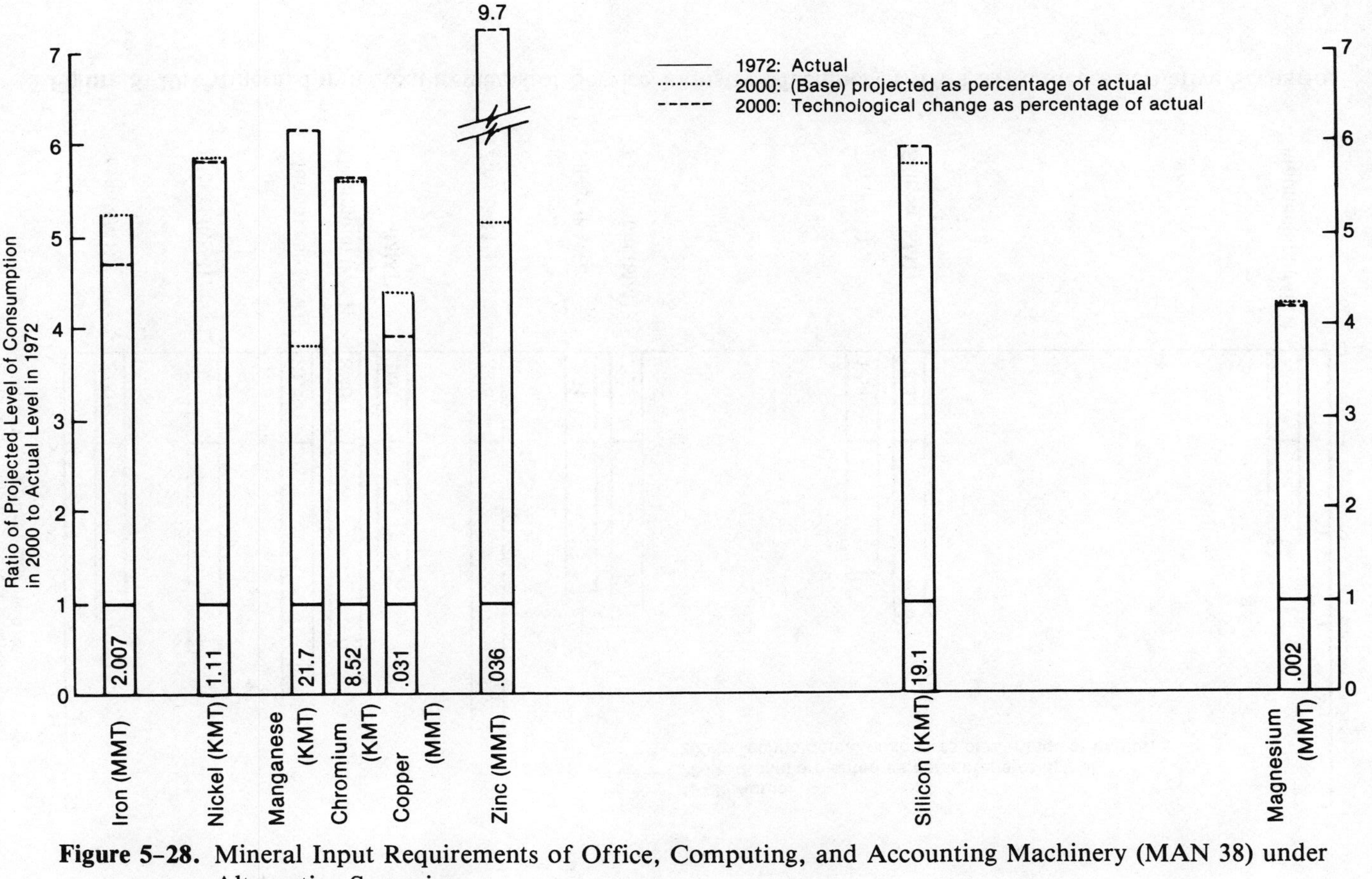

Figure 5–28. Mineral Input Requirements of Office, Computing, and Accounting Machinery (MAN 38) under Alternative Scenarios

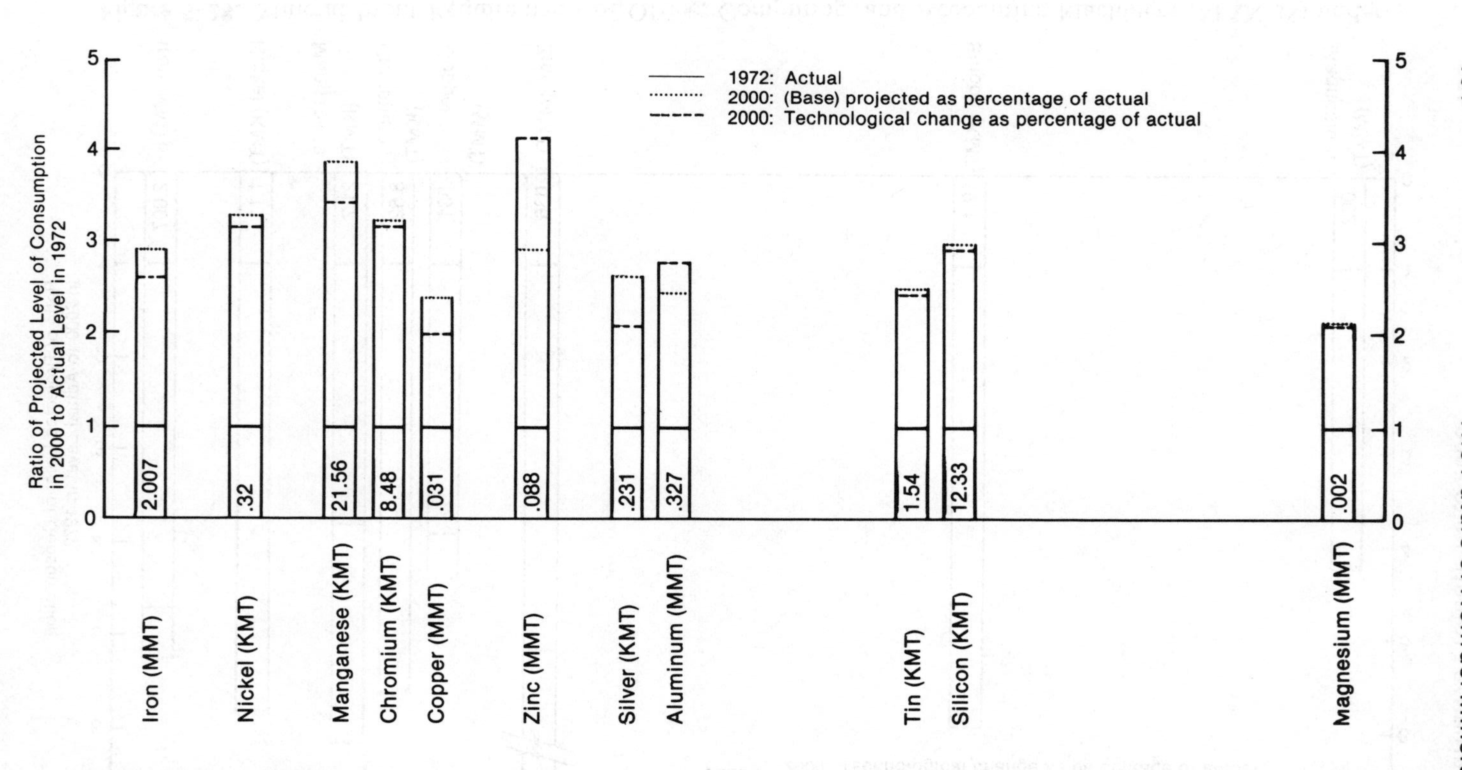

Figure 5–29. Mineral Input Requirements of Service Industry Machinery (MAN 39) under Alternative Scenarios

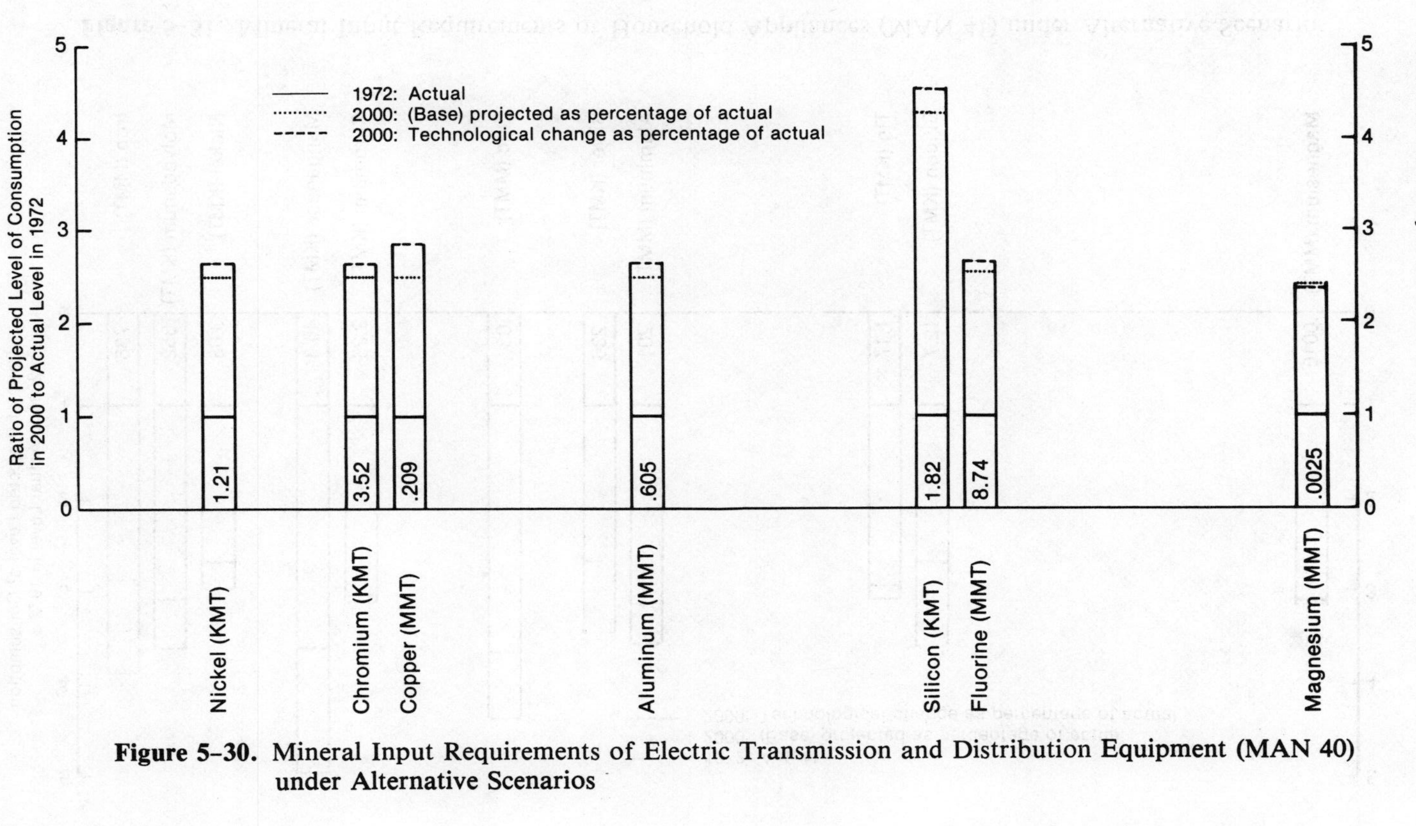

Figure 5–30. Mineral Input Requirements of Electric Transmission and Distribution Equipment (MAN 40) under Alternative Scenarios

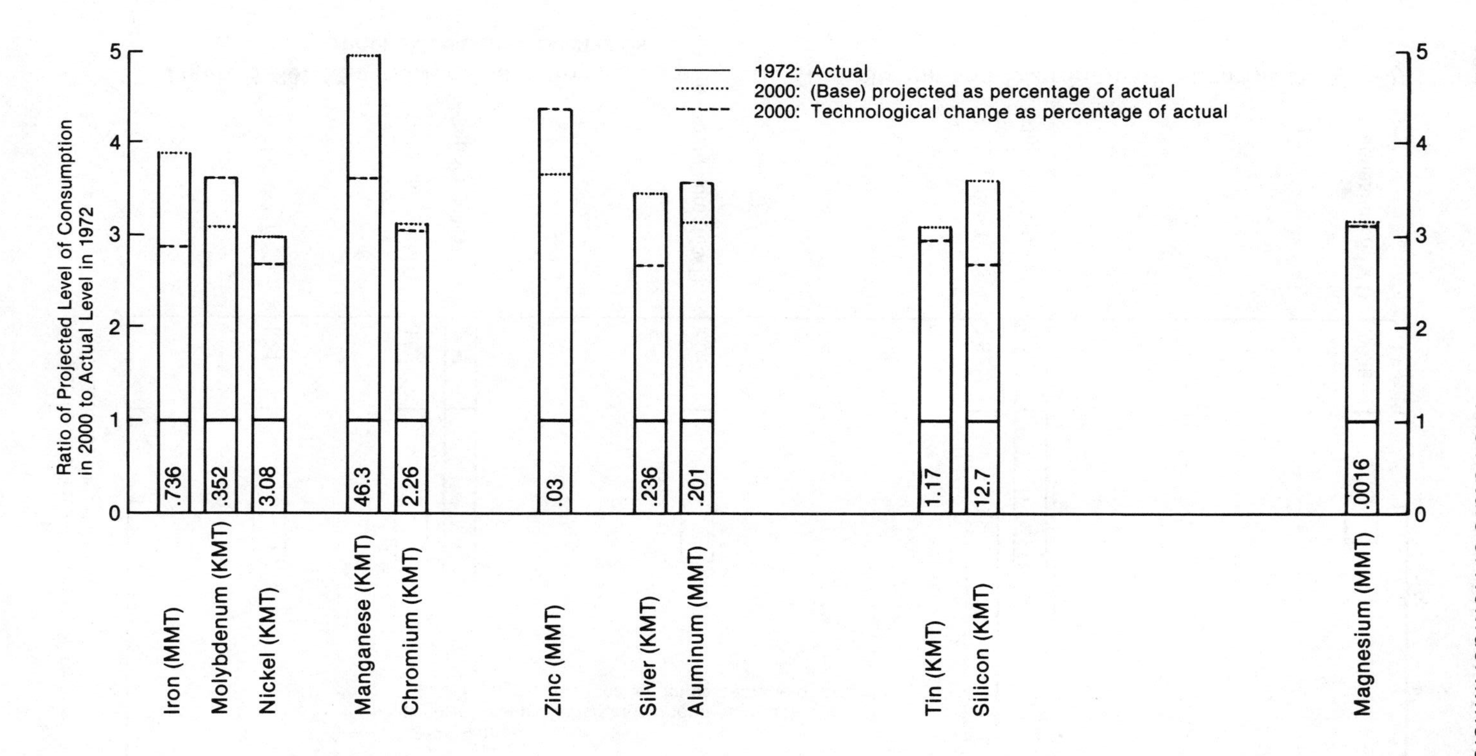

Figure 5-31. Mineral Input Requirements of Household Appliances (MAN 41) under Alternative Scenarios

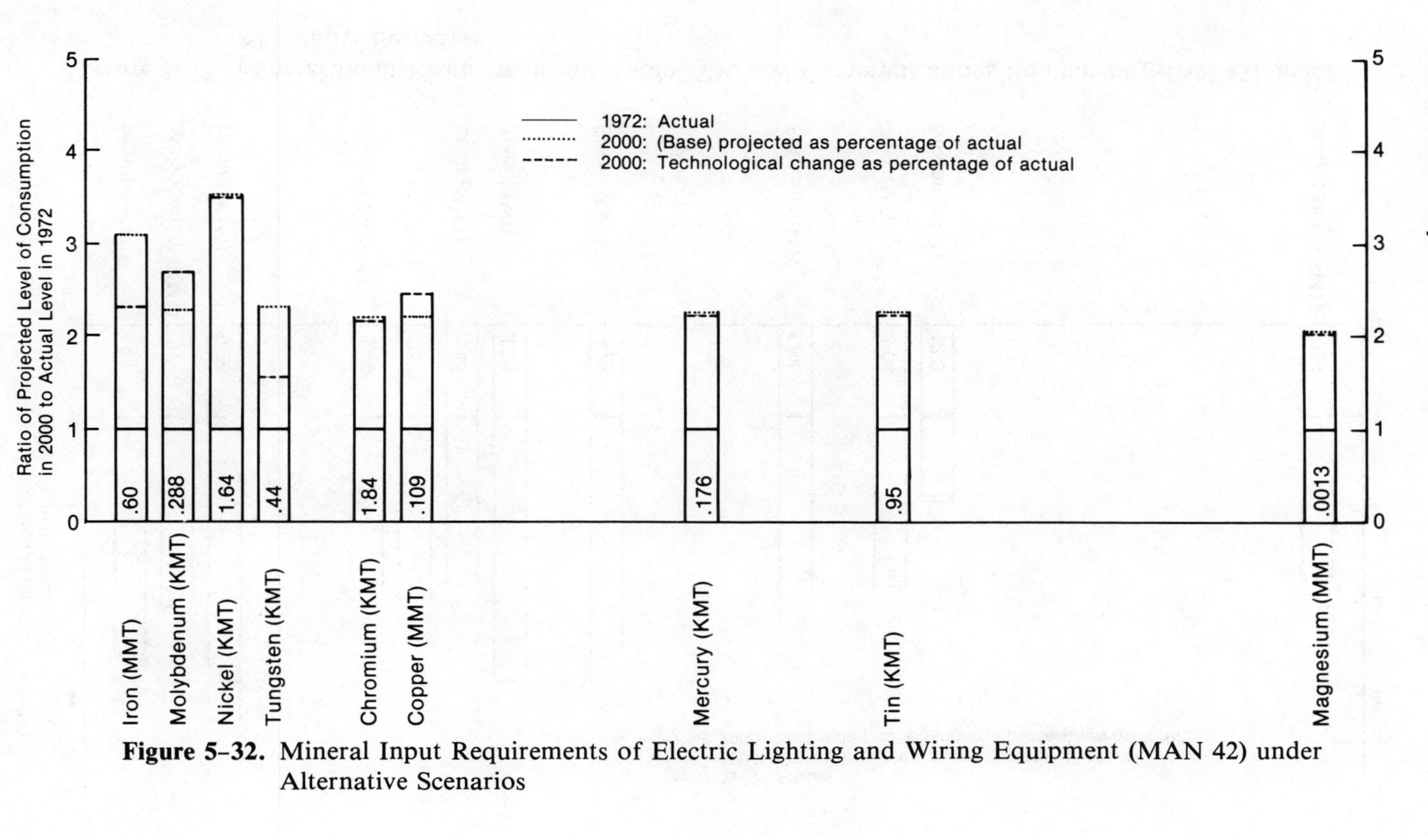

Figure 5–32. Mineral Input Requirements of Electric Lighting and Wiring Equipment (MAN 42) under Alternative Scenarios

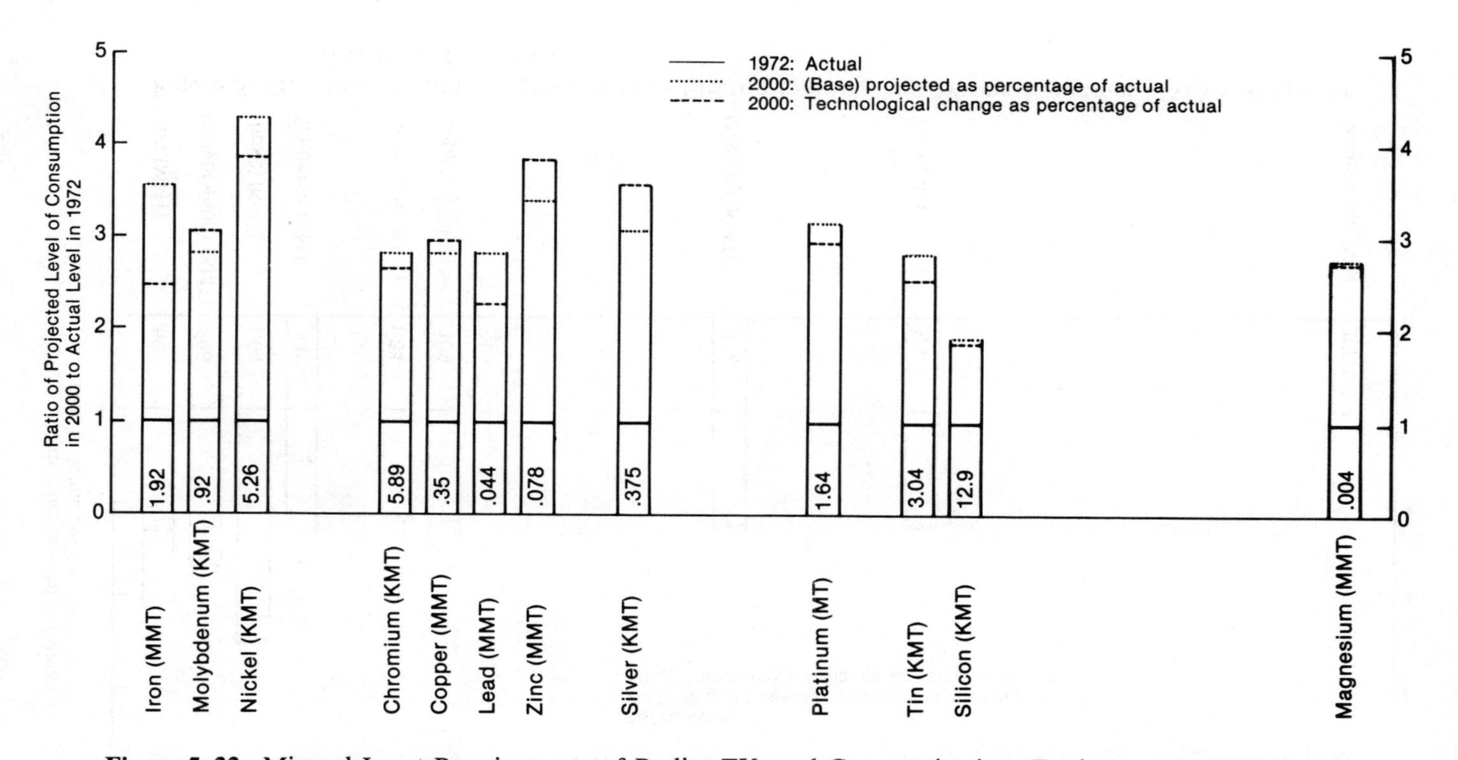

Figure 5–33. Mineral Input Requirements of Radio, TV, and Communications Equipment (MAN 43) under Alternative Scenarios

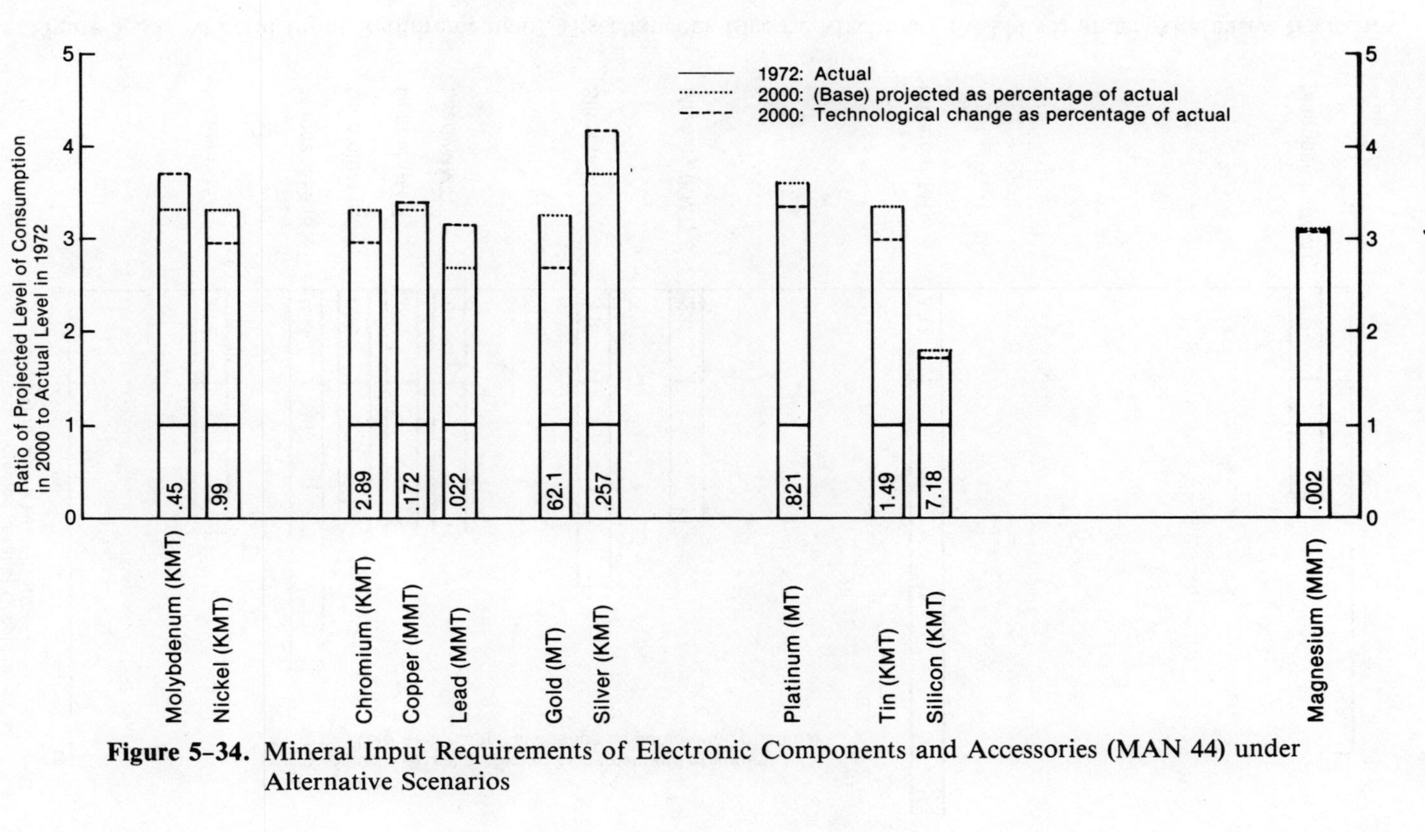

Figure 5–34. Mineral Input Requirements of Electronic Components and Accessories (MAN 44) under Alternative Scenarios

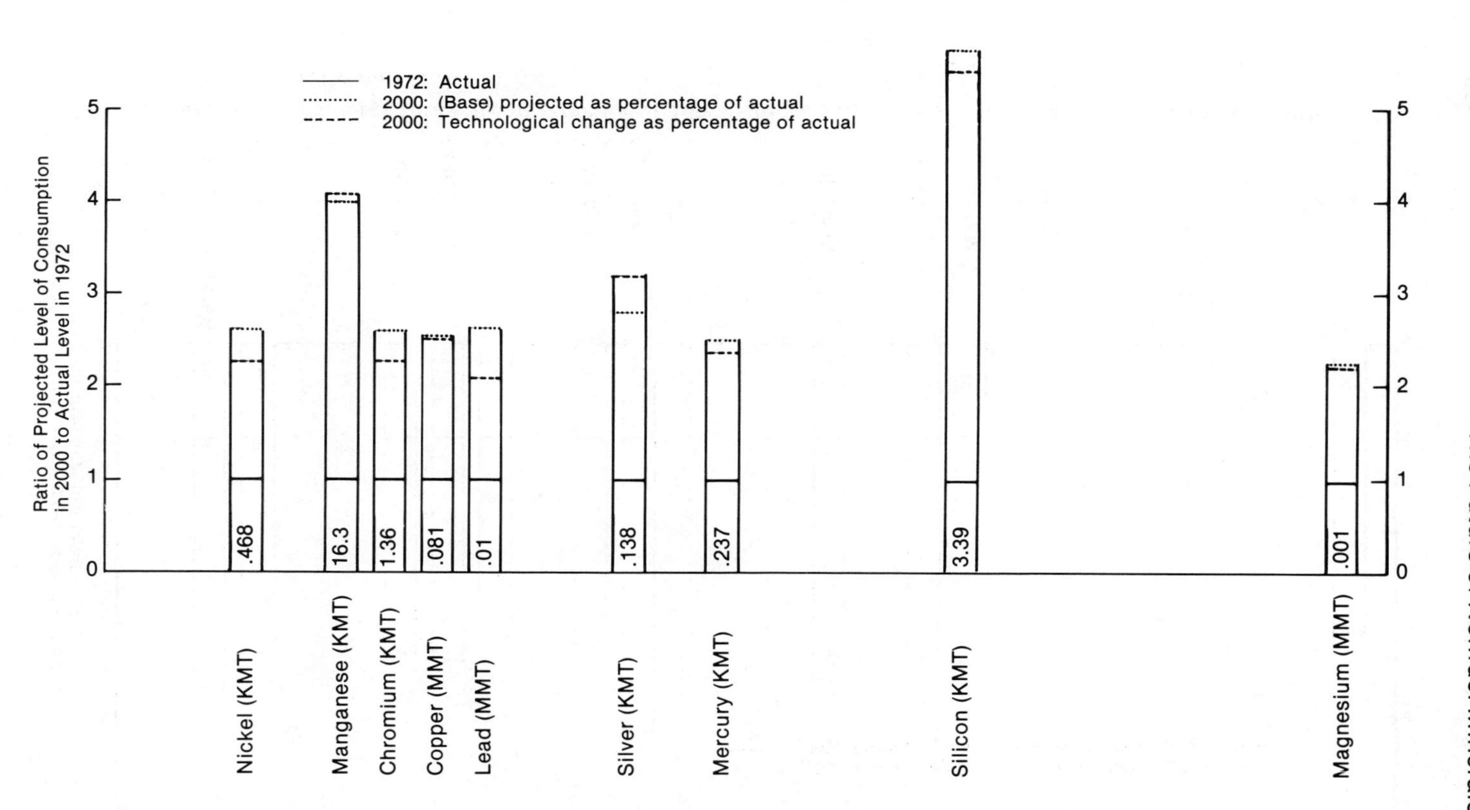

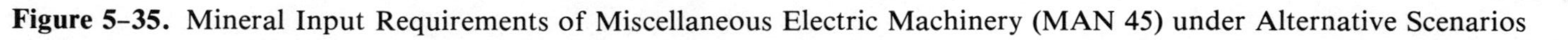
Figure 5–35. Mineral Input Requirements of Miscellaneous Electric Machinery (MAN 45) under Alternative Scenarios

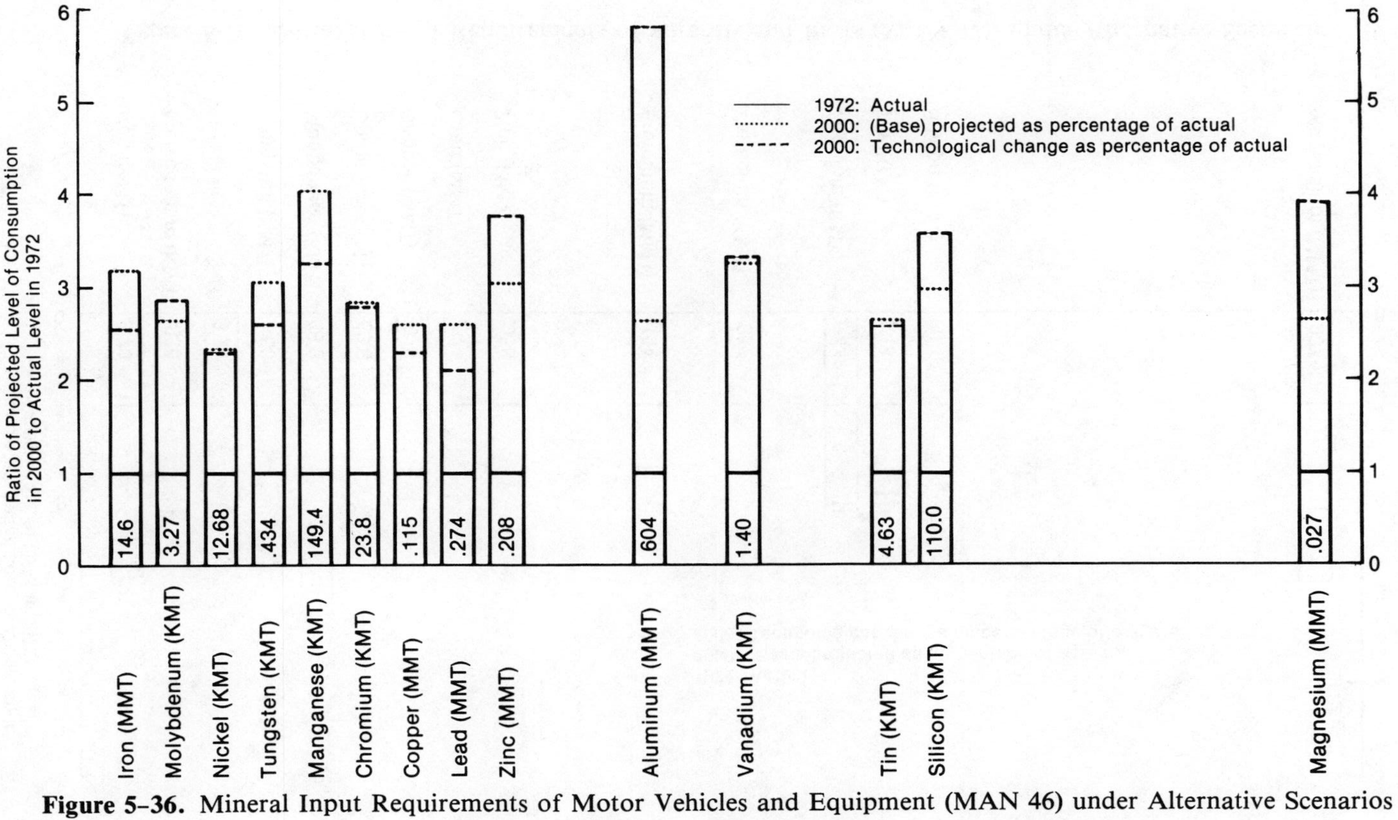

Figure 5–36. Mineral Input Requirements of Motor Vehicles and Equipment (MAN 46) under Alternative Scenarios

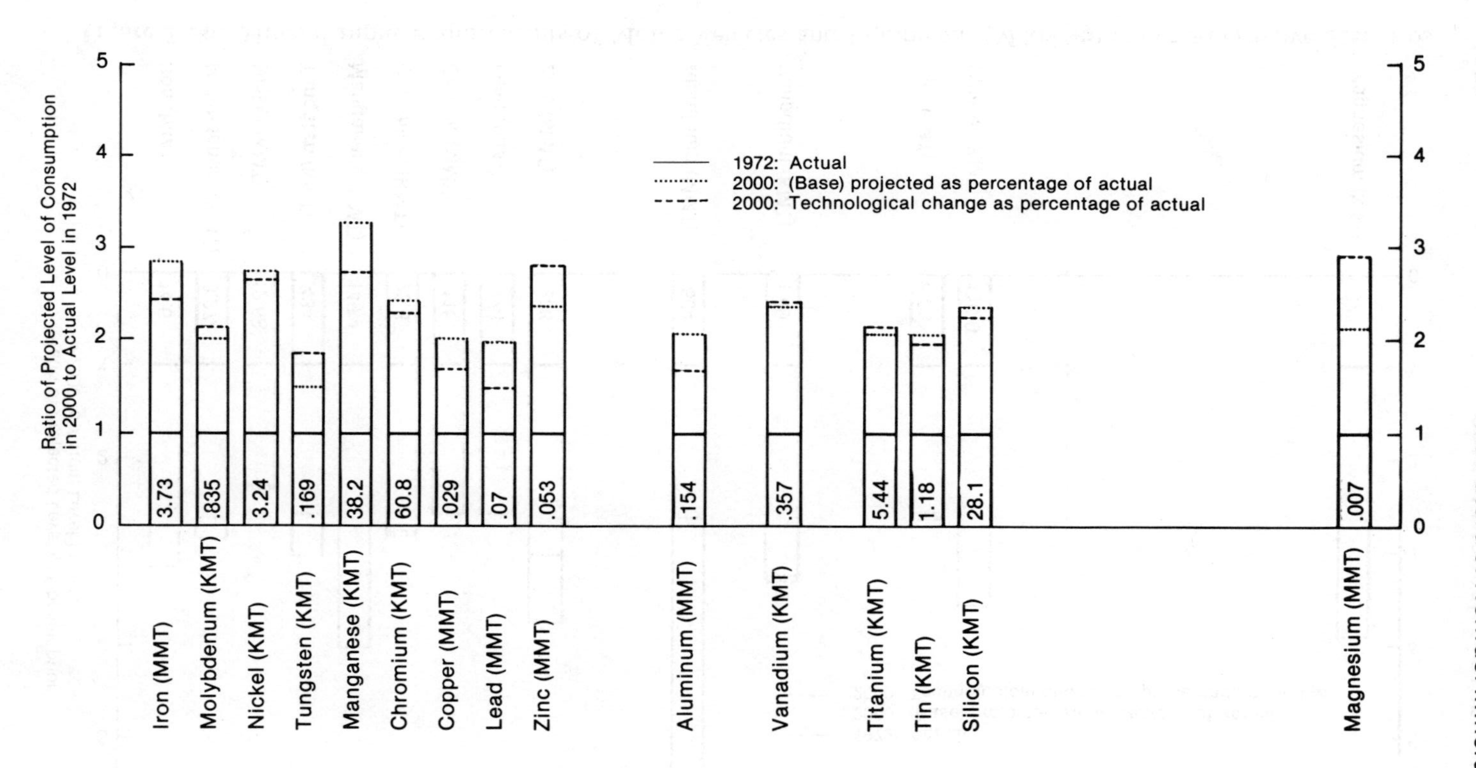

Figure 5-37. Mineral Input Requirements of Aircraft and Parts (MAN 47), under Alternative Scenarios

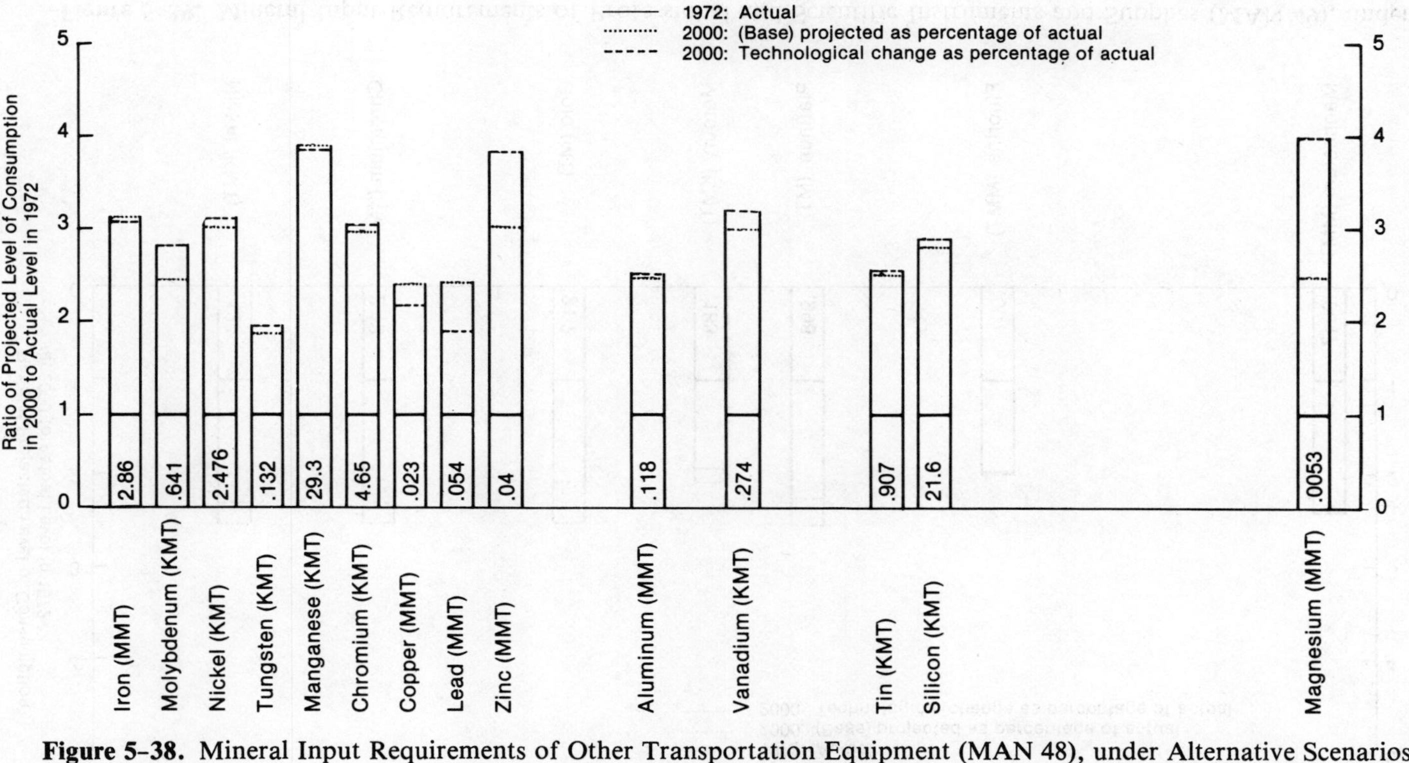

Figure 5–38. Mineral Input Requirements of Other Transportation Equipment (MAN 48), under Alternative Scenarios

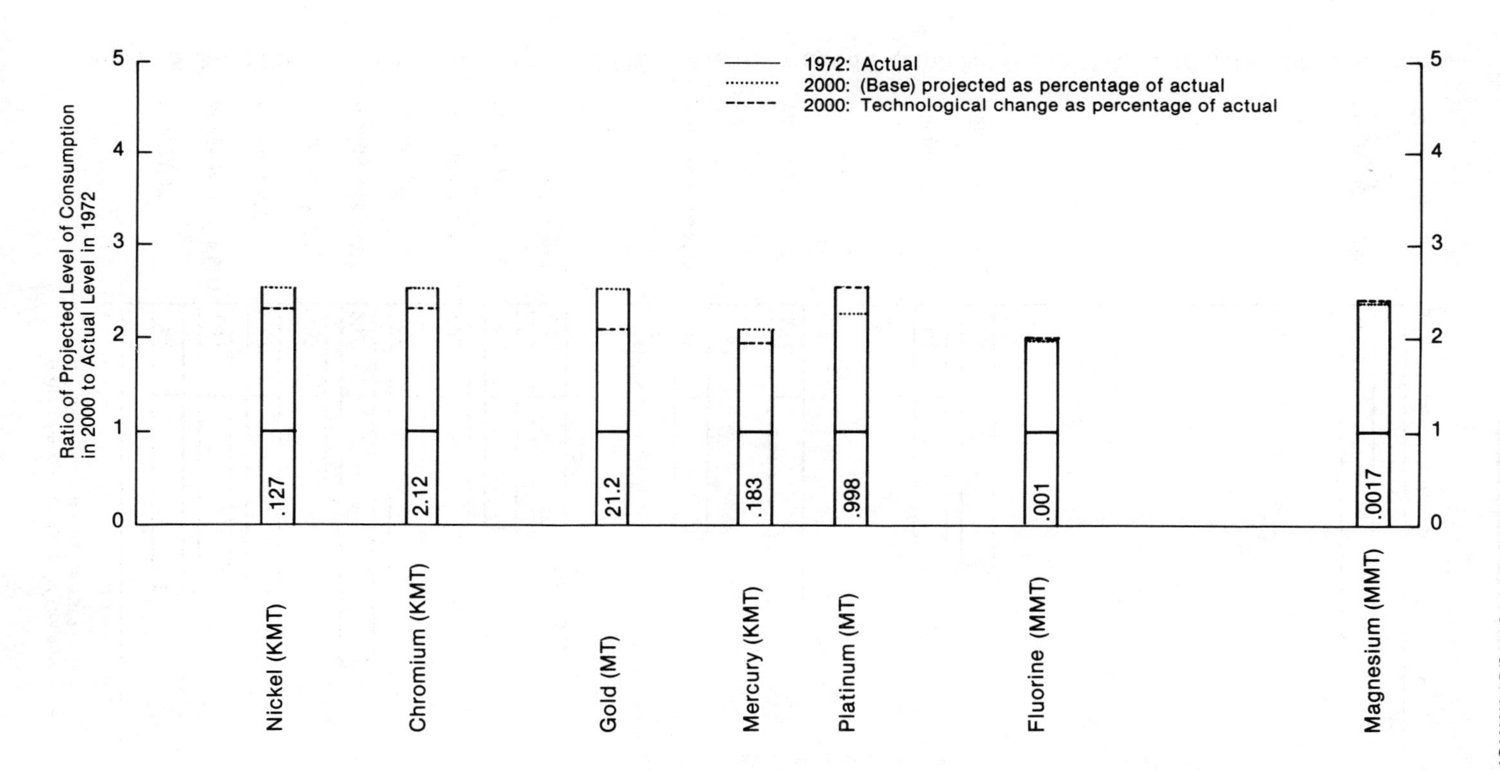

Figure 5–39. Mineral Input Requirements of Professional and Scientific Instruments and Supplies (MAN 49), under Alternative Scenarios

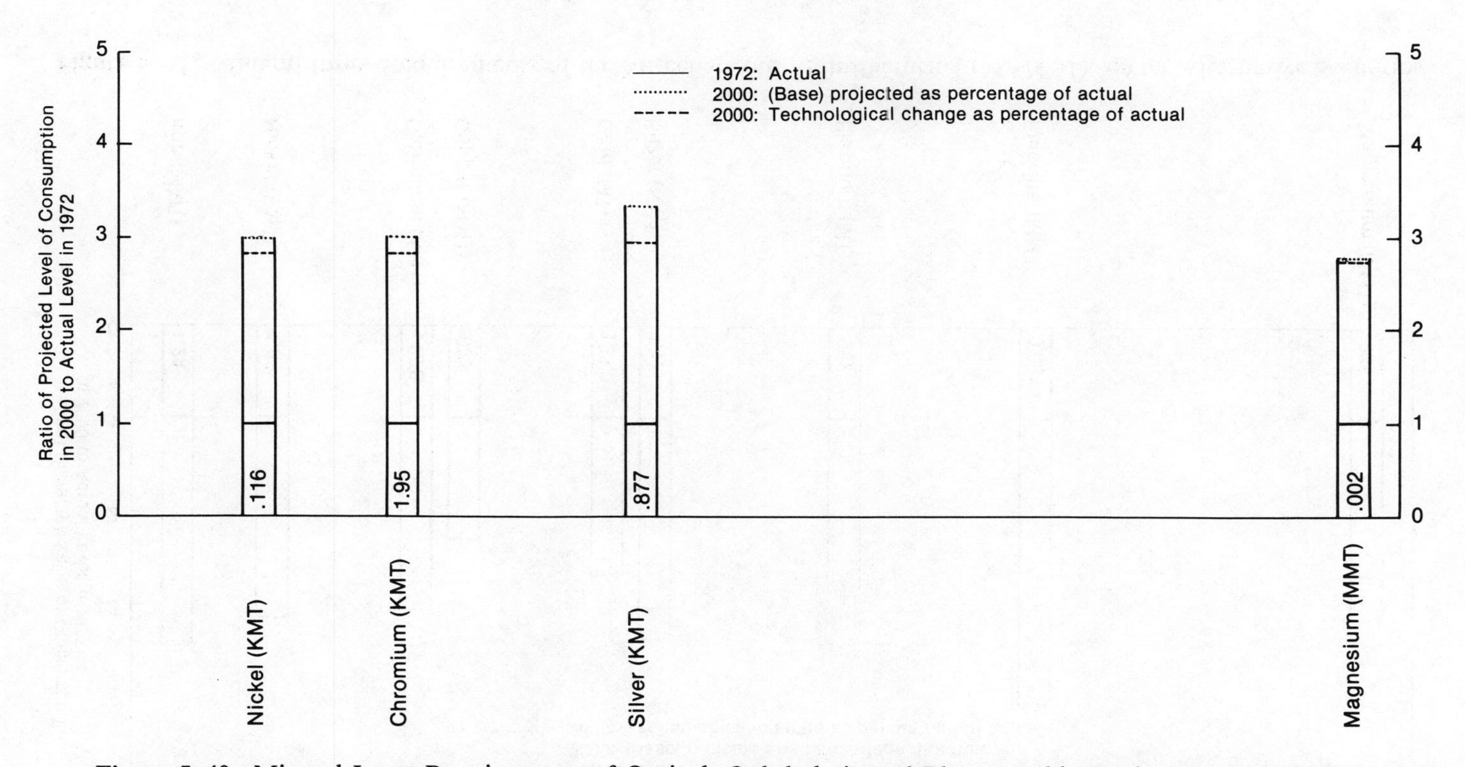

Figure 5–40. Mineral Input Requirements of Optical, Ophthalmic and Photographic Equipment (MAN 50), under Alternative Scenarios

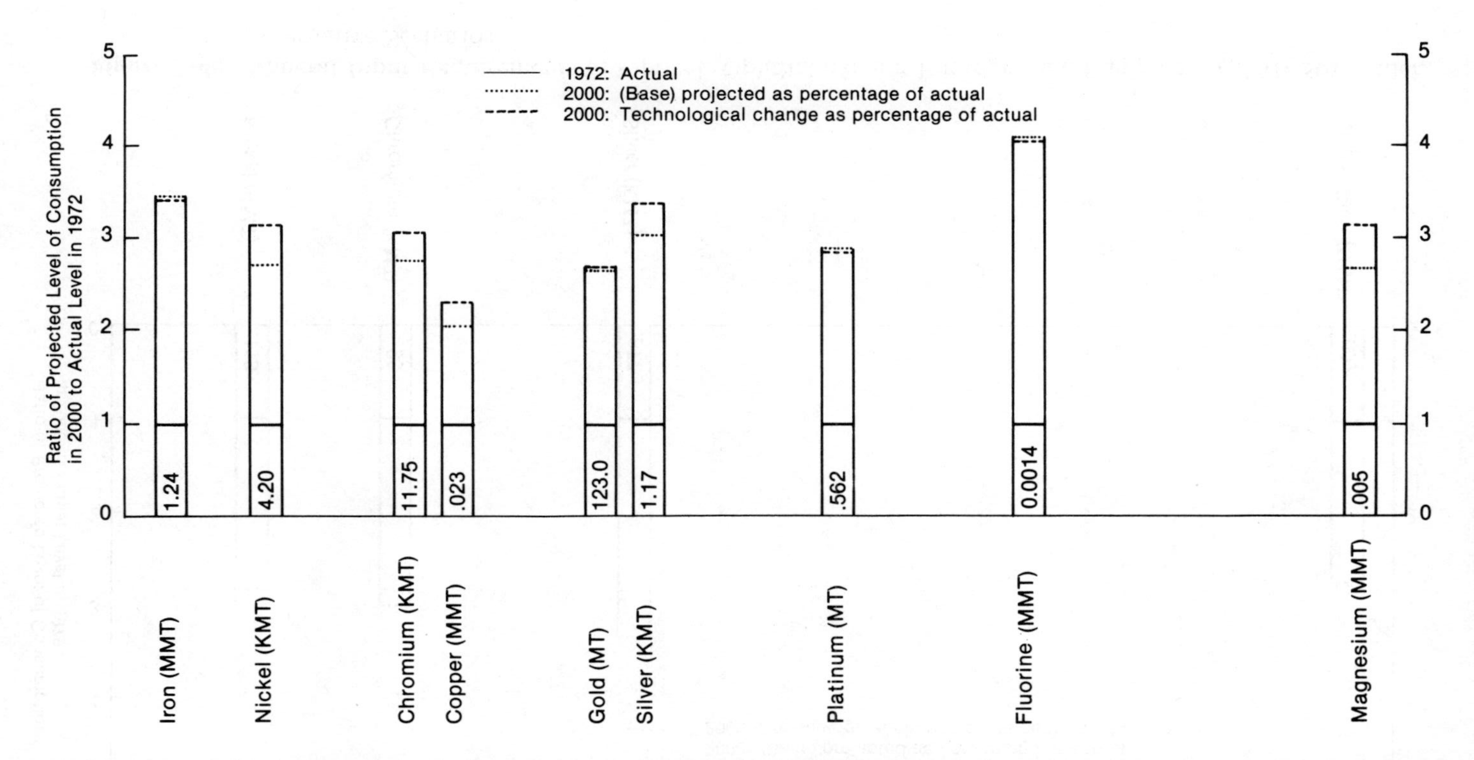

Figure 5–41. Mineral Input Requirements of the Miscellaneous Manufacturing (MAN 51), under Alternative Scenarios

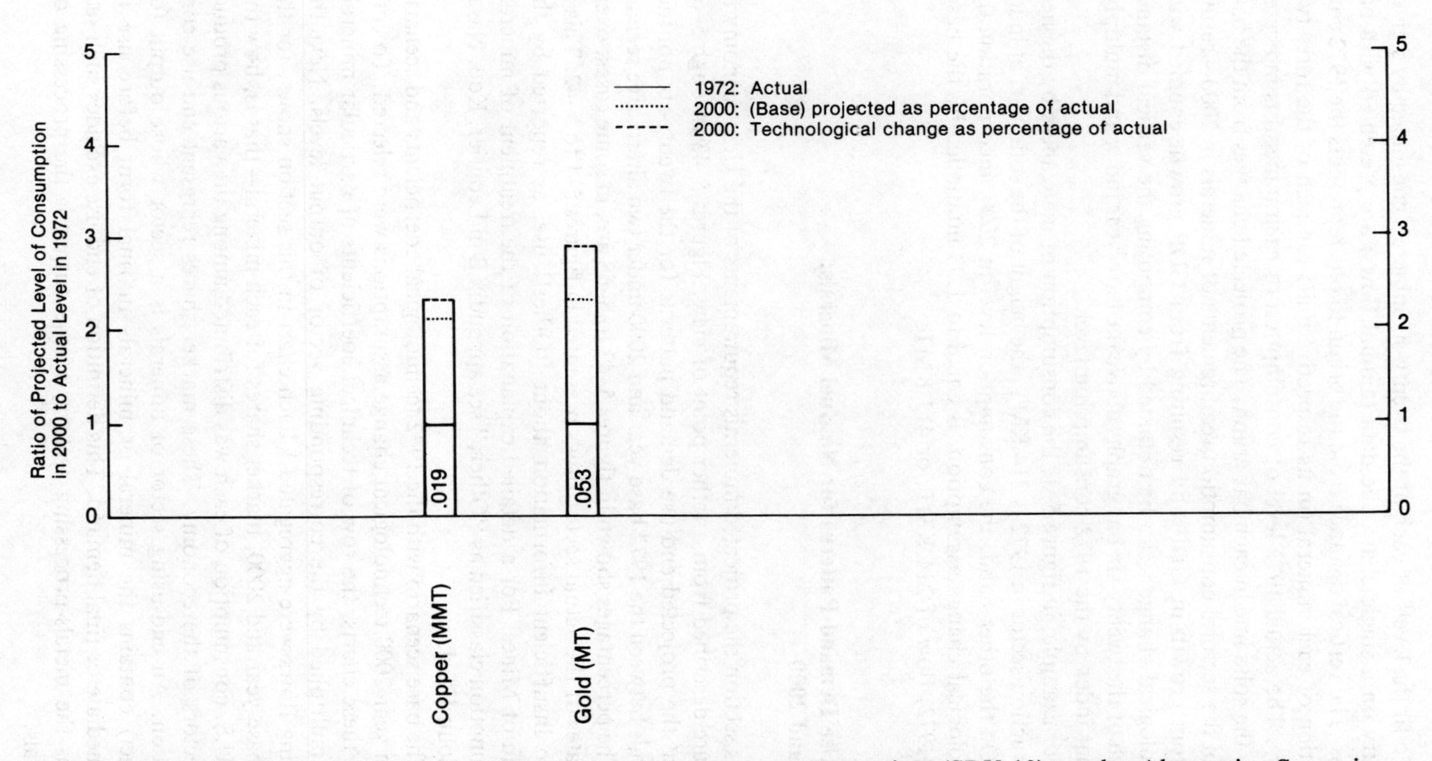

Figure 5-42. Mineral Input Requirements of Federal Government Enterprises (SRV 13), under Alternative Scenarios

To fit the levels of each of the twenty-six minerals consumed by a given industry on a single chart, the data (annual flows) are presented with bar graphs. The vertical distance from the origin to one represents the 1972 consumption of each mineral (in its respective unit) for each of the forty-two sectors. The actual 1972 level of consumption of each mineral is indicated below the solid line in each bar graph. The projected changes from the base year in the sectoral consumption requirements of minerals in 2000—resulting from growth in GDP and resulting from GDP growth coupled with technological change—can be obtained by computing the vertical distance of each of the twenty-six bar graphs above the base-year index and multiplying that index by the 1972 consumption level.

For example, in figure 5–1, the consumption of manganese by the new construction sector in 1972 is 214 KMT, the height of the solid bar, at index one. On the other hand, the consumption level in 2000, incorporating the technological change assumption, is equal to 1.93 multiplied by the base-year (1972) flow of 214 KMT, or 413 KMT.

End-Use Demand Patterns for Nonfuel Minerals: 1972 and 2000

In this section the projected minerals requirements of the U.S. economy to 2000 are described from another point of view. Figures 5–43 through 5–68 present the projected end-use demand patterns for the twenty-six nonfuel minerals between the 1972 base-year and 2000 under two alternative scenarios. The percentages shown in figures 5–43 to 5–68 are, in some cases, overestimates of the actual sector flows presented in figures 5–1 to 5–42. This is due to insufficient information about "Other" use, as reported by the Bureau of Mines. For a detailed explanation of the treatment of minerals consumption classified as "Other," see appendix B in Leontief, Koo, Nasar and Sohn [25a].

The base scenario with the 1972 technological coefficients and scenario 8 with year 2000 technological change assumptions were selected. To construct these charts, the row of technical coefficients of particular minerals were multiplied by the corresponding vector of output levels. Next the resulting flows were aggregated.[1] Each chart in this section shows, for the 1972 base-year and 2000, market shares for each mineral—that is, how the total U.S. consumption of each was distributed among the various producing sectors of the economy. These market shares represent end-use consumption. An end-using sector of minerals is a sector whose output (or product) contains the mineral or minerals in final form before use in intermediate or final demand. This definition of end-use excludes such sectors as the metals-processing sectors, whose output is the processing of minerals.

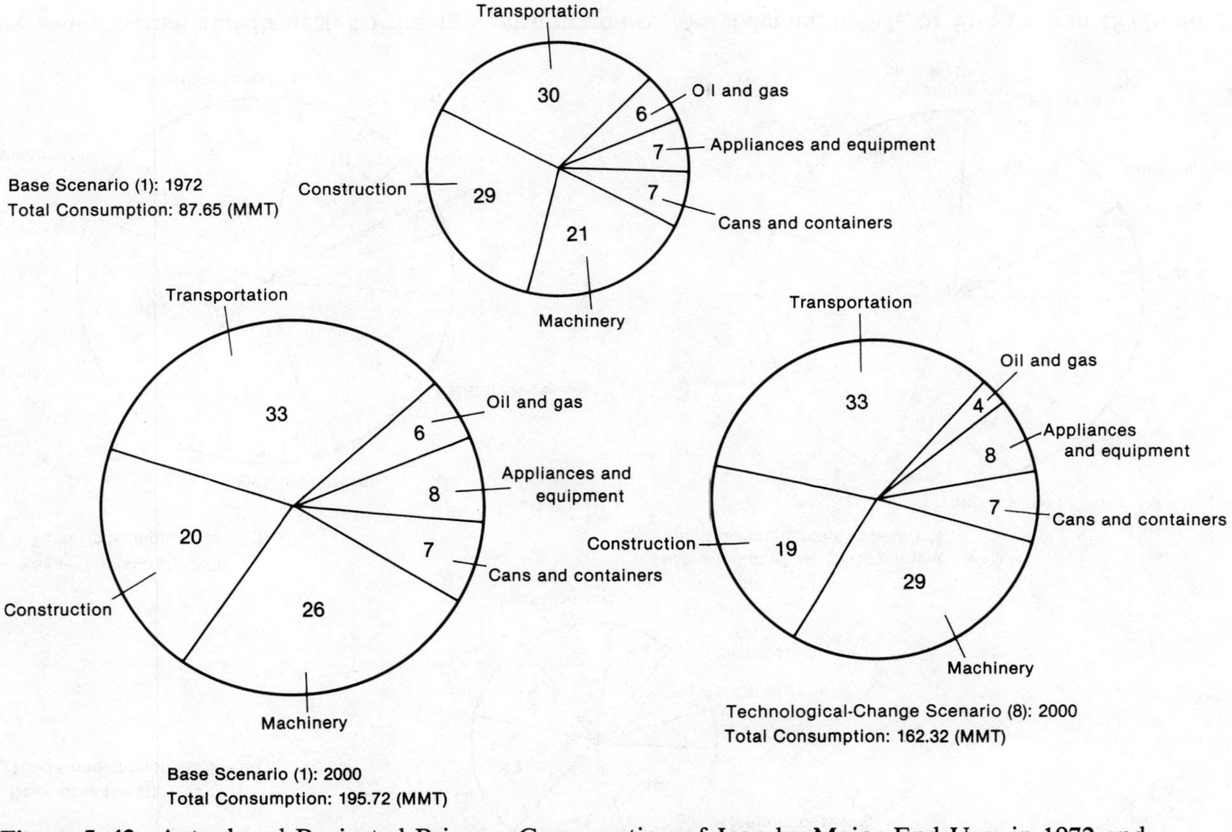

Figure 5-43. Actual and Projected Primary Consumption of Iron by Major End Uses in 1972 and 2000 (percentages)

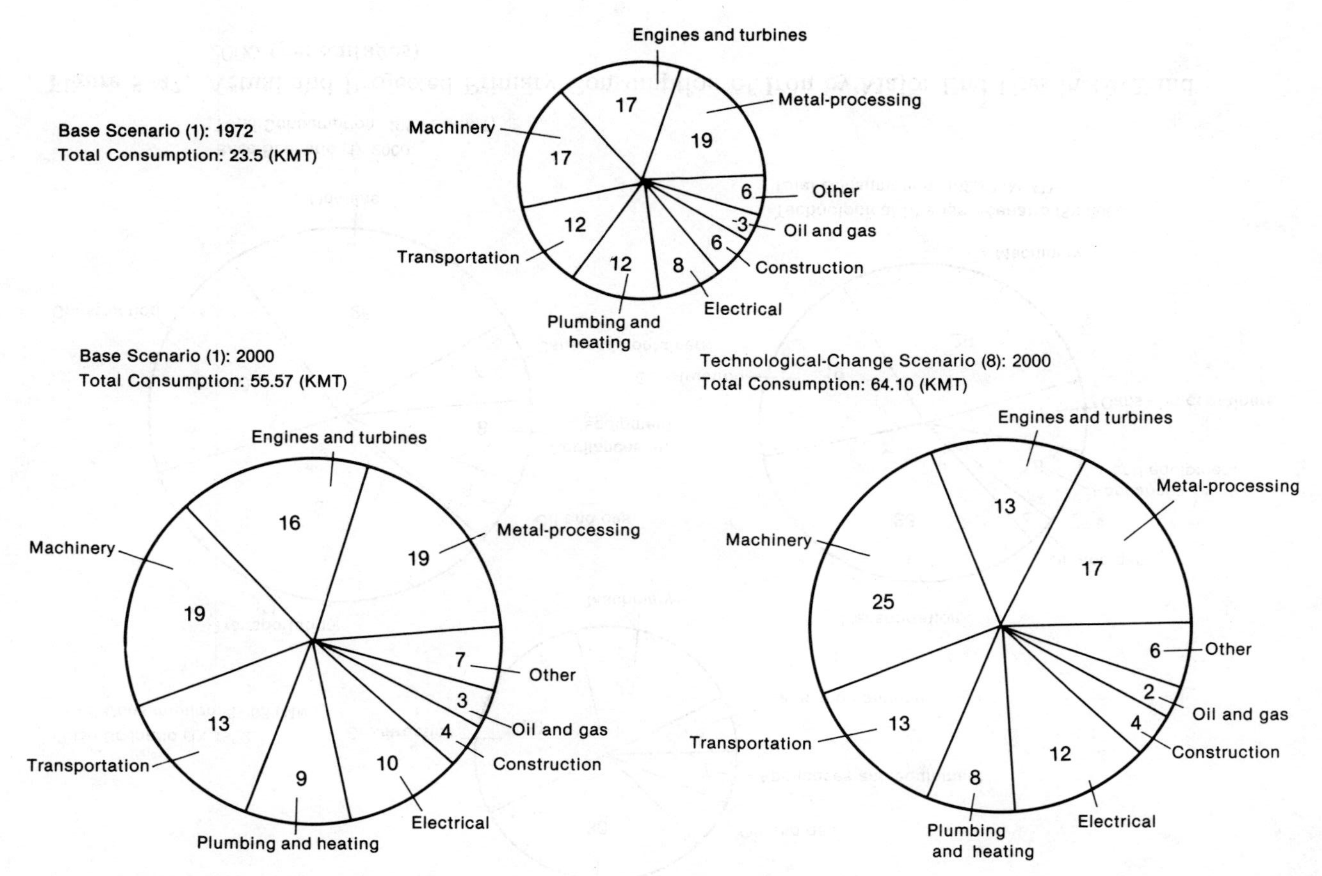

Figure 5–44. Actual and Projected Primary Consumption of Molybdenum by Major End Uses in 1972 and 2000 (percentages)

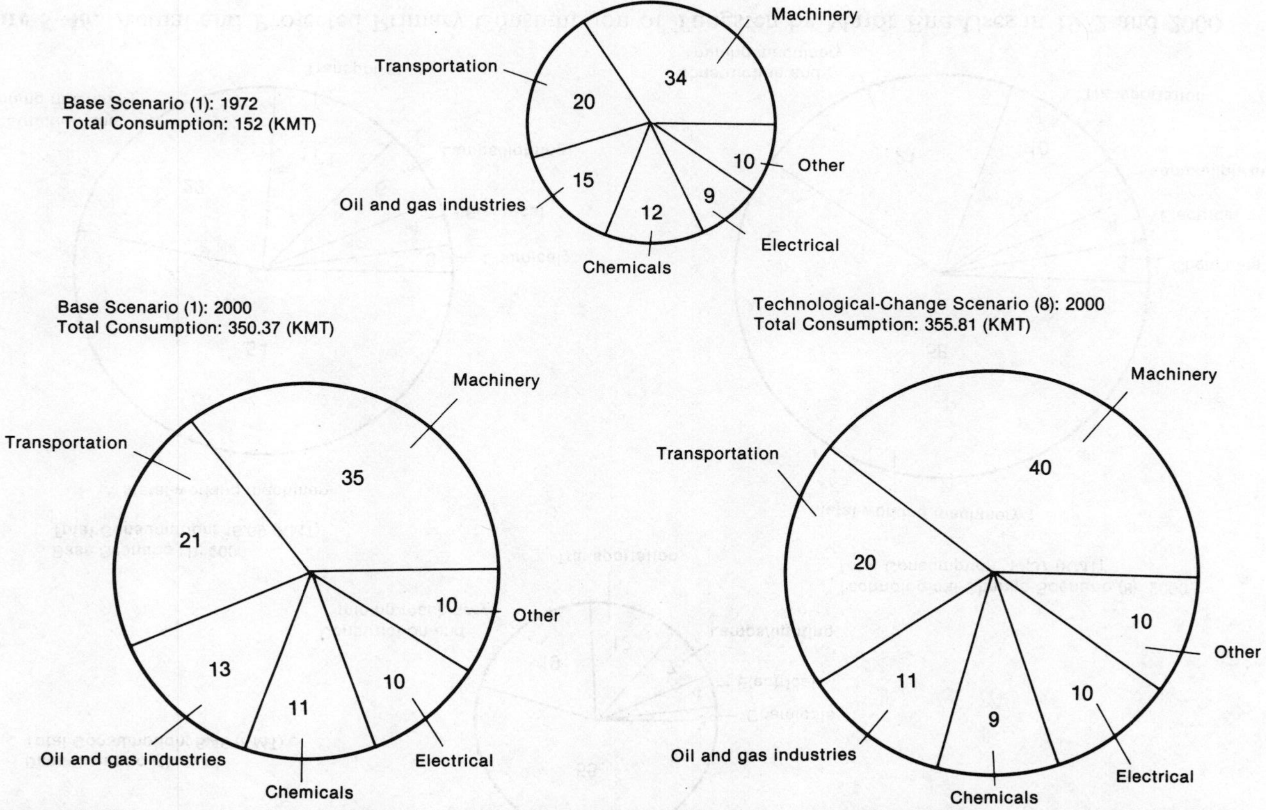

Figure 5–45. Actual and Projected Primary Consumption of Nickel by Major End Uses in 1972 and 2000 (percentages)

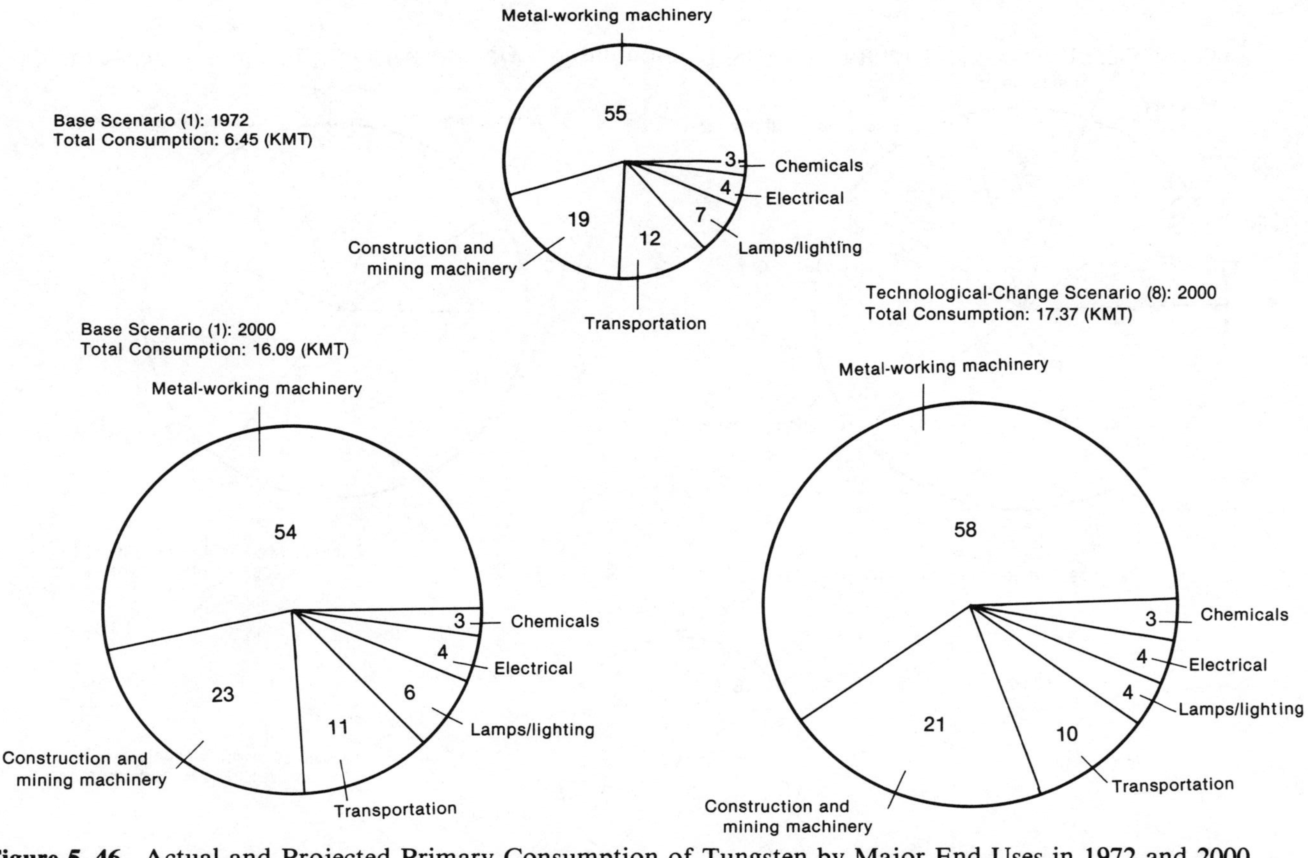

Figure 5-46. Actual and Projected Primary Consumption of Tungsten by Major End Uses in 1972 and 2000 (percentages)

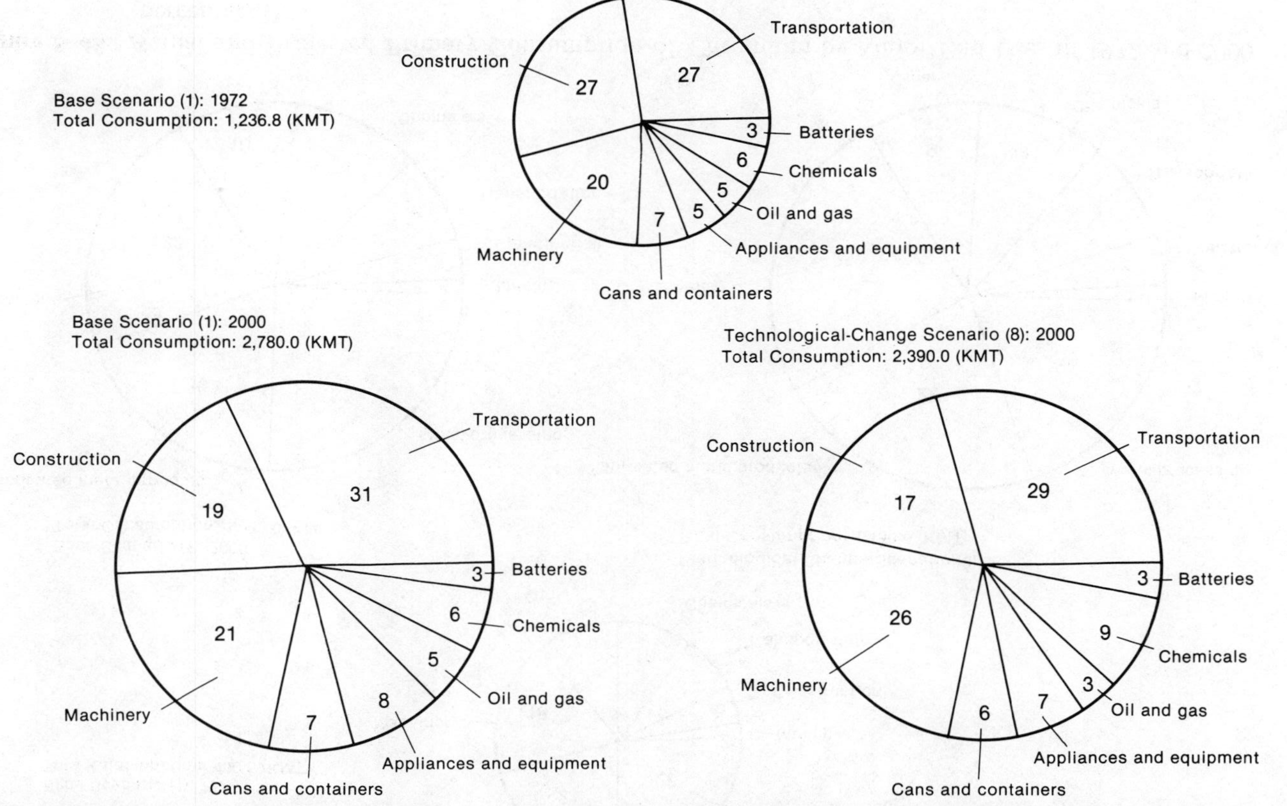

Figure 5–47. Actual and Projected Primary Consumption of Manganese by Major End Uses in 1972 and 2000 (percentages)

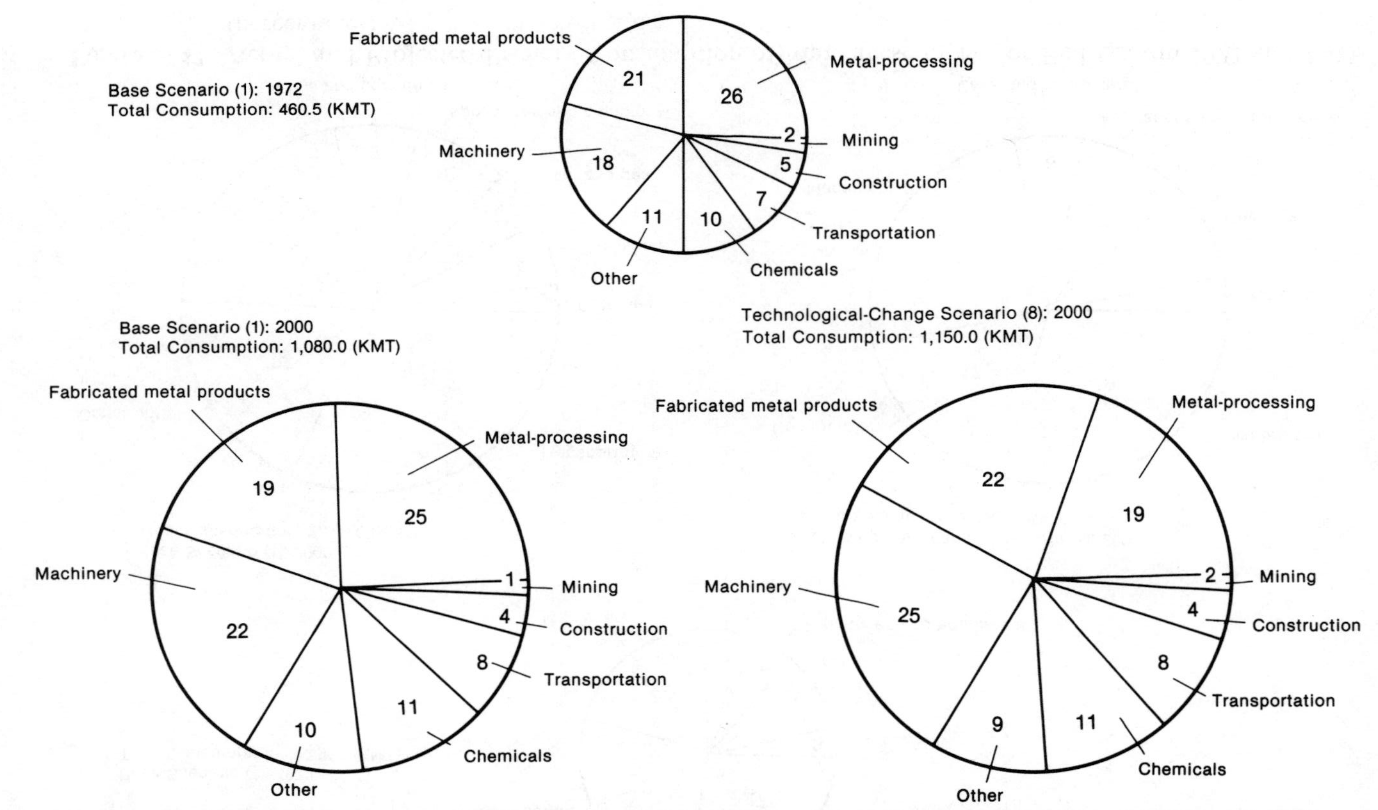

Figure 5–48. Actual and Projected Primary Consumption of Chromium by Major End Uses in 1972 and 2000 (percentages)

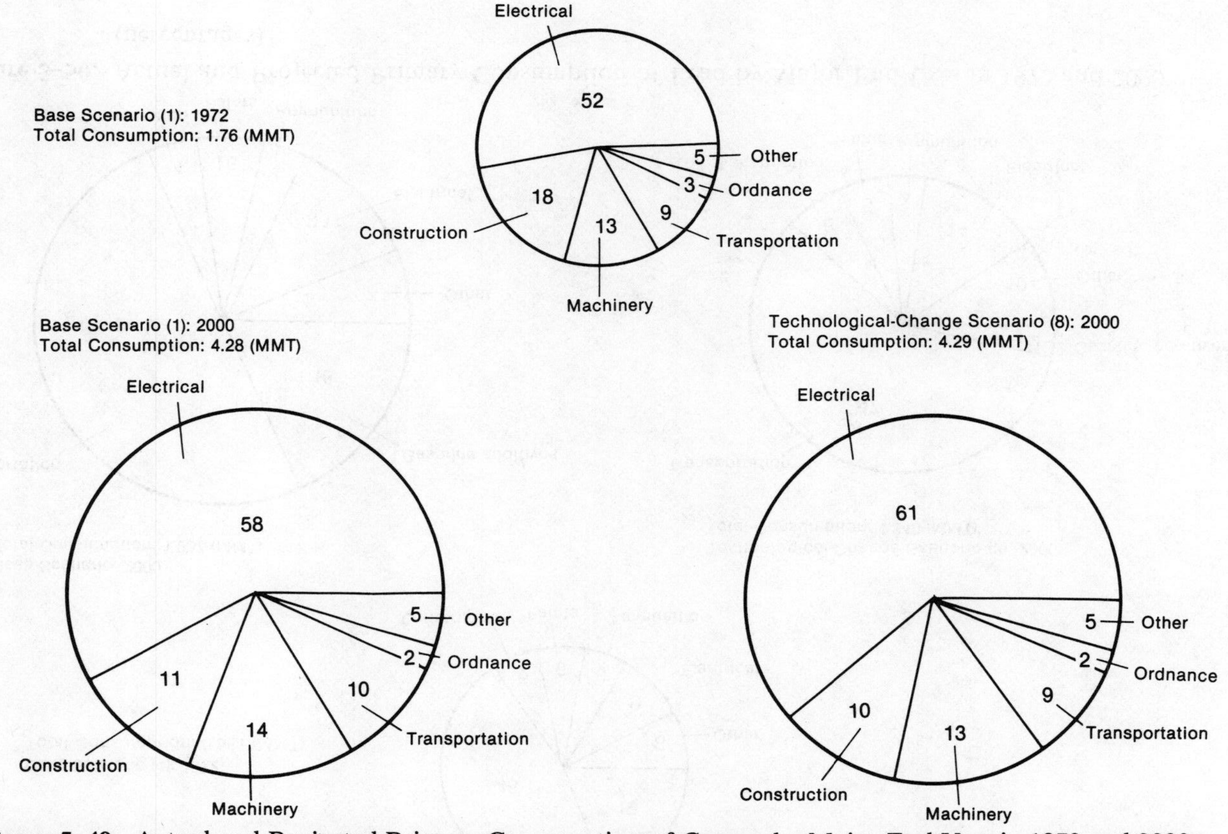

Figure 5–49. Actual and Projected Primary Consumption of Copper by Major End Uses in 1972 and 2000 (percentages)

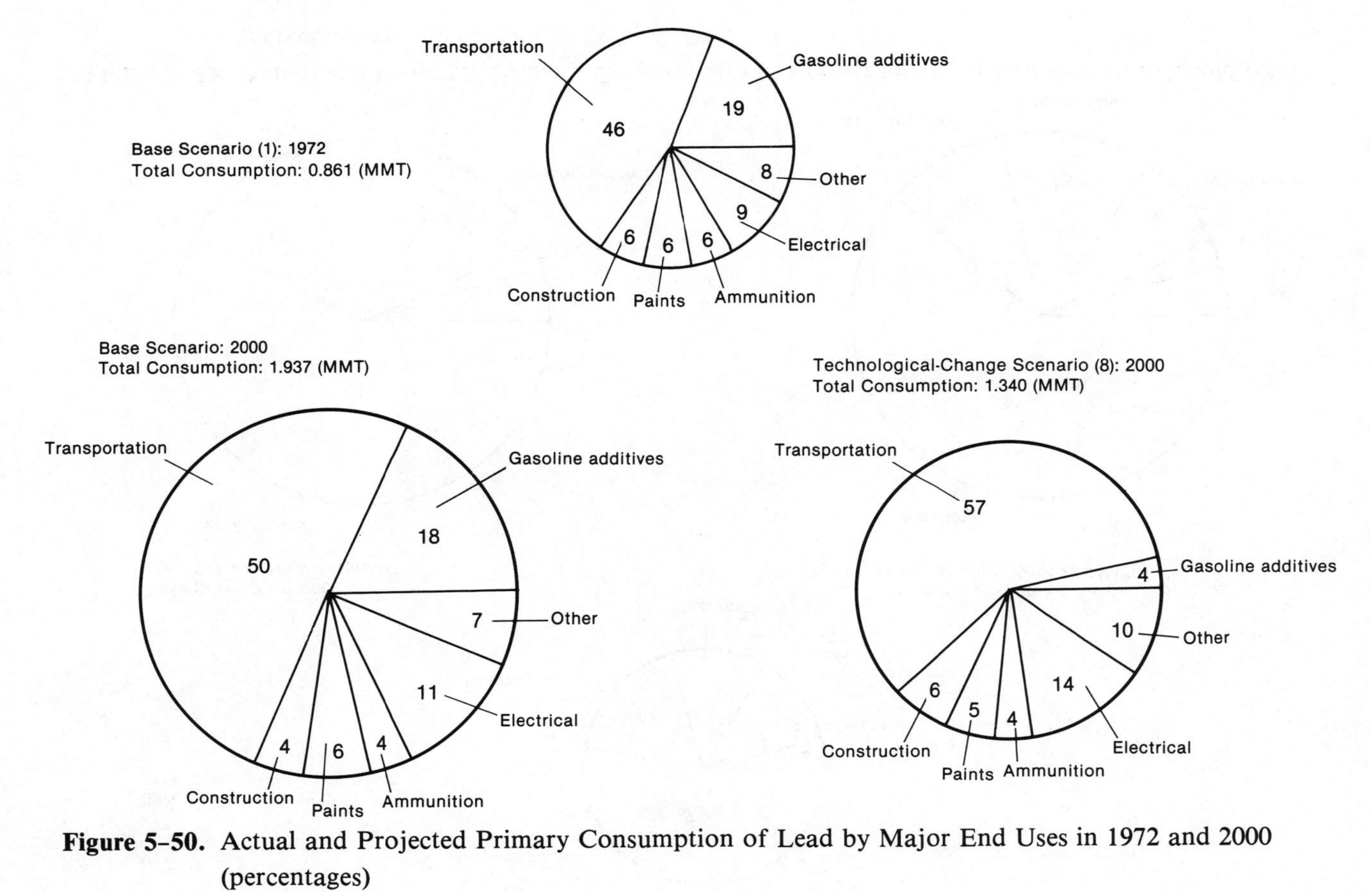

Figure 5–50. Actual and Projected Primary Consumption of Lead by Major End Uses in 1972 and 2000 (percentages)

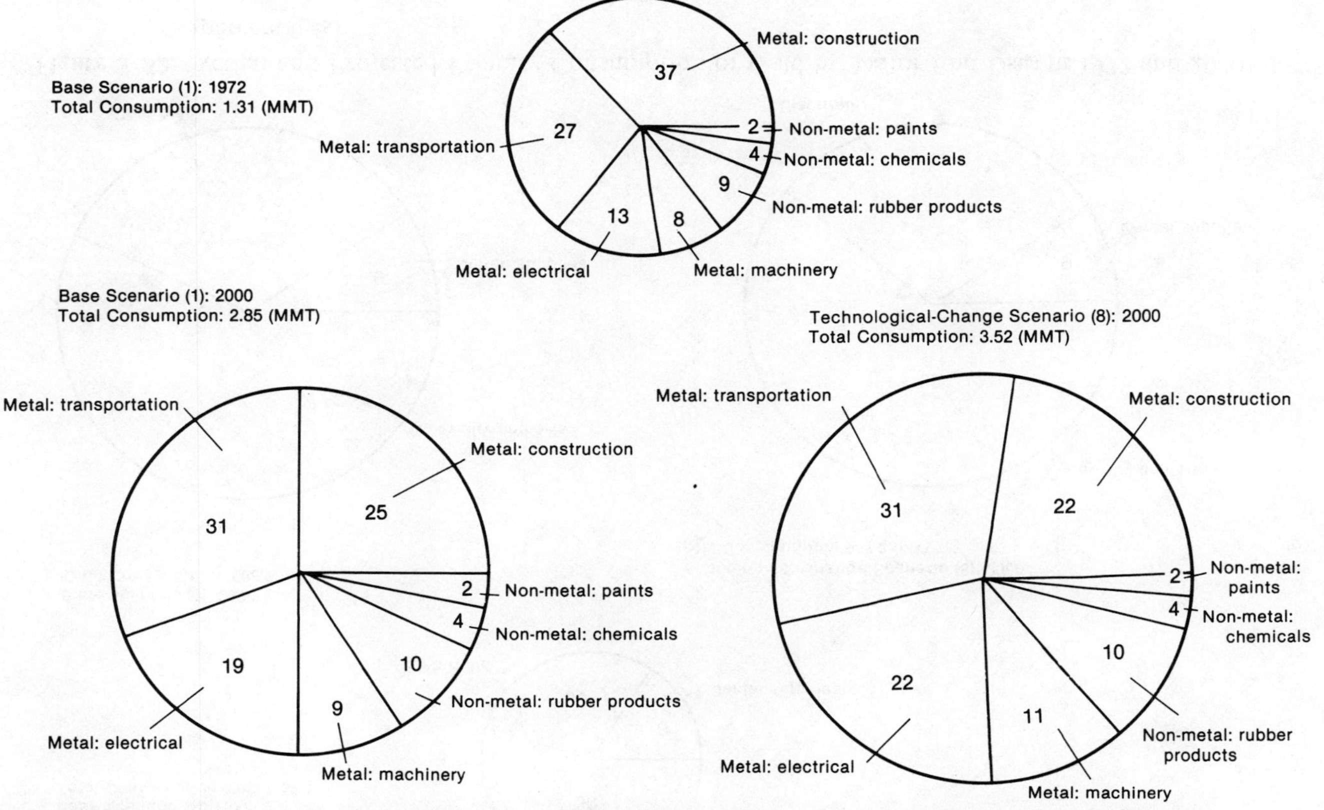

Figure 5–51. Actual and Projected Primary Consumption of Zinc by Major End Uses in 1972 and 2000 (percentages)

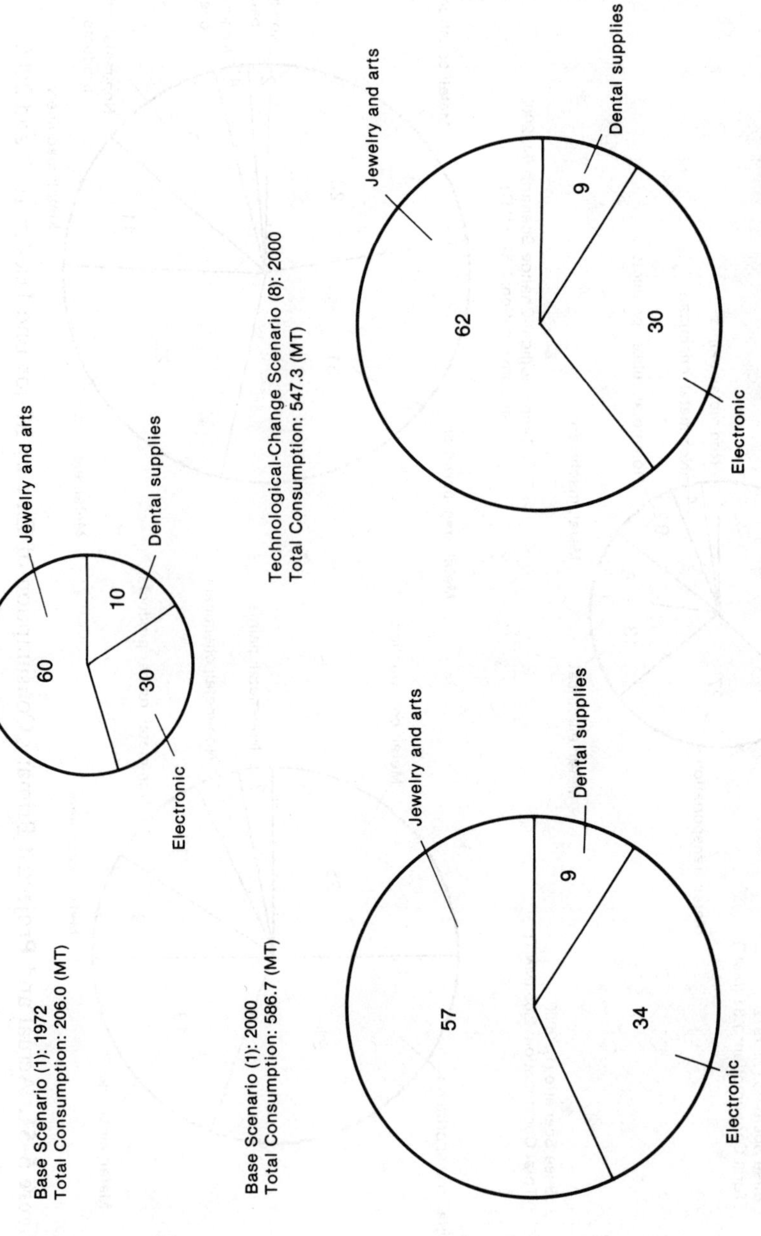

Figure 5–52. Actual and Projected Primary Consumption of Gold by Major End Uses in 1972 and 2000 (percentages)

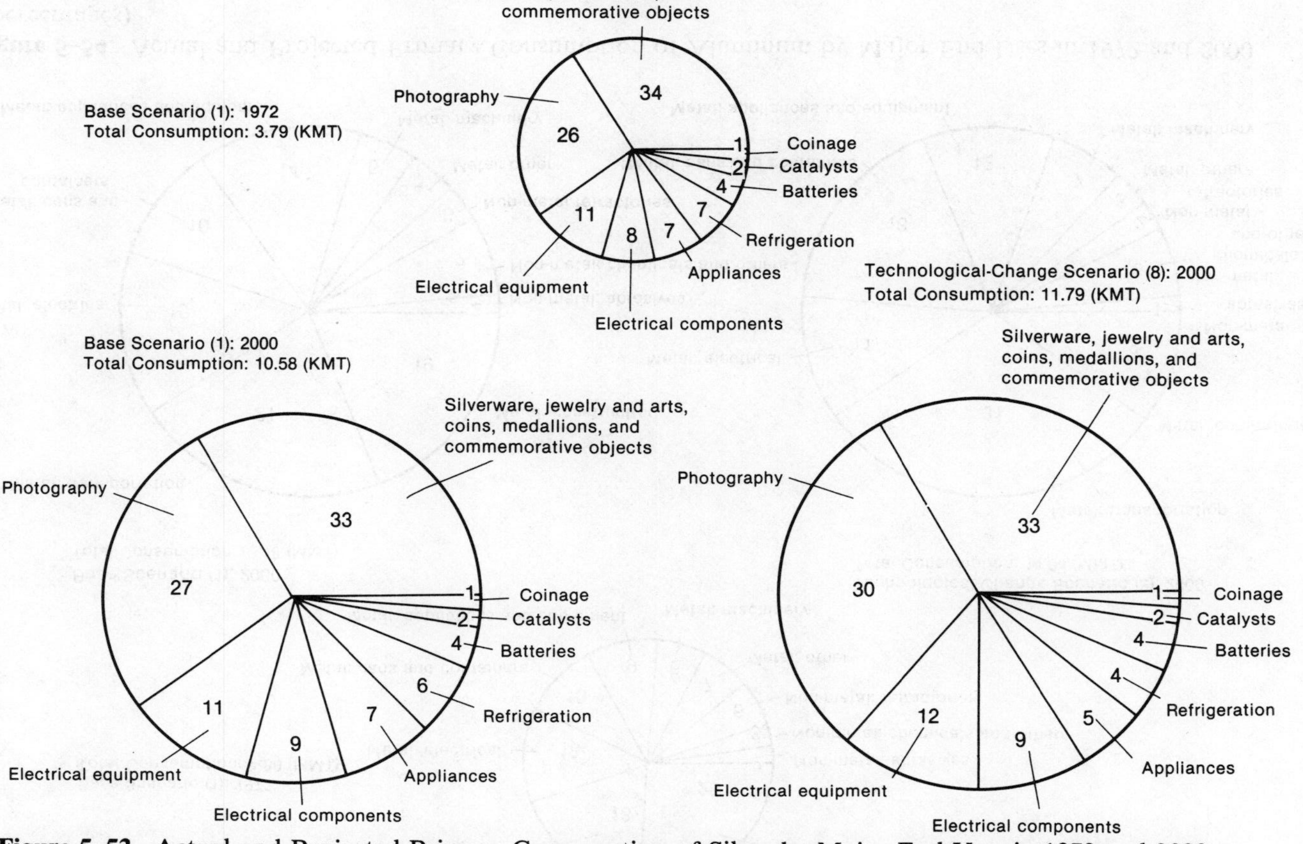

Figure 5–53. Actual and Projected Primary Consumption of Silver by Major End Uses in 1972 and 2000 (percentages)

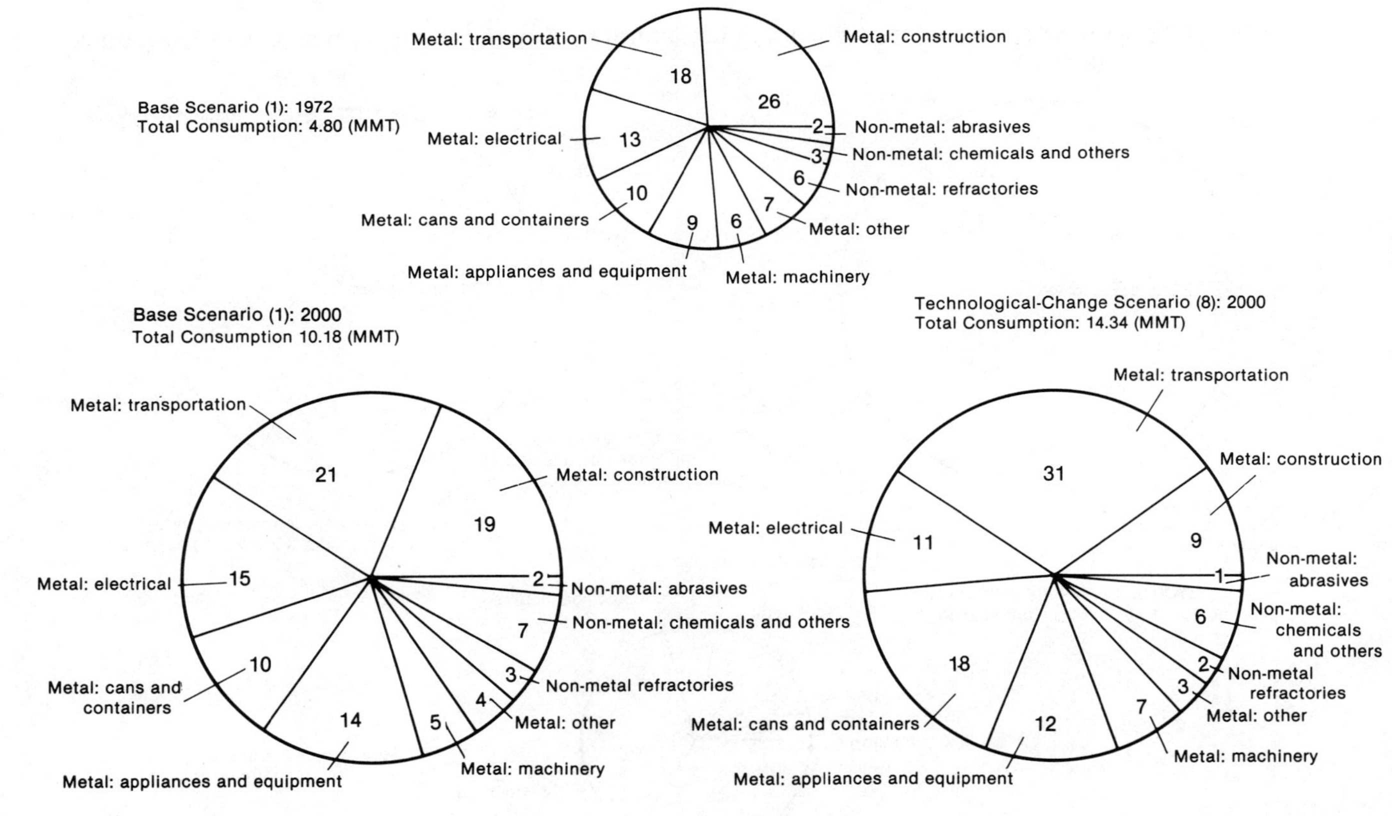

Figure 5–54. Actual and Projected Primary Consumption of Aluminum by Major End Uses in 1972 and 2000 (percentages)

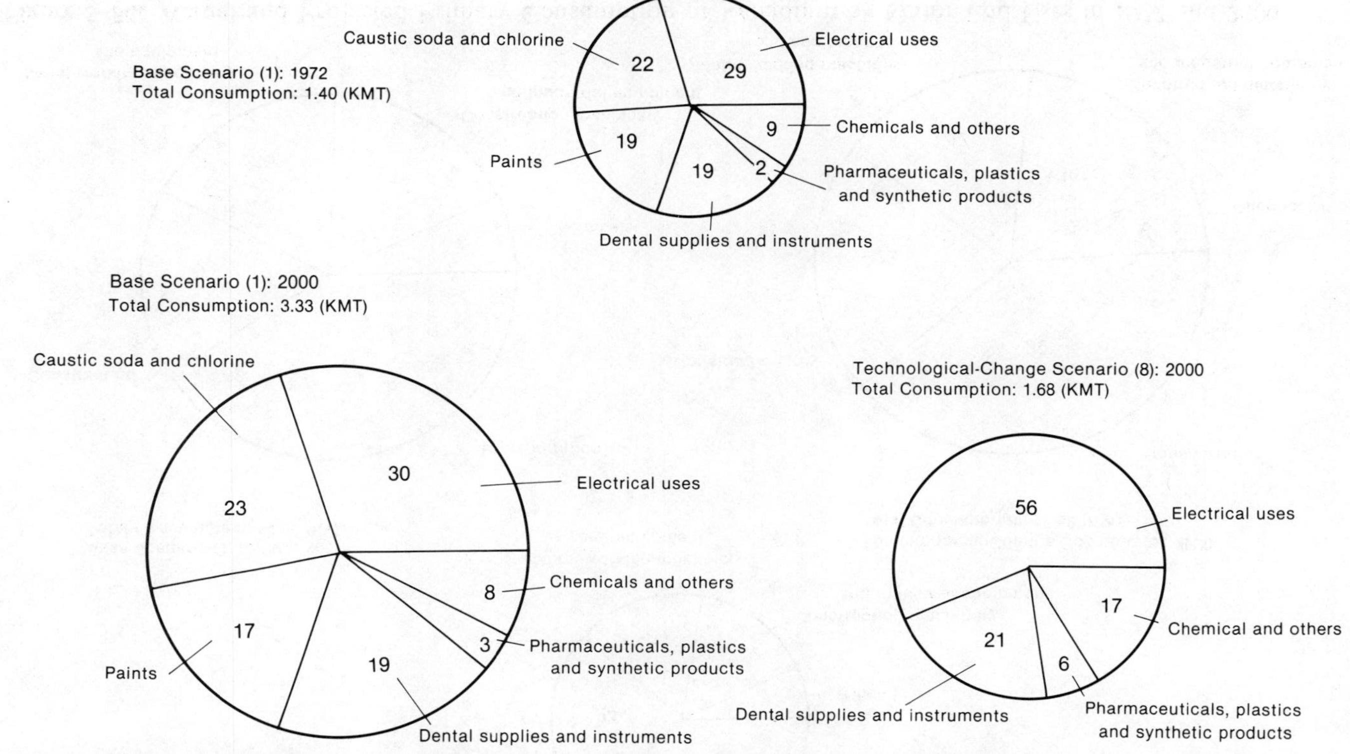

Figure 5–55. Actual and Projected Primary Consumption of Mercury by Major End Uses in 1972 and 2000 (percentages)

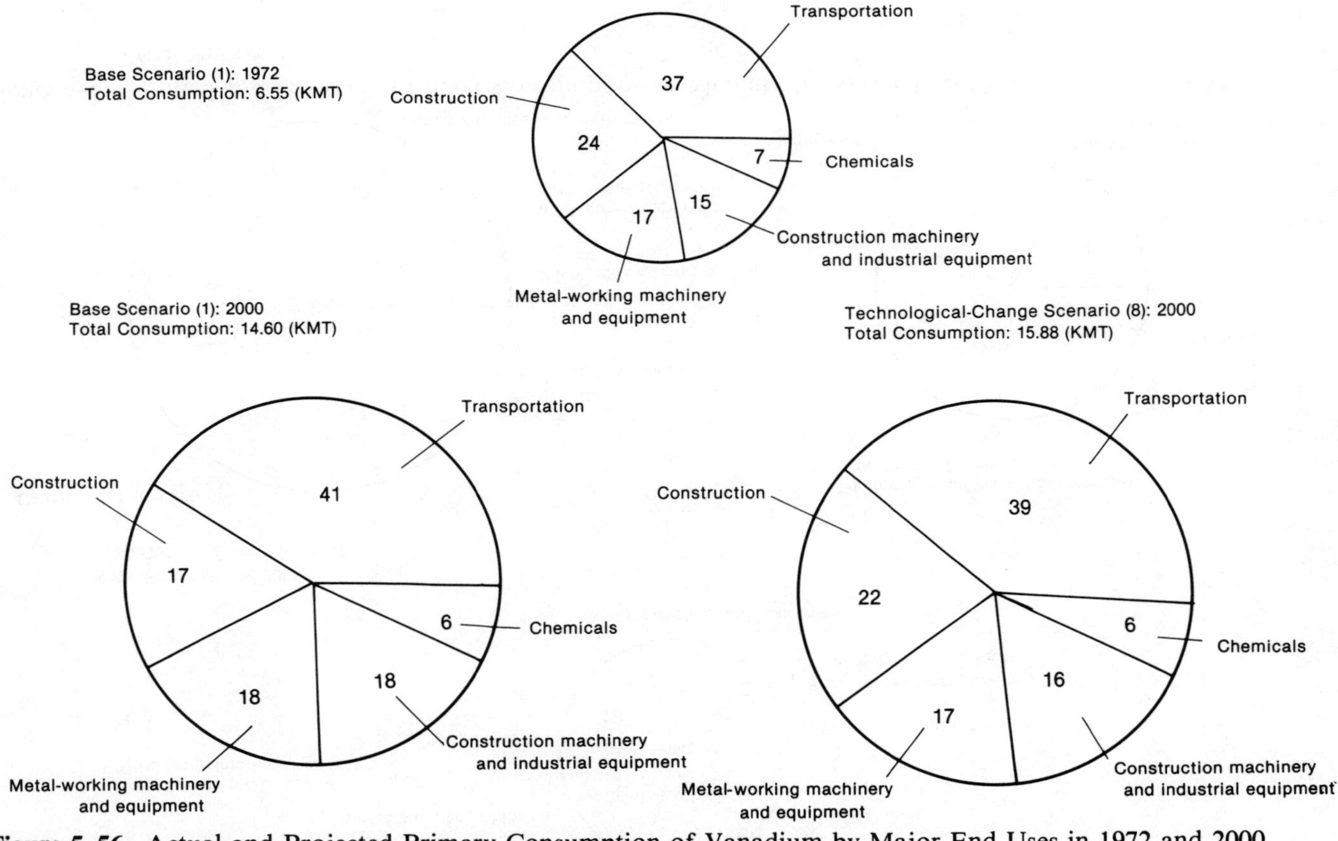

Figure 5–56. Actual and Projected Primary Consumption of Vanadium by Major End Uses in 1972 and 2000 (percentages)

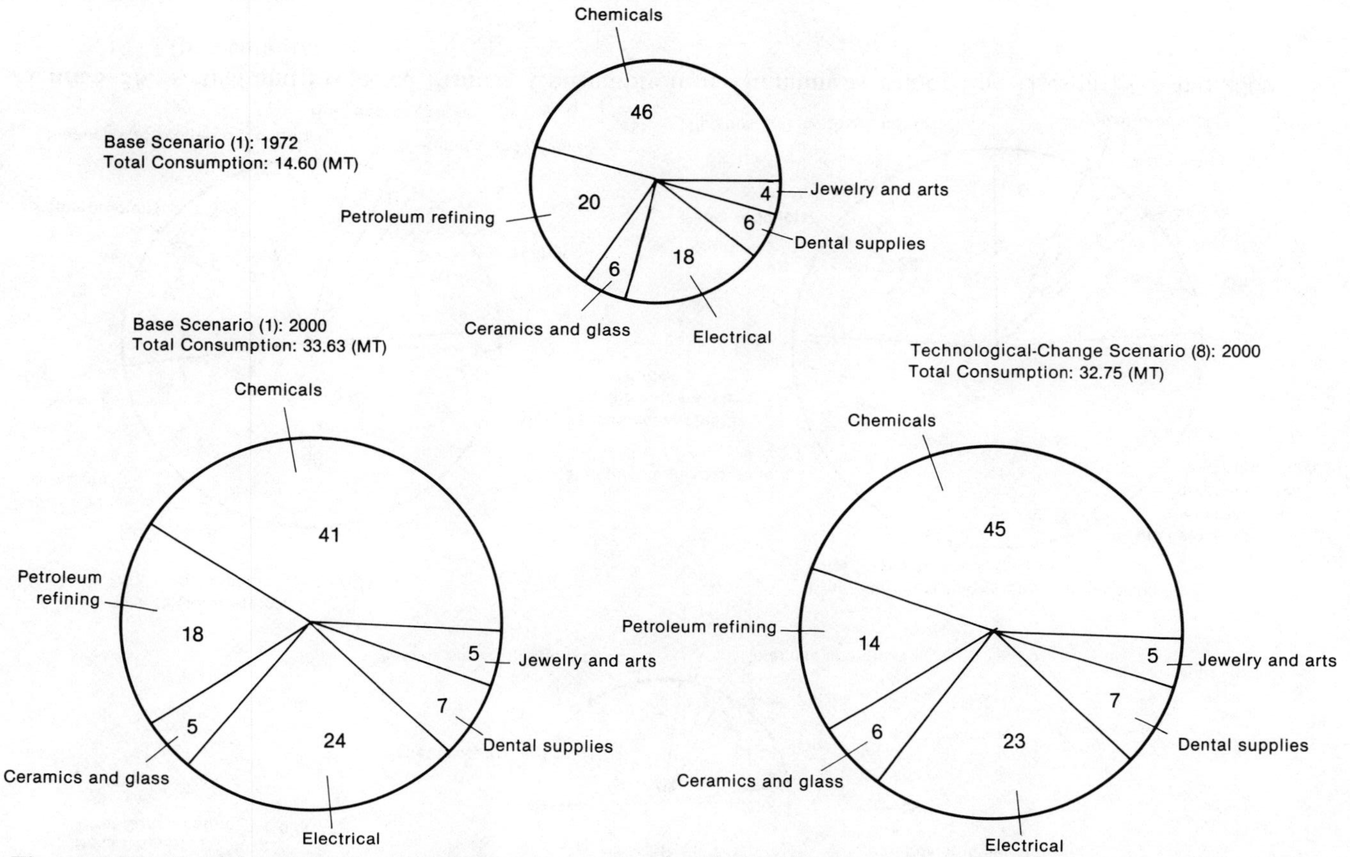

Figure 5–57. Actual and Projected Primary Consumption of Platinum by Major End Uses in 1972 and 2000 (percentages)

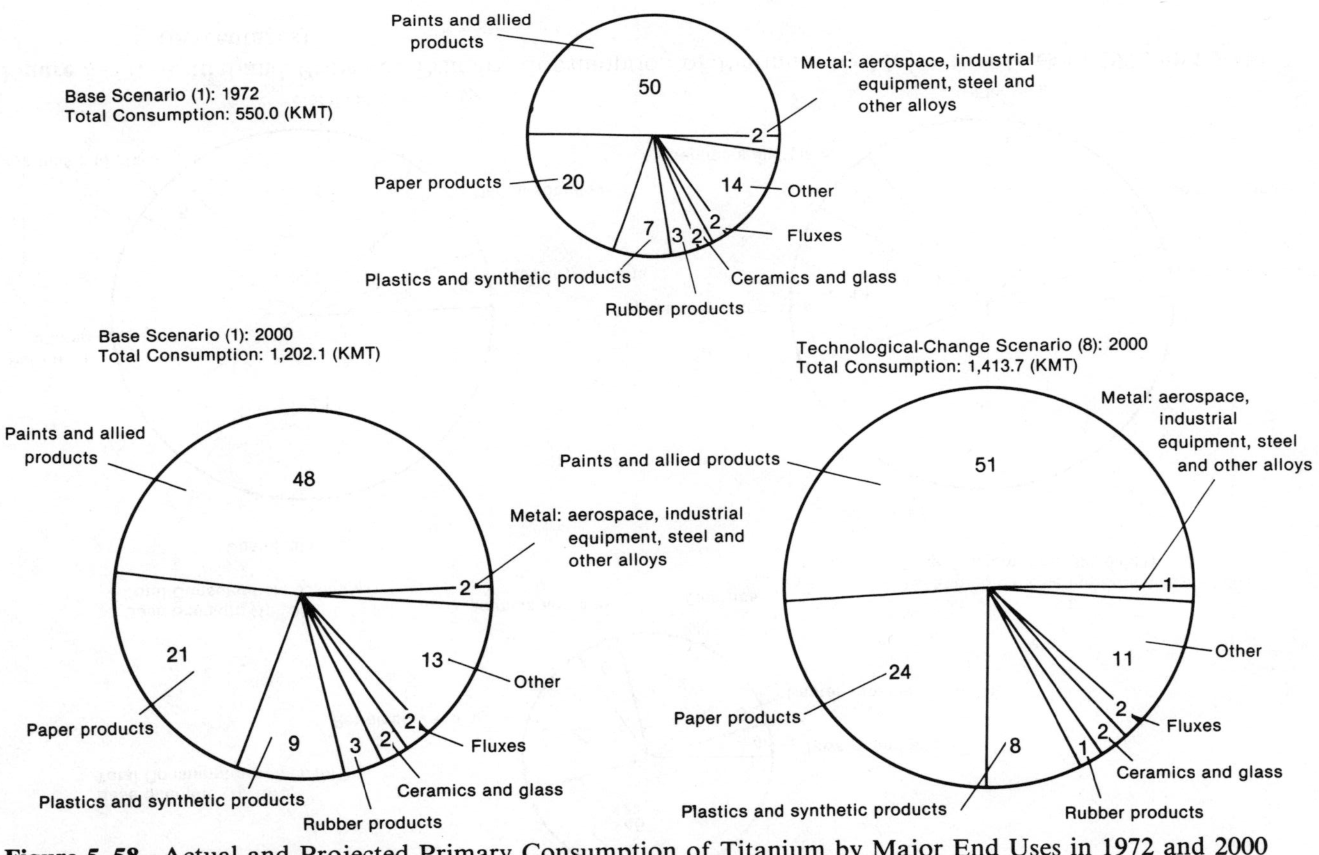

Figure 5–58. Actual and Projected Primary Consumption of Titanium by Major End Uses in 1972 and 2000 (percentages)

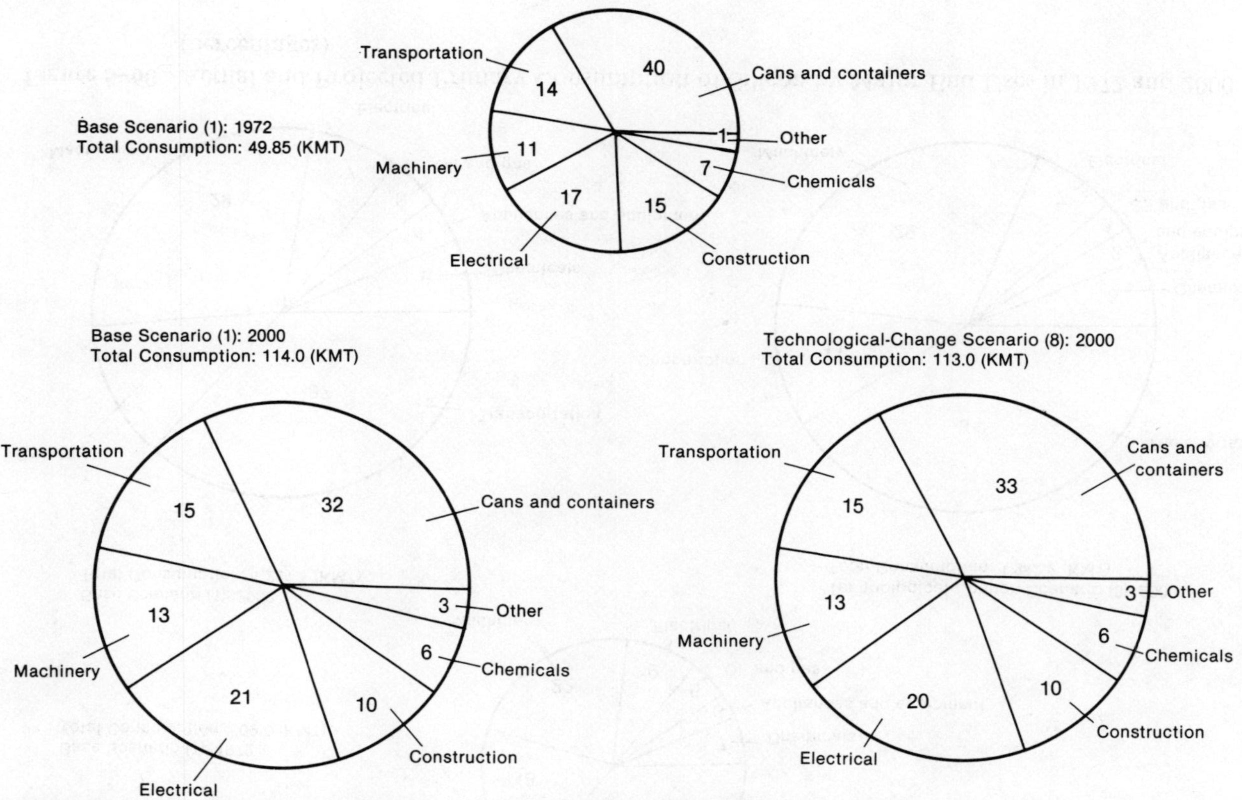

Figure 5–59. Actual and Projected Primary Consumption of Tin by Major End Uses in 1972 and 2000 (percentages)

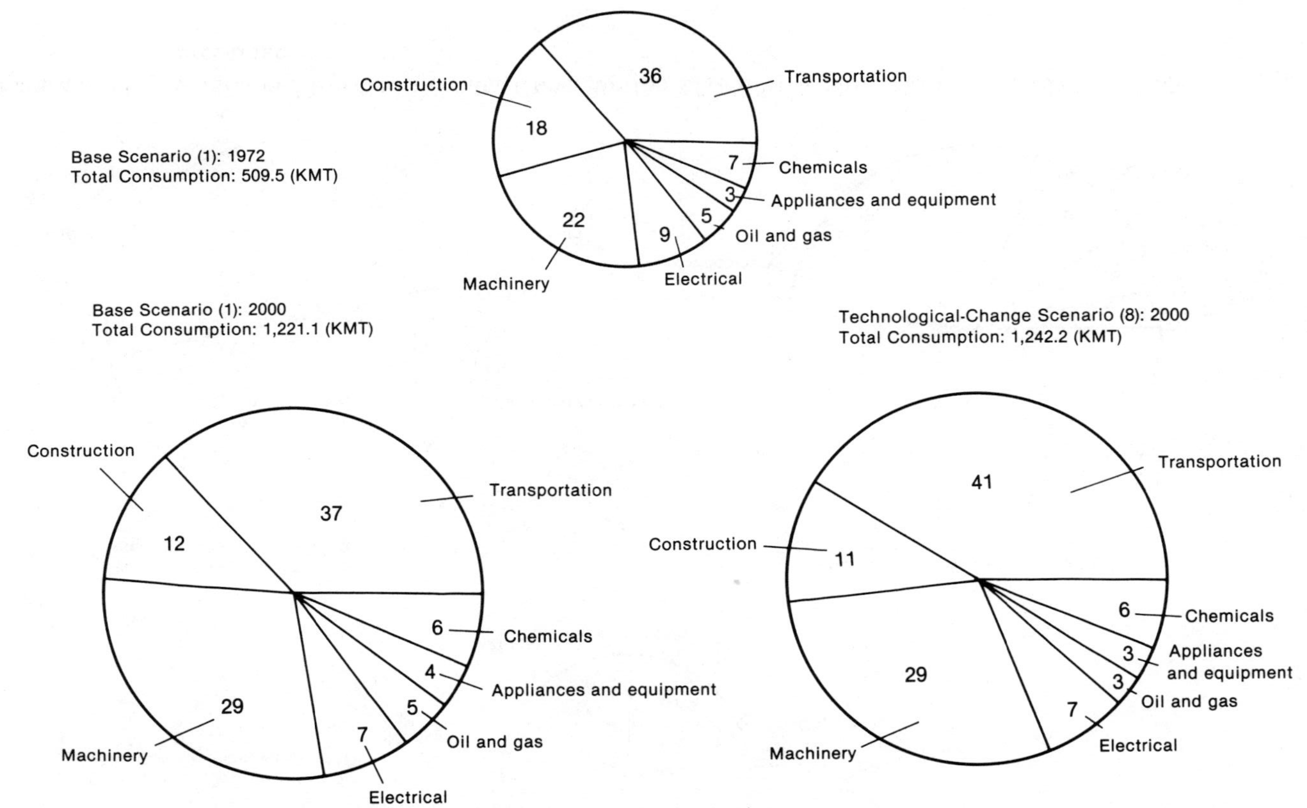

Figure 5–60. Actual and Projected Primary Consumption of Silicon by Major End Uses in 1972 and 2000 (percentages)

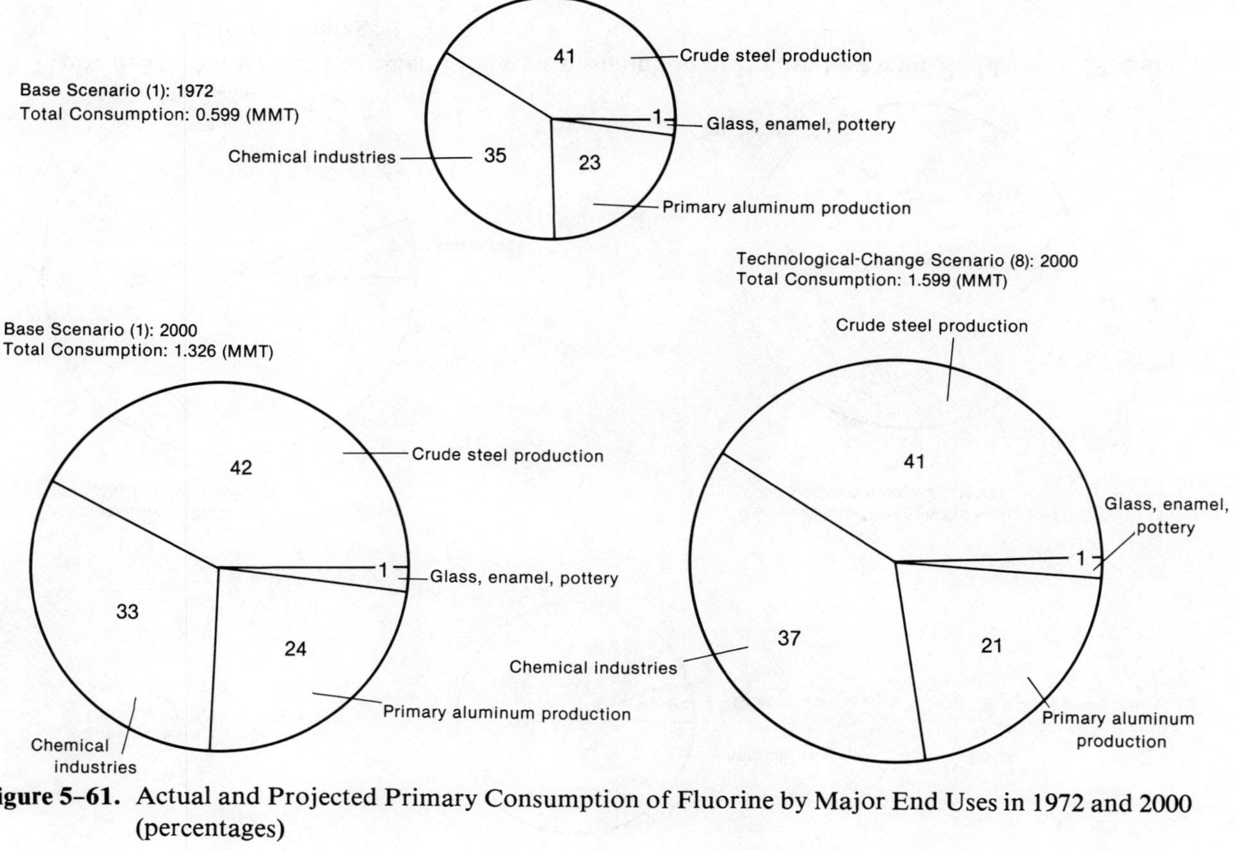

Figure 5–61. Actual and Projected Primary Consumption of Fluorine by Major End Uses in 1972 and 2000 (percentages)

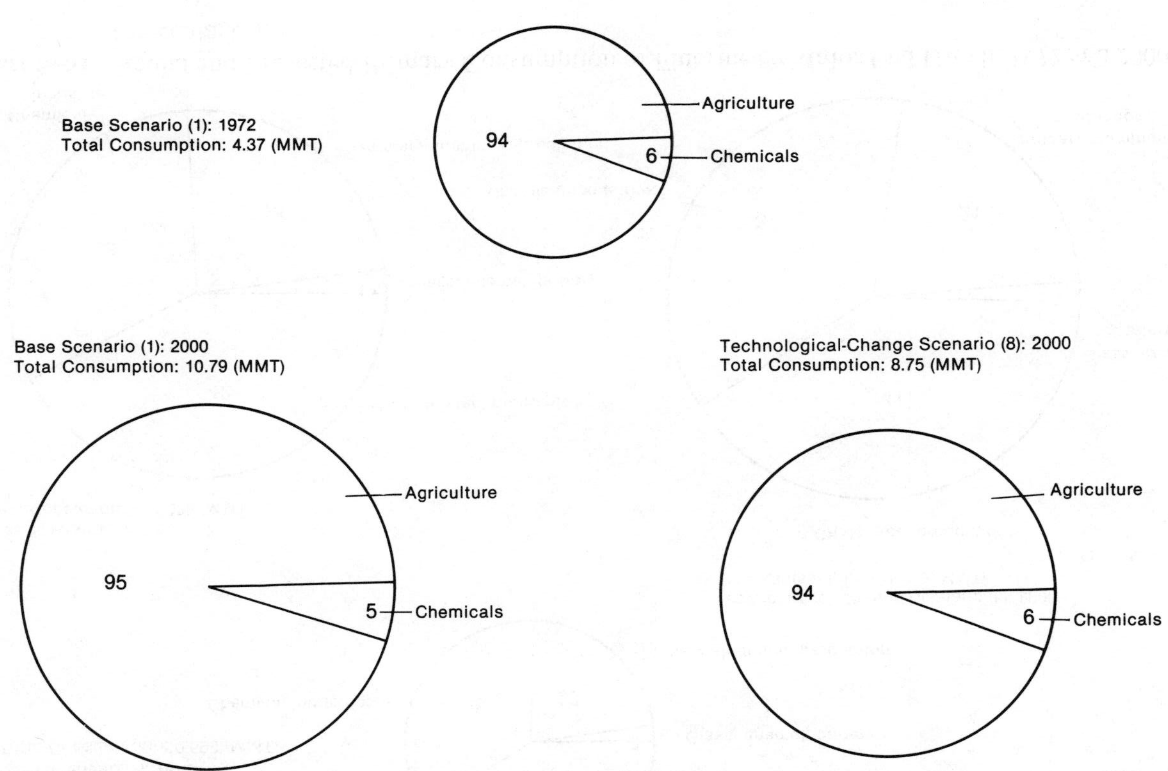

Figure 5-62. Actual and Projected Primary Consumption of Potash by Major End Uses in 1972 and 2000 (percentages)

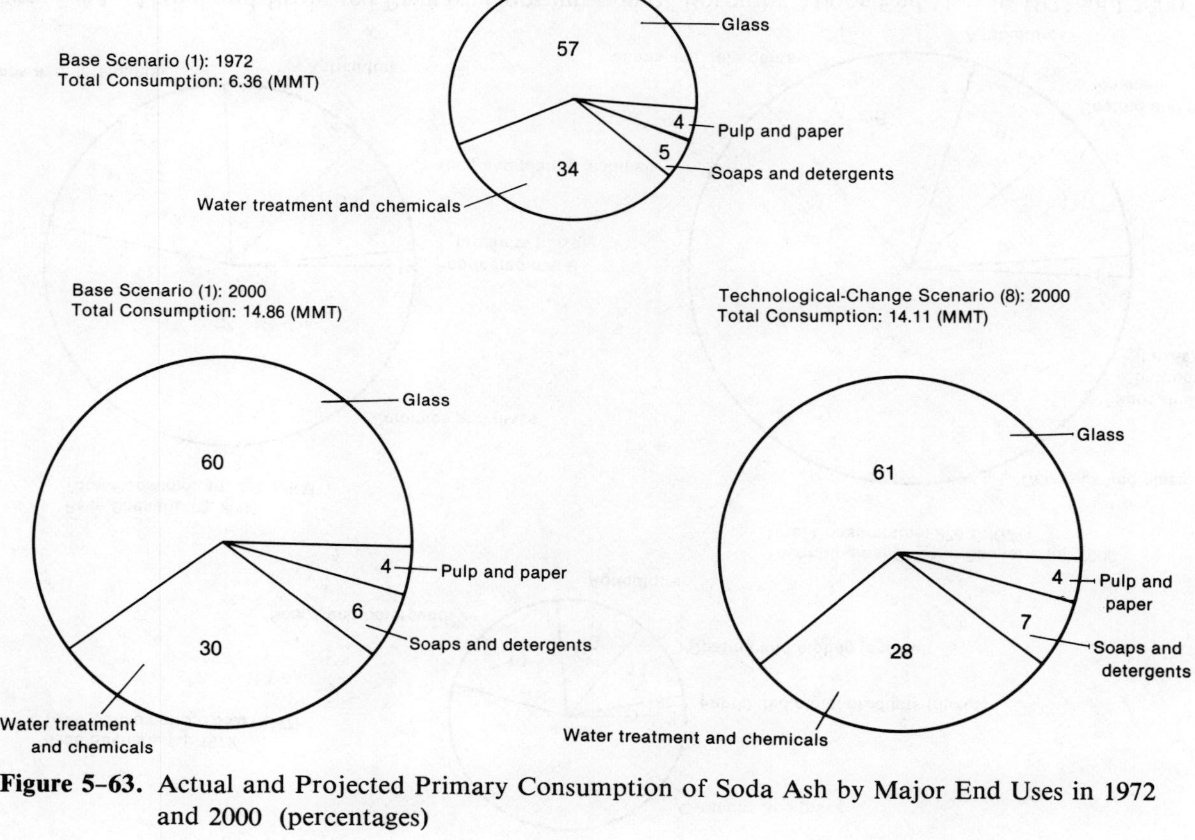

Figure 5–63. Actual and Projected Primary Consumption of Soda Ash by Major End Uses in 1972 and 2000 (percentages)

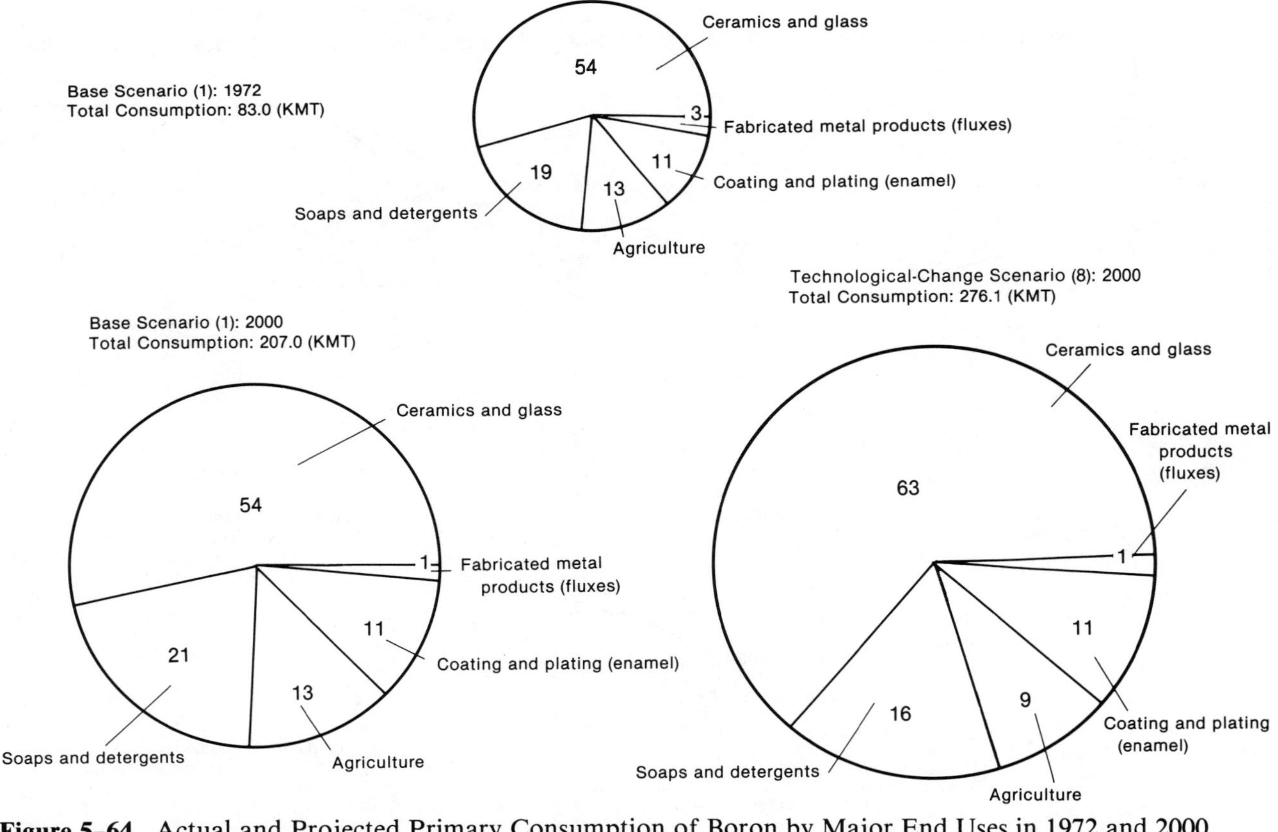

Figure 5–64. Actual and Projected Primary Consumption of Boron by Major End Uses in 1972 and 2000 (percentages)

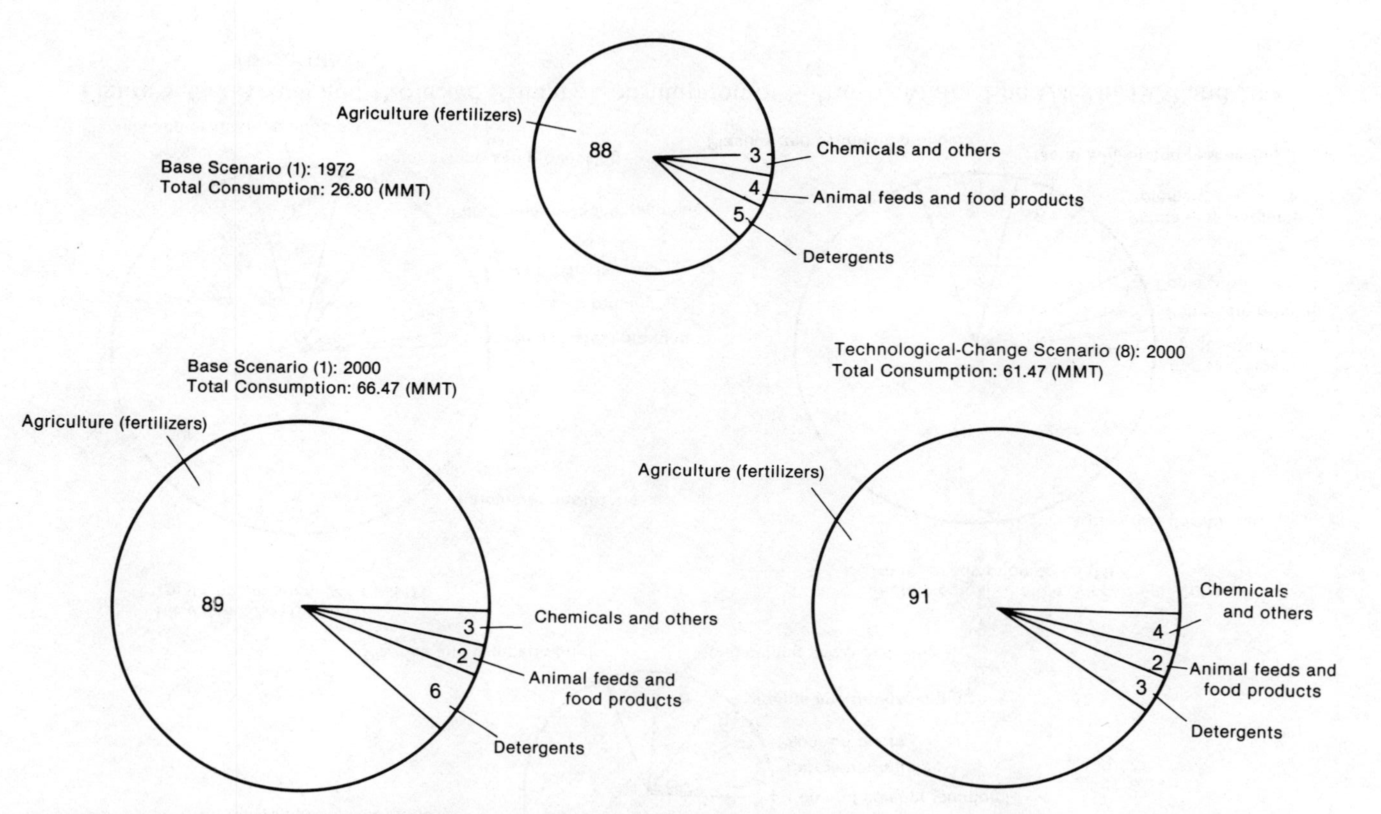

Figure 5–65. Actual and Projected Primary Consumption of Phosphate Rock by Major End Uses in 1972 and 2000 (percentages)

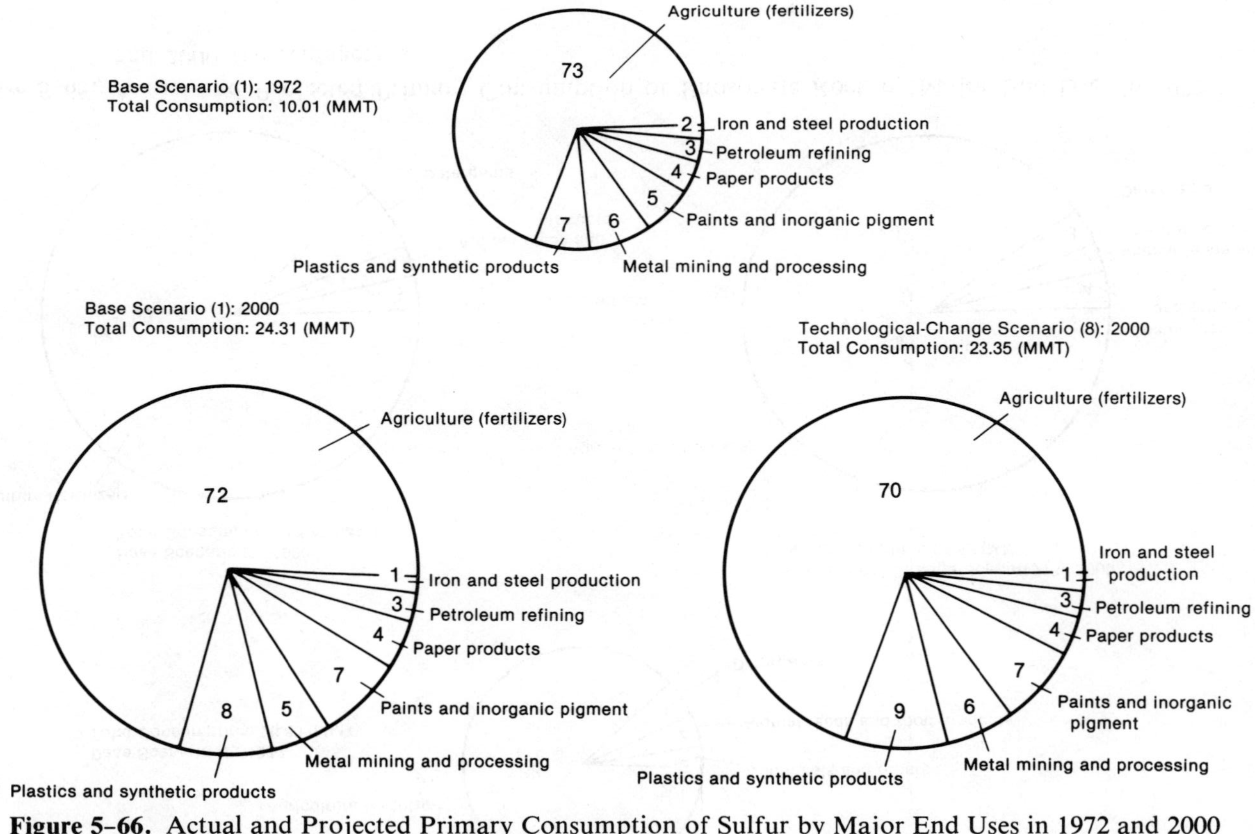

Figure 5–66. Actual and Projected Primary Consumption of Sulfur by Major End Uses in 1972 and 2000 (percentages)

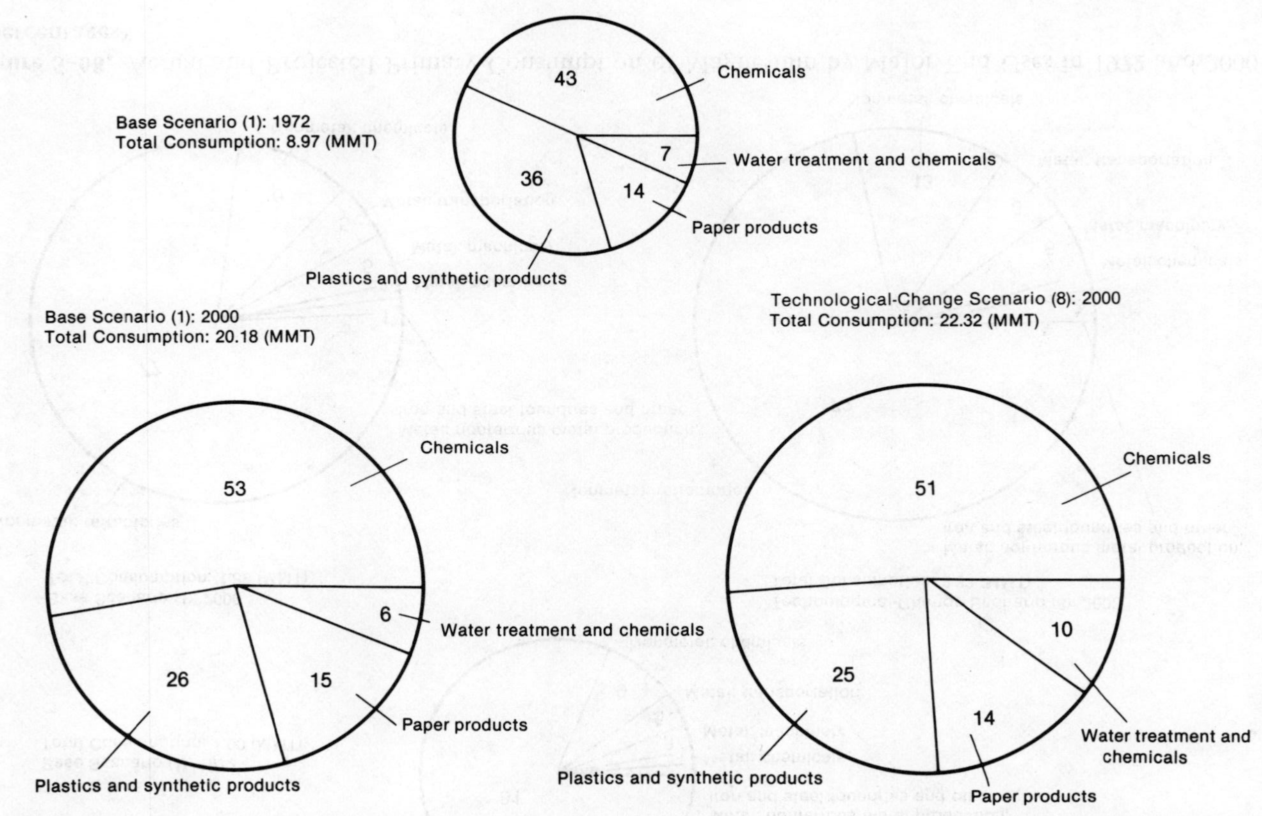

Figure 5-67. Actual and Projected Primary Consumption of Chlorine by Major End Uses in 1972 and 2000 (percentages)

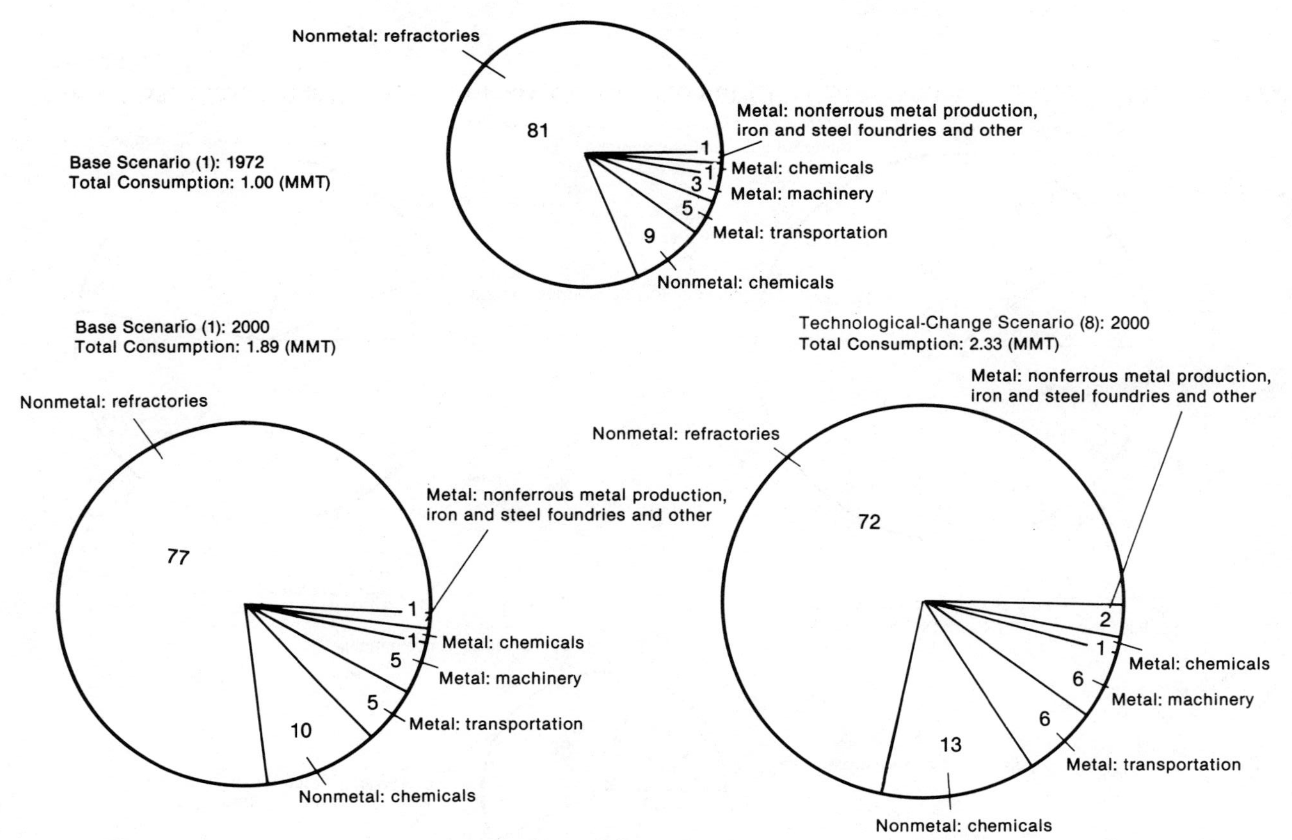

Figure 5–68. Actual and Projected Primary Consumption of Magnesium by Major End Uses in 1972 and 2000 (percentages)

The figures in each segment of the charts represent the quantity of minerals consumed by an end-use sector as a percentage of total U.S. consumption of that mineral. The sizes of the charts reflect the quantities of U.S. total consumption of that particular mineral. Taking Iron (MIN 1) as an example, 1972 U.S. consumption amounted to 87.7 million metric tons (MMT). Most of it was distributed among three major end-use sectors: transportation, construction, and machinery, accounting for 80 percent of the metal's total consumption. The remaining 20 percent was consumed by the cans and containers, appliances and equipment, and oil and gas sectors. In 2000, considering only the growth of the economy projected on the basis of an unchanged 1972 technology (that is, the base scenario), the three major end-use sectors would still account for 79 percent of the projected total iron consumption of 195.7 MMT. However, even though these three sectors will remain the dominant consuming sectors, their relative shares in demand are expected to change. This shift is even more pronounced in scenario 8 under which the growth of the economy is projected including the assumption of expected changes in the input structure of some producing sectors, particularly materials substitution brought about by ongoing technological changes.

As an analytical tool, figures 5–43 through 5–68 aid in visualizing the projected changes in the demand patterns for each of the twenty-six nonfuel minerals through 2000.

Consumption and Production Indexes: 1972 and 2000

Before describing the long-term effects of alternative future levels of production on U.S. reserves, we first summarize the results—both from the input side (demand) and from the output side (supply)—by comparing the effects of the alternative sets of assumptions described in the scenarios on future consumption and production levels.

Although the well-known balance between production and consumption cannot be violated for a closed economy, as shown in figure 4–30, the United States is a major importer of many of the twenty-six nonfuel minerals in this study. Some nonfuel minerals can be recycled to some degree, as shown in figure 4–29. These two additional sources of minerals supply require a modification of the restrictive-balance equation of production equal to consumption to a more general condition of total supply equal to total demand, permitting the inclusion of imports and recycled metals on the supply side of the equation. These comparisons are conveniently summarized in the consumption and production indexes, tables 5–1 through 5–39.

For example, it is possible to isolate the source of the expected growth

Table 5–1
Iron: Indexes for Comparing Changes in U.S. Consumption and Production Levels under Different Scenarios

Consumption		
Effect	*Year/Scenarios*	*Indexes*
Income growth	2000 [1] / 1972 [1]	2.233
Technological change	2000 [8] / 2000 [1]	0.829
Technological change and income growth	2000 [8] / 1972 [1]	1.852
High recycling	2000 [7] / 2000 [8]	0.569
Technological change, income growth, and high recycling	2000 [7] / 1972 [1]	1.054
Production		
Effect	*Year/Scenarios*	*Indexes*
Income growth	2000 [1] / 1972 [1]	2.398
Technological change	2000 [8] / 2000 [1]	0.837
Technological change and income growth	2000 [8] / 1972 [1]	2.006
High recycling	2000 [7] / 2000 [8]	0.591
Technological change, income growth, and high recycling	2000 [7] / 1972 [1]	1.186
High import dependence	2000 [6] / 2000 [8]	0.729
Low import dependence	2000 [11] / 2000 [8]	1.376

Table 5–2
Molybdenum: Indexes for Comparing Changes in U.S. Consumption and Production Levels under Different Scenarios

Consumption		
Effect	*Year/Scenarios*	*Indexes*
Income growth	2000 [1] / 1972 [1]	2.36
Technological change	2000 [8] / 2000 [1]	1.15
Technological change and income growth	2000 [8] / 1972 [1]	2.73
High recycling	2000 [7] / 2000 [8]	1.00
Technological change, income growth, and high recycling	2000 [7] / 1972 [1]	2.73

Table 5-2 continued

	Production	
Effect	*Year/Scenarios*	*Indexes*
Income growth	2000 [1] / 1972 [1]	2.00
Technological change	2000 [8] / 2000 [1]	1.08
Technological change and income growth	2000 [8] / 1972 [1]	2.17
High recycling	2000 [7] / 2000 [8]	1.00
Technological change, income growth, and high recycling	2000 [7] / 1972 [1]	2.17
High import dependence	2000 [6] / 2000 [8]	1.00
Low import dependence	2000 [11] / 2000 [8]	1.00

Table 5-3
Nickel: Indexes for Comparing Changes in U.S. Consumption and Production Levels under Different Scenarios

	Consumption	
Effect	*Year/Scenarios*	*Indexes*
Income growth	2000 [1] / 1972 [1]	2.305
Technological change	2000 [8] / 2000 [1]	1.016
Technological change and income growth	2000 [8] / 1972 [1]	2.341
High recycling	2000 [7] / 2000 [8]	0.675
Technological change, income growth, and high recycling	2000 [7] / 1972 [1]	1.581
	Production	
Effect	*Year/Scenarios*	*Indexes*
Income growth	2000 [1] / 1972 [1]	2.067
Technological change	2000 [8] / 2000 [1]	1.015
Technological change and income growth	2000 [8] / 1972 [1]	2.098
High recycling	2000 [7] / 2000 [8]	0.681
Technological change, income growth, and high recycling	2000 [7] / 1972 [1]	1.429
High import dependence	2000 [6] / 2000 [8]	1.038
Low import dependence	2000 [11] / 2000 [8]	8.661

Table 5-4
Tungsten: Indexes for Comparing Changes in U.S. Consumption and Production Levels under Different Scenarios

Consumption		
Effect	*Year/Scenarios*	*Indexes*
Income growth	2000 [1] / 1972 [1]	2.49
Technological change	2000 [8] / 2000 [1]	1.08
Technological change and income growth	2000 [8] / 1972 [1]	2.69
High recycling	2000 [7] / 2000 [8]	0.48
Technological change, income growth, and high recycling	2000 [7] / 1972 [1]	1.28
Production		
Effect	*Year/Scenarios*	*Indexes*
Income growth	2000 [1] / 1972 [1]	2.69
Technological change	2000 [8] / 2000 [1]	1.07
Technological change and income growth	2000 [8] / 1972 [1]	2.89
High recycling	2000 [7] / 2000 [8]	0.51
Technological change, income growth, and high recycling	2000 [7] / 1972 [1]	1.48
High import dependence	2000 [6] / 2000 [8]	0.08
Low import dependence	2000 [11] / 2000 [8]	0.94

Table 5-5
Manganese: Indexes for Comparing Changes in U.S. Consumption and Production Levels under Different Scenarios

Consumption		
Effect	*Year/Scenarios*	*Indexes*
Income growth	2000 [1] / 1972 [1]	2.25
Technological change	2000 [8] / 2000 [1]	0.86
Technological change and income growth	2000 [8] / 1972 [1]	1.93
High recycling	2000 [7] / 2000 [8]	1.00
Technological change, income growth, and high recycling	2000 [7] / 1972 [1]	1.93

Table 5-5 continued

Production		
Effect	*Year/Scenarios*	*Indexes*
Income growth	2000 [1] / 1972 [1]	0
Technological change	2000 [8] / 2000 [1]	0
Technological change and income growth	2000 [8] / 1972 [1]	0
High recycling	2000 [7] / 2000 [8]	0
Technological change, income growth, and high recycling	2000 [7] / 1972 [1]	0
High import dependence	2000 [6] / 2000 [8]	0
Low import dependence	2000 [11] / 2000 [8]	0

Table 5-6
Chromium: Indexes for Comparing Changes in U.S. Consumption and Production Levels under Different Scenarios

Consumption		
Effect	*Year/Scenarios*	*Indexes*
Income growth	2000 [1] / 1972 [1]	2.35
Technological change	2000 [8] / 2000 [1]	1.06
Technological change and income growth	2000 [8] / 1972 [1]	2.50
High recycling	2000 [7] / 2000 [8]	0.74
Technological change, income growth, and high recycling	2000 [7] / 1972 [1]	1.85
Production		
Effect	*Year/Scenarios*	*Indexes*
Income growth	2000 [1] / 1972 [1]	0
Technological change	2000 [8] / 2000 [1]	0
Technological change and income growth	2000 [8] / 1972 [1]	0
High recycling	2000 [7] / 2000 [8]	0
Technological change, income growth, and high recycling	2000 [7] / 1972 [1]	0
High import dependence	2000 [6] / 2000 [8]	0
Low import dependence	2000 [11] / 2000 [8]	0

Table 5-7
Copper: Indexes for Comparing Changes in U.S. Consumption and Production Levels under Different Scenarios

Consumption		
Effect	*Year/Scenarios*	*Indexes*
Income growth	2000 [1] / 1972 [1]	2.43
Technological change	2000 [8] / 2000 [1]	1.00
Technological change and income growth	2000 [8] / 1972 [1]	2.44
High recycling	2000 [7] / 2000 [8]	0.49
Technological change, income growth, and high recycling	2000 [7] / 1972 [1]	1.20

Production		
Effect	*Year/Scenarios*	*Indexes*
Income growth	2000 [1] / 1972 [1]	2.77
Technological change	2000 [8] / 2000 [1]	1.00
Technological change and income growth	2000 [8] / 1972 [1]	2.72
High recycling	2000 [7] / 2000 [8]	0.58
Technological change, income growth, and high recycling	2000 [7] / 1972 [1]	1.60
High import dependence	2000 [6] / 2000 [8]	0.97
Low import dependence	2000 [11] / 2000 [8]	1.24

Table 5-8
Lead: Indexes for Comparing Changes in U.S. Consumption and Production Levels under Different Scenarios

Consumption		
Effect	*Year/Scenarios*	*Indexes*
Income growth	2000 [1] / 1972 [1]	2.25
Technological change	2000 [8] / 2000 [1]	0.69
Technological change and income growth	2000 [8] / 1972 [1]	1.56
High recycling	2000 [7] / 2000 [8]	0.74
Technological change, income growth, and high recycling	2000 [7] / 1972 [1]	1.15

Table 5-8 continued

Production		
Effect	*Year/Scenarios*	*Indexes*
Income growth	2000 [1] / 1972 [1]	2.33
Technological change	2000 [8] / 2000 [1]	0.70
Technological change and income growth	2000 [8] / 1972 [1]	1.63
High recycling	2000 [7] / 2000 [8]	0.75
Technological change, income growth, and high recycling	2000 [7] / 1972 [1]	1.21
High import dependence	2000 [6] / 2000 [8]	0.92
Low import dependence	2000 [11] / 2000 [8]	1.13

Table 5-9
Zinc: Indexes for Comparing Changes in U.S. Consumption and Production Levels under Different Scenarios

Consumption		
Effect	*Year/Scenarios*	*Indexes*
Income growth	2000 [1] / 1972 [1]	2.17
Technological change	2000 [8] / 2000 [1]	1.23
Technological change and income growth	2000 [8] / 1972 [1]	2.68
High recycling	2000 [7] / 2000 [8]	0.82
Technological change, income growth, and high recycling	2000 [7] / 1972 [1]	2.20
Production		
Effect	*Year/Scenarios*	*Indexes*
Income growth	2000 [1] / 1972 [1]	2.49
Technological change	2000 [8] / 2000 [1]	1.23
Technological change and income growth	2000 [8] / 1972 [1]	3.07
High recycling	2000 [7] / 2000 [8]	0.82
Technological change, income growth, and high recycling	2000 [7] / 1972 [1]	2.53
High import dependence	2000 [6] / 2000 [8]	0.87
Low import dependence	2000 [11] / 2000 [8]	1.65

Table 5-10
Gold: Indexes for Comparing Changes in U.S. Consumption and Production Levels under Different Scenarios

	Consumption	
Effect	*Year/Scenarios*	*Indexes*
Income growth	2000 [1] / 1972 [1]	2.848
Technological change	2000 [8] / 2000 [1]	0.933
Technological change and income growth	2000 [8] / 1972 [1]	2.657
High recycling	2000 [7] / 2000 [8]	0.091
Technological change, income growth, and high recycling	2000 [7] / 1972 [1]	0.242
	Production	
Effect	*Year/Scenarios*	*Indexes*
Income growth	2000 [1] / 1972 [1]	2.820
Technological change	2000 [8] / 2000 [1]	0.942
Technological change and income growth	2000 [8] / 1972 [1]	2.657
High recycling	2000 [7] / 2000 [8]	0.226
Technological change, income growth, and high recycling	2000 [7] / 1972 [1]	0.600
High import dependence	2000 [6] / 2000 [8]	0.629
Low import dependence	2000 [11] / 2000 [8]	1.167

Table 5-11
Silver: Indexes for Comparing Changes in U.S. Consumption and Production Levels under Different Scenarios

	Consumption	
Effect	*Year/Scenarios*	*Indexes*
Income growth	2000 [1] / 1972 [1]	2.79
Technological change	2000 [8] / 2000 [1]	1.11
Technological change and income growth	2000 [8] / 1972 [1]	3.11
High recycling	2000 [7] / 2000 [8]	1.00
Technological change, income growth, and high recycling	2000 [7] / 1972 [1]	3.12

Table 5-11 continued

Production		
Effect	*Year/Scenarios*	*Indexes*
Income growth	2000 [1] / 1972 [1]	4.25
Technological change	2000 [8] / 2000 [1]	1.08
Technological change and income growth	2000 [8] / 1972 [1]	4.60
High recycling	2000 [7] / 2000 [8]	1.00
Technological change, income growth, and high recycling	2000 [7] / 1972 [1]	4.61
High import dependence	2000 [6] / 2000 [8]	0.56
Low import dependence	2000 [11] / 2000 [8]	0.67

Table 5-12
Aluminum: Indexes for Comparing Changes in U.S. Consumption and Production Levels under Different Scenarios

Consumption		
Effect	*Year/Scenarios*	*Indexes*
Income growth	2000 [1] / 1972 [1]	2.12
Technological change	2000 [8] / 2000 [1]	1.41
Technological change and income growth	2000 [8] / 1972 [1]	2.99
High recycling	2000 [7] / 2000 [8]	0.78
Technological change, income growth, and high recycling	2000 [7] / 1972 [1]	2.32
Production		
Effect	*Year/Scenarios*	*Indexes*
Income growth	2000 [1] / 1972 [1]	2.50
Technological change	2000 [8] / 2000 [1]	1.32
Technological change and income growth	2000 [8] / 1972 [1]	3.30
High recycling	2000 [7] / 2000 [8]	0.81
Technological change, income growth, and high recycling	2000 [7] / 1972 [1]	2.69
High import dependence	2000 [6] / 2000 [8]	0.31
Low import dependence	2000 [11] / 2000 [8]	1.23

Table 5–13
Mercury: Indexes for Comparing Changes in U.S. Consumption and Production Levels under Different Scenarios

	Consumption	
Effect	*Year/Scenarios*	*Indexes*
Income growth	2000 [1] / 1972 [1]	2.380
Technological change	2000 [8] / 2000 [1]	0.503
Technological change and income growth	2000 [8] / 1972 [1]	1.196
High recycling	2000 [7] / 2000 [8]	1.000
Technological change, income growth, and high recycling	2000 [7] / 1972 [1]	1.197

	Production	
Effect	*Year/Scenarios*	*Indexes*
Income growth	2000 [1] / 1972 [1]	2.715
Technological change	2000 [8] / 2000 [1]	0.511
Technological change and income growth	2000 [8] / 1972 [1]	1.387
High recycling	2000 [7] / 2000 [8]	1.000
Technological change, income growth, and high recycling	2000 [7] / 1972 [1]	1.387
High import dependence	2000 [6] / 2000 [8]	1.000
Low import dependence	2000 [11] / 2000 [8]	1.002

Table 5–14
Vanadium: Indexes for Comparing Changes in U.S. Consumption and Production Levels under Different Scenarios

	Consumption	
Effect	*Year/Scenarios*	*Indexes*
Income growth	2000 [1] / 1972 [1]	2.229
Technological change	2000 [8] / 2000 [1]	1.088
Technological change and income growth	2000 [8] / 1972 [1]	2.424
High recycling	2000 [7] / 2000 [8]	1.000
Technological change, income growth, and high recycling	2000 [7] / 1972 [1]	2.424

Table 5–14 continued

Production		
Effect	*Year/Scenarios*	*Indexes*
Income growth	2000 [1] / 1972 [1]	2.472
Technological change	2000 [8] / 2000 [1]	1.081
Technological change and income growth	2000 [8] / 1972 [1]	2.673
High recycling	2000 [7] / 2000 [8]	1.000
Technological change, income growth, and high recycling	2000 [7] / 1972 [1]	2.673
High import dependence	2000 [6] / 2000 [8]	0.351
Low import dependence	2000 [11] / 2000 [8]	0.849

Table 5–15
Platinum: Indexes for Comparing Changes in U.S. Consumption and Production Levels under Different Scenarios

Consumption		
Effect	*Year/Scenarios*	*Indexes*
Income growth	2000 [1] / 1972 [1]	2.303
Technological change	2000 [8] / 2000 [1]	0.974
Technological change and income growth	2000 [8] / 1972 [1]	2.243
High recycling	2000 [7] / 2000 [8]	0.999
Technological change, income growth, and high recycling	2000 [7] / 1972 [1]	2.241

Production		
Effect	*Year/Scenarios*	*Indexes*
Income growth	2000 [1] / 1972 [1]	3.370
Technological change	2000 [8] / 2000 [1]	0.989
Technological change and income growth	2000 [8] / 1972 [1]	3.331
High recycling	2000 [7] / 2000 [8]	1.000
Technological change, income growth, and high recycling	2000 [7] / 1972 [1]	3.331
High import dependence	2000 [6] / 2000 [8]	0.361
Low import dependence	2000 [11] / 2000 [8]	0.717

Table 5–16
Titanium: Indexes for Comparing Changes in U.S. Consumption and Production Levels under Different Scenarios

Consumption		
Effect	*Year/Scenarios*	*Indexes*
Income growth	2000 [1] / 1972 [1]	2.186
Technological change	2000 [8] / 2000 [1]	1.176
Technological change and income growth	2000 [8] / 1972 [1]	2.570
High recycling	2000 [7] / 2000 [8]	0.998
Technological change, income growth, and high recycling	2000 [7] / 1972 [1]	2.566
Production		
Effect	*Year/Scenarios*	*Indexes*
Income growth	2000 [1] / 1972 [1]	2.609
Technological change	2000 [8] / 2000 [1]	1.170
Technological change and income growth	2000 [8] / 1972 [1]	3.053
High recycling	2000 [7] / 2000 [8]	0.998
Technological change, income growth, and high recycling	2000 [7] / 1972 [1]	3.048
High import dependence	2000 [6] / 2000 [8]	0.715
Low import dependence	2000 [11] / 2000 [8]	1.149

Table 5–17
Tin: Indexes for Comparing Changes in U.S. Consumption and Production Levels under Different Scenarios

Consumption		
Effect	*Year/Scenarios*	*Indexes*
Income growth	2000 [1] / 1972 [1]	2.287
Technological change	2000 [8] / 2000 [1]	0.991
Technological change and income growth	2000 [8] / 1972 [1]	2.267
High recycling	2000 [7] / 2000 [8]	0.604
Technological change, income growth, and high recycling	2000 [7] / 1972 [1]	1.368

Table 5-17 continued

Production		
Effect	*Year/Scenarios*	*Indexes*
Income growth	2000 [1] / 1972 [1]	0
Technological change	2000 [8] / 2000 [1]	0
Technological change and income growth	2000 [8] / 1972 [1]	0
High recycling	2000 [7] / 2000 [8]	0
Technological change, income growth, and high recycling	2000 [7] / 1972 [1]	0
High import dependence	2000 [6] / 2000 [8]	0
Low import dependence	2000 [11] / 2000 [8]	0

Table 5-18
Silicon: Indexes for Comparing Changes in U.S. Consumption and Production Levels under Different Scenarios

Consumption		
Effect	*Year/Scenarios*	*Indexes*
Income growth	2000 [1] / 1972 [1]	2.397
Technological change	2000 [8] / 2000 [1]	1.017
Technological change and income growth	2000 [8] / 1972 [1]	2.438
High recycling	2000 [7] / 2000 [8]	1.000
Technological change, income growth, and high recycling	2000 [7] / 1972 [1]	2.438
Production		
Effect	*Year/Scenarios*	*Indexes*
Income growth	2000 [1] / 1972 [1]	2.454
Technological change	2000 [8] / 2000 [1]	1.017
Technological change and income growth	2000 [8] / 1972 [1]	2.495
High recycling	2000 [7] / 2000 [8]	1.000
Technological change, income growth, and high recycling	2000 [7] / 1972 [1]	2.495
High import dependence	2000 [6] / 2000 [8]	0.860
Low import dependence	2000 [11] / 2000 [8]	1.058

Table 5-19
Fluorine: Indexes for Comparing Changes in U.S. Consumption and Production Levels under Different Scenarios

Consumption		
Effect	*Year/Scenarios*	*Indexes*
Income growth	2000 [1] / 1972 [1]	2.2
Technological change	2000 [8] / 2000 [1]	1.2
Technological change and income growth	2000 [8] / 1972 [1]	2.7
High recycling	2000 [7] / 2000 [8]	1.0
Technological change, income growth, and high recycling	2000 [7] / 1972 [1]	2.7
Production		
Effect	*Year/Scenarios*	*Indexes*
Income growth	2000 [1] / 1972 [1]	2.0
Technological change	2000 [8] / 2000 [1]	1.2
Technological change and income growth	2000 [8] / 1972 [1]	2.5
High recycling	2000 [7] / 2000 [8]	1.0
Technological change, income growth, and high recycling	2000 [7] / 1972 [1]	2.5
High import dependence	2000 [6] / 2000 [8]	0.2
Low import dependence	2000 [11] / 2000 [8]	1.4

Table 5-20
Potash: Indexes for Comparing Changes in U.S. Consumption and Production Levels under Different Scenarios

Consumption		
Effect	*Year/Scenarios*	*Indexes*
Income growth	2000 [1] / 1972 [1]	2.47
Technological change	2000 [8] / 2000 [1]	0.81
Technological change and income growth	2000 [8] / 1972 [1]	2.00
High recycling	2000 [7] / 2000 [8]	1.00
Technological change, income growth, and high recycling	2000 [7] / 1972 [1]	2.00

Table 5-20 continued

Production		
Effect	*Year/Scenarios*	*Indexes*
Income growth	2000 [1] / 1972 [1]	2.76
Technological change	2000 [8] / 2000 [1]	0.85
Technological change and income growth	2000 [8] / 1972 [1]	2.36
High recycling	2000 [7] / 2000 [8]	1.00
Technological change, income growth, and high recycling	2000 [7] / 1972 [1]	2.36
High import dependence	2000 [6] / 2000 [8]	0.18
Low import dependence	2000 [11] / 2000 [8]	0.72

Table 5-21
Soda Ash: Indexes for Comparing Changes in U.S. Consumption and Production Levels under Different Scenarios

Consumption		
Effect	*Year/Scenarios*	*Indexes*
Income growth	2000 [1] / 1972 [1]	2.336
Technological change	2000 [8] / 2000 [1]	0.949
Technological change and income growth	2000 [8] / 1972 [1]	2.218
High recycling	2000 [7] / 2000 [8]	0.997
Technological change, income growth, and high recycling	2000 [7] / 1972 [1]	2.212

Production		
Effect	*Year/Scenarios*	*Indexes*
Income growth	2000 [1] / 1972 [1]	2.479
Technological change	2000 [8] / 2000 [1]	0.956
Technological change and income growth	2000 [8] / 1972 [1]	2.369
High recycling	2000 [7] / 2000 [8]	0.998
Technological change, income growth, and high recycling	2000 [7] / 1972 [1]	2.364
High import dependence	2000 [6] / 2000 [8]	0.702
Low import dependence	2000 [11] / 2000 [8]	1.000

Table 5-22
Boron: Indexes for Comparing Changes in U.S. Consumption and Production Levels under Different Scenarios

Consumption		
Effect	*Year/Scenarios*	*Indexes*
Income growth	2000 [1] / 1972 [1]	2.494
Technological change	2000 [8] / 2000 [1]	1.334
Technological change and income growth	2000 [8] / 1972 [1]	3.326
High recycling	2000 [7] / 2000 [8]	1.000
Technological change, income growth, and high recycling	2000 [7] / 1972 [1]	3.327
Production		
Effect	*Year/Scenarios*	*Indexes*
Income growth	2000 [1] / 1972 [1]	3.673
Technological change	2000 [8] / 2000 [1]	1.108
Technological change and income growth	2000 [8] / 1972 [1]	4.070
High recycling	2000 [7] / 2000 [8]	1.000
Technological change, income growth, and high recycling	2000 [7] / 1972 [1]	4.070
High import dependence	2000 [6] / 2000 [8]	1.016
Low import dependence	2000 [11] / 2000 [8]	1.107

Table 5-23
Phosphate Rock: Indexes for Comparing Changes in U.S. Consumption and Production Levels under Different Scenarios

Consumption		
Effect	*Year/Scenarios*	*Indexes*
Income growth	2000 [1] / 1972 [1]	2.480
Technological change	2000 [8] / 2000 [1]	0.925
Technological change and income growth	2000 [8] / 1972 [1]	2.293
High recycling	2000 [7] / 2000 [8]	1.000
Technological change, income growth, and high recycling	2000 [7] / 1972 [1]	2.293

Table 5-23 continued

Production		
Effect	*Year/Scenarios*	*Indexes*
Income growth	2000 [1] / 1972 [1]	3.432
Technological change	2000 [8] / 2000 [1]	0.961
Technological change and income growth	2000 [8] / 1972 [1]	3.297
High recycling	2000 [7] / 2000 [8]	1.000
Technological change, income growth, and high recycling	2000 [7] / 1972 [1]	3.297
High import dependence	2000 [6] / 2000 [8]	1.000
Low import dependence	2000 [11] / 2000 [8]	1.000

Table 5-24
Sulfur: Indexes for Comparing Changes in U.S. Consumption and Production Levels under Different Scenarios

Consumption		
Effect	*Year/Scenarios*	*Indexes*
Income growth	2000 [1] / 1972 [1]	2.43
Technological change	2000 [8] / 2000 [1]	0.96
Technological change and income growth	2000 [8] / 1972 [1]	2.33
High recycling	2000 [7] / 2000 [8]	1.00
Technological change, income growth, and high recycling	2000 [7] / 1972 [1]	2.33
Production		
Effect	*Year/Scenarios*	*Indexes*
Income growth	2000 [1] / 1972 [1]	2.86
Technological change	2000 [8] / 2000 [1]	0.97
Technological change and income growth	2000 [8] / 1972 [1]	2.78
High recycling	2000 [7] / 2000 [8]	1.00
Technological change, income growth, and high recycling	2000 [7] / 1972 [1]	2.78
High import dependence	2000 [6] / 2000 [8]	0.79
Low import dependence	2000 [11] / 2000 [8]	1.13

Table 5-25
Chlorine: Indexes for Comparing Changes in U.S. Consumption and Production Levels under Different Scenarios

	Consumption	
Effect	*Year/Scenarios*	*Indexes*
Income growth	2000 [1] / 1972 [1]	2.25
Technological change	2000 [8] / 2000 [1]	1.11
Technological change and income growth	2000 [8] / 1972 [1]	2.49
High recycling	2000 [7] / 2000 [8]	1.00
Technological change, income growth, and high recycling	2000 [7] / 1972 [1]	2.48
	Production	
Effect	*Year/Scenarios*	*Indexes*
Income growth	2000 [1] / 1972 [1]	2.26
Technological change	2000 [8] / 2000 [1]	1.11
Technological change and income growth	2000 [8] / 1972 [1]	2.50
High recycling	2000 [7] / 2000 [8]	1.00
Technological change, income growth, and high recycling	2000 [7] / 1972 [1]	2.49
High import dependence	2000 [6] / 2000 [8]	0.55
Low import dependence	2000 [11] / 2000 [8]	1.00

Table 5-26
Magnesium: Indexes for Comnaring Changes in U.S. Consumption and Production Levels under Different Scenarios

	Consumption	
Effect	*Year/Scenarios*	*Indexes*
Income growth	2000 [1] / 1972 [1]	1.889
Technological change	2000 [8] / 2000 [1]	1.235
Technological change and income growth	2000 [8] / 1972 [1]	2.333
High recycling	2000 [7] / 2000 [8]	0.701
Technological change, income growth, and high recycling	2000 [7] / 1972 [1]	1.636

Table 5-26 continued

Production		
Effect	*Year/Scenarios*	*Indexes*
Income growth	2000 [1] / 1972 [1]	1.889
Technological change	2000 [8] / 2000 [1]	1.235
Technological change and income growth	2000 [8] / 1972 [1]	2.333
High recycling	2000 [7] / 2000 [8]	0.701
Technological change, income growth, and high recycling	2000 [7] / 1972 [1]	1.636
High import dependence	2000 [6] / 2000 [8]	0.996
Low import dependence	2000 [11] / 2000 [8]	1.000

Table 5-27
Iron and Steel Scrap: Indexes for Comparing Changes in U.S. Consumption and Production Levels under Different Scenarios

Consumption		
Effect	*Year/Scenarios*	*Indexes*
Income growth	2000 [1] / 1972 [1]	2.242
Technological change	2000 [8] / 2000 [1]	0.827
Technological change and income growth	2000 [8] / 1972 [1]	1.853
High recycling	2000 [7] / 2000 [8]	1.776
Technological change, income growth, and high recycling	2000 [7] / 1972 [1]	3.291

Production		
Effect	*Year/Scenarios*	*Indexes*
Income growth	2000 [1] / 1972 [1]	2.253
Technological change	2000 [8] / 2000 [1]	0.853
Technological change and income growth	2000 [8] / 1972 [1]	1.923
High recycling	2000 [7] / 2000 [8]	1.635
Technological change, income growth, and high recycling	2000 [7] / 1972 [1]	3.145
High import dependence	2000 [6] / 2000 [8]	0.999
Low import dependence	2000 [11] / 2000 [8]	1.000

Table 5–28
Copper Scrap: Indexes for Comparing Changes in U.S. Consumption and Production Levels under Different Scenarios

Consumption		
Effect	*Year/Scenarios*	*Indexes*
Income growth	2000 [1] / 1972 [1]	2.49
Technological change	2000 [8] / 2000 [1]	0.94
Technological change and income growth	2000 [8] / 1972 [1]	2.30
High recycling	2000 [7] / 2000 [8]	1.69
Technological change, income growth, and high recycling	2000 [7] / 1972 [1]	3.90
Production		
Effect	*Year/Scenarios*	*Indexes*
Income growth	2000 [1] / 1972 [1]	2.57
Technological change	2000 [8] / 2000 [1]	0.95
Technological change and income growth	2000 [8] / 1972 [1]	2.44
High recycling	2000 [7] / 2000 [8]	1.63
Technological change, income growth, and high recycling	2000 [7] / 1972 [1]	3.97
High import dependence	2000 [6] / 2000 [8]	0.78
Low import dependence	2000 [11] / 2000 [8]	1.00

Table 5–29
Lead Scrap: Indexes for Comparing Changes in U.S. Consumption and Production Levels under Different Scenarios

Consumption		
Effect	*Year/Scenarios*	*Indexes*
Income growth	2000 [1] / 1972 [1]	2.248
Technological change	2000 [8] / 2000 [1]	0.692
Technological change and income growth	2000 [8] / 1972 [1]	1.555
High recycling	2000 [7] / 2000 [8]	1.256
Technological change, income growth, and high recycling	2000 [7] / 1972 [1]	1.953

Table 5-29 continued

Production		
Effect	*Year/Scenarios*	*Indexes*
Income growth	2000 [1] / 1972 [1]	2.371
Technological change	2000 [8] / 2000 [1]	0.720
Technological change and income growth	2000 [8] / 1972 [1]	1.708
High recycling	2000 [7] / 2000 [8]	1.223
Technological change, income growth, and high recycling	2000 [7] / 1972 [1]	2.088
High import dependence	2000 [6] / 2000 [8]	0.999
Low import dependence	2000 [11] / 2000 [8]	1.000

Table 5-30
Zinc Scrap: Indexes for Comparing Changes in U.S. Consumption and Production Levels under Different Scenarios

Consumption		
Effect	*Year/Scenarios*	*Indexes*
Income growth	2000 [1] / 1972 [1]	2.168
Technological change	2000 [8] / 2000 [1]	1.182
Technological change and income growth	2000 [8] / 1972 [1]	2.563
High recycling	2000 [7] / 2000 [8]	1.691
Technological change, income growth, and high recycling	2000 [7] / 1972 [1]	4.332
Production		
Effect	*Year/Scenarios*	*Indexes*
Income growth	2000 [1] / 1972 [1]	2.276
Technological change	2000 [8] / 2000 [1]	1.180
Technological change and income growth	2000 [8] / 1972 [1]	2.685
High recycling	2000 [7] / 2000 [8]	1.687
Technological change, income growth, and high recycling	2000 [7] / 1972 [1]	4.531
High import dependence	2000 [6] / 2000 [8]	1.000
Low import dependence	2000 [11] / 2000 [8]	1.002

Table 5-31
Aluminum Scrap: Indexes for Comparing Changes in U.S. Consumption and Production Levels under Different Scenarios

Consumption		
Effect	*Year/Scenarios*	*Indexes*
Income growth	2000 [1] / 1972 [1]	2.231
Technological change	2000 [8] / 2000 [1]	1.386
Technological change and income growth	2000 [8] / 1972 [1]	3.092
High recycling	2000 [7] / 2000 [8]	2.143
Technological change, income growth, and high recycling	2000 [7] / 1972 [1]	6.625
Production		
Effect	*Year/Scenarios*	*Indexes*
Income growth	2000 [1] / 1972 [1]	2.361
Technological change	2000 [8] / 2000 [1]	1.343
Technological change and income growth	2000 [8] / 1972 [1]	3.171
High recycling	2000 [7] / 2000 [8]	2.047
Technological change, income growth, and high recycling	2000 [7] / 1972 [1]	6.493
High import dependence	2000 [6] / 2000 [8]	1.000
Low import dependence	2000 [11] / 2000 [8]	1.000

Table 5-32
Nickel Scrap: Indexes for Comparing Changes in U.S. Consumption and Production Levels under Different Scenarios

Consumption		
Effect	*Year/Scenarios*	*Indexes*
Income growth	2000 [1] / 1972 [1]	2.296
Technological change	2000 [8] / 2000 [1]	0.972
Technological change and income growth	2000 [8] / 1972 [1]	2.232
High recycling	2000 [7] / 2000 [8]	1.579
Technological change, income growth, and high recycling	2000 [7] / 1972 [1]	3.525

Table 5-32 continued

Production		
Effect	*Year/Scenarios*	*Indexes*
Income growth	2000 [1] / 1972 [1]	2.856
Technological change	2000 [8] / 2000 [1]	0.977
Technological change and income growth	2000 [8] / 1972 [1]	2.790
High recycling	2000 [7] / 2000 [8]	1.478
Technological change, income growth, and high recycling	2000 [7] / 1972 [1]	4.123
High import dependence	2000 [6] / 2000 [8]	0.986
Low import dependence	2000 [11] / 2000 [8]	1.017

Table 5-33
Chromium Scrap: Indexes for Comparing Changes in U.S. Consumption and Production Levels under Different Scenarios

Consumption		
Effect	*Year/Scenarios*	*Indexes*
Income growth	2000 [1] / 1972 [1]	2.336
Technological change	2000 [8] / 2000 [1]	1.138
Technological change and income growth	2000 [8] / 1972 [1]	2.659
High recycling	2000 [7] / 2000 [8]	2.197
Technological change, income growth, and high recycling	2000 [7] / 1972 [1]	5.843

Production		
Effect	*Year/Scenarios*	*Indexes*
Income growth	2000 [1] / 1972 [1]	2.325
Technological change	2000 [8] / 2000 [1]	1.124
Technological change and income growth	2000 [8] / 1972 [1]	2.613
High recycling	2000 [7] / 2000 [8]	2.087
Technological change, income growth, and high recycling	2000 [7] / 1972 [1]	5.453
High import dependence	2000 [6] / 2000 [8]	0.992
Low import dependence	2000 [11] / 2000 [8]	1.012

Table 5-34
Gold Scrap: Indexes for Comparing Changes in U.S. Consumption and Production Levels under Different Scenarios

Consumption		
Effect	*Year/Scenarios*	*Indexes*
Income growth	2000 [1] / 1972 [1]	2.847
Technological change	2000 [8] / 2000 [1]	1.041
Technological change and income growth	2000 [8] / 1972 [1]	2.965
High recycling	2000 [7] / 2000 [8]	0.505
Technological change, income growth, and high recycling	2000 [7] / 1972 [1]	1.498
Production		
Effect	*Year/Scenarios*	*Indexes*
Income growth	2000 [1] / 1972 [1]	3.151
Technological change	2000 [8] / 2000 [1]	1.027
Technological change and income growth	2000 [8] / 1972 [1]	3.237
High recycling	2000 [7] / 2000 [8]	0.667
Technological change, income growth, and high recycling	2000 [7] / 1972 [1]	2.160
High import dependence	2000 [6] / 2000 [8]	0.990
Low import dependence	2000 [11] / 2000 [8]	1.000

Table 5-35
Silver Scrap: Indexes for Comparing Changes in U.S. Consumption and Production Levels under Different Scenarios

Consumption		
Effect	*Year/Scenarios*	*Indexes*
Income growth	2000 [1] / 1972 [1]	2.795
Technological change	2000 [8] / 2000 [1]	1.007
Technological change and income growth	2000 [8] / 1972 [1]	2.815
High recycling	2000 [7] / 2000 [8]	1.000
Technological change, income growth, and high recycling	2000 [7] / 1972 [1]	2.815

Table 5-35 continued

Effect	Year/Scenarios	Indexes
Production		
Income growth	2000 [1] / 1972 [1]	2.795
Technological change	2000 [8] / 2000 [1]	1.007
Technological change and income growth	2000 [8] / 1972 [1]	2.815
High recycling	2000 [7] / 2000 [8]	1.000
Technological change, income growth, and high recycling	2000 [7] / 1972 [1]	2.815
High import dependence	2000 [6] / 2000 [8]	0.996
Low import dependence	2000 [11] / 2000 [8]	1.004

Table 5-36
Tungsten Scrap: Indexes for Comparing Changes in U.S. Consumption and Production Levels under Different Scenarios

Effect	Year/Scenarios	Indexes
Consumption		
Income growth	2000 [1] / 1972 [1]	2.49
Technological change	2000 [8] / 2000 [1]	1.13
Technological change and income growth	2000 [8] / 1972 [1]	2.82
High recycling	2000 [7] / 2000 [8]	12.50
Technological change, income growth, and high recycling	2000 [7] / 1972 [1]	35.40
Production		
Income growth	2000 [1] / 1972 [1]	2.49
Technological change	2000 [8] / 2000 [1]	1.13
Technological change and income growth	2000 [8] / 1972 [1]	2.82
High recycling	2000 [7] / 2000 [8]	12.50
Technological change, income growth, and high recycling	2000 [7] / 1972 [1]	35.40
High import dependence	2000 [6] / 2000 [8]	1.00
Low import dependence	2000 [11] / 2000 [8]	1.00

Table 5-37
Mercury Scrap: Indexes for Comparing Changes in U.S. Consumption and Production Levels under Different Scenarios

Consumption		
Effect	*Year/Scenarios*	*Indexes*
Income growth	2000 [1] / 1972 [1]	2.373
Technological change	2000 [8] / 2000 [1]	0.504
Technological change and income growth	2000 [8] / 1972 [1]	1.196
High recycling	2000 [7] / 2000 [8]	0.998
Technological change, income growth, and high recycling	2000 [7] / 1972 [1]	1.194
Production		
Effect	*Year/Scenarios*	*Indexes*
Income growth	2000 [1] / 1972 [1]	2.373
Technological change	2000 [8] / 2000 [1]	0.504
Technological change and income growth	2000 [8] / 1972 [1]	1.196
High recycling	2000 [7] / 2000 [8]	0.998
Technological change, income growth, and high recycling	2000 [7] / 1972 [1]	1.194
High import dependence	2000 [6] / 2000 [8]	0.998
Low import dependence	2000 [11] / 2000 [8]	1.012

Table 5-38
Tin Scrap: Indexes for Comparing Changes in U.S. Consumption and Production Levels under Different Scenarios

Consumption		
Effect	*Year/Scenarios*	*Indexes*
Income growth	2000 [1] / 1972 [1]	2.382
Technological change	2000 [8] / 2000 [1]	0.886
Technological change and income growth	2000 [8] / 1972 [1]	2.111
High recycling	2000 [7] / 2000 [8]	1.805
Technological change, income growth, and high recycling	2000 [7] / 1972 [1]	3.811

Table 5–38 continued

Production		
Effect	*Year/Scenarios*	*Indexes*
Income growth	2000 [1] / 1972 [1]	2.655
Technological change	2000 [8] / 2000 [1]	0.926
Technological change and income growth	2000 [8] / 1972 [1]	2.457
High recycling	2000 [7] / 2000 [8]	1.505
Technological change, income growth, and high recycling	2000 [7] / 1972 [1]	3.698
High import dependence	2000 [6] / 2000 [8]	1.000
Low import dependence	2000 [11] / 2000 [8]	1.000

Table 5–39
Magnesium Scrap: Indexes for Comparing Changes in U.S. Consumption and Production Levels under Different Scenarios

Consumption		
Effect	*Year/Scenarios*	*Indexes*
Income growth	2000 [1] / 1972 [1]	1.887
Technological change	2000 [8] / 2000 [1]	1.224
Technological change and income growth	2000 [8] / 1972 [1]	2.310
High recycling	2000 [7] / 2000 [8]	22.226
Technological change, income growth, and high recycling	2000 [7] / 1972 [1]	51.338
Production		
Effect	*Year/Scenarios*	*Indexes*
Income growth	2000 [1] / 1972 [1]	1.852
Technological change	2000 [8] / 2000 [1]	1.222
Technological change and income growth	2000 [8] / 1972 [1]	2.263
High recycling	2000 [7] / 2000 [8]	21.926
Technological change, income growth, and high recycling	2000 [7] / 1972 [1]	49.623
High import dependence	2000 [6] / 2000 [8]	0.981
Low import dependence	2000 [11] / 2000 [8]	1.004

in the production and consumption of each mineral from 1972 to 2000. That is, all else equal, what would be the effect of projected income growth on the production and consumption of minerals? Or, on the other hand, how would the production and consumption of a particular mineral be affected by the assumed changes in technology, all other parameters being held constant? A third comparison is the combined effects of technological change and the growth in income.

The expected effect both on production and consumption of minerals—as well as measures aimed at reducing (or permitting an increase in) the import dependency of the United States—are shown in tables 5-1 through 5-39. In the case of iron, comparing the projected consumption of iron brought about only by the expected economic growth of the U.S. economy as measured by the growth in GDP, we find it would be approximately 2.2 times as high in 2000 as in 1972. On the other hand, under the high recycling assumption, with the projected growth in GDP, iron production in 2000 would be only 20 percent higher than the 1972 base-year figure (assuming the same ratio of imports to domestic output).

The Effects of Projected U.S. Cumulative Production through 2000 on 1980 U.S. Reserves

A study that projects future nonfuel minerals requirements, as any other study of nonrenewable resources, must address the question of future minerals available from domestic and foreign reserves. Changes in such important factors as income, consumption patterns, technology, recycling rates, and import and export trade patterns can affect the rates of domestic mine production and influence the size of yet-unused domestic and foreign reserves. To allow a closer examination of this question, computations based on eleven different scenarios were carried out, with each scenario representing a different combination of changes in these factors. The total effect of these combinations on U.S. mine output of each mineral could then be analyzed by comparing cumulative production of that mineral from 1980 to 2000 with its estimated domestic reserve base. Both global and U.S. reserve and resource estimates for 1980 are listed in table 5-40.[2]

For such minerals as silicon, chlorine, and magnesium, where resource availabilities were considered virtually unlimited even over the long run, no comparisons were made of domestic reserve estimates to projected U.S. cumulative production levels. Similarly, this analysis bypasses minerals such as chromium and manganese, for which there was either no or negligible amounts of U.S. production, and platinum, for which no information was available on its reserve base. Each mineral's reserve and resource estimates were provided by the Bureau of Mines, and the projected U.S.

Table 5–40
U.S. and World Reserves and Resources[a]: 1980 Estimates for 26 Nonfuel Minerals

IEA Code	Mineral Name (Unit)	Reserves		Resources	
		U.S.	*World Total*	*U.S.*	*World Total*
MIN 1	Iron (MMT)	3,628.0	93,421.0	17,777.0	19,6819.0
MIN 2	Molybdenum (KMT)	4,983.0	9,467.7	8,652.3	20,906.4
MIN 3	Nickel (KMT)[b]	326.5	54,420.0	13,877.1	206,796.0
MIN 4	Tungsten (KMT)	124.5	2,582.1	450.7	6,749.7
IMM 5	Manganese (KMT)[b]	0.0	1,360,500.0	66,755.2	2,811,700.0
IMM 6	Chromium (KMT)	0.0	1,038,600.0	2,294.7	10,205,000.0
MIN 7	Copper (MMT)[b]	92.0	494.0	382.0	1,627.0
MIN 8	Lead (MMT)	27.0	127.0	74.0	288.0
MIN 9	Zinc (MMT)	15.0	162.0	65.0	325.0
MIN 10	Gold (MT)	1,399.5	32,344.0	7,464.0	61,267.0
MIN 11	Silver (KMT)	47.0	252.7	177.6	770.0
MIN 12	Aluminum (MMT)	9.1	4,716.0	45.4	8,026.0
MIN 13	Mercury (KMT)	12.1	152.6	27.6	570.3
MIN 14	Vanadium (KMT)	104.3	15,781.0	9,101.0	56,234.0
MIN 15	Platinum (MT)	(c)	16,172.0	(c)	(c)
MIN 16	Titanium (KMT)	16,326.0	273,007.0	94,872.2	697,483.0
IMM 17	Tin (KMT)	50.0	10,000.0	200.0	37,000.0
MIN 21	Silicon (KMT)	(d)	(d)	(d)	(d)
MIN 23	Fluorine (MMT)	12.4	114.3	43.3	422.7
MIN 24	Potash (MMT)	300.0	9,100.0	6,000.0	144,500.0
MIN 25	Soda Ash (MMT)	26,937.0	27,391.4	(d)	(d)

Table 5-40 continued

IEA Code	Mineral Name (Unit)	Reserves		Resources	
		U.S.	World Total	U.S.	World Total
MIN 26	Boron (KMT)	18,100.0	90,700.0	(c)	(c)
MIN 27	Phosphate Rock (MMT)	1,800.0	34,500.0	9,250.0	129,500.0
MIN 28	Sulfur (MMT)	175.0	1,765.0	330.0	6,385.0
CHM 1	Chlorine (MMT)	(d)	(d)	(d)	(d)
CHM 2	Magnesium (MMT)	9.1	2,785.0	(d)	(d)

Source: *Mineral Facts and Problems,* 1980. Bureau of Mines, United States Department of Interior, 1980.

[a]Includes reserves.

[b]Excludes deposits contained in seabed nodules.

[c]Not available.

[d]Virtually unlimited.

Table 5–41
U.S. Cumulative Production (1980 to 2000) under Different Scenarios as a Percentage of Estimated Reserves (1980)

IEA Code	*Mineral Name*	*Scenario* 1	2	3	4	5	6	7	8	9	10	11
MIN 1	Iron	.49	.41	.60	.37	.69	.36	.33	.43	.29	.40	.53
MIN 2	Molybdenum	.36	.36	.36	.36	.36	.38	.38	.38	.38	.38	.38
MIN 3	Nickel	1.40	1.53	7.47	1.13	1.98	1.55	1.15	1.41	1.21	5.71	7.51
MIN 4	Tungsten	1.07	.40	1.13	.79	1.11	.42	.81	1.12	.35	.86	1.18
IMM 5	Manganese	(a)	(a)	(a)	(a)	(a)	(a)	(a)	(a)	(a)	(a)	(a)
IMM 6	Chromium	(a)	(a)	(a)	(a)	(a)	(a)	(a)	(a)	(a)	(a)	(a)
MIN 7	Copper	.70	.68	.81	.52	1.05	.68	.53	.70	.51	.60	.80
MIN 8	Lead	.71	.71	.80	.61	1.12	.55	.48	.56	.48	.54	.62
MIN 9	Zinc	1.13	1.05	1.66	1.01	1.38	1.22	1.18	1.32	1.09	1.72	1.94
MIN 10	Gold	1.37	1.10	1.64	1.37	1.54	1.03	.64	1.29	.55	.74	1.54
MIN 11	Silver	1.59	1.03	1.16	1.59	2.09	1.08	1.67	1.67	1.08	1.22	1.22
MIN 12	Aluminum	1.82	1.01	2.34	1.64	2.07	1.21	1.99	2.23	1.11	2.57	2.86
MIN 13	Mercury	.88	.93	2.07	.88	1.09	.57	.56	.56	.57	1.25	1.25
MIN 14	Vanadium	1.76	1.01	1.80	1.76	1.77	1.06	1.87	1.87	1.06	1.90	1.90
MIN 15	Platinum	(b)	(b)	(b)	(b)	(b)	(b)	(b)	(b)	(b)	(b)	(b)
MIN 16	Titanium	.54	.45	.61	.54	.55	.50	.61	.61	.50	.69	.69
IMM 17	Tin	(a)	(a)	(a)	(a)	(a)	(a)	(a)	(a)	(a)	(a)	(a)
MIN 21	Silicon	(c)	(c)	(c)	(c)	(c)	(c)	(c)	(c)	(c)	(c)	(c)
MIN 23	Fluorine	.40	.15	.47	.40	.40	.17	.48	.48	.17	.55	.55
MIN 24	Potash	.34	.15	.28	.34	.34	.14	.30	.30	.14	.25	.25
MIN 25	Soda Ash	.01	.01	.01	.01	.01	.01	.01	.01	.01	.01	.01
MIN 26	Boron	.53	.53	.53	.53	.53	.57	.57	.57	.57	.57	.57

Table 5-41 continued

IEA Code	Mineral Name	Scenario 1	2	3	4	5	6	7	8	9	10	11
MIN 27	Phosphate rock	1.07	1.08	1.08	1.07	1.08	1.04	1.04	1.04	1.04	1.04	1.04
MIN 28	Sulfur	2.60	2.39	2.85	2.60	2.61	2.35	2.54	2.55	2.35	2.79	2.79
CHM 1	Chlorine	(c)	(c)	(c)	(c)	(c)	(c)	(c)	(c)	(c)	(c)	(c)
CHM 2	Magnesium	(c)	(c)	(c)	(c)	(c)	(c)	(c)	(c)	(c)	(c)	(c)

[a] Noncompetitive import.
[b] Reserve estimate not available
[c] Virtually unlimited reserves.

cumulative production figure under each individual scenario was calculated by interpolating linearly the projected U.S. production level over the years 1980, 1990, and 2000.

Taking the U.S. cumulative production of a mineral as a ratio of its estimated reserve, one could see whether that mineral would still be available for extraction under each of the different projections in 2000. If that ratio were smaller than one it would be; if the ratio were larger than one, it would not.

For example, from table 5-41, U.S. tungsten reserves are expected to be exhausted under scenarios 1, 3, 5, 8, and 11, unless additional reserves become available through new discoveries, changes in prices and in institutional factors, or if recycling rates, trade patterns, or use of the mineral are modified. In this case, ratios in both scenarios 1 and 8 are greater than one, indicating that projected technological changes alone would not save enough to avoid the depletion of the current U.S. reserve. In addition, the rate of depletion would certainly accelerate if policies for reducing import dependence were instituted, as the two ratios in scenarios 3 and 11 clearly indicate. However, the ratios in the remaining scenarios imply that either increasing import dependency or increasing recycling alone would preserve the limited domestic reserves of tungsten up to 2000. As a result, increasing import dependency and higher recycling would obviously enhance this conservation effort.

Notes

1. These aggregations were made in appendix B, tables B.1 through B.26 of Leontief, Koo, Nasar, and Sohn [25a].

2. See appendix D for a discussion of the definition of reserves and resources.

6 Minerals Use by Final Demand Components

Chapter 5 presented the results of the computations based on the alternative sets of assumptions about income growth, technological change and materials substitution, the recycling rates of metals, and the degree of import dependence of the United States for twenty-six minerals through 2000. A consideration of the effects of variations in the rate of growth of the economy as represented by a rising GDP is deferred until chapter 7, where the U.S. economy is embedded within the wider framework of the world economy. In this chapter, it is useful to exploit yet another capability of the input-output technique, the separability characteristic.

It is by now well known that if a structural matrix has certain properties it can be used to solve not one particular bill of goods, but any bill of goods, including the individual parts of a given bill of goods [21]. Because of the linearity property, the sum of the solutions from the various parts of a particular bill of goods is equal to the solution of the total bill of goods. With this property of separability in mind, we can impute the amounts of the twenty-six minerals needed to satisfy the different component parts of final demand—that is, personal consumption expenditures, investment, exports, imports, and military and civilian government expenditures.[1] In addition, by calculating the direct and indirect mineral requirements needed to produce domestically goods that are now or will be imported, we are in a position to ask a critical question. By engaging in international trade, is the United States economizing on its use of minerals, and in particular, imported minerals? These and similar issues are considered in this chapter.

The method used to answer these questions is an algebraic manipulation of the structural matrix used in the computations, which isolates the minerals portion of the inverse matrix—that is, a submatrix with 39 rows and 106 columns. A variant of figure 3-1 is reproduced as figure 6-1, showing, by the shaded portion, the rows and columns in the coefficients matrix that were isolated once the larger inverse matrix was created.

Any element within a particular row of the inverse matrix has the following interpretation. It represents the total—that is, the direct and indirect—minerals requirements needed to deliver a billion dollars worth of the good identified by the column in which this element is located. Before describing the method by which the calculations were performed and the results presented, we include a short digression on the component parts of final demand and changes in their levels.

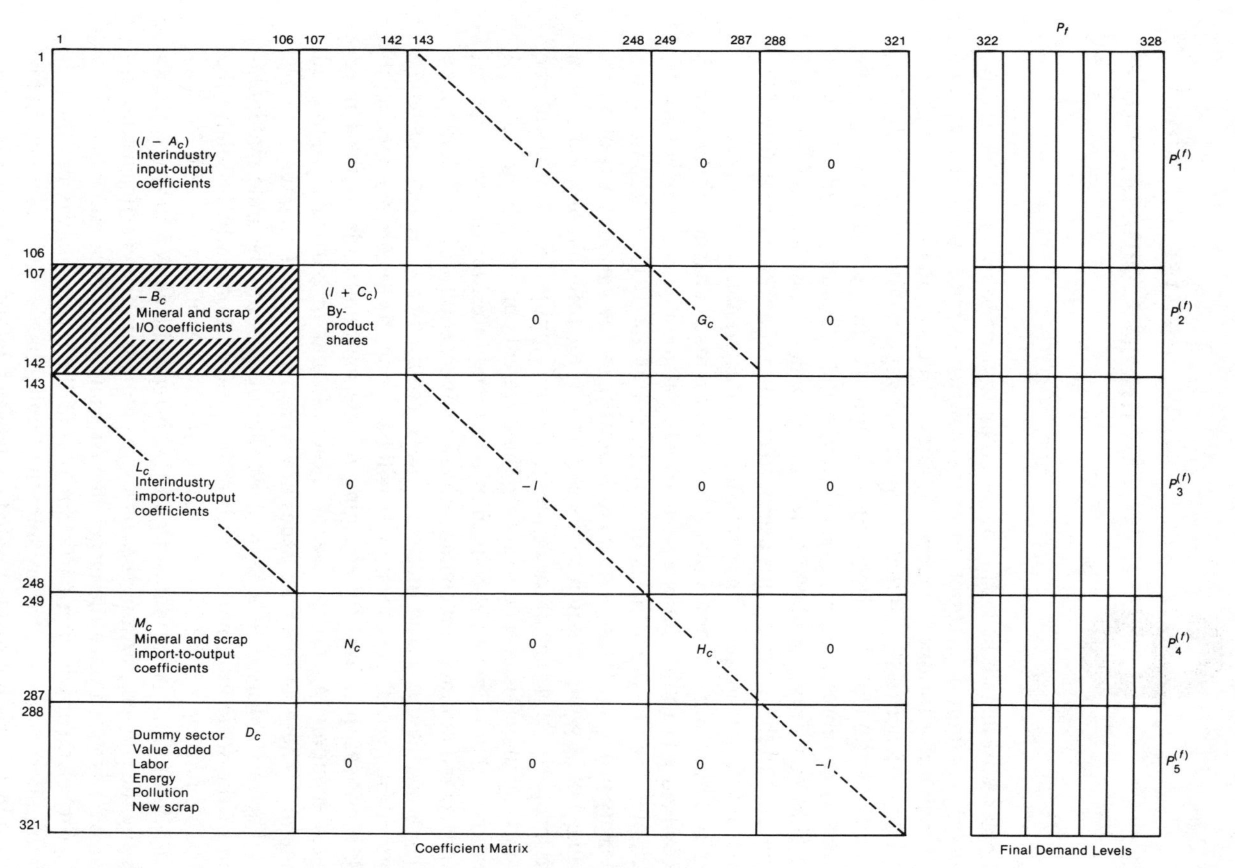

Figure 6–1. Rows and Columns Required to Construct the Minerals Portion of the Inverse Matrix

The Component Parts of Final Demand

The final demand projections used in this study are based on projections made from the Bureau of Labor Statistics' macroeconomic model [55]. This model takes into consideration detailed projections of demographic trends and corresponding changes in the pattern of consumption, investment, exports, imports, and labor productivity.

Because long-range projections of future final demand (GDP), even as detailed as these, are highly imperfect estimates, it would be useful for policy purposes to determine how the solution values would vary under sets of alternative bills of final demand. Variations in the growth rate are introduced in chapter 7, where the U.S. economy is incorporated into the wider framework of the world economy.

Minerals Use by Final Demand Component

Because the input-output system is additive, the solution of the model need not aggregate all production requirements into a single set of total outputs. Instead, the model can be solved with various sets of characteristic bundles of final demands—for example, for consumption, investment, and government expenditures. The solution provides the output levels necessary to sustain one unit—say a billion dollars' worth—of each component of GDP. The difference between these output vectors are easy to distinguish and provide a clearer understanding than a solution in which the differences have been obscured by the aggregation of consumption, investment, and government spending into a single vector of final demand.

In the present model, this approach has the further advantage that it provides results that address frequently raised policy questions. It is possible, for example, to determine the effects on minerals requirements of a dollar-for-dollar shift from consumption to investment or from government nonmilitary to military spending in specific, detailed terms.

A number of such minerals-use computations were performed for this study. Each computation, in turn, examines minerals requirements at a quite detailed level of disaggregation.[2]

The first computation yields the levels of minerals needed to satisfy the direct and indirect requirements generated by each of the separate GDP components given in the base year (1972) and the projected final demand, using the 1972 technical coefficients. The results of this first computation are entered in tables 6-1 and 6-2.

In the second computation, each column of final demand was divided by its column total (in billions), thereby restating the GDP accounts in terms of $1 billion bundles for each component of final demand. These $1

Table 6–1
1972 Levels of Direct and Indirect Minerals Requirements

Mineral (Unit)	*Personal Consumption Expenditures*	*Gross Private Domestic Fixed Investment*	*Exports*	*Imports*	*Government Purchases, Nondefense*	*Government Purchases, Defense*	*Total, Excluding Changes in Business Inventories*
Iron (MMT)	30.1	42.5	7.12	8.30	9.23	5.39	86.0
Molybdenum (KMT)	10.5	9.69	2.54	3.33	1.49	1.80	22.7
Nickel (KMT)	53.1	71.8	15.5	20.0	15.6	10.4	146.
Tungsten (KMT)	1.37	3.77	.987	.543	.326	.318	6.23
Manganese (KMT)	457.	575.	103.	115.	124.	66.1	.121D+04
Chromium (KMT)	189.	195.	54.9	73.2	46.1	29.2	441.
Copper (MMT)	.500	.788	.177	.184	.193	.251	1.73
Lead (MMT)	.423	.256	.703D−01	.973D−01	.787D−01	.114	.844
Zinc (MMT)	.405	.633	.848D−01	.100	.162	.927D−01	1.28
Gold (MT)	132.	39.6	26.0	35.3	16.7	19.0	198.
Silver (KMT)	2.26	.944	.370	.524	.311	.296	3.66
Aluminum (MMT)	1.54	2.14	.353	.333	.582	.400	4.68
Mercury (KMT)	.692	.399	.181	.162	.156	.803D−01	1.35
Vanadium (KMT)	1.81	3.48	.689	.604	.581	.432	6.39
Platinum (MT)	8.16	3.06	1.76	1.87	1.40	1.55	14.1
Titanium (KMT)	298.	134.	49.1	54.0	75.7	23.3	526.
Tin (KMT)	24.8	16.4	4.07	4.19	4.23	3.13	48.4
Silicon (KMT)	182.	235.	53.6	56.1	45.5	37.5	497.

Fluorine (MMT)	.283	.232	.927D-01	.151	.665D−01	.462D−01	.570
Potash (MMT)	3.23	.284	1.04	.678	.106	.640D−01	4.04
Soda ash (MMT)	4.32	.979	.702	.789	.574	.281	6.06
Boron (KMT)	57.0	14.0	8.53	9.29	7.27	2.05	79.7
Phosphate rock (MMT)	20.1	1.60	5.99	3.93	.717	.360	24.9
Sulfur (MMT)	7.09	1.17	2.16	1,81	.465	.252	9.33
Chlorine (MMT)	5.69	1.40	1.27	1.16	.839	.523	8.55
Magnesium (MMT)	.258	.519	.637D−01	.104	.190	.355D−01	.963
Iron and steel scrap (MMT)	13.0	18.3	3.07	3.58	3.98	2.32	37.0
Copper scrap (MMT)	.328	.534	.117	.123	.130	.170	1.16
Lead scrap (MMT)	.274	.166	.456D−01	.631D−01	.510D−01	.737D−01	.548
Zinc scrap (MMT)	.121	.156	.270D−01	.287D−01	.425D−*01*	*.253D*−01	.343
Aluminum scrap (MMT)	.272	.377	.575D−01	.620D−01	.105	.911D−01	.841
Nickel scrap (KMT)	22.4	30.3	6.54	8.44	6.59	4.40	61.8
Chromium scrap (KMT)	31.8	39.8	10.3	13.2	8.31	6.27	83.3
Gold scrap (MT)	14.7	4.40	2.89	3.92	1.86	2.11	22.0
Silver scrap (KMT)	1.16	.486	.190	.270	.160	.152	1.88
Tungsten scrap (KMT)	.590D−01	.163	.427D−01	.235D−01	.141D−01	.137D−01	.269
Mercury scrap (KMT)	.206	.119	.538D−01	.482D−01	.463D−01	.239D−01	.401
Tin scrap (KMT)	7.85	5.20	1.29	1.33	1.34	.992	15.3
Magnesium scrap (MMT)	.363D−02	.734D−02	.884D−03	.138D−02	.269D−02	.501D−03	.137D−01

Table 6–2
2000 Levels of Direct and Indirect Minerals Requirements

Mineral (Unit)	*Personal Consumption Expenditures*	*Gross Private Domestic Fixed Investment*	*Exports*	*Imports*	*Government Purchases, Nondefense*	*Government Purchases, Defense*	*Total, Excluding Changes in Business Inventories*
Iron (MMT)	82.3	91.5	24.5	25.5	10.9	8.31	192.
Molybdenum (KMT)	27.4	22.9	8.39	10.4	2.08	2.79	53.2
Nickel (KMT)	151.	160.	53.1	59.6	19.0	16.1	340.
Tungsten (KMT)	3.68	9.25	3.53	1.68	.513	.458	15.8
Manganese (KMT)	.124D+04	.123D+04	342.	343.	150.	111.	.272D+04
Chromium (KMT)	520.	441.	193.	201.	58.7	46.2	.106D+04
Copper (MMT)	1.59	1.90	.674	.695	.249	.377	4.09
Lead (MMT)	1.13	.560	.226	.275	.102	.164	1.91
Zinc (MMT)	1.18	1.31	.294	.322	.190	.144	2.79
Gold (MT)	405.	108.	122.	128.	21.6	29.0	557.
Silver (KMT)	6.91	2.66	1.59	1.87	.430	.450	10.2
Aluminum (MMT)	4.17	4.36	1.10	.913	.689	.647	10.0
Mercury (KMT)	1.84	.877	.609	.564	.258	.144	3.17
Vanadium (KMT)	4.97	7.53	2.23	1.67	.708	.678	14.4
Platinum (MT)	21.2	7.10	5.09	5.33	1.79	2.80	32.6
Titanium (KMT)	766.	266.	159.	128.	99.9	40.0	.120D+04
Tin (KMT)	64.2	36.0	13.5	12.2	5.26	5.07	112.
Silicon (KMT)	506.	549.	185.	171.	58.3	61.6	.119D+04

Fluorine (MMT)	.738	.505	.275	.307	.821D−01	.845D−01	1.38
Potash (MMT)	7.66	.541	3.65	2.29	.385	.137	10.1
Soda ash (MMT)	11.1	2.07	2.20	1.87	.843	.607	14.9
Boron (KMT)	150.	29.6	32.9	27.1	12.8	3.57	202.
Phosphate rock (MMT)	48.1	3.04	21.1	13.2	2.54	.759	62.3
Sulfur (MMT)	17.2	2.39	7.43	5.23	1.00	.488	23.3
Chlorine (MMT)	14.3	2.84	3.55	2.60	1.09	1.19	20.4
Magnesium (MMT)	.675	.931	.230	.294	.206	.647D−01	1.81
Iron and steel scrap (MMT)	35.5	39.4	10.6	11.0	4.69	3.58	82.8
Copper scrap (MMT)	1.06	1.29	.452	.469	.167	.253	2.75
Lead scrap (MMT)	.734	.363	.147	.178	.662D−01	.107	1.24
Zinc scrap (MMT)	.337	.322	.850D−01	.865D−01	.503D−01	.425D−01	.750
Aluminum scrap (MMT)	.791	.821	.204	.205	.124	.133	1.87
Nickel scrap (KMT)	63.8	67.6	22.3	25.1	8.00	6.82	143.
Chromium scrap (KMT)	89.6	91.8	35.2	35.2	10.4	9.66	201.
Gold scrap (MT)	45.0	12.0	13.5	14.3	2.40	3.22	61.9
Silver scrap (KMT)	3.56	1.37	.820	.961	.221	.232	5.24
Tungsten scrap (KMT)	.159	.400	.153	.726D−01	.222D−01	.198D−01	.681
Mercury scrap (KMT)	.549	.261	.181	.168	.767D−01	.430D−01	.944
Tin scrap (KMT)	20.3	11.4	4.27	3.87	1.67	1.61	35.4
Magnesium scrap (MMT)	.947D-02	.132D-01	.321D−02	.404D−02	.291D−02	.914D−03	.256D−01

billion bundles were premultiplied one by one by the minerals portion of the inverse. The solution provided the direct and indirect minerals requirements needed to produce one $1 billion unit of each of the different categories of GDP. The results of this computation appear in tables 6–3 and 6–4.

Table 6–5 presents the projected percentage change in the levels of minerals requirements from 1972 to 2000 by GDP components. For example, the 1972 direct and indirect requirements of iron needed to satisfy total household demand for the same year was 30.1 MMT. In the year 2000, these same requirements increased to 82.3 MMT, an increase of 174 percent over the base-year.

A similar calculation was carried out to describe the percentage change in the intensity of minerals use, mineral by mineral, per billion dollars of GDP component from 1972 to 2000. These results are presented in table 6–6. Because the matrix of technological coefficients was unchanged for this computation, the figures in table 6–6 reflect only changes in the composition of the goods delivered to the various GDP components from 1972 to 2000. Tables 6–7 and 6–8 describe the percentage distribution of the total use of each mineral by each GDP component for 1972 and 2000.

To determine the effect that technological change can be expected to have on the levels of minerals needed to satisfy the different GDP components, the same computations were performed using the matrix incorporating the technological-change assumptions. For example, table 6–9 presents the same information as table 6–2. The difference in values results from the technological-change and materials-substitution assumptions embedded in the 2000 structural matrix. For example, comparing the use of iron and steel necessary to satisfy the same bill of final demands, the requirements are significantly lower after incorporating the assumptions regarding a substitution of aluminum and other metals and materials for steel. Tables 6–10 and 6–11 present the same information as tables 6–4 and 6–8 respectively, however, incorporating the assumptions concerning technological change used in the construction of the 2000 structural matrix.

The minerals portion of the inverse matrices sheds some light on the perennial controversy regarding the relative resource intensity of agricultural, manufacturing, and service activities. The minerals portion of the inverse, both for 1972 and 2000, on which all the computations in this section were based, are reproduced in appendix G.

Throughout chapters 4 and 5, the solutions of the eleven scenarios and the calculations performed on the minerals portion of the inverse matrix were determined, in part, by the same projected rate of growth in economic activity—the BLS estimates [53]—which assumed an annual rate of growth from 1972 to 2000 of 3.1 percent.[3] The level of economic activity is a major determinant of the level of minerals consumption; therefore, changes in the rate of economic growth over time are bound to have significant effects on

Table 6–3
1972 Intensity of Use of Minerals by GNP Account

Mineral (Unit)	*Personal Consumption Expenditures*	*Gross Private Domestic Fixed Investment*	*Exports*	*Imports*	*Government Purchases, Nondefense*	*Government Purchases, Defense*	*Total, Excluding Changes in Business Inventories*
Iron (MMT)	.412D-01	.230	.123	.146	.573D-01	.7950-01	.832D-01
Molybdenum (KMT)	.144D-01	.524D-01	.438D-01	.586D-01	.925D-02	.265D-01	.220D-01
Nickel (KMT)	.727D-01	.388	.268	.352	.968D-01	.154	.142
Tungsten (KMT)	.187D-02	.204D-01	.170D-01	.956D-02	.202D-02	.469D-02	.603D-02
Manganese (KMT)	.626	3.11	1.77	2.03	.771	.975	1.17
Chromium (KMT)	.259	1.06	.947	1.29	.286	.431	.427
Copper (MMT)	.685D-03	.426D-02	.306D-02	.323D-02	.120D-02	.371D-02	.167D-02
Lead (MMT)	.580D-03	.138D-02	.121D-02	.171D-02	.488D-03	.168D-02	.817D-03
Zinc (MMT)	.556D-03	.342D-02	.146D-02	.176D-02	.100D-02	.137D-02	.124D-02
Gold (MT)	.181	.214	.448	.621	.104	.281	.192
Silver (KMT)	.310D-02	.510D-02	.638D-02	.922D-02	.193D-02	.437D-02	.354D-02
Aluminum (MMT)	.211D-02	.116D-01	.610D-02	.586D-02	.361D-02	.590D-02	.453D-02
Mercury (KMT)	.949D-03	.216D-02	.312D-02	.285D-02	.965D-03	.119D-02	.130D-02
Vanadium (KMT)	.248D-02	.188D-01	.119D-01	.106D-01	.361D-02	.638D-02	.618D-02
Platinum (MT)	.112D-01	.165D-01	.303D-01	.329D-01	.871D-02	.229D-01	.136D-01
Titanium (KMT)	.408	.723	.848	.949	.470	.344	.509
Tin (KMT)	.339D-01	.888D-01	.703D-01	.737D-01	.263D-01	.462D-01	.469D-01
Silicon (KMT)	.249	1.27	.926	.987	.282	.553	.481
Fluorine (MMT)	.388D-03	.126D-02	.160D-02	.266D-02	.412D-03	.682D-03	.551D-03
Potash (MMT)	.442D-02	.154D-02	.179D-01	.119D-01	.660D-03	.944D-03	.391D-02

Table 6–3 continued

Mineral (Unit)	*Personal Consumption Expenditures*	*Gross Private Domestic Fixed Investment*	*Exports*	*Imports*	*Government Purchases, Nondefense*	*Government Purchases, Defense*	*Total, Excluding Changes in Business Inventories*
Soda ash (MMT)	.592D–02	.530D–02	.121D–01	.139D–01	.356D–02	.414D–02	.587D–02
Boron (KMT)	.782D–01	.759D–01	.147	.163	.451D–01	.303D–01	.771D–01
Phosphate rock (MMT)	.276D–01	.865D–02	.103	.691D–01	.445D–02	.531D–02	.241D–01
Sulfur (MMT)	.971D–02	.635D–02	.372D–01	.318D–01	.289D–02	.372D–02	.903D–02
Chlorine (MMT)	.780D–02	.755D–02	.219D–01	.205D–01	.521D–02	.772D–02	.828D–02
Magnesium (MMT)	.354D–03	.281D–02	.110D–02	.183D–02	.118D–02	.524D–03	.932D–03
Iron and steel scrap (MMT)	.178D–01	.989D–01	.529D–01	.629D–01	.247D–01	.342D–01	.358D–01
Copper scrap (MMT)	.449D–03	.289D–02	.203D–02	.216D–02	.805D–03	.250D–02	.112D–02
Lead scrap (MMT)	.376D–03	.897D–03	.787D–03	.111D–02	.317D–03	.109D–02	.530D–03
Zinc scrap (MMT)	.166D–03	.844D–03	.466D–03	.505D–03	.264D–03	.373D–03	.332D–03
Aluminum scrap (MMT)	.373D–03	.204D–02	.992D–03	.109D–02	.653D–03	.134D–02	.814D–03
Nickel scrap (KMT)	.307D–01	.164	.113	.148	.409D–01	.650D–01	.598D–01
Chromium scrap (KMT)	.436D–01	.215	.178	.233	.516D–01	.925D–01	.806D–01
Gold scrap (MT)	.201D–01	.238D–01	.498D–01	.690D–01	.115D–01	.312D–01	.213D–01
Silver scrap (KMT)	.159D–02	.263D–02	.329D–02	.475D–02	.933D–03	.225D–02	.182D–02
Tungsten scrap (KMT)	.809D–04	.882D–03	.736D–03	.413D–03	.874D–04	.203D–03	.260D–03
Mercury scrap (KMT)	.282D–03	.642D–03	.929D–03	.848D–03	.287D–03	.353D–03	.388D–03
Tin scrap (KMT)	.108D–01	.281D–01	.223D–01	.233D–01	.832D–02	.146D–01	.149D–01
Magnesium scrap (MMT)	.497D–05	.397D–04	.153D–04	.243D–04	.167D–04	.740D–05	.132D–04

Table 6–4
2000 Intensity of Use of Minerals by GNP Account

Mineral (Unit)	*Personal Consumption Expenditures*	*Gross Private Domestic Fixed Investment*	*Exports*	*Imports*	*Government Purchases, Nondefense*	*Government Purchases, Defense*	*Total, Excluding Changes in Business Inventories*
Iron (MMT)	.441D–01	.226	.133	.173	.475D–01	.798D–01	.774D–01
Molybdenum (KMT)	.147D–01	.566D–01	.457D–01	.711D–01	.906D–02	.268D–01	.214D–01
Nickel (KMT)	.809D–01	.395	.289	.406	.827D–01	.155	.137
Tungsten (KMT)	.197D–02	.229D–01	.192D–01	.114D–01	.224D–02	.440D–02	.635D–02
Manganese (KMT)	.662	3.03	1.86	2.34	.654	1.07	1.10
Chromium (KMT)	.278	1.09	1.05	1.37	.256	.443	.427
Copper (MMT)	.853D–03	.469D–02	.367D–02	.473D–02	.109D–02	.362D–02	.165D–02
Lead (MMT)	.605D–03	.138D–02	.123D–02	.187D–02	.445D–03	.158D–02	.769D–03
Zinc (MMT)	.630D–03	.323D–02	.160D–02	.219D–02	.831D–03	.138D–02	.112D–02
Gold (MT)	.217	.267	.663	.874	.943D–01	.278	.225
Silver (KMT)	.370D–02	.657D–02	.868D–02	.127D–01	.187D–02	.432D–02	.410D–02
Aluminum (MMT)	.223D–02	.108D–01	.601D–02	.622D–02	.301D–02	.621D–02	.405D–02
Mercury (KMT)	.988D–03	.217D–02	.332D–02	.384D–02	.112D–02	.139D–02	.128D–02
Vanadium (KMT)	.266D–02	.186D–01	.121D–01	.114D–01	.309D–02	.652D–02	.583D–02
Platinum (MT)	.113D–01	.176D–01	.277D–01	.363D–01	.781D–02	.269D–01	.132D–01
Titanium (KMT)	.410	.657	.866	.869	.436	.384	.485
Tin (KMT)	.344D–01	.889D–01	.734D–01	.833D–01	.229D–01	.487D–01	.451D–01
Silicon (KMT)	.271	1.36	1.01	1.17	.254	.592	.479
Fluorine (MMT)	.395D–03	.125D–02	.150D–02	.209D–02	.358D–03	.812D–03	.555D–03
Potash (MMT)	.410D–02	.134D–02	.199D–01	.156D–01	.168D–02	.132D–02	.406D–02

Table 6–4 continued

Mineral (Unit)	*Personal Consumption Expenditures*	*Gross Private Domestic Fixed Investment*	*Exports*	*Imports*	*Government Purchases, Nondefense*	*Government Purchases, Defense*	*Total, Excluding Changes in Business Inventories*
Soda ash (MMT)	.593D-02	.511D-02	.120D-01	.127D-01	.368D-02	.583D-02	.602D-02
Boron (KMT)	.802D-01	.733D-01	.179	.184	.560D-01	.342D-01	.814D-01
Phosphate rock (MMT)	.258D-01	.752D-02	.115	.901D-01	.111D-01	.729D-02	.251D-01
Sulfur (MMT)	.923D-02	.590D-02	.404D-01	.356D-01	.436D-02	.469D-02	.940D-02
Chlorine (MMT)	.767D-02	.703D-02	.193D-01	.177D-01	.477D-02	.114D-01	.822D-02
Magnesium (MMT)	.361D-03	.230D-02	.125D-02	.200D-02	.898D-03	.622D-03	.731D-03
Iron and steel scrap (MMT)	.190D-01	.975D-01	.575D-01	.747D-01	.204D-01	.344D-01	.334D-01
Copper scrap (MMT)	.565D-03	.318D-02	.246D-02	.319D-02	.729D-03	.243D-02	.111D-02
Lead scrap (MMT)	.393D-03	.898D-03	.798D-03	.122D-02	.289D-03	.103D-02	.499D-03
Zinc scrap (MMT)	.181D-03	.795D-03	.463D-03	.589D-03	.219D-03	.408D-03	.302D-03
Aluminum scrap (MMT)	.423D-03	.203D-02	.111D-02	.140D-02	.542D-03	.128D-02	.753D-03
Nickel scrap (KMT)	.342D-01	.167	.122	.171	.349D-01	.655D-01	.578D-01
Chromium scrap (KMT)	.480D-01	.227	.191	.240	.454D-01	.928D-01	.812D-01
Gold scrap (MT)	.241D-01	.297D-01	.736D-01	.971D-01	.105D-01	.309D-01	.250D-01
Silver scrap (KMT)	.190D-02	.338D-02	.447D-02	.655D-02	.965D-03	.222D-02	.211D-02
Tungsten scrap (KMT)	.851D-04	.989D-03	.832D-03	.494D-03	.967D-04	.190D-03	.275D-03
Mercury scrap (KMT)	.294D-03	.646D-03	.988D-03	.114D-02	.335D-03	.413D-03	.380D-03
Tin scrap (KMT)	.109D-01	.282D-01	.232D-01	.264D-01	.727D-02	.154D-01	.143D-01
Magnesium scrap (MMT)	.507D-05	.326D-04	.175D-04	.275D-04	.127D-04	.878D-05	.103D-04

Table 6–5
Percentage Change in Levels of Minerals Requirements from 1972 to 2000

Mineral (Unit)	*Personal Consumption Expenditures*	*Gross Private Domestic Fixed Investment*	*Exports*	*Imports*	*Government Purchases, Nondefense*	*Government Purchases, Defense*	*Total, Excluding Changes in Business Inventories*
Iron (MMT)	1.74	1.16	2.44	2.07	.178	.543	1.23
Molybdenum (KMT)	1.60	1.36	2.31	2.13	.394	.552	1.34
Nickel (KMT)	1.85	1.23	2.42	1.98	.215	.549	1.32
Tungsten (KMT)	1.69	1.45	2.58	2.09	.575	.442	1.53
Manganese (KMT)	1.70	1.13	2.33	1.98	.207	.686	1.25
Chromium (KMT)	1.75	1.26	2.52	1.74	.274	.580	1.40
Copper (MMT)	2.19	1.41	2.80	2.78	.287	.499	1.37
Lead (MMT)	1.67	1.19	2.21	1.83	.297	.447	1.26
Zinc (MMT)	1.90	1.06	2.47	2.22	.177	.551	1.18
Gold (MT)	2.07	1.73	3.69	2.64	.294	.523	1.81
Silver (KMT)	2.06	1.81	3.31	2.56	.382	.520	1.78
Aluminum (MMT)	1.70	1.04	2.12	1.74	.184	.616	1.15
Mercury (KMT)	1.67	1.20	2.37	2.49	.656	.799	1.35
Vanadium (KMT)	1.75	1.16	2.24	1.76	.219	.569	1.26
Platinum (MT)	1.60	1.32	1.90	1.85	.274	.805	1.32
Titanium (KMT)	1.57	.988	2.24	1.36	.320	.716	1.29
Tin (KMT)	1.59	1.19	2.31	1.92	.242	.618	1.31
Silicon (KMT)	1.78	1.34	2.46	2.06	.281	.645	1.39
Fluorine (MMT)	1.61	1.17	1.97	1.04	.235	.827	1.42
Potash (MMT)	1.37	.903	2.51	2.37	2.62	1.14	1.50

Table 6–5 continued

Mineral (Unit)	*Personal Consumption Expenditures*	*Gross Private Domestic Fixed Investment*	*Exports*	*Imports*	*Government Purchases, Nondefense*	*Government Purchases, Defense*	*Total, Excluding Changes in Business Inventories*
Soda ash (MMT)	1.57	1.11	2.14	1.37	.469	1.16	1.46
Boron (KMT)	1.63	1.11	2.86	1.91	.766	.737	1.53
Phosphate rock (MMT)	1.39	.903	2.53	2.37	2.54	1.11	1.51
Sulfur (MMT)	1.43	1.03	2.44	1.90	1.15	.939	1.50
Chlorine (MMT)	1.52	1.04	1.80	1.23	.301	1.26	1.39
Magnesium (MMT)	1.61	.795	2.61	1.84	.813D–01	.824	.882
Iron and steel scrap (MMT)	1.74	1.16	2.44	2.07	.178	.543	1.23
Copper scrap (MMT)	2.22	1.41	2.85	2.82	.288	.489	1.38
Lead scrap (MMT)	1.67	1.19	2.21	1.83	.297	.447	1.26
Zinc scrap (MMT)	1.79	1.06	2.15	2.02	.184	.678	1.19
Aluminum scrap (MMT)	1.91	1.18	2.55	2.30	.181	.459	1.22
Nickel scrap (KMT)	1.85	1.23	2.41	1.98	.215	.549	1.32
Chromium scrap (KMT)	1.81	1.31	2.41	1.66	.252	.541	1.42
Gold scrap (MT)	2.07	1.73	3.69	2.64	.294	.523	1.81
Silver scrap (KMT)	2.06	1.81	3.31	2.56	.382	.520	1.78
Tungsten scrap (KMT)	1.69	1.45	2.58	2.09	.575	.442	1.53
Mercury scrap (KMT)	1.67	1.20	2.37	2.49	.656	.799	1.35
Tin scrap (KMT)	1.59	1.19	2.31	1.92	.242	.618	1.31
Magnesium scrap (MMT)	1.61	.794	2.63	1.92	.815D–01	.822	.876

Table 6–6
Percentage Change in Intensities of Minerals Requirements from 1972 to 2000

Mineral (Unit)	*Personal Consumption Expenditures*	*Gross Private Domestic Fixed Investment*	*Exports*	*Imports*	*Government Purchases, Nondefense*	*Government Purchases, Defense*	*Total, Excluding Changes in Business Inventories*
Iron (MMT)	.694D–01	−.147D–01	.866D–01	.187	−.171	.472D–02	−.692D–01
Molybdenum (KMT)	.171D–01	.807D–01	.432D–01	.212	−.199D–01	.105D–01	−.252D–01
Nickel (KMT)	.113	.185D–01	.780D–01	.153	−.146	.854D–02	−.335D–01
Tungsten (KMT)	.514D–01	.121	.130	.197	.107	−.611D–01	.542D–01
Manganese (KMT)	.568D–01	−.246D–01	.502D–01	.154	−.151	.975D–01	−.623D–01
Chromium (KMT)	.756D–01	.331D–01	.110	.619D–01	−.104	.288D–01	−.484D–03
Copper (MMT)	.246	.101	.200	.464	−.948D–01	−.238D–01	−.115D–01
Lead (MMT)	.445D–01	.615D–03	.145D–01	.940D–01	−.878D–01	−.579D–01	−.582D–01
Zinc (MMT)	.133	−.564D–01	.942D–01	.246	−.173	.979D–02	−.907D–01
Gold (MT)	.198	.249	.478	.408	−.902D–01	−.848D–02	.172
Silver (KMT)	.195	.286	.359	.380	−.282D–01	−.105D–01	.159
Aluminum (MMT)	.556D–01	−.688D–01	−.148D–01	.620D–01	−.168	.522D–01	−.106
Mercury (KMT)	.413D–01	.587D–02	.632D–01	.350	.165	.171	−.189D–01
Vanadium (KMT)	.745D–01	−.121D–01	.213D–01	.700D–01	−.143	.217D–01	−.580D–01
Platinum (MT)	.141D–01	.609D–01	−.861D–01	.102	−.104	.175	−.331D–01
Titanium (KMT)	.517D–02	−.912D–01	.214D–01	−.850D–01	−.719D–01	.117	−.463D–01
Tin (KMT)	.128D–01	.687D–03	.437D–01	.131	−.126	.536D–01	−.391D–01
Silicon (KMT)	.877D–01	.679D–01	.906D–01	.183	−.989D–01	.710D–01	−.401D–02
Fluorine (MMT)	.183D–01	−.648D–02	−.622D–01	−.212	−.132	.189	.715D–02
Potash (MMT)	−.724D–01	−.130	.109	.305	1.55	.393	.398D–01

Table 6–6 continued

Mineral (Unit)	*Personal Consumption Expenditures*	*Gross Private Domestic Fixed Investment*	*Exports*	*Imports*	*Government Purchases, Nondefense*	*Government Purchases, Defense*	*Total, Excluding Changes in Business Inventories*
Soda ash (MMT)	.251D-02	−.350D-01	−.946D-02	−.824D-01	.333D-01	.408	.256D-01
Boron (KMT)	.264D-01	−.353D-01	.218	.128	.242	.131	.554D-01
Phosphate rock (MMT)	−.668D-01	−.130	.114	.304	1.49	.373	.445D-01
Sulfur (MMT)	−.499D-01	−.712D-01	.867D-01	.121	.512	.262	.407D-01
Chlorine (MMT)	−.162D-01	−.689D-01	−.117	−.137	−.847D-01	.474	−.628D-02
Magnesium (MMT)	.205D-01	−.179	.140	.982D-01	−.240	.188	−.216
Iron and steel scrap (MMT)	.694D-01	−.147D-01	.866D-01	.187	−.171	.472D-02	−.692D-01
Copper scrap (MMT)	.259	.102	.215	.479	−.943D-01	−.307D-01	−.101D-01
Lead scrap (MMT)	.445D-01	.615D-03	.145D-01	.940D-01	−.878D-01	−.579D-01	−.582D-01
Zinc scrap (MMT)	.890D-01	−.577D-01	−.619D-02	.167	−.167	.927D-01	−.892D-01
Aluminum scrap (MMT)	.136	−.416D-02	.120	.279	−.170	−.504D-01	−.743D-01
Nickel scrap (KMT)	.113	.186D-01	.770D-01	.152	−.146	.840D-02	−.334D-01
Chromium scrap (KMT)	.992D-01	.549D-01	.752D-01	.302D-01	−.119	.346D-02	.737D-02
Gold scrap (MT)	.198	.249	.478	.408	−.902D-01	−.848D-02	.172
Silver scrap (KMT)	.195	.286	.359	.380	−.282D-01	−.105D-01	.159
Tungsten scrap (KMT)	.514D-01	.121	.130	.197	.107	−.611D-01	.542D-01
Mercury scrap (KMT)	.413D-01	.587D-02	.632D-01	.350	.165	.171	−.189D-01
Tin scrap (KMT)	.128D-01	.686D-03	.437D-01	.131	−.126	.536D-01	−.391D-01
Magnesium scrap (MMT)	.196D-01	−.180	.145	.131	−.239	.186	−.218

Table 6–7
1972 Distribution of Each Minerals Use Level by GNP Account

Mineral (Unit)	*Personal Consumption Expenditures*	*Gross Private Domestic Fixed Investment*	*Exports*	*Imports*	*Government Purchases, Nondefense*	*Government Purchases, Defense*	*Total, Excluding Changes in Business Inventories*
Iron (MMT)	.350	.494	.828D–01	.965D–01	.107	.626D–01	1.00
Molybdenum (KMT)	.464	.427	.112	.147	.656D–01	.790D–01	1.00
Nickel (KMT)	.362	.490	.106	.137	.107	.712D–01	1.00
Tungsten (KMT)	.219	.606	.158	.872D–01	.523D–01	.510D–01	1.00
Manganese (KMT)	.378	.475	.849D–01	.951D–01	.103	.546D–01	1.00
Chromium (KMT)	.428	.443	.124	.166	.105	.662D–01	1.00
Copper (MMT)	.290	.457	.103	.107	.112	.146	1.00
Lead (MMT)	.501	.303	.833D–01	.115	.932D–01	.135	1.00
Zinc (MMT)	.317	.495	.663D–01	.782D–01	.127	.725D–01	1.00
Gold (MT)	.667	.200	.131	.178	.843D–01	.961D–01	1.00
Silver (KMT)	.618	.258	.101	.143	.851D–01	.810D–01	1.00
Aluminum (MMT)	.329	.457	.755D–01	.711D–01	.124	.854D–01	1.00
Mercury (KMT)	.514	.296	.134	.120	.116	.597D–01	1.00
Vanadium (KMT)	.283	.545	.108	.945D–01	.909D–01	.676D–01	1.00
Platinum (MT)	.580	.218	.125	.133	.999D–01	.110	1.00
Titanium (KMT)	.566	.254	.935D–01	.103	.144	.443D–01	1.00
Tin (KMT)	.511	.339	.841D–01	.864D–01	.874D–01	.646D–01	1.00
Silicon (KMT)	.365	.473	.108	.113	.915D–01	.753D–01	1.00
Fluorine (MMT)	.497	.407	.163	.265	.117	.812D–01	1.00
Potash (MMT)	.799	.704D–01	.257	.168	.263D–01	.158D–01	1.00

Table 6-7 continued

Mineral (Unit)	*Personal Consumption Expenditures*	*Gross Private Domestic Fixed Investment*	*Exports*	*Imports*	*Government Purchases, Nondefense*	*Government Purchases, Defense*	*Total, Excluding Changes in Business Inventories*
Soda ash (MMT)	.712	.161	.116	.130	.946D-01	.463D-01	1.00
Boron (KMT)	.716	.176	.107	.117	.913D-01	.258D-01	1.00
Phosphate rock (MMT)	.810	.643D-01	.241	.158	.288D-01	.145D-01	1.00
Sulfur (MMT)	.760	.126	.231	.194	.499D-01	.270D-01	1.00
Chlorine (MMT)	.665	.163	.148	.136	.982D-01	.612D-01	1.00
Magnesium (MMT)	.268	.539	.661D-01	.108	.198	.369D-01	1.00
Iron and steel scrap (MMT)	.350	.494	.828D-01	.965D-01	.107	.626D-01	1.00
Copper scrap (MMT)	.283	.462	.102	.106	.112	.147	1.00
Lead scrap (MMT)	.501	.303	.833D-01	.115	.932D-01	.135	1.00
Zinc scrap (MMT)	.353	.455	.786D-01	.836D-01	.124	.737D-01	1.00
Aluminum scrap (MMT)	.323	.449	.684D-01	.738D-01	.125	.108	1.00
Nickel scrap (KMT)	.362	.491	.106	.136	.107	.712D-01	1.00
Chromium scrap (KMT)	.382	.478	.124	.159	.998D-01	.753D-01	1.00
Gold scrap (MT)	.667	.200	.131	.178	.843D-01	.961D-01	1.00
Silver scrap (KMT)	.618	.258	.101	.143	.851D-01	.810D-01	1.00
Tungsten scrap (KMT)	.219	.606	.158	.872D-01	.523D-01	.510D-01	1.00
Mercury scrap (KMT)	.514	.296	.134	.120	.116	.597D-01	1.00
Tin scrap (KMT)	.511	.339	.841D-01	.864D-01	.874D-01	.646D-01	1.00
Magnesium scrap (MMT)	.266	.537	.647D-01	.101	.197	.367D-01	1.00

Table 6–8
2000 Distribution of Each Minerals Use Level by GNP Account

Mineral (Unit)	*Personal Consumption Expenditures*	*Gross Private Domestic Fixed Investment*	*Exports*	*Imports*	*Government Purchases, Nondefense*	*Government Purchases, Defense*	*Total, Excluding Changes in Business Inventories*
Iron (MMT)	.429	.476	.128	.133	.566D–01	.433D–01	1.00
Molybdenum (KMT)	.516	.431	.158	.196	.391D–01	.524D–01	1.00
Nickel (KMT)	.445	.471	.156	.176	.558D–01	.475D–01	1.00
Tungsten (KMT)	.233	.587	.224	.107	.326D–01	.291D–01	1.00
Manganese (KMT)	.454	.451	.126	.126	.551D–01	.409D–01	1.00
Chromium (KMT)	.491	.417	.182	.190	.555D–01	.436D–01	1.00
Copper (MMT)	.389	.463	.165	.170	.608D–01	.920D–01	1.00
Lead (MMT)	.593	.293	.118	.144	.535D–01	.862D–01	1.00
Zinc (MMT)	.422	.469	.105	.115	.683D–01	.516D–01	1.00
Gold (MT)	.727	.194	.218	.230	.388D–01	.520D–01	1.00
Silver (KMT)	.679	.261	.157	.184	.422D–01	.442D–01	1.00
Aluminum (MMT)	.415	.433	.110	.909D–01	.685D–01	.644D–01	1.00
Mercury (KMT)	.582	.277	.192	.178	.813D–01	.456D–01	1.00
Vanadium (KMT)	.344	.521	.154	.116	.490D–01	.470D–01	1.00
Platinum (MT)	.649	.218	.156	.163	.548D–01	.859D–01	1.00
Titanium (KMT)	.637	.221	.132	.106	.831D–01	.332D–01	1.00
Tin (KMT)	.575	.322	.121	.109	.471D–01	.454D–01	1.00
Silicon (KMT)	.425	.462	.156	.144	.490D–01	.518D–01	1.00
Fluorine (MMT)	.536	.366	.200	.223	.596D–01	.613D–01	1.00
Potash (MMT)	.760	.537D–01	.362	.227	.382D–01	.136D–01	1.00

Table 6–8 continued

Mineral (Unit)	*Personal Consumption Expenditures*	*Gross Private Domestic Fixed Investment*	*Exports*	*Imports*	*Government Purchases, Nondefense*	*Government Purchases, Defense*	*Total, Excluding Changes in Business Inventories*
Soda ash (MMT)	.742	.138	.148	.125	.565D-01	.407D-01	1.00
Boron (KMT)	.743	.147	.163	.134	.636D-01	.177D-01	1.00
Phosphate rock (MMT)	.771	.488D-01	.339	.212	.407D-01	.122D-01	1.00
Sulfur (MMT)	.739	.102	.319	.224	.429D-01	.209D-01	1.00
Chlorine (MMT)	.702	.139	.174	.127	.536D-01	.581D-01	1.00
Magnesium (MMT)	.372	.514	.127	.162	.114	.357D-01	1.00
Iron and steel scrap (MMT)	.429	.476	.128	.133	.566D-01	.433D-01	1.00
Copper scrap (MMT)	.384	.469	.165	.171	.608D-01	.920D-01	1.00
Lead scrap (MMT)	.593	.293	.118	.144	.535D-01	.862D-01	1.00
Zinc scrap (MMT)	.449	.429	.113	.115	.671D-01	.566D-01	1.00
Aluminum scrap (MMT)	.423	.440	.109	.110	.665D-01	.711D-01	1.00
Nickel scrap (KMT)	.445	.471	.156	.175	.558D-01	.475D-01	1.00
Chromium scrap (KMT)	.445	.456	.175	.175	.517D-01	.480D-01	1.00
Gold scrap (MT)	.727	.194	.218	.230	.388D-01	.520D-01	1.00
Silver scrap (KMT)	.679	.261	.157	.184	.422D-01	.442D-01	1.00
Tungsten scrap (KMT)	.233	.587	.224	.107	.326D-01	.291D-01	1.00
Mercury scrap (KMT)	.582	.277	.192	.178	.813D-01	.456D-01	1.00
Tin scrap (KMT)	.575	.322	.121	.109	.471D-01	.454D-01	1.00
Magnesium scrap (MMT)	.369	.514	.125	.158	.114	.356D-01	1.00

Table 6–9
2000 Levels of Direct and Indirect Minerals Requirements with Technological Change Assumptions

Mineral (Unit)	*Personal Consumption Expenditures*	*Gross Private Domestic Fixed Investment*	*Exports*	*Imports*	*Government Purchases, Nondefense*	*Government Purchases, Defense*	*Total, Excluding Changes in Business Inventories*
Iron (MMT)	67.8	77.9	21.5	20.8	8.95	7.26	163.
Molybdenum (KMT)	31.6	29.0	10.3	12.5	2.47	3.30	64.2
Nickel (KMT)	153.	168.	58.5	62.5	18.8	18.4	355.
Tungsten (KMT)	3.92	10.4	3.84	1.80	.548	.534	17.4
Manganese (KMT)	.107D+04	.107D+04	315.	293.	130.	107.	.239D+04
Chromium (KMT)	547.	492.	212.	212.	63.3	50.3	.115D+04
Copper (MMT)	1.72	1.93	.718	.751	.259	.421	4.30
Lead (MMT)	.726	.447	.167	.197	.744D−01	.125	1.34
Zinc (MMT)	1.48	1.65	.401	.414	.228	.179	3.52
Gold (MT)	21.1	20.6	8.78	9.11	4.67	3.34	49.4
Silver (KMT)	7.71	3.36	1.98	2.26	.524	.583	11.9
Aluminum (MMT)	7.75	5.10	1.56	1.54	.767	.730	14.4
Mercury (KMT)	.989	.531	.282	.330	.130	.828D-01	1.69
Vanadium (KMT)	5.28	8.46	2.29	1.74	.860	.724	15.9
Platinum (MT)	20.5	7.55	5.22	5.39	1.83	2.86	32.6
Titanium (KMT)	897.	315.	184.	146.	119.	46.0	.141D+04
Tin (KMT)	65.0	36.3	13.7	12.4	5.31	5.17	113.
Silicon (KMT)	535.	566.	193.	175.	58.4	62.4	.124D+04
Fluorine (MMT)	.904	.550	.312	.357	.924D-01	.987D-01	1.60
Potash (MMT)	6.66	.495	3.13	1.97	.338	.128	8.78

Table 6–9 continued

Mineral (Unit)	*Personal Consumption Expenditures*	*Gross Private Domestic Fixed Investment*	*Exports*	*Imports*	*Government Purchases, Nondefense*	*Government Purchases, Defense*	*Total, Excluding Changes in Business Inventories*
Soda ash (MMT)	10.4	2.05	2.11	1.83	.793	.556	14.1
Boron (KMT)	202.	45.5	45.0	39.2	17.1	5.47	276.
Phosphate rock (MMT)	47.1	3.17	21.1	13.3	2.41	.777	61.3
Sulfur (MMT)	17.4	2.39	7.49	5.31	.985	.507	23.5
Chlorine (MMT)	15.7	3.27	3.75	2.81	1.18	1.24	22.4
Magnesium (MMT)	.911	1.16	.296	.372	.257	.878D–01	2.34
Iron and steel scrap (MMT)	29.2	33.6	9.25	8.98	3.86	3.13	70.1
Copper scrap (MMT)	1.07	1.24	.452	.470	.166	.266	2.72
Lead scrap (MMT)	.471	.290	.108	.128	.483D–01	.812D–01	.871
Zinc scrap (MMT)	.405	.389	.105	.105	.583D–01	.506D–01	.904
Aluminum scrap (MMT)	1.48	.932	.280	.337	.136	.157	2.65
Nickel scrap (KMT)	64.3	67.3	22.5	25.3	7.97	6.87	144.
Chromium scrap (KMT)	99.4	108.	41.3	39.2	11.9	11.3	232.
Gold scrap (MT)	48.1	14.7	14.8	15.7	2.71	3.64	68.3
Silver scrap (KMT)	3.70	1.49	.879	1.03	.236	.247	5.53
Tungsten scrap (KMT)	.191	.462	.171	.852D–01	.270D–01	.240D–01	.789
Mercury scrap (KMT)	.295	.158	.840D–01	.984D–01	.386D–01	.247D–01	.502
Tin scrap (KMT)	17.3	11.1	4.02	3.59	1.57	1.53	31.9
Magnesium scrap (MMT)	.127D–01	.164D–01	.410D–02	.507D–02	.363D–02	.123D–02	.330D–01

Table 6–10
2000 Intensity of Use of Minerals by GNP Account with Technological Change Assumptions

Mineral (Unit)	*Personal Consumption Expenditures*	*Gross Private Domestic Fixed Investment*	*Exports*	*Imports*	*Government Purchases, Nondefense*	*Government Purchases, Defense*	*Total, Excluding Changes in Business Inventories*
Iron (MMT)	.363D–01	.193	.117	.142	.391D–01	.698D–01	.656D–01
Molybdenum (KMT)	.169D–01	.717D–01	.562D–01	.851D–01	.108D–01	.317D–01	.259D–01
Nickel (KMT)	.822D–01	.416	.319	.425	.821D–01	.177	.143
Tungsten (KMT)	.210D–02	.256D–01	.209D–01	.123D–01	.239D–02	.513D–02	.701D–02
Manganese (KMT)	.574	2.63	1.71	2.00	.567	1.03	.966
Chromium (KMT)	.293	1.22	1.16	1.45	.276	.484	.465
Copper (MMT)	.924D–03	.476D–02	.391D–02	.512D–02	.113D–02	.404D–02	.173D–02
Lead (MMT)	.389D–03	.111D–02	.908D–03	.134D–02	.325D–03	.120D–02	.541D–03
Zinc (MMT)	.794D–03	.407D–02	.219D–02	.282D–02	.995D–03	.172D–02	.142D–02
Gold (MT)	.113D–01	.510D–01	.478D–01	.620D–01	.204D–01	.321D–01	.199D–01
Silver (KMT)	.413D–02	.830D–02	.108D–01	.154D–01	.229D–02	.560D–02	.479D–02
Aluminum (MMT)	.415D–02	.126D–01	.850D–02	.105D–01	.334D–02	.701D–02	.579D–02
Mercury (KMT)	.530D–03	.131D–02	.154D–02	.225D–02	.566D–03	.796D–03	.679D–03
Vanadium (KMT)	.283D–02	.209D–01	.125D–01	.119D–01	.375D–02	.696D–02	.640D–02
Platinum (MT)	.110D–01	.187D–01	.284D–01	.367D–01	.797D–02	.274D–01	.131D–01
Titanium (KMT)	.480	.778	.999	.994	.519	.441	.570
Tin (KMT)	.348D–01	.897D–01	.745D–01	.846D–01	.232D–01	.497D–01	.456D–01
Silicon (KMT)	.287	1.40	1.05	1.19	.255	.599	.500
Fluorine (MMT)	.484D–03	.136D–02	.170D–02	.244D–02	.403D–03	.949D–03	.645D–03
Potash (MMT)	.357D–02	.122D–02	.170D–01	.134D–01	.148D–02	.123D–02	.354D–02

Table 6–10 continued

Mineral (Unit)	*Personal Consumption Expenditures*	*Gross Private Domestic Fixed Investment*	*Exports*	*Imports*	*Government Purchases, Nondefense*	*Government Purchases, Defense*	*Total, Excluding Changes in Business Inventories*
Soda ash (MMT)	.558D–02	.508D–02	.115D–01	.125D–01	.346D–02	.534D–02	.569D–02
Boron (KMT)	.108	.113	.245	.267	.747D–01	.526D–01	.111
Phosphate rock (MMT)	.252D–01	.784D–02	.115	.904D–01	.105D–01	.746D–02	.247D–01
Sulfur (MMT)	.934D–02	.592D–02	.408D–01	.362D–01	.430D–02	.487D–02	.947D–02
Chlorine (MMT)	.842D–02	.808D–02	.204D–01	.191D–01	.514D–02	.119D–01	.901D–02
Magnesium (MMT)	.488D–03	.288D–02	.161D–02	.253D–02	.112D–02	.844D–03	.945D–03
Iron and steel scrap (MMT)	.156D–01	.830D–01	.504D–01	.611D–01	.168D–01	.301D–01	.282D–01
Copper scrap (MMT)	.572D–03	.307D–02	.246D–02	.320D–02	.724D–03	.255D–02	.110D–02
Lead scrap (MMT)	.252D–03	.717D–03	.589D–03	.870D–03	.211D–03	.780D–03	.351D–03
Zinc scrap (MMT)	.217D–03	.963D–03	.574D–03	.715D–03	.254D–03	.486D–03	.364D–03
Aluminum scrap (MMT)	.793D–03	.230D–02	.152D–02	.229D–02	.593D–03	.151D–02	.107D–02
Nickel scrap (KMT)	.344D–01	.166	.123	.172	.348D–01	.660D–01	.579D–01
Chromium scrap (KMT)	.532D–01	.266	.225	.267	.518D–01	.108	.937D–01
Gold scrap (MT)	.257D–01	.364D–01	.809D–01	.107	.118D–01	.349D–01	.276D–01
Silver scrap (KMT)	.198D–02	.370D–02	.479D–02	.699D–02	.103D–02	.237D–02	.223D–02
Tungsten scrap (KMT)	.102D–03	.114D–02	.930D–03	.581D–03	.118D–03	.230D–03	.318D–03
Mercury scrap (KMT)	.158D–03	.391D–03	.458D–03	.670D–03	.169D–03	.237D–03	.202D–03
Tin scrap (KMT)	.927D–02	.273D–01	.219D–01	.244D–01	.686D–02	.147D–01	.129D–01
Magnesium scrap (MMT)	.680D–05	.406D–04	.223D–04	.346D–04	.158D–04	.118D–04	.133D–04

Table 6–11
2000 Distribution of Each Minerals Use Level by GNP Account with Technological Change Assumptions

	Personal Consumption Expenditures	*Gross Private Domestic Fixed Investment*	*Exports*	*Imports*	*Government Purchases, Nondefense*	*Government Purchases, Defense*	*Total, Excluding Changes in Business Inventories*
Iron (MMT)	.417	.479	.132	.128	.551D–01	.447D–01	1.00
Molybdenum (KMT)	.492	.452	.161	.195	.385D–01	.514D–01	1.00
Nickel (KMT)	.432	.474	.165	.176	.530D–01	.518D–01	1.00
Tungsten (KMT)	.225	.595	.221	.104	.315D–01	.307D–01	1.00
Manganese (KMT)	.447	.445	.131	.122	.543D–01	.447D–01	1.00
Chromium (KMT)	.474	.427	.184	.184	.549D–01	.437D–01	1.00
Copper (MMT)	.401	.448	.167	.175	.603D–01	.979D–01	1.00
Lead (MMT)	.541	.333	.124	.147	.554D–01	.933D–01	1.00
Zinc (MMT)	.421	.467	.114	.118	.647D–01	.509D–01	1.00
Gold (MT)	.427	.418	.178	.184	.944D–01	.676D–01	1.00
Silver (KMT)	.649	.282	.166	.190	.441D–01	.491D–01	1.00
Aluminum (MMT)	.539	.355	.109	.107	.533D–01	.508D–01	1.00
Mercury (KMT)	.587	.315	.167	.196	.770D–01	.492D–01	1.00
Vanadium (KMT)	.333	.533	.144	.110	.542D–01	.456D–01	1.00
Platinum (MT)	.630	.232	.160	.165	.561D–01	.876D–01	1.00
Titanium (KMT)	.634	.223	.130	.103	.842D–01	.325D–01	1.00
Tin (KMT)	.575	.321	.121	.110	.470D–01	.458D–01	1.00
Silicon (KMT)	.432	.457	.156	.141	.471D–01	.503D–01	1.00
Fluorine (MMT)	.565	.344	.195	.223	.577D–01	.617D–01	1.00
Potash (MMT)	.759	.564D–01	.356	.224	.385D–01	.146D–01	1.00

Table 6–11 continued

Mineral (Unit)	Personal Consumption Expenditures	Gross Private Domestic Fixed Investment	Exports	Imports	Government Purchases, Nondefense	Government Defense	Total, Excluding Changes in Business Inventories
Soda ash (MMT)	.739	.146	.149	.130	.562D–01	.394D–01	1.00
Boron (KMT)	.732	.165	.163	.142	.620D–01	.198D–01	1.00
Phosphate rock (MMT)	.769	.518D–01	.344	.217	.394D–01	.127D–01	1.00
Sulfur (MMT)	.742	.102	.319	.226	.419D–01	.216D–01	1.00
Chlorine (MMT)	.704	.146	.168	.126	.527D–01	.554D–01	1.00
Magnesium (MMT)	.389	.496	.126	.159	.110	.375D–01	1.00
Iron and steel scrap (MMT)	.417	.479	.132	.128	.551D–01	.447D–01	1.00
Copper scrap (MMT)	.392	.456	.166	.173	.609D–01	.976D–01	1.00
Lead scrap (MMT)	.541	.333	.124	.147	.554D–01	.933D–01	1.00
Zinc scrap (MMT)	.448	.431	.117	.116	.645D–01	.560D–01	1.00
Aluminum scrap (MMT)	.559	.352	.106	.127	.513D–01	.593D–01	1.00
Nickel scrap (KMT)	.448	.468	.157	.176	.555D–01	.478D–01	1.00
Chromium scrap (KMT)	.428	.464	.178	.169	.511D–01	.486D–01	1.00
Gold scrap (MT)	.704	.216	.217	.229	.397D–01	.532D–01	1.00
Silver scrap (KMT)	.669	.270	.159	.186	.427D–01	.446D–01	1.00
Tungsten scrap (KMT)	.242	.585	.216	.108	.343D–01	.304D–01	1.00
Mercury scrap (KMT)	.587	.315	.167	.196	.770D–01	.492D–01	1.00
Tin scrap (KMT)	.543	.346	.126	.112	.493D–01	.479D–01	1.00
Magnesium scrap (MMT)	.385	.497	.124	.154	.110	.373D–01	1.00

the level of minerals consumption. In chapter 7, which integrates the U.S. economy into the World Input-Output Model, the question of alternative annual growth rates in GDP is considered and the resulting changes in the levels of minerals consumption and production are presented.

Notes

1. Figure 6-1 shows the precise position and structure of the component parts of final demand within IEA/USMIN. Each of these column vectors represents a final user of the goods and services produced by the economy. Aggregating the final demand vectors into one column vector and then adding together the values of the goods and services in the rows of this column vector represents the value of final output of the economy for a particular year—GDP.

2. To compare the results of 1972 and 2000, the 1972 column vector of final demand of changes in business inventory was excluded from this computation. Changes in business inventories being a cyclical term in the national accounts, no attempt was made to project that bill of final demand for 2000.

3. The assumed growth rates by decade, taken from the Bureau of Labor Statistics base-scenario projections, were: an annual growth rate of 4.3 percent from 1972 to 1980; an annual growth rate of 3.4 percent from 1980 to 1990; and an annual growth rate of 2.55 percent from 1990 to 2000.

Part II
Nonfuel Minerals and the World Economy

7 Nonfuel Minerals and the World Economy through 2030

Judging by the events that we have witnessed throughout the 1970s, the U.S. economy has become increasingly reliant on foreign sources of some natural resources—particularly petroleum—as well as many consumer durables such as automobiles, televisions, cameras, and other household appliances. On the other hand, the United States is playing a critical role in the world economy not only in providing computers, aircraft, and other high-technology products including military hardware, but also in the production and export of many basic foodstuffs to both the developed and the developing countries. In addition, the United States is projected to become the major exporter of coal by the end of this century and into the first quarter of the next century [1,11,23,26].

Consequently, it is apparent that the United States is only a part—albeit a large one—of the wider world economy. Although decisions on energy policy made in other developed countries will affect the U.S. economy, increasing population levels in the developing countries will also make demands on the U.S. economy over the next fifty years. The ability of other developed and developing countries to compete with U.S. producers and the implementation of general agreements to reduce tariff and nontariff barriers have helped to create competitive conditions on a global level. As a result, the important role played by the United States both as a producer and as a consumer of minerals and other goods and services requires that the U.S. economy be incorporated within the broader framework of the world economy.

In 1973 the United Nations, with special financial support from the Netherlands, commissioned the construction of a general-purpose model of the world economy. To transform the vast collection of microeconomic facts that describe the world economy into an organized system from which macroeconomic projections of future growth could be made, the model relied on the method of input-output analysis. The World Input-Output Model [22] is capable of tracking many economic and environmental interdependencies throughout the world.

Despite its global scope, the model incorporates an unusual degree of detail. The world economy is subdivided into fifteen regions that fall into three main groups: the developed regions, characterized by relatively high per capita income (North America, Europe, the Soviet Union, Oceania, South Africa, and Japan); the less-developed regions rich in natural resources (the Middle East, some of the South American countries, and some countries in tropical Africa); and the less-developed countries with few resources.[1] The model describes each region in terms of 45 sectors of economic activity, including various types of agriculture, mining, manufacturing, utilities, construction, services, transportation, communication, and pollution abatement. Though each region is initially treated separately, the model provides a linkage mechanism that permits its user to trace the complex interconnections of trade, foreign investment, loans, interest payments, and foreign aid.[2]

The World Model has yielded several different projections of the future world economy, depending on the combination of several alternative assumptions about the rates of growth of population and of GDP per capita that is selected. For example, rather than a single rate of population growth for each region, alternative projections are based on a high, a medium, and a low rate. The degree of detail in the model permits the use of very specialized data—for example, about specific industries in specific regions—and results in relatively specific conclusions.

These conclusions, however, are not based solely on the alternative assumptions about the growth of population and per capita GDP; they are also subject to modification by other variables. For example, estimates of available reserves of various mineral resources differ widely. Instead of choosing only one estimate, alternative computations on the basis of different estimates were made, even though no attempts were made to predict possible discoveries of mineral resources. As a result, each scenario yields a hypothetical picture of the world economy in coming decades, a picture based on different combination of assumptions about some of the variables in the model.

Since its publication in 1976, the World Input-Output Model has been used in a number of specialized studies [23,24,25,26]. Each of these studies including this work necessitated extensive data collection and the incorporation of the prospective structural changes in the equations that describe the economic relationships in the model.

For example, for this study it was necessary to represent the United States and Canada separately.[3] We also incorporated the twenty new minerals into the structural framework of the World Model.[4] In addition, we have updated the original nine (fuel and nonfuel) minerals to ensure compatability with published regional production and consumption totals for 1980. Furthermore, to gain a first glimpse of the prospective regional and global minerals requirements for the world economy in the first thirty years of the next century, we extended the time frame of the model to the year 2030, with intermediate computations for 2010 and 2020.

Disaggregating the North American Region into Canada and the United States

To conform with the structural description of the other World Model regions, we prepared a set of basic data—an interindustry structural matrix, import coefficients, export shares, and other macroeconomic exogenous variables—for each of the two new regions, Canada and the United States.

The basic data sources for the interindustry part of the Canadian regional matrix was *The Input-Output Structure of the Canadian Economy, 1961–71* [40] and *The Input-Output Structure of the Canadian Economy in Constant 1961 Prices* [41]. The dominant constituent of the North American region was the United States. Therefore, the latest available input-output table of the U.S. economy [43] was aggregated to the World Model classification scheme to represent the 1970 base-year U.S. regional matrix. The interindustry matrix of coefficients for Canada was constructed from *Canada and the Future of the International Economy* [35].[5]

In addition to embedding their own interindustry matrices into each of the two new World Model regions, some region-specific exogeneously set macroeconomic variables had to be provided for the two new regions. For example, labor force and population projections had to be split from the North American region to reflect the U.S. and Canadian labor force estimates. At the same time, some activities, such as abatement or emissions levels or base-year resource output levels, had to be made compatible with the new regional classification. The regional import coefficients and export shares were also modified to reflect the import dependency of each country in the traded-goods sectors and the respective shares of the United States and Canada in the world commodity export pools.

The production, consumption, and net trade data for the nine fuel and nonfuel minerals (SE5–13) already represented in the World Model for the base-year were treated as follows: since the production figures for 1970 for all these minerals were known, they were set exogenously. Consumption and net trade figures for 1970 in both regions were also known, so any positive (negative) residuals were eliminated by scaling up (down) the consumption coefficients along the minerals rows so that the required amounts computed by the model agreed with the published amounts.

The following equation represents the balance of one commodity for one region:

$$\text{Output} = \text{Consumption} + \text{Exports} - \text{Imports}$$

Obviously, if values are known for any three terms, it is possible to solve for the fourth. Moreover, given the regional matrix of technology, minerals consumption can be computed if the vector of regional-sectoral outputs is known. In this case, any two of the three remaining variables—minerals output, exports, and imports—must be specified. Because projected mineral output levels are provided by the solution vector of IEA/USMIN, and World Model interregional resource trade is described on a net basis (that is, for each mineral, a region would have either zero net imports, zero net exports, or both), the equation can be solved.

To specify output levels, IEA/USMIN solution values for each decade were inserted for the new USA region in the World Model as exogenous variables, while the values of the difference—that is, North America less IEA/USMIN solution values—were entered for the new Canada region. This method ensured that the output levels given in the respecified World Model regional classification were consistent with both the old World Model North America totals and IEA/USMIN projections of U.S. production from 1980 to 2000. Because there was no IEA/USMIN solution corresponding to the World Model base-year (1970), output levels for that year were taken from empirical data as reported in the U.S. Bureau of Mines *Mineral Yearbook,* 1972 edition. (See table 8-2 for the corresponding export shares of these six nonfuel minerals that were assumed for the U.S. and Canada in the computations.)

Specification of trade data for the two new regions was facilitated by the World Model's net accounting system. Because the United States is a net importer of all nine of the World Model resources except coal, its World Model export shares could be set equal to zero. Similarly, the export share of coal for Canada was set to zero. Where both countries had net imports, a zero was assigned to each export share corresponding to that mineral. According to the previous equation, the level of imports could then be found from the solution of the model.

The 1980, 1990, and 2000 matrices for Canada and the United States were projected using the respective coefficient updaters from the North American region for 1980, 1990, and 2000 in the original World Model.

Incorporating the New Nonfuel Minerals within the Framework of the World Model

In addition to the projected levels of production, consumption, and net trade of the six nonfuel minerals originally represented in the World Input-Output Model, this study also traces the prospective levels of production, consumption, and trade for the twenty nonfuel minerals discussed in chapters 3 to 6. These new minerals will be needed to satisfy the requirements for future global economic growth projected under alternative assumptions until 2030.[6] The computations also track the cumulative outputs of the producing regions and possible regional exhaustion of resources of one or more minerals over the fifty-year time span.

In the World Model, a region can be a net importer, a net exporter, or self-sufficient in minerals. If a region is a net importer, its regional output is set exogenously for the decade in question and its export share is set to zero. If a region is a net exporter (to the world economy) a positive export share is set, regional imports are exogenously set to zero, and regional output is determined endogenously. Because of the long time frame of these projections, the following considerations were taken into account. First, regions that have been net exporters or self-sufficient in the past could become net importers because of resource exhaustion over the next fifty-years; second, regions that are or have been self-sufficient or net importers could become net exporters in the future as geological explorations identify new resource endowments. The first of these two possibilities is incorporated in the projections appearing in this study. The second, more speculative, possibility is not considered at this time.

Estimating the stock of resources to be worked over a fifty-year interval is a highly speculative undertaking. For example, as Tilton reports [42], the additions to the world reserve base over the past twenty-five years have been formidable. For example, world reserves of bauxite increased by 1,103 percent, lead by 433 percent, and potash by 1,525 percent. The reasons for this are well known: increased prices or lower costs of production brought about by technological change in mining, processing, or transporting the minerals, coupled with the discoveries of heretofore unknown deposits. With the already substantial amounts of mineral deposits identified on the ocean floor and the development of an appropriate technology to exploit these deposits profitably, continued supplements to the world's reserve base can be expected to occur in the future.

Consequently, the regional resource estimates chosen in these projections were the least restrictive estimate. That is, the figure included economic and subeconomic reserves and hypothetical resources.[7] Even under these optimistic estimates of endowments, some producing regions are expected to exhaust their endowments and turn to imports to satisfy their resource requirements.

Global resource estimates on a regional basis were obtained from the 1981 edition of *Mineral Facts and Problems* [47]. They are reproduced in table 5-40.

Updating the Fuel and Nonfuel Minerals Sectors for 1980 Regional and Global Balance

To bring the 1980 projections of minerals use and production in line with the actual levels of regional and global production, consumption, and net trade, we adjusted the exogenous variable file and the technical coefficients on a regional level to achieve regional as well as global balance for 1980. Data for 1979 provided the last set of complete information on the original three fuel and six nonfuel minerals available when the computations were made (July 1981).

The regional data for the nonfuel nonferrous minerals were taken from the 1980 edition of *Metal Statistics* [31).[8] Data for regional iron ore production, consumption, and trade were available from various Bureau of Mines reports. The 1979 data for regional energy production, consumption and trade were taken from the 1980 *Exxon World Energy Outlook* [7] and the *Statistical Review of the World Oil Industry for 1979,* published by the British Petroleum Company [2]. Energy conservation and substitution efforts already realized and those expected to occur by the end of the century in the developed countries were also incorporated into the coefficient matrices of the various regions for 1980, 1990, and 2000.[9]

Extending the World Model to 2030 with Intermediate Computations for 2010 and 2020

Because of the crucial role played by minerals in the world economy, they, like food production, have historically been given center stage as far as long-term projections are concerned [3,22,38]. Because of technological, economic, political, and institutional factors, long lead times are required to increase productive capacity for extracting and processing minerals. With the steady exhaustion of higher-grade deposits, the search for alternative sources or possible substitutes, and conservation efforts, twenty-year minerals supply-and-demand projections under alternative assumptions are appropriate and fifty-year projections, albeit under conditions of considerably more uncertainty, are not uncommon [20,26]. To address the need for long-term projections of nonfuel minerals, we extended the time frame of the World Model to 2030 on a region-by-region and mineral-by-mineral basis.

The rates of regional or world economic growth as determined by the regional rates of population growth, technological change, and savings, will, to a great extent, determine the global long-term requirements for nonfuel minerals. In projecting nonfuel minerals requirements from 2000 to 2030, no attempt was made to account for prospective changes in technology beyond the year 2000. Consequently, after that year, the projections are based solely on assumptions regarding population growth and the increase in regional GDP.

Two sets of assumptions portray an optimistic and a pessimistic future state of the world economy. For example, in the optimistic scenario, relatively high levels of economic activity in the developed countries were assumed by setting annual GDP growth at a relatively high rate of 3 percent per year from 2000 to 2030. For the less-developed countries, the optimistic scenario incorporates the low UN population projections from 1980 to

2030. For agriculture, the optimistic scenario is based on the assumption that several of the less-developed regions will become self-sufficient in food production by the year 2030. Such self-sufficiency eases one of the major constraints on the development process—the balance of payments.

The pessimistic scenario assumes that the GDP in the developed regions will increase at a modest 2 percent per year from 2000 to 2030. As far as the less-developed regions are concerned, pessimism is manifested by assuming that the high UN population projections from 1980 to 2030 will prevail, in addition to continued problems in achieving self-sufficiency in food production.

Table 7-1 displays the different combinations of assumptions used in the alternative projections. Each describes a possible path that the world economy can follow to the year 2030. Table 7-2 displays the different regional rates of GDP growth either assumed (in the case of the developed countries after the year 2000) or (in the case of most of the developing countries) resulting from the different sets of assumptions.

Table 7-3 presents the export shares assumed for the six nonfuel minerals (SE5-SE10) in 1980, 2000, and 2030. Tables 7-4 to 7-25 present the projected regional and global output, cumulative output (from a 1970 base year), net exports, and consumption from 1970 to 2030, decade by decade, for the six nonfuel minerals originally represented in the World Model under the alternative sets of assumptions listed in table 7-1.

The results of the alternative projections for future production and consumption of these six nonfuel minerals, yield several interesting findings. Only insignificant changes in the production and consumption of nonfuel minerals in the major consuming regions—that is, the developed countries—would be caused by a shift of the electricity feedstock from coal to electricity generated by nuclear power. In addition, when the conditions for the developing countries in the pessimistic scenario were replaced by those in the optimistic scenario, the levels of nonfuel minerals consumption remained practically unchanged. Consequently, only one set of figures for regional production, consumption, net exports and cumulative resource output is presented for the 1970-2000 period. However, two sets—one based on optimistic growth and the other based on pessimistic growth in the developed countries—are presented for the 2010-2030 period.

Furthermore, in the case of lead and zinc, the projections indicate that by the year 2010, the world's current resource base of these minerals will be exhausted.[10] This global exhaustion was accounted for in the model by making production of lead and zinc in the last two decades of the model's time frame endogenous for all regions. Presumably, by that time, new discoveries, greater efficiency in the use of the remaining stock of resources, substitution, or increased recycling of these minerals will provide a solution to this problem.

Tables 7-26 to 7-65 present the results of the computations for the

Table 7-1
Alternative Scenarios

Scenario	*1*	*2*	*3*	*4*
Population	Developing regions: Low U.N. population projections		Developing regions: High U.N. population projections	
	Developed regions: Medium U.N. population projections			
GDP	Developed regions: Endogenous until 2000 and set at 3 percent annual rate of growth from 2010 to 2030		Developed regions: Endogenous until 2000 then set at 2 percent annual rate growth from 2010 to 2030	
	Developing regions: Asia Centrally Planned, Latin America Low Income, and Middle East set exogenously; Asia Low Income set exogenously until 2000. Other developing regions: Endogenous		Same as in scenarios 1 and 2 except Tropical Africa set exogenously from 2010 to 2030	
Employment	Developed regions: Exogenous until 2000; endogenous from 2010 to 2030			
Investment	Endogenous			
Surplus or deficit in balance of payments	LDC/II (Excluding Asia Centrally Planned and Asia Low Income) and Tropical Africa: set to zero Asia Low Income: Endogenous until 2000; then set to zero from 2010 to 2030 Other regions: Endogenous		Same as in scenarios 1 and 2 except Tropical Africa set to zero until 2000 and then endogenous from 2010 to 2030	
Food imports	Asia Centrally Planned, Asia Low Income, Latin America Low Income, Latin America Medium Income: import coefficients decline to zero by the year 2030 All other regions: Import coefficients retained at 2000 values		All regions: Import coefficients retained at 2000 values	
Energy feedstock (from 2000 to 2030)	Increased use of nuclear energy in electricity generation in developed countries	Increased use of coal in electricity generation in developed countries	Same as in scenario 1	Same as in scenario 2

Note: See appendix H for the aggregated regional classification.

twenty new minerals for the 1970–2030 period, for both the optimistic and pessimistic scenarios for the last three decades. For each of these decades, production, consumption, and net trade levels are presented for the twenty minerals in the sixteen regions, in addition to a world total for each mineral in each decade. A table presenting the assumed export shares for 1970, 1980, 2000 and 2030 is also included with the projections.

Table 7-2
Average Annual Regional GDP Growth Rates from 1970 to 2030
(percentages)

				2000–2010		*2010–2020*		*2020–2030*	
Regions	*1970–1980*	*1980–1990*	*1990–2000*	*Optimistic*	*Pessimistic*	*Optimistic*	*Pessimistic*	*Optimistic*	*Pessimistic*
Africa, arid (AAF)	−1.43	2.35	3.41	3.61	3.03	3.23	2.58	3.17	2.49
Asia, centrally planned (ASC)	4.26	3.75	4.57	3.05	3.05	3.00	3.00	3.00	3.00
Asia, low income (ASL)	2.47	0.01	1.92	3.71	2.50	3.75	2.88	3.48	2.87
Canada (CAN)	3.83	4.39	2.50	3.00	2.00	3.00	2.00	3.00	2.00
Eastern Europe (EEM)	5.06	4.91	3.41	3.00	2.00	3.00	2.00	3.00	2.00
Asia, high income (JAP)	6.56	5.67	3.42	3.00	2.00	3.00	2.00	3.00	2.00
Latin America, resource rich (LAL)	6.37	7.99	4.37	5.62	5.62	2.70	2.70	2.70	2.70
Latin America, medium income (LAM)	1.80	5.51	6.14	3.98	2.94	3.63	2.85	3.43	2.73
Oceania (OCH)	4.15	5.34	2.00	3.00	2.00	3.00	2.00	3.00	2.00
Middle East-Africa (oil producers) (OIL)	17.42	10.78	7.95	2.41	2.16	5.11	5.14	4.18	4.16
Soviet Union (RUH)	5.47	5.18	4.14	3.00	2.00	3.00	2.00	3.00	2.00
Southern Africa (SAF)	2.86	2.76	8.42	3.00	2.00	3.00	2.00	3.00	2.00
Africa, tropical (TAF)	4.24	4.43	3.53	1.48	2.50	2.49	2.50	2.97	2.49
United States (USA)	3.36	2.63	2.50	3.00	2.00	3.00	2.00	3.00	2.00
Western Europe, high income (WEH)	3.15	4.56	2.20	3.00	2.00	3.00	2.00	3.00	2.00
Western Europe, medium income (WEM)	1.99	2.38	2.85	3.00	2.00	3.00	2.00	3.00	2.00

Note: The countries included in each of the sixteen regions are listed in appendix H.

Table 7–3
Export Shares for the Original Six Nonfuel Minerals Represented in the World Model

	Copper			Aluminum			Nickel			Zinc			Lead			Iron		
Regions	*1980*	*2000*	*2030*	*1980*	*2000*	*2030*	*1980*	*2000*	*2030*	*1980*	*2000*	*2030*[a]	*1980*	*2000*	*2030*[a]	*1980*	*2000*	*2030*
AAF	0	0	0	0	0	0	0	0	0	0	0	0	.07	.05	0	.02	0	0
ASC	0	0	0	0	0	0	0	0	0	0	0	0	0	0	0	0	0	0
ASL	.05	0	0	.03	.03	.03	.13	.13	.14	0	0	0	0	0	0	.15	.17	.17
CAN	.07	.10	.17	0	0	0	.30	.30	.31	.33	.37	0	.15	.23	0	.04	.06	.06
EEM	0	0	0	.02	0	0	0	0	0	0	0	0	0	0	0	0	0	0
JAP	0	0	0	0	0	0	0	0	0	0	0	0	0	0	0	0	0	0
LAL	.10	.16	.16	.31	.30	.25	0	0	0	.21	.19	0	.25	.27	0	.10	.10	.10
LAM	.22	.33	.32	0	.02	.09	.07	.07	.07	.11	.10	0	.06	0	0	.17	.23	.23
OCH	.03	.02	.09	.35	.36	.30	.36	.36	.38	.23	.24	0	.41	.44	0	.13	.02	.02
OIL	0	0	0	0	0	0	0	0	0	.02	0	0	.03	0	0	0	0	0
RUH	0	0	0	0	0	0	.07	.07	.07	0	0	0	0	0	0	.23	.28	.28
SAF	.03	0	0	0	0	0	.04	.04	0	0	0	0	0	0	0	.10	0	0
TAF	.50	.39	.20	.27	.27	.24	.03	.03	.03	.09	.10	0	.02	.01	0	.15	.14	.14
USA	0	0	.06	0	0	0	0	0	0	0	0	0	0	0	0	0	0	0
WEH	0	0	0	0	0	0	0	0	0	0	0	0	0	0	0	0	0	0
WEM	0	0	0	.02	.02	.09	0	0	0	.01	0	0	.01	0	0	0	0	0

Note: The countries included in each of the sixteen regions are listed in appendix H.

[a]Regional export shares of zinc and lead for the year 2030 were set to zero due to global exhaustion of these two minerals.

Table 7-4
1970 Nonfuel Minerals Production, Cumulative Output, and Net Exports

		AAF70	ASC70	ASL70	CAN70	EEM70	JAP70	LAL70	LAM70
RESOURCE OUTPUTS									
COPPER	MT	0.0*	0.1	0.2	1.0	0.1	0.1	0.2	0.8
BAUXITE	MT	0.0	0.1	0.7	0.0	0.5	0.0	5.4	0.1
NICKEL	KT	0.1	0.0	225.0	274.2	4.7	0.0	1.8	37.7
ZINC	MT	0.0*	0.2	0.0*	1.3	0.3	0.3	0.4	0.3
LEAD	MT	0.1	0.2	0.0*	0.5	0.2	0.1	0.2	0.2
IRON	MT	8.2	26.0	22.9	28.7	4.3	0.9	20.4	28.8
CUMULATIVE RESOURCE OUTPUT AT END OF PERIOD									
COPPER	MT	0.	0.	0.	0.	0.	0.	0.	0.
BAUXITE	MT	0.	0.	0.	0.	0.	0.	0.	0.
NICKEL	KT	0.	0.	0.	0.	0.	0.	0.	0.
ZINC	MT	0.	0.	0.	0.	0.	0.	0.	0.
LEAD	MT	0.	0.	0.	0.	0.	0.	0.	0.
IRON	MT	0.	0.	0.	0.	0.	0.	0.	0.
NET EXPORTS OF RESOURCES									
COPPER	MT	0.0*	-0.0*	0.2	0.4	-0.1	-0.7	0.2	0.4
BAUXITE	MT	-0.0*	-0.0*	0.2	-0.2	0.2	-1.1	5.2	-0.1
NICKEL	KT	0.0	-0.3	162.7	160.5	0.0	-103.1	0.0	-1.6
ZINC	MT	0.0*	-0.0*	-0.2	1.0	-0.0*	-0.5	0.4	0.2
LEAD	MT	0.1	-0.0*	-0.0*	0.4	-0.0*	-0.1	0.2	0.1
IRON	MT	7.2	-0.0*	19.4	4.4	-22.4	-57.6	14.7	23.8

		OCH70	OIL70	RUH70	SAF70	TAF70	USA70	WEH70	WEM70
RESOURCE OUTPUTS									
COPPER	MT	0.1	0.0*	0.9	0.1	1.1	1.4	0.1	0.2
BAUXITE	MT	1.9	0.0	0.9	0.0	0.7	0.4	0.7	0.9
NICKEL	KT	29.8	0.0	110.0	11.6	18.9	14.3	5.1	8.6
ZINC	MT	0.5	0.1	0.5	0.0	0.2	0.4	0.5	0.2
LEAD	MT	0.4	0.1	0.5	0.0	0.1	0.5	0.3	0.2
IRON	MT	32.8	1.5	117.0	5.5	23.2	48.8	49.0	6.8
CUMULATIVE RESOURCE OUTPUT AT END OF PERIOD									
COPPER	MT	0.	0.	0.	0.	0.	0.	0.	0.
BAUXITE	MT	0.	0.	0.	0.	0.	0.	0.	0.
NICKEL	KT	0.	0.	0.	0.	0.	0.	0.	0.
ZINC	MT	0.	0.	0.	0.	0.	0.	0.	0.
LEAD	MT	0.	0.	0.	0.	0.	0.	0.	0.
IRON	MT	0.	0.	0.	0.	0.	0.	0.	0.
NET EXPORTS OF RESOURCES									
COPPER	MT	0.1	-0.0*	0.1	0.1	1.1	-0.1	-1.7	-0.1
BAUXITE	MT	1.7	-0.0*	-0.2	-0.1	0.2	-3.8	-2.4	0.6
NICKEL	KT	-1.0	0.0	0.0	6.2	18.8	-130.4	-108.0	-3.8
ZINC	MT	0.4	0.1	0.0*	-0.0*	0.2	-0.7	-1.0	0.0
LEAD	MT	0.3	0.0*	0.0*	0.0*	0.0	-0.2	-0.8	0.0
IRON	MT	24.2	1.4	32.2	2.6	22.3	-24.1	-47.1	-1.0

Notes: The countries included in each of the sixteen regions are listed in appendix H.
All values appearing in the bauxite row represent aluminum content therein.
*Rounded off to zero.

Table 7-5
1970 Regional Nonfuel Minerals Consumption

		AAF70	ASC70	ASL70	CAN70	EEM70	JAP70	LAL70	LAM70	OCH70
COPPER	MT	.0	.1	.0	.6	.2	.8	.0	.4	.0
BAUXITE	MT	.0	.1	.5	.2	.3	1.1	.2	.2	.2
NICKEL	MT	.1	.3	62.3	113.7	4.7	103.1	1.8	39.3	30.8
ZINC	MT	.0	.2	.2	.3	.3	.8	.0	.1	.1
LEAD	MT	.0	.2	.0	.1	.2	.2	.0	.1	.1
IRON	MT	1.0	26.0	3.5	24.3	26.7	58.5	5.7	5.0	8.6

Notes: The countries included in each of the sixteen regions are listed in appendix H.
All values appearing in the bauxite row represent aluminum content therein.

Table 7-6
1980 Nonfuel Minerals Production, Cumulative Output, and Net Trade

		AAF80	ASC80	ASL80	CAN80	EEM80	JAP80	LAL80
RESOURCE OUTPUTS								
COPPER	MT	0.0*	0.2	0.5	0.5	0.5	0.1	0.5
BAUXITE	MT	0.0	0.4	1.1	0.0	1.0	0.0	3.2
NICKEL	KT	0.0	11.0	87.6	168.7	10.6	0.0	1.5
ZINC	MT	0.0*	0.3	0.1	0.9	0.4	0.2	0.5
LEAD	MT	0.2	0.2	0.1	0.5	0.3	0.1*	0.7
IRON	MT	4.1	33.4	32.4	34.9	12.0	0.6	21.5
CUMULATIVE RESOURCE OUTPUT AT END OF PERIOD								
COPPER	MT	0.*	2.	3.	7.	3.	1.	4.
BAUXITE	MT	0.	3.	9.	0.	8.	0.	43.
NICKEL	KT	1.*	55.	1563.	2215.	76.	0.	17.
ZINC	MT	0.*	3.	1.	11.	4.	3.	4.
LEAD	MT	1.	2.	0.*	5.	3.	1.	4.
IRON	MT	62.	297.	276.	318.	81.	8.	210.
NET EXPORTS OF RESOURCES								
COPPER	MT	0.0	-0.2	0.2	0.3	-0.0*	-1.3	0.4
BAUXITE	MT	-0.0*	-0.3	0.3	-0.3	0.2	-1.8	3.1
NICKEL	KT	-0.5	-4.9	1.9	156.5	-18.0	-132.4	0.0
ZINC	MT	0.0	0.1	-0.4	0.7	-0.0*	-0.6	0.4
LEAD	MT	0.2	0.0	-0.2	0.4	-0.1	-0.2	0.6
IRON	MT	3.5	0.0	26.1	7.0	-33.5	-111.1	17.4

Notes: The countries included in each of the sixteen regions are listed in appendix H.
All values appearing in the bauxite row represent aluminum content therein.
*Rounded off to zero.

OIL70	RUH70	SAF70	TAF70	USA70	WEH70	WEM70	WORLD
.0	.8	.0	.0	1.5	1.8	.3	6.5
.0	1.1	.1	.5	4.2	3.1	.3	12.1
.0	110.0	5.4	.1	144.7	113.1	12.4	741.8
.0	.5	.0	.0	1.1	1.5	.2	5.3
.1	.5	.0	.2	.6	1.3	.2	3.8
.1	84.8	2.9	.9	72.9	96.1	7.8	424.8

LAM80	OCH80	OIL80	RUH80	SAF80	TAF80	USA80	WEH80	WEM80
1.3	0.3	0.0*	1.3	0.1	2.1	2.2	0.1	0.1
0.4	3.7	0.0	1.6	0.0	2.7	0.7	0.5	0.9
45.7	195.0	0.0	178.1	32.9	15.4	16.0	7.0	16.1
0.5	0.6	0.2	1.0	0.1	0.2	0.6	0.7	0.2
0.4	1.1	0.2	0.6	0.0*	0.1	0.6	0.3	0.3
46.7	45.2	7.4	170.8	5.0	26.8	59.7	97.6	11.4
10.	2.	0.*	11.	1.	16.	18.	1.	2.
3.	28.	0.	12.	0.	17.	5.	6.	9.
417.	1124.	0.	1440.	223.	171.	152.	61.	124.
4.	5.	1.	8.	0.*	2.	5.	6.	2.
3.	8.	1.	6.	0.*	1.	6.	3.	2.
377.	390.	45.	1439.	53.	250.	542.	733.	91.
0.9	0.1	-0.1	0.0	0.1	2.1	-0.0	-2.4	-0.2
0.0	3.5	-0.1	-0.3	-0.1	2.7	-4.4	-2.9	0.2
35.9	189.3	-2.0	35.9	23.8	15.3	-167.6	-207.7	4.4
0.2	0.5	0.0*	0.0*	0.0*	0.2	-0.4	-0.8	0.0
0.2	1.0	0.1	-0.2	0.0*	0.1	-0.7	-1.1	0.0
31.4	22.6	0.0	41.8	1.7	26.1	-25.9	0.0	-7.2

Table 7–7
1980 Regional Nonfuel Minerals Consumption

		AAF70	ASC80	ASL80	CAN80	EEM80	JAP80	LAL80	LAM80	OCH80
COPPER	MT	.0	.4	.3	.2	.5	1.4	.1	.4	.2
BAUXITE	MT	.0	.7	.8	.3	.8	1.8	.1	.4	.2
NICKEL	MT	.5	15.9	15.7	12.2	28.6	132.4	1.5	9.8	5.7
ZINC	MT	.0	.2	.5	.2	.4	.8	.1	.3	.1
LEAD	MT	.0	.2	.3	.1	.4	.3	.1	.2	.1
IRON	MT	.6	33.4	6.3	27.9	45.5	111.7	4.1	15.3	22.6

Notes: The countries included in each of the sixteen regions are listed in appendix H.
All values appearing in the bauxite row represent aluminum content therein.

Table 7–8
1990 Nonfuel Minerals Production, Cumulative Output, and Net Trade

		AAF90	ASC90	ASL90	CAN90	EEM90	JAP90	LAL90
RESOURCE OUTPUTS								
COPPER	MT	0.0*	0.3	0.6	1.2	0.3	0.2	1.1
BAUXITE	MT	0.0	0.1	1.5	0.0	1.4	0.0	5.5
NICKEL	KT	0.7	0.0	142.8	281.0	0.0	0.0	2.8
ZINC	MT	0.0*	0.2	0.1	1.9	0.6	0.3	1.0
LEAD	MT	0.3	0.3	0.0*	1.1	0.0	0.1	1.4
IRON	MT	5.2	51.3	70.1	61.1	12.3	1.5	50.2
CUMULATIVE RESOURCE OUTPUT AT END OF PERIOD								
COPPER	MT	0.*	4.	9.	16.	7.	2.	11.
BAUXITE	MT	0.	5.	23.	0.	20.	0.	87.
NICKEL	KT	4.	110.	2715.	4464.	130.	0.	38.
ZINC	MT	0.*	5.	2.	24.	9.	5.	12.
LEAD	MT	4.	4.	1.	12.	4.	1.	15.
IRON	MT	108.	721.	788.	798.	203.	18.	568.
NET EXPORTS OF RESOURCES								
COPPER	MT	0.0	-0.3	0.2	0.9	-0.7	-2.3	1.0
BAUXITE	MT	-0.0*	-0.9	0.6	-0.5	0.0	-3.4	5.4
NICKEL	KT	0.0	-23.2	121.5	264.4	-50.2	-246.8	0.0
ZINC	MT	0.0	-0.1	-0.5	1.6	-0.1	-1.1	0.9
LEAD	MT	0.3	0.0*	-0.2	1.0	-0.8	-0.3	1.3
IRON	MT	4.0	0.0	63.9	20.0	-64.0	-178.1	39.9

Notes: The countries included in each of the sixteen regions are listed in appendix H.
All values appearing in the bauxite row represent aluminum content therein.
*Rounded off to zero.

OIL80	RUH80	SAF80	TAF80	USA80	WEH80	WEM80	WORLD
.1	1.3	.0	.0	2.2	2.5	.3	9.9
.1	1.9	.1	.0	5.1	3.4	.7	16.4
2.0	142.2	9.1	.1	183.6	214.7	11.7	785.7
.2	1.0	.1	.0	1.0	1.5	.2	6.6
.1	.8	.0	.0	1.3	1.4	.3	5.6
7.4	129.0	3.3	.7	85.6	97.6	18.6	609.6

LAM90	OCH90	OIL90	RUH90	SAF90	TAF90	USA90	WEH90	WEM90
2.9	0.5	0.0	1.7	0.1	4.7	2.9	0.2	0.3
1.2	6.9	0.0	1.3	0.0	5.0	1.0	0.8	1.4
79.6	330.7	0.0	296.9	53.7	26.0	23.0	0.0	0.0
1.0	1.3	0.0	1.0	0.0	0.5	1.0	1.0	0.4
0.6	2.3	0.0	1.5	0.0	0.1	0.8	0.0	0.4
114.2	63.3	9.4	307.8	4.7	57.0	78.9	56.0	26.8
31.	6.	0.*	26.	3.	50.	44.	3.	4.
11.	81.	0.	27.	0.	56.	14.	12.	20.
1044.	3752.	0.	3815.	656.	378.	347.	96.	204.
12.	15.	5.	18.	1.	5.	14.	14.	5.
8.	25.	6.	16.	0.*	2.	13.	4.	6.
1182.	932.	128.	3832.	101.	669.	1235.	1501.	282.
2.1	0.3	-0.4	-0.7	0.0	4.7	-0.0	-4.3	-0.4
0.4	6.5	-0.7	-2.0	-0.1	5.0	-5.5	-5.1	0.4
60.7	319.8	-8.2	60.7	40.2	25.9	-179.3	-364.9	-20.6
0.5	1.1	0.0	-0.7	-0.1	0.5	-0.4	-1.5	0.1
0.2	2.1	0.0	-0.1	-0.0*	0.1	-1.0	-2.5	0.0
83.9	24.0	-21.4	107.8	0.0	55.9	-31.5	-104.3	0.0

Table 7-9
1990 Regional Nonfuel Minerals Consumption

		AAF90	ASC90	ASL90	CAN90	EEM90	JAP90	LAL90	LAM90	OCH90
COPPER	MT	.0	.6	.4	.3	1.0	2.5	.1	.8	.2
BAUXITE	MT	.0	1.0	.9	.5	1.4	3.4	.1	.8	.4
NICKEL	MT	.7	23.2	21.3	16.6	50.2	246.8	2.8	18.9	10.9
ZINC	MT	.0	.3	.6	.3	.7	1.4	.1	.5	.2
LEAD	MT	.0	.3	.2	.1	.8	.4	.1	.4	.2
IRON	MT	1.2	51.3	6.2	41.1	76.3	179.6	10.3	30.3	39.3

Notes: The countries included in each of the sixteen regions are listed in appendix H.
All values appearing in the bauxite row represent aluminum content therein.

Table 7-10
2000 Nonfuel Minerals Production, Cumulative Output, and Net Trade

		AAF20	ASC20	ASL20	CAN20	EEM20	JAP20	LAL20
RESOURCE OUTPUTS								
COPPER	MT	0.0*	0.0	0.6	1.7	0.4	0.3	2.1
BAUXITE	MT	0.0	0.2	2.1	0.0	2.0	0.0	7.1
NICKEL	KT	0.9	0.0	184.4	359.2	0.0	0.0	6.1
ZINC	MT	0.0*	0.6	0.9	2.2	1.1	0.3	1.2
LEAD	MT	0.2	0.5	0.4	0.9	1.1	0.0	1.1
IRON	MT	1.6	99.9	86.6	78.1	20.1	2.0	63.1
CUMULATIVE RESOURCE OUTPUT AT END OF PERIOD								
COPPER	MT	0.*	5.	15.	31.	10.	5.	27.
BAUXITE	MT	0.	7.	41.	0.	37.	0.	150.
NICKEL	KT	12.	110.	4351.	7665.	130.	0.	82.
ZINC	MT	1.	9.	6.	45.	17.	8.	23.
LEAD	MT	7.	8.	3.	22.	10.	2.	27.
IRON	MT	142.	1477.	1572.	1494.	365.	36.	1134.
NET EXPORTS OF RESOURCES								
COPPER	MT	0.0	-1.1	0.0	1.3	-1.0	-3.1	2.0
BAUXITE	MT	-0.0*	-1.5	0.7	-0.6	0.0	-4.5	6.9
NICKEL	KT	0.0	-38.6	156.7	341.0	-69.2	-306.1	0.0
ZINC	MT	0.0	0.0	0.0	2.0	0.0	-1.6	1.0
LEAD	MT	0.2	0.0	0.0	0.8	0.0	-0.6	0.9
IRON	MT	0.0	0.0	77.6	27.4	-82.7	-241.2	45.6

Notes: The countries included in each of the sixteen regions are listed in appendix H.
All values appearing in the bauxite row represent aluminum content therein.
*Rounded off to zero.

OIL90	RUH90	SAF90	TAF90	USA90	WEH90	WEM90	WORLD
.4	2.4	.1	.0	2.9	4.5	.7	16.9
.7	3.3	.1	.0	6.5	5.9	1.0	26.0
8.2	236.2	13.5	.1	202.3	364.9	20.6	1237.2
.0	1.7	.1	.0	1.4	2.5	.3	10.1
.0	1.6	.0	.0	1.8	2.5	.4	8.8
30.8	200.0	4.7	1.1	110.4	160.3	26.8	969.7

LAM20	OCH20	OIL20	RUH20	SAF20	TAF20	USA20	WEH20	WEM20
6.1	0.6	0.0	2.3	0.3	5.0	3.7	0.2	0.4
2.3	8.7	0.3	2.1	0.0	6.5	1.4	1.0	2.2
120.6	423.7	0.0	417.8	90.6	33.5	29.8	0.0	0.0
1.9	1.5	1.0	2.5	0.2	0.6	1.3	0.0	0.0
1.0	1.6	0.0	2.2	0.1	0.1	0.9	3.0	0.6
171.8	52.5	24.3	432.0	14.0	65.8	93.7	140.2	45.8
76.	11.	0.*	47.	4.	99.	77.	5.	7.
29.	159.	2.	44.	0.	113.	26.	21.	38.
2044.	7524.	0.	7389.	1378.	675.	611.	96.	204.
27.	29.	10.	36.	2.	10.	25.	19.	7.
16.	44.	6.	35.	1.	2.	21.	19.	11.
2612.	1512.	297.	7531.	195.	1283.	2098.	2482.	645.
4.2	0.3	-0.7	-1.2	0.0	5.0	0.0	-5.1	-0.8
0.5	8.3	-1.0	-2.7	-0.3	6.4	-6.8	-5.9	0.5
78.3	412.4	-14.5	78.3	51.8	33.4	-288.8	-399.4	-35.4
0.5	1.3	0.0	0.0	0.0	0.5	-0.4	-2.9	-0.5
0.0	1.5	-1.2	0.0	0.0	0.0	-1.6	0.0	0.0
105.0	9.1	-33.9	127.8	0.0	63.9	-47.1	-51.4	0.0

Table 7-11
2000 Regional Nonfuel Minerals Consumption

		AAF20	ASC20	ASL20	CAN20	EEM20	JAP20	LAL20	LAM20	OCH20
COPPER	MT	.0	1.1	.6	.4	1.4	3.4	.1	1.9	.3
BAUXITE	MT	.0	1.7	1.4	.6	2.0	4.5	.2	1.8	.4
NICKEL	MT	.9	38.6	27.7	18.2	69.2	306.1	6.1	42.3	11.3
ZINC	MT	.0	.6	.9	.2	1.1	1.9	.2	1.4	.2
LEAD	MT	.0	.5	.4	.1	1.1	.6	.2	1.0	.1
IRON	MT	1.6	99.9	9.0	50.7	102.8	243.2	15.5	66.8	43.4

Notes: The countries included in each of the sixteen regions are listed in appendix H.
All values appearing in the bauxite row represent aluminum content therein.

Table 7-12
2010 Nonfuel Minerals Production, Cumulative Output, and Net

		AAF21	ASC21	ASL21	CAN21	EEM21	JAP21	LAL21
RESOURCE OUTPUTS								
COPPER	MT	0.0*	0.0	0.8	3.1	0.5	0.3	2.6
BAUXITE	MT	0.0	0.4	2.8	0.0	2.7	0.0	8.7
NICKEL	KT	1.4	0.0	253.5	503.6	0.0	0.0	10.2
ZINC	MT	0.0*	0.6	1.2	5.4	0.0	0.0	2.9
LEAD	MT	1.0	0.6	0.0	4.5	1.4	0.0	0.2
IRON	MT	2.5	116.3	121.8	109.9	32.7	2.7	93.9
CUMULATIVE RESOURCE OUTPUT AT END OF PERIOD								
COPPER	MT	1.	5.	22.	55.	15.	8.	50.
BAUXITE	MT	0.	10.	66.	0.	61.	0.	229.
NICKEL	KT	24.	110.	6543.	11981.	130.	0.	163.
ZINC	MT	1.	12.	12.	83.	17.	16.	43.
LEAD	MT	12.	11.	3.	49.	17.	5.	28.
IRON	MT	162.	2054.	2618.	2435.	629.	59.	1919.
NET EXPORTS OF RESOURCES								
COPPER	MT	0.0	-1.3	0.0	2.6	-1.3	-4.2	2.4
BAUXITE	MT	-0.1	-1.8	0.9	-0.8	0.0	-5.9	8.3
NICKEL	KT	0.0	-49.4	219.4	477.5	-90.4	-404.5	0.0
ZINC	MT	0.0	0.0	0.0	5.1	-1.4	-2.5	2.6
LEAD	MT	0.9	0.0	-0.6	4.3	0.0	-0.7	0.0
IRON	MT	0.0	0.0	109.0	38.5	-101.9	-319.5	64.1

Notes: The countries included in each of the sixteen regions are listed in appendix H.
All values appearing in the bauxite row represent aluminum content therein.
*Rounded off to zero.

OIL20	RUH20	SAF20	TAF20	USA20	WEH20	WEM20	WORLD
.7	3.5	.3	.0	3.7	5.3	1.2	23.9
1.3	4.8	.3	.1	8.2	6.9	1.7	35.9
14.5	339.5	38.8	.1	318.6	399.4	35.4	1666.7
1.0	2.5	.2	.1	1.7	2.9	.5	15.4
1.2	2.2	.1	.1	2.5	3.0	.6	13.7
58.2	304.2	14.0	1.9	140.8	191.6	45.8	1389.4

Trade—Optimistic Scenario

LAM21	OCH21	OIL21	RUH21	SAF21	TAF21	USA21	WEH21	WEM21
7.4	1.7	0.0	4.5	0.3	3.0	6.1	0.0	0.5
5.3	10.6	0.0	2.3	0.0	7.7	1.3	1.2	5.0
165.2	592.9	0.0	541.1	46.5	46.9	36.0	0.0	0.0
3.1	3.6	0.0	0.0	0.3	1.4	1.3	0.0	0.0
1.3	0.2	0.0	0.0	2.0	0.2	1.4	4.0	0.8
237.9	73.3	24.3	567.8	17.0	91.6	137.8	140.2	57.7
144.	23.	0.*	80.	7.	139.	126.	6.	12.
67.	256.	3.	65.	0.	184.	39.	32.	74.
3474.	12609.	0.	12185.	2064.	1077.	940.	96.	204.
52.	55.	10.	36.	3.	20.	39.	34.	10.
22.	46.	12.	35.	11.	4.	32.	39.	15.
4666.	2141.	541.	12534.	350.	2072.	3256.	3884.	1163.
5.0	1.4	-0.7	0.0	0.0	3.0	1.0	-7.0	-0.9
2.9	10.0	-1.3	-3.9	-0.4	7.7	-10.3	-8.2	2.9
109.7	577.6	-14.6	109.7	0.0	46.8	-399.0	-538.0	-44.8
1.4	3.3	-0.9	-3.2	0.0	1.4	-1.1	-3.9	-0.7
0.0	0.0	-1.3	-2.8	1.9	0.2	-1.8	0.0	0.0
147.4	12.8	-38.4	179.5	0.0				

Table 7-13
2010 Regional Nonfuel Minerals Consumption—Optimistic Scenario

		AAF21	ASC21	ASL21	CAN21	EEM21	JAP21	LAL21	LAM21	OCH21	OIL21
COPPER	MT	.0	1.3	.8	.5	1.8	4.5	.2	2.4	.3	.7
BAUXITE	MT	.1	2.2	1.9	.8	2.7	5.9	.4	2.4	.6	1.3
NICKEL	KT	1.4	49.4	34.1	26.1	90.4	404.5	10.2	55.5	15.3	14.6
ZINC	MT	.0	.6	1.2	.3	1.4	2.5	.3	1.7	.3	.9
LEAD	MT	.1	.6	.6	.2	1.4	.7	.2	1.3	.2	1.3
IRON	MT	2.5	116.3	12.8	71.4	134.6	322.2	29.8	90.5	60.5	62.7

Notes: The countries included in each of the sixteen regions are listed in appendix H.
All values appearing in the bauxite row represent aluminum content therein.

Table 7-14
2010 Nonfuel Minerals Production, Cumulative Output, and Net

RESOURCE OUTPUTS		AAF21	ASC21	ASL21	CAN21	EEM21	JAP21	LAL21
COPPER	MT	0.0*	0.0	0.7	2.7	0.5	0.3	2.3
BAUXITE	MT	0.0	0.4	2.4	0.0	2.3	0.0	7.5
NICKEL	KT	1.3	0.0	223.2	444.4	0.0	0.0	9.9
ZINC	MT	0.0*	0.6	0.9	4.7	0.0	0.0	2.5
LEAD	MT	0.9	0.6	0.0	3.9	1.3	0.0	0.2
IRON	MT	2.3	113.8	97.7	93.4	32.7	2.7	81.6
CUMULATIVE RESOURCE OUTPUT AT END OF PERIOD								
COPPER	MT	1.	5.	21.	53.	15.	8.	49.
BAUXITE	MT	0.	10.	64.	0.	59.	0.	223.
NICKEL	KT	23.	110.	6390.	11682.	130.	0.	162.
ZINC	MT	1.	12.	11.	79.	17.	16.	41.
LEAD	MT	11.	11.	3.	46.	16.	5.	28.
IRON	MT	161.	2046.	2494.	2351.	629.	59.	1858.
NET EXPORTS OF RESOURCES								
COPPER	MT	0.0	-1.2	0.0	2.3	-1.0	-3.6	2.1
BAUXITE	MT	-0.1	-1.8	0.7	-0.7	0.0	-5.2	7.1
NICKEL	KT	0.0	-49.2	193.6	421.4	-79.7	-355.5	0.0
ZINC	MT	0.0	0.0	0.0	4.4	-1.2	-2.2	2.3
LEAD	MT	0.8	0.0	-0.5	3.8	0.0	-0.6	0.0
IRON	MT	0.0	0.0	88.0	31.1	-84.3	-277.6	51.8

Notes: The countries included in each of the sixteen regions are listed in appendix H.
All values appearing in the bauxite row represent aluminum content therein.
*Rounded off to zero.

RUH21	SAF21	TAF21	USA21	WEH21	WEM21	WORLD
4.5	.3	.0	5.1	7.0	1.4	30.8
6.2	.4	.0	11.6	9.4	2.1	48.0
431.4	46.5	.1	435.0	538.0	44.8	2197.3
3.2	.3	.0	2.4	3.9	.7	19.7
2.8	.1	.0	3.2	4.0	.8	17.5
388.3	17.0	1.9	198.7	260.5	57.7	1827.4

Trade—Pessimistic Scenario

LAM21	OCH21	OIL21	RUH21	SAF21	TAF21	USA21	WEH21	WEM21
6.5	1.5	0.0	3.9	0.3	2.6	5.3	0.0	0.5
4.6	9.1	0.0	2.3	0.0	6.7	1.3	1.2	4.4
145.0	523.1	0.0	478.9	39.7	41.4	36.0	0.0	0.0
2.7	3.1	0.0	0.0	0.2	1.2	1.3	0.0	0.0
1.1	0.2	0.0	0.0	1.8	0.2	1.4	3.6	0.7
197.4	62.2	24.3	482.5	14.5	74.6	137.8	140.2	50.6
139.	22.	0.*	78.	7.	138.	122.	6.	12.
63.	248.	3.	65.	0.	179.	39.	32.	71.
3372.	12258.	0.	11872.	2029.	1050.	940.	96.	204.
49.	53.	10.	36.	3.	19.	39.	34.	10.
21.	45.	12.	35.	10.	4.	32.	37.	14.
4458.	2085.	541.	12103.	338.	1985.	3256.	3684.	1127.
4.4	1.2	-0.7	0.0	0.0	2.6	0.9	-6.2	-0.7
2.5	8.6	-1.2	-3.1	-0.3	6.6	-8.7	-7.0	2.5
96.8	509.7	-13.7	96.8	0.0	41.3	-347.0	-475.0	-39.6
1.2	2.8	-0.9	-2.8	0.0	1.2	-0.8	-3.4	-0.6
0.0	0.0	-1.2	-2.5	1.6	0.2	-1.5	0.0	0.0
119.1	10.4	-36.1	145.0	0.0	72.5	-32.7	-87.1	0.0

Table 7-15
2010 Regional Nonfuel Minerals Consumption—Pessimistic Scenario

		AAF21	ASC21	ASL21	CAN21	EEM21	JAP21	LAL21	LAM21	OCH21	OIL21
COPPER	MT	.0	1.2	.7	.4	1.5	3.9	.2	2.1	.3	.7
BAUXITE	MT	.1	2.2	1.7	.7	2.3	5.2	.4	2.1	.5	1.2
NICKEL	MT	1.3	49.2	29.6	23.0	79.7	355.5	9.9	48.2	13.4	13.7
ZINC	MT	.0	.6	.9	.3	1.2	2.2	.2	1.5	.3	.9
LEAD	MT	.1	.6	.5	.1	1.3	.6	.2	1.1	.2	1.2
IRON	MT	2.3	113.8	9.7	62.3	117.0	280.3	29.8	78.3	51.8	60.4

Notes: The countries included in each of the sixteen regions are listed in appendix H.
All values appearing in the bauxite row represent aluminum content therein.

Table 7-16
2020 Nonfuel Minerals Production, Cumulative Output, and Net

		AAF22	ASC22	ASL22	CAN22	EEM22	JAP22	LAL22
RESOURCE OUTPUTS								
COPPER	MT	0.0*	0.0	1.2	4.3	0.7	0.3	3.6
BAUXITE	MT	0.0	0.6	4.3	0.0	0.0	0.0	13.8
NICKEL	KT	1.8	0.0	349.5	692.9	0.0	0.0	12.1
ZINC	MT	0.0*	0.9	1.7	0.4	1.9	3.4	0.3
LEAD	MT	0.1	0.9	0.9	0.2	1.9	1.0	0.3
IRON	MT	2.9	163.5	183.0	148.3	44.0	3.6	132.1
CUMULATIVE RESOURCE OUTPUT AT END OF PERIOD								
COPPER	MT	1.	5.	32.	92.	20.	11.	81.
BAUXITE	MT	0.	16.	102.	0.	74.	0.	341.
NICKEL	KT	39.	110.	9557.	17962.	130.	0.	275.
ZINC	MT	1.	16.	21.	112.	34.	46.	59.
LEAD	MT	12.	16.	11.	73.	26.	13.	30.
IRON	MT	189.	2872.	4142.	3726.	1012.	91.	3049.
NET EXPORTS OF RESOURCES								
COPPER	MT	0.0	-1.7	0.0	3.7	-1.8	-5.9	3.4
BAUXITE	MT	-0.1	-2.4	1.4	-1.0	-3.7	-8.2	13.3
NICKEL	KT	0.0	-66.6	303.0	659.4	-123.9	-556.2	0.0
ZINC	MT	0.0	0.0	0.0	0.0	0.0	0.0	0.0
LEAD	MT	-0.0*	0.0	0.0	-0.0*	0.0	0.0	0.0
IRON	MT	0.0	0.0	163.3	57.6	-142.1	-441.2	96.0

Notes: The countries included in each of the sixteen regions are listed in appendix H.
All values appearing in the bauxite row represent aluminum content therein.
*Rounded off to zero.

RUH21	SAF21	TAF21	USA21	WEH21	WEM21	WORLD
3.9	.3	.0	4.4	6.2	1.2	27.0
5.4	.3	.1	10.0	8.2	1.9	42.3
382.1	39.7	.1	383.0	475.0	39.6	1943.0
2.8	.2	.0	2.1	3.4	.6	17.2
2.5	.2	.0	2.9	3.6	.7	15.8
337.5	14.5	2.1	170.5	227.3	50.6	1608.2

Trade—Optimistic Scenario

LAM22	OCH22	OIL22	RUH22	SAF22	TAF22	USA22	WEH22	WEM22
10.5	2.4	0.0	6.1	0.4	4.3	8.4	0.0	0.7
8.1	16.8	0.0	0.0	0.0	12.4	1.7	1.4	7.6
229.5	818.1	0.0	734.8	59.9	64.7	45.3	0.0	0.0
2.4	0.4	1.8	4.4	0.3	0.0	3.4	5.4	1.0
1.9	0.3	2.4	3.8	0.2	0.0	4.3	5.5	1.1
350.1	100.3	24.3	799.9	22.3	137.0	206.7	140.2	79.7
233.	44.	0.*	133.	11.	176.	198.	6.	18.
134.	393.	3.	77.	0.	285.	55.	45.	137.
5446.	19662.	0.	18563.	2594.	1635.	1347.	96.	204.
79.	75.	23.	73.	5.	27.	62.	80.	18.
32.	47.	31.	68.	12.	5.	61.	67.	20.
7606.	3008.	784.	19371.	547.	3215.	4978.	5286.	1850.
7.1	2.0	-1.4	0.0	0.0	4.3	1.4	-9.8	-1.3
4.6	16.1	-2.3	-8.4	-0.5	12.4	-14.3	-11.5	4.6
151.5	797.6	-27.8	151.5	0.0	64.6	-551.1	-740.3	-61.8
0.0	0.0	0.0	0.0	0.0	0.0	0.0	0.0	0.0
0.0	0.0	0.0	0.0	0.0	-0.0	0.0	0.0	0.0
220.9	19.2	-89.5	268.9	0.0	134.5	-67.6	-220.0	0.0

Table 7-17
2020 Regional Nonfuel Minerals Consumption—Optimistic Scenario

		AAF22	ASC22	ASL22	CAN22	EEM22	JAP22	LAL22	LAM22	OCH22	OIL22
COPPER	MT	.0	1.7	1.2	.6	2.5	6.2	.2	3.4	.4	1.4
BAUXITE	MT	.1	3.0	2.9	1.0	3.7	8.2	.5	3.5	.7	2.3
NICKEL	KT	1.8	66.6	46.5	33.5	123.9	556.2	12.1	78.0	20.5	27.8
ZINC	MT	.0	.9	1.7	.4	1.9	3.4	.3	2.4	.4	1.8
LEAD	MT	.1	.9	.9	.2	1.9	1.0	.3	1.9	.3	2.4
IRON	MT	2.9	163.5	19.7	90.7	186.1	444.8	36.1	129.2	81.1	113.8

Notes: The countries included in each of the sixteen regions are listed in appendix H.
All values appearing in the bauxite row represent aluminum content therein.

Table 7-18
2020 Nonfuel Minerals Production, Cumulative Output, and Net

		AAF22	ASC22	ASL22	CAN22	EEM22	JAP22	LAL22
RESOURCE OUTPUTS								
COPPER	MT	0.0*	0.0	1.0	3.5	0.7	0.3	3.0
BAUXITE	MT	0.0	0.6	3.4	0.0	0.0	0.0	11.0
NICKEL	KT	1.6	0.0	283.5	559.5	0.0	0.0	11.7
ZINC	MT	0.0*	0.9	1.4	0.4	1.5	2.8	0.3
LEAD	MT	0.1	0.9	0.7	0.2	1.6	0.8	0.3
IRON	MT	2.3	161.8	131.9	113.2	44.0	3.6	105.4
CUMULATIVE RESOURCE OUTPUT AT END OF PERIOD								
COPPER	MT	1.	5.	29.	84.	20.	11.	75.
BAUXITE	MT	0.	16.	93.	0.	71.	0.	316.
NICKEL	KT	38.	110.	8923.	16702.	130.	0.	271.
ZINC	MT	1.	16.	18.	105.	31.	41.	56.
LEAD	MT	11.	16.	10.	67.	24.	12.	29.
IRON	MT	184.	2855.	3642.	3384.	1012.	91.	2792.
NET EXPORTS OF RESOURCES								
COPPER	MT	0.0	-1.7	0.0	3.0	-1.3	-4.7	2.8
BAUXITE	MT	-0.1	-2.3	1.1	-0.8	-2.9	-6.5	10.6
NICKEL	KT	0.0	-66.4	244.7	532.6	-100.0	-446.0	0.0
ZINC	MT	0.0	0.0	0.0	0.0	0.0	0.0	0.0
LEAD	MT	0.0	0.0	0.0	0.0	0.0	0.0	0.0
IRON	MT	0.0	0.0	117.3	41.4	-103.9	-348.3	69.0

Notes: The countries included in each of the sixteen regions are listed in appendix H.
All values appearing in the bauxite row represent aluminum content therein.
*Rounded off to zero.

RUH22	SAF22	TAF22	USA22	WEH22	WEM22	WORLD
6.1	.4	.0	7.0	9.8	2.0	42.9
8.4	.5	.0	16.0	12.9	3.0	66.7
583.3	59.9	.1	596.4	740.3	61.8	3008.7
4.4	.3	.0	3.4	5.4	1.0	27.7
3.8	.2	.0	4.3	5.5	1.1	24.8
531.0	22.3	2.5	274.3	360.2	79.7	2537.9

Trade—Pessimistic Scenario

LAM22	OCH22	OIL22	RUH22	SAF22	TAF22	USA22	WEH22	WEM22
8.6	2.0	0.0	4.9	0.3	3.5	6.7	0.0	0.7
6.4	13.3	0.0	0.0	0.0	9.9	1.7	1.4	6.1
186.1	660.4	0.0	592.2	48.1	52.3	45.3	0.0	0.0
2.0	0.3	1.7	3.5	0.3	0.0	2.6	4.3	0.8
1.5	0.2	2.3	3.0	0.2	0.0	3.5	4.6	0.9
263.4	76.4	24.3	610.6	17.7	99.2	206.7	140.2	63.8
214.	40.	0.*	122.	10.	168.	182.	6.	18.
118.	360.	3.	77.	0.	262.	55.	45.	123.
5027.	18175.	0.	17228.	2480.	1518.	1347.	96.	204.
72.	70.	22.	67.	4.	25.	58.	72.	17.
29.	47.	30.	62.	11.	5.	57.	60.	19.
6763.	2778.	784.	17569.	503.	2854.	4978.	5286.	1698.
5.8	1.6	-1.3	0.0	0.0	3.5	1.2	-8.0	-0.9
3.6	12.7	-2.2	-6.7	-0.4	9.8	-10.6	-8.9	3.6
122.4	644.1	-26.3	122.4	0.0	52.2	-430.5	-598.7	-50.5
0.0	0.0	0.0	0.0	0.0	0.0	0.0	0.0	0.0
0.0	0.0	0.0	0.0	0.0	0.0	0.0	0.0	0.0
158.6	13.8	-86.7	193.1	0.0	96.6	-4.2	-146.7	0.0

Table 7-19
2020 Regional Nonfuel Minerals Consumption—Pessimistic Scenario

		AAF22	ASC22	ASL22	CAN22	EEM22	JAP22	LAL22	LAM22	OCH22	OIL22
COPPER	MT	.0	1.7	1.0	.5	2.0	5.0	.2	2.8	.4	1.3
BAUXITE	MT	.1	2.9	2.3	.8	2.9	6.5	.4	2.8	.6	2.2
NICKEL	KT	1.6	66.4	38.8	26.9	100.0	446.0	11.7	63.7	16.3	26.3
ZINC	MT	.0	.9	1.4	.4	1.5	2.8	.3	2.0	.3	1.7
LEAD	MT	.1	.9	.7	.2	1.6	.8	.3	1.5	.2	2.3
IRON	MT	2.3	161.8	14.6	71.8	147.9	351.9	36.4	104.8	62.6	111.0

Notes: The countries included in each of the sixteen regions are listed in appendix H.
All values appearing in the bauxite row represent aluminum content therein.

Table 7-20
2030 Nonfuel Minerals Production, Cumulative Output, and Net

		AAF23	ASC23	ASL23	CAN23	EEM23	JAP23	LAL23
RESOURCE OUTPUTS								
COPPER	MT	0.1	0.0	1.7	5.9	0.9	0.3	4.9
BAUXITE	MT	0.0	1.1	5.9	0.0	0.0	0.0	18.8
NICKEL	KT	2.5	0.0	471.2	939.7	0.0	0.0	16.1
ZINC	MT	0.1	1.2	2.4	0.6	2.6	4.7	0.4
LEAD	MT	0.1	1.2	1.2	0.3	2.6	1.4	0.4
IRON	MT	4.4	226.0	265.1	211.0	59.5	4.9	187.0
CUMULATIVE RESOURCE OUTPUT AT END OF PERIOD								
COPPER	MT	1.	5.	46.	143.	28.	14.	124.
BAUXITE	MT	0.	24.	152.	0.	74.	0.	504.
NICKEL	KT	61.	110.	13661.	26125.	130.	0.	416.
ZINC	MT	2.	22.	33.	117.	46.	69.	63.
LEAD	MT	13.	22.	17.	75.	39.	20.	32.
IRON	MT	226.	4002.	6383.	5523.	1530.	133.	4645.
NET EXPORTS OF RESOURCES								
COPPER	MT	0.0	-2.4	0.0	5.0	-2.4	-8.1	4.7
BAUXITE	MT	-0.1	-3.1	1.9	-1.4	-5.0	-11.1	18.2
NICKEL	KT	0.0	-89.7	410.4	893.2	-167.3	-751.9	0.0
ZINC	MT	0.0	0.0	0.0	0.0	0.0	0.0	0.0
LEAD	MT	0.0*	0.0	0.0	0.0*	0.0	0.0	0.0
IRON	MT	0.0	0.0	237.1	83.7	-192.9	-597.3	139.5

Notes: The countries included in each of the sixteen regions are listed in appendix H.
All values appearing in the bauxite row represent aluminum content therein.
*Rounded off to zero.

RUH22	SAF22	TAF22	USA22	WEH22	WEM22	WORLD
4.9	.3	.0	5.5	8.0	1.6	35.2
6.7	.4	.1	12.3	10.3	2.5	53.8
469.8	48.1	.1	475.8	598.7	50.5	2440.7
3.5	.3	.0	2.6	4.3	.8	22.8
3.0	.2	.0	3.5	4.6	.9	20.8
417.5	17.7	2.6	210.9	286.9	63.8	2064.5

Trade—Optimistic Scenario

LAM23	OCH23	OIL23	RUH23	SAF23	TAF23	USA23	WEH23	WEM23
14.5	3.3	0.0	8.1	0.6	5.8	11.4	0.0	0.9
11.1	23.0	0.0	0.0	0.0	17.0	2.2	1.7	10.3
313.5	1108.1	0.0	986.4	81.2	87.7	57.1	0.0	0.0
3.4	0.6	2.6	5.9	0.4	0.0	4.6	7.3	1.3
2.6	0.4	3.5	5.0	0.3	0.0	5.8	7.4	1.6
501.8	138.5	24.3	1104.9	30.4	199.0	258.4	140.2	107.3
358.	73.	0.*	204.	16.	227.	298.	6.	26.
229.	592.	3.	77.	0.	432.	74.	61.	227.
8162.	29293.	0.	27169.	3300.	2397.	1859.	96.	204.
108.	81.	36.	103.	7.	27.	102.	116.	25.
45.	49.	48.	93.	13.	5.	111.	104.	28.
11865.	4201.	1028.	28894.	810.	4895.	7304.	6688.	2785.
9.7	2.7	-1.9	0.0	0.0	5.8	1.9	-13.2	-1.8
6.3	21.9	-3.3	-11.3	-0.7	16.9	-19.6	-15.8	6.3
205.2	1080.3	-39.1	205.2	0.0	87.5	-748.1	-1002.5	-83.1
0.0	0.0	0.0	0.0	0.0	0.0	0.0	0.0	0.0
0.0	0.0	0.0	0.0	0.0	0.0	0.0	0.0	0.0
320.8	27.9	-141.1	390.5	0.0	195.2	-114.6	-348.7	0.0

Table 7-21
2030 Regional Nonfuel Minerals Consumption—Optimistic Scenario

		AAF23	ASC23	ASL23	CAN23	EEM23	JAP23	LAL23	LAM23	OCH23	OIL23
COPPER	MT	.1	2.4	1.7	.9	3.3	8.4	.2	4.8	.6	1.9
BAUXITE	MT	.1	4.2	4.0	1.4	5.0	11.1	.6	4.8	1.1	3.3
NICKEL	KT	2.5	89.7	60.8	46.5	167.3	751.9	16.1	108.3	27.8	39.1
ZINC	MT	.1	1.2	2.4	.6	2.6	4.7	.4	3.4	.6	2.6
LEAD	MT	.1	1.2	1.2	.3	2.6	1.4	.4	2.6	.4	3.5
IRON	MT	4.4	226.0	28.0	127.3	252.4	602.2	47.5	181.0	110.6	165.4

Notes: The countries included in each of the sixteen regions are listed in appendix H.
All values appearing in the bauxite row represent aluminum content therein.

Table 7-22
2030 Nonfuel Minerals Production, Cumulative Output, and Net

		AAF23	ASC23	ASL23	CAN23	EEM23	JAP23	LAL23
RESOURCE OUTPUTS								
COPPER	MT	0.0*	0.0	1.3	4.5	0.9	0.3	3.8
BAUXITE	MT	0.0	1.1	4.4	0.0	0.0	0.0	13.8
NICKEL	KT	2.1	0.0	352.1	695.9	0.0	0.0	15.2
ZINC	MT	0.0*	1.2	1.8	0.5	1.9	3.4	0.4
LEAD	MT	0.1	1.2	0.9	0.2	1.9	1.0	0.4
IRON	MT	3.3	223.4	173.8	146.5	59.5	4.9	138.6
CUMULATIVE RESOURCE OUTPUT AT END OF PERIOD								
COPPER	MT	1.	5.	40.	124.	28.	14.	109.
BAUXITE	MT	0.	24.	132.	0.	71.	0.	440.
NICKEL	KT	56.	110.	12101.	22979.	130.	0.	405.
ZINC	MT	2.	22.	27.	109.	40.	58.	60.
LEAD	MT	12.	22.	14.	69.	33.	17.	31.
IRON	MT	212.	3972.	5171.	4682.	1530.	133.	4012.
NET EXPORTS OF RESOURCES								
COPPER	MT	0.0	-2.4	0.0	3.8	-1.5	-6.0	3.5
BAUXITE	MT	-0.1	-3.0	1.4	-1.0	-3.6	-8.0	13.2
NICKEL	KT	0.0	-89.5	304.1	661.8	-123.5	-551.4	0.0
ZINC	MT	0.0	0.0	0.0	0.0	0.0	0.0	0.0
LEAD	MT	0.0	0.0	0.0	0.0	0.0	0.0	0.0
IRON	MT	0.0	0.0	154.2	54.4	-123.8	-429.6	90.7

Notes: The countries included in each of the sixteen regions are listed in appendix H.
All values appearing in the bauxite row represent aluminum content therein.
*Rounded off to zero.

RUH23	SAF23	TAF23	USA23	WEH23	WEM23	WORLD
8.1	.6	.0	9.5	13.2	2.7	58.4
11.3	.7	.1	21.8	17.5	4.0	91.0
781.2	81.2	.2	805.2	1002.5	83.1	4063.4
5.9	.4	.0	4.6	7.3	1.3	38.1
5.0	.3	.0	5.8	7.4	1.6	33.8
714.4	30.4	3.8	373.0	488.9	107.3	3462.6

Trade—Pessimistic Scenario

LAM23	OCH23	OIL23	RUH23	SAF23	TAF23	USA23	WEH23	WEM23
11.0	2.5	0.0	5.9	0.4	4.5	8.3	0.0	0.9
8.2	16.6	0.0	0.0	0.0	12.3	2.2	1.7	7.6
235.1	820.4	0.0	725.2	59.7	65.0	57.1	0.0	0.0
2.6	0.4	2.4	4.2	0.3	0.0	3.2	5.4	1.0
2.0	0.3	3.3	3.7	0.2	0.0	4.3	5.7	1.2
345.8	95.5	24.3	763.2	21.8	130.3	258.4	140.2	78.1
312.	62.	0.*	176.	14.	208.	257.	6.	26.
191.	510.	3.	77.	0.	373.	74.	61.	191.
7133.	25579.	0.	23815.	3019.	2105.	1859.	96.	204.
95.	74.	35.	88.	6.	25.	87.	99.	22.
39.	48.	46.	81.	12.	5.	96.	89.	25.
9809.	3637.	1028.	24438.	701.	4001.	7304.	6688.	2408.
7.4	2.1	-1.8	0.0	0.0	4.4	1.5	-9.9	-1.1
4.5	15.9	-3.2	-8.2	-0.5	12.3	-13.0	-11.1	4.5
152.0	800.4	-36.5	152.0	0.0	64.8	-527.5	-744.3	-62.6
0.0	0.0	0.0	0.0	0.0	0.0	0.0	0.0	0.0
0.0	0.0	0.0	0.0	0.0	0.0	0.0	0.0	0.0
208.6	18.1	-136.4	253.9	0.0	127.0	-1.1	-215.9	0.0

Table 7-23
2030 Regional Nonfuel Minerals Consumption—Pessimistic Scenario

		AAF23	ASC23	ASL23	CAN23	EEM23	JAP23	LAL23	LAM23	OCH23	OIL23	EUH23	SAF23	TAF23	USA23	WEH23	WEM23	WORLD
COPPER	MT	.0	2.4	1.3	.7	2.4	6.3	.3	3.6	.4	1.8	5.9	.4	.1	6.8	9.9	2.0	44.3
BAUXITE	MT	.1	4.1	3.0	1.0	3.6	8.0	.6	3.7	.7	3.2	8.2	.5	.0	15.2	12.8	3.1	67.8
NICKEL	KT	2.1	89.5	48.0	34.1	123.5	551.4	15.2	83.1	20.0	36.5	573.2	59.7	.2	584.6	744.3	62.6	3028.0
ZINC	MT	.0	1.2	1.8	.5	1.9	3.4	.4	2.6	.4	2.4	4.2	.3	.0	3.2	5.4	1.0	28.7
LEAD	MT	.1	1.2	.9	.2	1.9	1.0	.4	2.0	.3	3.3	3.7	.2	.0	4.3	5.7	1.2	26.4
IRON	MT	3.3	223.4	19.6	92.1	183.3	434.5	47.9	137.2	77.4	160.7	509.3	21.8	3.3	259.5	356.1	78.1	2607.5

Notes: The countries included in each of the sixteen regions are listed in appendix H.
All values appearing in the bauxite row represent aluminum content therein.

Table 7–24
World Total—Optimistic Scenario

Resource Outputs	*1970*	*1980*	*1990*	*2000*	*2010*	*2020*	*2030*
Copper (MT)	6.6	9.9	16.9	23.7	30.9	43.0	58.5
Bauxite (MT)	12.2	16.3	26.2	36.0	48.0	66.8	91.0
Nickel (KT)	741.8	785.8	1,237.4	1,667.2	2,196.4	3,008.8	4,063.6
Zinc (MT)	5.4	6.6	10.2	15.3	19.8	27.9	38.0
Lead (MT)	3.7	5.5	8.9	13.8	17.7	24.8	33.8
Iron (MT)	424.8	609.4	970.2	1,392.3	1,826.9	2,538.1	3,462.5
Cumulative Resource Output at End of Period							
Copper (MT)	0	82.0	216.0	420.0	693.0	1,062.0	1,570.0
Bauxite (MT)	0	143.0	355.0	666.0	1,087.0	1,661.0	2,450.0
Nickel (KT)	0	7,638.0	17,754.0	32,277.0	51,595.0	77,620.0	112,982.0
Zinc (MT)	0	60.0	146.0	273.0	441.0	732.0	958.0
Lead (MT)	0	46.0	121.0	234.0	341.0	523.0	715.0
Iron (MT)	0	5,171.0	13,070.0	24,882.0	40,482.0	61,726.0	90,911.0

Note: All values appearing in the bauxite row represent aluminum content therein.

Table 7–25
World Total—Pessimistic Scenario

Resource Outputs	*1970*	*1980*	*1990*	*2000*	*2010*	*2020*	*2030*
Copper (MT)	6.6	9.9	16.9	23.7	27.2	35.2	44.3
Bauxite (MT)	12.2	16.3	26.2	36.0	42.2	53.9	67.9
Nickel (KT)	741.8	785.8	1,237.1	1,666.7	1,944.1	2,440.6	3,027.8
Zinc (MT)	5.4	6.6	10.1	15.3	17.4	22.7	28.8
Lead (MT)	3.7	5.5	8.9	13.8	15.9	20.9	26.5
Iron (MT)	424.8	609.4	969.8	1,391.5	1,608.8	2,064.6	2,607.7
Cumulative Resource Output at End of Period							
Copper (MT)	0	82.0	216.0	419.0	674.0	986.0	1,383.0
Bauxite (MT)	0	143.0	355.0	666.0	1,057.0	1,538.0	2,147.0
Nickel (KT)	0	7,638.0	17,752.0	32,270.0	50,324.0	72,248.0	99,590.0
Zinc (MT)	0	60.0	146.0	273.0	429.0	676.0	848.0
Lead (MT)	0	46.0	121.0	234.0	332.0	488.0	639.0
Iron (MT)	0	5,171.0	13,067.0	24,874.0	39,376.0	57,173.0	79,726.0

Note: All values appearing in the bauxite row represent aluminum content therein.

Table 7–26
Production 1970

	AAF	ASC	ASL	CAN	EEM	JAP	LAL	LAM	OCH	OIL
MOLYB. K	.0	1.51	.0	12.6	.0	.500	.728	5.93	.0	.0
TUNGS. K	.0	9.36	4.28	1.70	.0	.841	2.64	.841	1.08	.0
MANGAN K	65.5	339.	483.	.0	50.9	44.7	.0	991.	621.	816.
CHROM. K	.0	.0	231.	.0	194.	.0	.0	9.07	.0	57.1
GOLD	.0	.0	.0	60.0	.0	.0	.0	.0	24.0	.0
SILVER K	.559D-01	.0	.174	1.50	.217	.310	1.95	1.64	.890	.0
MERC. K	.0	.687	.795D-01	.450	.699	.138	.114	1.30	.0	.515
VANAD. K	.0	.0	.0	.0	.0	.0	.0	.0	.0	.0
PLATIN	.0	.0	.0	3.00	.0	.0	.606	.0	.0	.0
TITAN. K	.0	.0	124.	410.	.0	9.20	.0	.0	472.	.0
TIN K	.0	22.3	118.	.0	1.20	.700	31.0	5.00	8.80	8.00
SILIC. K	.0	.0	17.0	71.0	7.64	110.	.0	18.7	6.79	.0
FLOUR. M	.205D-01	.193	.167	.580D-01	.401D-01	.0	.0	.480	.0	.0
POTASH M	.437	.248	.0	3.33	2.19	.0	.0	.210D-01	.0	.0
SD.ASH M	.0	.0	.458	.0	1.55	1.39	.0	.376D-02	.0	.0
BORATE K	.0	3.41	.0	.0	.0	.0	.0	6.83	.0	.0
PHOSPH M	20.9	1.72	.0	.0	.0	.0	.0	.0	2.28	.0
SULFUR M	.0	1.20	.0	7.10	3.61	3.01	.0	1.66	.0	1.26
CHLOR. M	.0	.0	.0	.740	.0	2.60	.0	.0	.221	.632D-01
MAGNES M	.0	1.28	.129	.610D-01	.290	.178D-01	.0	.0	.0	.0

	RUH	SAF	TAF	USA	WEH	WEM	WD TOTAL
MOLYB. K	8.55	.0	.0	50.4	.0	.0	80.2
TUNGS. K	6.52	.0	.0	4.36	.600	1.32	33.6
MANGAN K	.306D+04	.147D+04	247.	59.9	.0	.0	.824D+04
CHROM. K	668.	500.	228.	.0	35.4	192.	.211D+04
GOLD	267.	.103D+04	27.1	54.2	.0	.0	.146D+04
SILVER K	1.57	.174	.621D-01	1.40	.345	.161	10.5
MERC. K	1.91	.0	.0	.940	1.40	3.10	11.3
VANAD. K	8.24	2.80	.0	5.07	1.80	.0	17.9
PLATIN	17.8	35.5	.0	.249	.0	.0	57.1
TITAN. K	176.	.0	.0	250.	283.	.0	.172D+04
TIN K	10.0	2.74	9.40	.0	3.75	.800	222.
SILIC. K	450.	25.5	.0	481.	406.	39.1	.163D+04
FLOUR. M	.196	.927D-01	.0	.110	.335	.170	1.86
POTASH M	4.73	.0	.263	2.48	4.98	.0	18.7
SD.ASH M	4.24	.0	.0	6.41	4.01	.394	18.5
BORATE K	30.1	.0	.0	159.	.0	54.6	254.
PHOSPH M	19.0	1.68	1.66	35.1	.0	.0	82.4
SULFUR M	7.53	.0	.0	8.67	3.87	1.00	38.9
CHLOR. M	2.66	.589D-01	.0	8.85	5.80	.119	21.1
MAGNES M	.822	.165D-02	.0	1.03	.750	.778	5.16

Note: The countries included in each of the sixteen regions are listed in appendix H.

Table 7–27
Consumption 1970

		AAF	ASC	ASL	CAN	EEM	JAP	LAL	LAM	OCH	OIL
MOLYB.	K	.502	.740	1.28	2.00	4.26	11.1	2.05	1.57	.505	8.07
TUNGS.	K	.675D-01	3.11	.188	1.20	5.24	2.11	.455D-01	.329	.139	.154D-01
MANGAN	K	17.0	415.	426.	89.0	796.	.101D+04	49.0	74.5	64.2	156.
CHROM.	K	5.48	34.8	20.2	20.0	213.	331.	9.54	26.2	2.99	17.7
GOLD		15.6	78.3	67.6	22.0	146.	39.5	18.4	72.8	14.4	7.87
SILVER	K	.737D-01	.372	.325	.262	.682	.968	.868D-01	.349	.923D-01	.368D-01
MERC.	K	.395D-01	1.61	.132	.105	2.25	.612	.390D-01	.191	.553D-01	.128D-01
VANAD.	K	.127D-01	.914D-01	.450D-01	.601	.215	2.11	.146D-01	.667D-01	.130	.904D-02
PLATIN		.242	1.39	.870	5.09	4.27	14.4	.337	.971	.419	.333
TITAN.	K	10.9	69.1	40.3	55.2	121.	89.6	14.3	65.3	72.2	3.25
TIN	K	.429	12.6	7.84	5.11	13.6	28.9	1.65	4.19	4.63	.276
SILIC.	K	6.91	37.4	20.8	51.0	86.0	152.	20.1	28.6	9.67	68.6
FLOUR.	M	.177D-02	.142D-01	.645D-02	.128	.232D-01	.547	.241D-02	.101D-01	.130D-01	.675D-03
POTASH	M	.197	1.11	.799	.200	1.74	1.20	.112	.734	.141	.733D-01
SD.ASH	M	.644D-01	.408	.258	.301	.749	1.40	.878D-01	.364	.195	.383D-01
BORATE	K	1.42	2.46	5.63	5.73	16.4	30.1	1.58	7.39	1.21	.649
PHOSPH	M	.787	1.39	3.19	4.27	7.01	2.99	.457	2.97	1.72	.291
SULFUR	M	.306	.512	1.24	7.12	2.73	1.84	.189	1.16	.201	.127
CHLOR.	M	.834D-01	.509	.326	.740	.686	2.61	.810D-01	.367	.222	.438D-01
MAGNES	M	.382D-01	.192	.121	.627D-01	.407	.234	.599D-01	.176	.357D-01	.109D-01

		RUH	SAF	TAF	USA	WEH	WEM	WD TOTAL
MOLYB.	K	9.15	.104	.961D-01	22.1	15.8	.840	80.2
TUNGS.	K	8.70	.276D-01	.847D-02	7.35	4.82	.214	33.6
MANGAN	K	.248D+04	583.	6.59	.121D+04	826.	48.5	.824D+04
CHROM.	K	246.	128.	2.78	491.	548.	16.9	.211D+04
GOLD		67.5	3.55	15.9	188.	657.	49.7	.146D+04
SILVER	K	1.57	.784D-01	.759D-01	2.27	2.98	.228	10.5
MERC.	K	1.91	.592D-01	.127D-01	1.87	2.30	.133	11.3
VANAD.	K	5.24	2.05	.544D-02	6.41	.865	.454D-01	17.9
PLATIN		5.05	.121	.116	12.2	10.6	.696	57.1
TITAN.	K	15.4	5.59	3.64	497.	619.	41.7	.172D+04
TIN	K	17.1	2.10	1.75	53.9	62.2	5.40	222.
SILIC.	K	300.	2.15	2.21	482.	347.	17.8	.163D+04
FLOUR.	M	.391	.477D-01	.138D-02	.549	.122	.514D-02	1.86
POTASH	M	3.36	.193	.510D-01	4.29	3.89	.591	18.7
SD.ASH	M	4.43	.615D-01	.368D-01	6.13	3.69	.250	18.5
BORATE	K	30.1	1.32	.515	80.0	63.5	5.52	254.
PHOSPH	M	13.7	.346	.204	24.6	16.0	2.38	82.4
SULFUR	M	7.31	.198	.847D-01	8.36	6.63	.919	38.9
CHLOR.	M	2.69	.588D-01	.466D-01	8.86	3.57	.223	21.1
MAGNES	M	.602	.114D-01	.445D-02	1.05	2.02	.138	5.16

Note: The countries included in each of the sixteen regions are listed in appendix H.

Table 7–28
Net Trade 1970

	AAF	ASC	ASL	CAN	EEM	JAP	LAL	LAM	OCH	OIL
MOLYB. K	-.502	.770	-1.28	10.6	-4.26	-10.6	-1.32	4.36	-.505	-8.07
TUNGS. K	-.675D-01	6.26	4.09	.501	-5.24	-1.27	2.60	.512	.941	-.154D-01
MANGAN K	48.5	-75.8	56.1	-89.0	-746.	-963.	-49.0	916.	557.	659.
CHROM. K	-5.48	-34.8	211.	-20.0	-18.8	-331.	-9.54	-17.1	-2.99	39.5
GOLD	-15.6	-78.3	-67.6	38.0	-146.	-39.5	-18.4	-72.8	9.56	-7.87
SILVER K	-.178D-01	-.372	-.151	1.24	-.465	-.657	1.87	1.29	.798	-.368D-01
MERC. K	-.395D-01	-.918	-.527D-01	.345	-1.56	-.474	.751D-01	1.11	-.553D-01	.503
VANAD. K	-.127D-01	-.914D-01	-.450D-01	-.601	-.215	-2.11	-.146D-01	-.667D-01	-.130	-.904D-02
PLATIN	-.242	-1.39	-.870	-2.09	-4.27	-14.4	.269	-.971	-.419	-.333
TITAN. K	-10.9	-69.1	83.2	355.	-121.	-80.4	-14.3	-65.3	400.	-3.25
TIN K	-.429	9.65	110.	-5.11	-12.4	-28.2	29.4	.813	4.17	7.72
SILIC. K	-6.91	-37.4	-3.86	20.0	-78.4	-42.2	-20.1	-9.88	-2.88	-68.6
FLOUR. M	.187D-01	.178	.160	-.699D-01	.169D-01	-.547	-.241D-02	.470	-.130D-01	-.675D-03
POTASH M	.240	-.858	-.799	3.13	.445	-1.20	-.112	-.713	-.141	-.733D-01
SD.ASH M	-.644D-01	-.408	.200	-.301	.805	-.979D-02	-.878D-01	-.361	-.195	-.383D-01
BORATE K	-1.42	.954	-5.63	-5.73	-16.4	-30.1	-1.58	-.566	-1.21	-.649
PHOSPH M	20.1	.329	-3.19	-4.27	-7.01	-2.99	-.457	-2.97	.560	-.291
SULFUR M	-.306	.693	-1.24	-.163D-01	.887	1.18	-.189	.497	-.201	1.13
CHLOR. M	-.834D-01	-.509	-.326	.410D-03	-.686	-.151D-01	-.810D-01	-.367	-.530D-03	.194D-01
MAGNES M	-.382D-01	1.09	.784D-02	-.171D-02	-.117	-.217	-.599D-01	-.176	-.357D-01	-.109D-01

	RUH	SAF	TAF	USA	WEH	WEM	WD TOTAL
MOLYB. K	-.594	-.104	-.961D-01	28.3	-15.8	-.840	.0
TUNGS. K	-2.17	-.276D-01	-.847D-02	-2.99	-4.22	1.11	.0
MANGAN K	583.	882.	240.	-.115D+04	-826.	-48.5	.0
CHROM. K	422.	372.	225.	-491.	-513.	175.	.0
GOLD	199.	.103D+04	11.2	-134.	-657.	-49.7	.0
SILVER K	.274D-02	.959D-01	-.138D-01	-.872	-2.64	-.671D-01	.0
MERC. K	.364D-02	-.592D-01	-.127D-01	-.933	-.900	2.96	.0
VANAD. K	3.00	.747	-.544D-02	-1.34	.937	-.454D-01	.0
PLATIN	12.8	35.3	-.116	-11.9	-10.6	-.696	.0
TITAN. K	161.	-5.59	-3.64	-247.	-336.	-41.7	.0
TIN K	-7.06	.634	7.65	-53.9	-58.5	-4.60	.0
SILIC. K	150.	23.3	-2.21	-.625	58.5	21.3	.0
FLOUR. M	-.195	.450D-01	-.138D-02	-.439	.213	.165	.0
POTASH M	1.37	-.193	.212	-1.81	1.09	-.591	.0
SD.ASH M	-.188	-.615D-01	-.368D-01	.282	.320	.144	.0
BORATE K	-.950D-02	-1.32	-.515	78.7	-63.5	49.1	.0
PHOSPH M	5.26	1.33	1.46	10.5	-16.0	-2.38	.0
SULFUR M	.220	-.198	-.847D-01	.305	-2.76	.852D-01	.0
CHLOR. M	-.268D-01	.573D-04	-.466D-01	-.781D-02	2.23	-.104	.0
MAGNES M	.220	-.979D-02	-.445D-02	-.103D-01	-1.27	.640	.0

Note: The countries included in each of the sixteen regions are listed in appendix H.

Table 7–29
Export Shares 1970

	AAF	ASC	ASL	CAN	EEM	JAP	LAL	LAM	OCH	OIL
MOLYB. K	.0	.175D-01	.0	.240	.0	.0	.0	.992D-01	.0	.0
TUNGS. K	.0	.391	.256	.313D-01	.0	.0	.162	.320D-01	.588D-01	.0
MANGAN K	.123D-01	.0	.142D-01	.0	.0	.0	.0	.232	.141	.167
CHROM. K	.0	.0	.146	.0	.0	.0	.0	.0	.0	.273D-01
GOLD	.0	.0	.0	.295D-01	.0	.0	.0	.0	.743D-02	.0
SILVER K	.0	.0	.0	.234	.0	.0	.353	.244	.151	.0
MERC. K	.0	.0	.0	.690D-01	.0	.0	.150D-01	.222	.0	.101
VANAD. K	.0	.0	.0	.0	.0	.0	.0	.0	.0	.0
PLATIN	.0	.0	.0	.0	.0	.0	.557D-02	.0	.0	.0
TITAN. K	.0	.0	.833D-01	.355	.0	.0	.0	.0	.401	.0
TIN K	.0	.567D-01	.647	.0	.0	.0	.173	.478D-02	.245D-01	.454D-01
SILIC. K	.0	.0	.0	.731D-01	.0	.0	.0	.0	.0	.0
FLOUR. M	.148D-01	.141	.126	.0	.134D-01	.0	.0	.371	.0	.0
POTASH M	.370D-01	.0	.0	.482	.686D-01	.0	.0	.0	.0	.0
SD.ASH M	.0	.0	.114	.0	.460	.0	.0	.0	.0	.0
BORATE K	.0	.741D-02	.0	.0	.0	.0	.0	.0	.0	.0
PHOSPH M	.509	.831D-02	.0	.0	.0	.0	.0	.0	.141D-01	.0
SULFUR M	.0	.139	.0	.0	.178	.236	.0	.995D-01	.0	.226
CHLOR. M	.0	.0	.0	.182D-03	.0	.0	.0	.0	.0	.861D-02
MAGNES M	.0	.557	.401D-02	.0	.0	.0	.0	.0	.0	.0

	RUH	SAF	TAF	USA	WEH	WEM
MOLYB. K	.0	.0	.0	.643	.0	.0
TUNGS. K	.0	.0	.0	.0	.0	.691D-01
MANGAN K	.148	.224	.609D-01	.0	.0	.0
CHROM. K	.292	.258	.156	.0	.0	.121
GOLD	.155	.800	.872D-02	.0	.0	.0
SILVER K	.519D-03	.181D-01	.0	.0	.0	.0
MERC. K	.728D-03	.0	.0	.0	.0	.593
VANAD. K	.641	.159	.0	.0	.200	.0
PLATIN	.264	.731	.0	.0	.0	.0
TITAN. K	.161	.0	.0	.0	.0	.0
TIN K	.0	.372D-02	.450D-01	.0	.0	.0
SILIC. K	.549	.854D-01	.0	.0	.214	.779D-01
FLOUR. M	.0	.355D-01	.0	.0	.168	.130
POTASH M	.211	.0	.326D-01	.0	.168	.0
SD.ASH M	.0	.0	.0	.161	.183	.821D-01
BORATE K	.0	.0	.0	.611	.0	.381
PHOSPH M	.133	.337D-01	.368D-01	.265	.0	.0
SULFUR M	.440D-01	.0	.0	.612D-01	.0	.171D-01
CHLOR. M	.0	.254D-04	.0	.0	.991	.0
MAGNES M	.112	.0	.0	.0	.0	.327

Note: The countries included in each of the sixteen regions are listed in appendix H.

Table 7-30
Import Coefficients 1970

	AAF	ASC	ASL	CAN	EEM	JAP	LAL	LAM	OCH	OIL
MOLYB. K	.0	.0	.0	.0	.0	-21.2	-1.82	.0	.0	.0
TUNGS. K	.0	.0	.0	.0	.0	-1.51	.0	.0	.0	.0
MANGAN K	.0	-.223	.0	.0	-14.7	-21.5	.0	.0	.0	.0
CHROM. K	.0	.0	.0	.0	-.972D-01	.0	.0	-1.89	.0	.0
GOLD	.0	.0	.0	.0	.0	.0	.0	.0	.0	.0
SILVER K	-.319	.0	-.868	.0	-2.14	-2.12	.0	.0	.0	.0
MERC. K	.0	-1.34	-.663	.0	-2.23	-3.44	.0	.0	.0	.0
VANAD. K	.0	.0	.0	.0	.0	.0	.0	.0	.0	.0
PLATIN	.0	.0	.0	-.696	.0	.0	.0	.0	.0	.0
TITAN. K	.0	.0	.0	.0	.0	-8.74	.0	.0	.0	.0
TIN K	.0	.0	.0	.0	-10.4	-40.2	.0	.0	.0	.0
SILIC. K	.0	.0	-.227	.0	-10.3	-.385	.0	-.529	-.424	.0
FLOUR. M	.0	.0	.0	-1.21	.0	.0	.0	.0	.0	.0
POTASH M	.0	-3.46	.0	.0	.0	.0	.0	-33.9	.0	.0
SD.ASH M	.0	.0	.0	.0	.0	-.705D-02	.0	-95.9	.0	.0
BORATE K	.0	.0	.0	.0	.0	.0	.0	-.829D-01	.0	.0
PHOSPH M	.0	.0	.0	.0	.0	.0	.0	.0	.0	.0
SULFUR M	.0	.0	.0	-.229D-02	.0	.0	.0	.0	.0	.0
CHLOR. M	.0	.0	.0	.0	.0	-.582D-02	.0	.0	-.240D-02	.0
MAGNES M	.0	.0	.0	-.280D-01	-.403	-12.2	.0	.0	.0	.0

	RUH	SAF	TAF	USA	WEH	WEM
MOLYB. K	-.694D-01	.0	.0	.0	.0	.0
TUNGS. K	-.333	.0	.0	-.685	-7.03	.0
MANGAN K	.0	.0	.0	-19.1	.0	.0
CHROM. K	.0	.0	.0	.0	-14.5	.0
GOLD	.0	.0	.0	-2.47	.0	.0
SILVER K	.0	.0	-.223	-.623	-7.66	-.416
MERC. K	.0	.0	.0	-.992	-.641	.0
VANAD. K	.0	.0	.0	-.264	.0	.0
PLATIN	.0	.0	.0	-48.0	.0	.0
TITAN. K	.0	.0	.0	-.989	-1.19	.0
TIN K	-.706	.0	.0	.0	-15.6	-5.75
SILIC. K	.0	.0	.0	-.130D-02	.0	.0
FLOUR. M	-.992	.0	.0	-3.99	.0	.0
POTASH M	.0	.0	.0	-.730	.0	.0
SD.ASH M	-.442D-01	.0	.0	.0	.0	.0
BORATE K	-.316D-03	.0	.0	.0	.0	.0
PHOSPH M	.0	.0	.0	.0	.0	.0
SULFUR M	.0	.0	.0	.0	-.714	.0
CHLOR. M	-.101D-01	.0	.0	-.883D-03	.0	-.878
MAGNES M	.0	-5.94	.0	-.991D-02	-1.70	.0

Note: The countries included in each of the sixteen regions are listed in appendix H.

Table 7–31
Production 1980

	AAF	ASC	ASL	CAN	EEM	JAP	LAL	LAM	OCH	OIL
MOLYB. K	.0	2.77	.0	19.0	.0	.500	1.15	9.48	.0	.0
TUNGS. K	.0	14.9	6.49	2.25	.0	.840	4.08	1.17	1.63	.0
MANGAN K	87.6	608.	598.	.0	87.3	44.7	.0	.157D+04	991.	.148D+04
CHROM. K	.0	.0	348.	.0	338.	.0	.0	12.1	.0	126.
GOLD	.0	.0	.0	88.4	.0	.0	.0	.0	33.3	.0
SILVER K	.443D-01	.0	.236	2.62	.365	.300	2.38	2.58	1.61	.0
MERC. K	.0	.467	.411D-01	.279	.469	.107	.779D-01	.815	.0	.386
VANAD. K	.0	.154	.0	.0	.0	.0	.0	.0	.0	.0
PLATIN	.0	.0	.0	3.49	.0	.0	1.11	.0	.0	.0
TITAN. K	.0	.0	140.	489.	.0	13.8	.0	.0	564.	.0
TIN K	.0	31.2	127.	.0	1.60	.0	35.2	5.95	11.6	12.6
SILIC. K	.0	.0	20.1	114.	13.1	201.	.0	24.7	10.0	.0
FLOUR. M	.145D-01	.144	.120	.463D-01	.397D-01	.0	.0	.351	.0	.0
POTASH M	.819	1.08	.0	7.30	3.95	.0	.0	.785D-01	.0	.0
SD.ASH M	.0	.0	.695	.0	2.76	2.88	.0	.484D-02	.0	.0
BORATE K	.0	8.61	.0	.0	.0	.0	.0	14.3	.0	.0
PHOSPH M	31.7	6.52	.0	.0	.0	.0	.0	.0	2.45	.0
SULFUR M	.0	3.01	.0	6.80	3.93	2.69	.0	4.77	.0	2.68
CHLOR. M	.731D-01	.933	.423	1.15	1.09	5.15	.146	.519	.297	.322
MAGNES M	.0	1.41	.160	.130	.530	.380D-01	.0	.0	.0	.0

	RUH	SAF	TAF	USA	WEH	WEM	WD TOTAL
MOLYB. K	14.3	.0	.0	74.1	.0	.0	121.
TUNGS. K	10.0	.0	.0	6.13	.736	1.92	50.2
MANGAN K	.523D+04	.206D+04	393.	87.2	.0	.0	.132D+05
CHROM. K	.107D+04	741.	349.	.0	47.2	287.	.332D+04
GOLD	403.	.151D+04	37.5	80.5	.0	.0	.215D+04
SILVER K	2.59	.256	.823D-01	2.08	.434	.210	15.8
MERC. K	1.30	.0	.0	.550	.810	1.97	7.27
VANAD. K	13.7	3.49	.0	7.37	2.66	.0	27.3
PLATIN	29.0	56.4	.0	.358	.0	.0	90.4
TITAN. K	212.	.0	.0	279.	327.	.0	.202D+04
TIN K	30.5	3.20	10.7	.0	3.99	.933	275.
SILIC. K	798.	46.5	.0	688.	582.	58.9	.256D+04
FLOUR. M	.228	.229	.0	.110	.166	.634D-01	1.51
POTASH M	9.31	.0	.451	2.50	7.95	.0	33.4
SD.ASH M	7.81	.0	.0	9.89	6.00	.575	30.6
BORATE K	56.2	.0	.0	251.	.0	88.0	419.
PHOSPH M	33.7	2.46	2.16	48.3	.0	.0	127.
SULFUR M	13.9	.0	.0	12.0	4.54	1.09	55.4
CHLOR. M	4.67	.777D-01	.531D-01	13.0	4.82	.302	33.1
MAGNES M	.975	.198D-02	.0	1.04	.382	.797	5.46

Note: The countries included in each of the sixteen regions are listed in appendix H.

Table 7-32
Consumption 1980

	AAF	ASC	ASL	CAN	EEM	JAP	LAL	LAM	OCH	OIL
MOLYB. K	.327	1.57	1.45	2.54	6.79	18.7	3.23	2.67	.656	16.5
TUNGS. K	.286D-01	5.33	.217	1.48	8.59	3.36	.108	.388	.185	.610
MANGAN K	9.73	744.	508.	143.	.137D+04	.187D+04	88.0	96.0	97.6	421.
CHROM. K	4.07	60.2	26.4	28.7	371.	611.	19.5	35.0	4.76	65.7
GOLD	12.5	126.	92.0	33.0	248.	74.4	35.5	89.6	19.3	82.3
SILVER K	.584D-01	.612	.441	.423	1.15	1.76	.168	.424	.123	.399
MERC. K	.122D-01	1.09	.684D-01	.574D-01	1.51	.475	.297D-01	.103	.323D-01	.639D-01
VANAD. K	.649D-02	.154	.547D-01	.959	.364	3.76	.367D-01	.772D-01	.194	.168
PLATIN	.178	2.42	1.11	5.93	6.80	26.7	.683	1.43	.551	2.86
TITAN. K	6.51	88.0	40.3	65.6	153.	135.	20.6	63.3	86.5	23.4
TIN K	.268	20.9	9.86	8.00	24.2	42.5	3.83	5.09	7.11	4.36
SILIC. K	4.09	67.0	24.7	76.4	147.	279.	34.9	37.7	14.2	172.
FLOUR. M	.117D-02	.174D-01	.565D-02	.102	.276D-01	.144	.472D-02	.106D-01	.178D-01	.856D-02
POTASH M	.292	4.82	1.05	.443	2.98	1.62	.301	2.74	.127	.743
SD.ASH M	.478D-01	.711	.333	.487	1.31	2.90	.185	.469	.305	.474
BORATE K	1.49	7.03	7.36	10.4	23.8	60.3	3.68	15.5	1.80	8.85
PHOSPH M	1.16	6.02	4.19	9.24	6.10	4.19	1.23	10.9	1.60	2.96
SULFUR M	.443	2.14	1.62	5.98	2.48	2.69	.499	4.14	.198	1.27
CHLOR. M	.731D-01	.933	.423	1.15	1.09	5.15	.146	.519	.297	.322
MAGNES M	.309D-01	.328	.152	.133	.744	.501	.132	.217	.590D-01	.130

	RUH	SAF	TAF	USA	WEH	WEM	WD TOTAL
MOLYB. K	15.3	.125	.643D-01	29.9	20.7	.811	121.
TUNGS. K	13.3	.314D-01	.152D-01	10.3	5.91	.227	50.2
MANGAN K	.429D+04	647.	8.23	.175D+04	.114D+04	50.4	.132D+05
CHROM. K	429.	174.	5.77	733.	732.	21.1	.332D+04
GOLD	113.	4.89	21.1	279.	852.	64.9	.215D+04
SILVER K	2.58	.108	.101	3.37	3.76	.297	15.8
MERC. K	1.30	.303D-01	.638D-02	1.10	1.33	.689D-01	7.27
VANAD. K	8.77	2.27	.767D-02	9.32	1.14	.475D-01	27.3
PLATIN	8.69	.153	.123	17.5	14.5	.745	90.4
TITAN. K	20.6	5.54	3.21	555.	715.	42.6	.202D+04
TIN K	30.5	2.53	2.54	40.6	66.4	6.30	275.
SILIC. K	515.	2.43	2.75	689.	472.	18.7	.256D+04
FLOUR. M	.454	.491D-01	.318D-02	.551	.110	.479D-02	1.51
POTASH M	6.30	.247	-.137D-01	5.61	5.55	.627	33.4
SD.ASH M	8.16	.807D-01	.415D-01	9.38	5.42	.315	30.6
BORATE K	56.2	1.71	.348	121.	92.7	6.51	419.
PHOSPH M	25.8	.441	-.515D-01	32.4	18.6	2.54	127.
SULFUR M	13.6	.254	.436D-03	11.3	7.79	.982	55.4
CHLOR. M	4.67	.777D-01	.531D-01	13.0	4.82	.302	33.1
MAGNES M	.757	.137D-01	.156D-01	1.05	1.03	.164	5.46

Note: The countries included in each of the sixteen regions are listed in appendix H.

Table 7–33
Net Trade 1980

	AAF	ASC	ASL	CAN	EEM	JAP	LAL	LAM	OCH	OIL
MOLYB. K	-.327	1.20	-1.45	16.5	-6.79	-18.2	-2.08	6.81	-.656	-16.5
TUNGS. K	-.286D-01	9.58	6.27	.766	-8.59	-2.52	3.97	.783	1.44	-.610
MANGAN K	77.9	-136.	90.0	-143.	-.128D+04	-.183D+04	-88.0	.147D+04	894.	.106D+04
CHROM. K	-4.07	-60.2	321.	-28.7	-32.8	-611.	-19.5	-22.9	-4.76	60.2
GOLD	-12.5	-126.	-92.0	55.4	-248.	-74.4	-35.5	-89.6	13.9	-82.3
SILVER K	-.141D-01	-.612	-.205	2.20	-.781	-1.46	2.21	2.16	1.48	-.399
MERC. K	-.122D-01	-.624	-.273D-01	.222	-1.04	-.368	.482D-01	.712	-.323D-01	.323
VANAD. K	-.649D-02	-.138D-07	-.547D-01	-.959	-.364	-3.76	-.367D-01	-.772D-01	-.194	-.168
PLATIN	-.178	-2.42	-1.11	-2.43	-6.80	-26.7	.429	-1.43	-.551	-2.86
TITAN. K	-6.51	-88.0	99.3	423.	-153.	-121.	-20.6	-63.3	477.	-23.4
TIN K	-.268	10.3	118.	-8.00	-22.6	-42.5	31.3	.867	4.45	8.25
SILIC. K	-4.09	-67.0	-4.57	37.7	-134.	-77.4	-34.9	-13.1	-4.24	-172.
FLOUR. M	.133D-01	.127	.114	-.558D-01	.120D-01	-.144	-.472D-02	.340	-.178D-01	-.856D-02
POTASH M	.527	-3.74	-1.05	6.86	.976	-1.62	-.301	-2.66	-.127	-.743
SD.ASH M	-.478D-01	-.711	.361	-.487	1.45	-.203D-01	-.185	-.464	-.305	-.474
BORATE K	-1.49	1.58	-7.36	-10.4	-23.8	-60.3	-3.68	-1.18	-1.80	-8.85
PHOSPH M	30.5	.499	-4.19	-9.24	-6.10	-4.19	-1.23	-10.9	.849	-2.96
SULFUR M	-.443	.869	-1.62	.818	1.46	.0	-.499	.623	-.198	1.41
CHLOR. M	-.550D-07	-.135D-07	-.550D-07	-.634D-06	-.695D-06	-.368D-06	-.371D-07	-.241D-07	-.593D-07	-.695D-08
MAGNES M	-.309D-01	1.08	.776D-02	-.363D-02	-.214	-.463	-.132	-.217	-.590D-01	-.130

	RUH	SAF	TAF	USA	WEH	WEM	WD TOTAL
MOLYB. K	-.994	-.125	-.643D-01	44.2	-20.7	-.811	.0
TUNGS. K	-3.34	-.314D-01	-.152D-01	-4.20	-5.17	1.69	.0
MANGAN K	935.	.142D+04	385.	-.167D+04	-.114D+04	-50.4	.0
CHROM. K	643.	567.	343.	-733.	-685.	266.	.0
GOLD	290.	.150D+04	16.4	-199.	-852.	-64.9	.0
SILVER K	.425D-02	.148	-.183D-01	-1.29	-3.32	-.873D-01	.541D-04
MERC. K	.234D-02	-.303D-01	-.638D-02	-.545	-.519	1.90	.0
VANAD. K	4.88	1.21	-.767D-02	-1.95	1.52	-.475D-01	.0
PLATIN	20.3	56.3	-.123	-17.2	-14.5	-.745	.0
TITAN. K	192.	-5.54	-3.21	-276.	-388.	-42.6	.0
TIN K	.0	.676	8.17	-40.6	-62.4	-5.37	.0
SILIC. K	283.	44.0	-2.75	-.895	110.	40.2	.0
FLOUR. M	-.226	.180	-.318D-02	-.441	.559D-01	.586D-01	.0
POTASH M	3.01	-.247	.464	-3.11	2.40	-.627	.0
SD.ASH M	-.346	-.807D-01	-.415D-01	.510	.577	.260	.0
BORATE K	-.177D-01	-1.71	-.348	131.	-92.7	81.5	.0
PHOSPH M	7.97	2.02	2.21	15.9	-18.6	-2.54	.0
SULFUR M	.276	-.254	-.436D-03	.695	-3.25	.107	.0
CHLOR. M	-.873D-06	-.271D-07	-.337D-08	-.213D-06	-.939D-06	-.446D-07	.0
MAGNES M	.218	-.117D-01	-.156D-01	-.103D-01	-.649	.634	.0

Note: The countries included in each of the sixteen regions are listed in appendix H.

Table 7–34
Production 1990

		AAF	ASC	ASL	CAN	EEM	JAP	LAL	LAM	OCH	OIL
MOLYB.	K	.0	4.58	.0	33.0	.0	.500	1.50	17.6	.0	.0
TUNGS.	K	.0	25.7	11.6	3.58	.0	.840	7.46	2.36	2.97	.0
MANGAN	K	155.	997.	786.	.0	161.	44.7	.0	.273D+04	.173D+04	.286D+04
CHROM.	K	.0	.0	559.	.0	600.	.0	.0	25.1	.0	334.
GOLD		.0	.0	.0	142.	.0	.0	.0	.0	59.5	.0
SILVER	K	.595D-01	.0	.236	5.05	.365	.0	5.00	3.41	3.42	.0
MERC.	K	.0	.722	.543D-01	.460	.856	.208	.157	1.40	.0	.875
VANAD.	K	.0	.253	.0	.0	.0	.0	.0	.0	.0	.0
PLATIN		.0	.0	.0	4.89	.0	.0	1.09	.0	.0	.0
TITAN.	K	.0	.0	224.	850.	.0	27.0	.0	.0	995.	.0
TIN	K	.0	51.8	220.	.0	1.60	.0	66.2	12.7	22.9	35.9
SILIC.	K	.0	.0	26.3	187.	24.3	381.	.0	54.7	20.4	.0
FLOUR.	M	.240D-01	.128	.192	.900D-01	.675D-01	.0	.0	.573	.0	.0
POTASH	M	2.54	1.45	.0	14.2	4.79	.0	.0	.121	.0	.0
SD.ASH	M	.0	.0	1.24	.0	5.65	5.66	.0	.956D-02	.0	.0
BORATE	K	.0	13.5	.0	.0	.0	.0	.0	24.9	.0	.0
PHOSPH	M	65.0	9.84	.0	.0	.0	.0	.0	.0	4.11	.0
SULFUR	M	.0	4.85	.0	7.94	5.88	6.44	.0	7.84	.0	9.99
CHLOR.	M	.122	1.38	.592	1.77	1.77	9.03	.339	.918	.507	1.57
MAGNES	M	.0	3.14	.287	.214	1.01	.817D-01	.0	.0	.0	.0

		RUH	SAF	TAF	USA	WEH	WEM	WD TOTAL
MOLYB.	K	23.6	.0	.0	117.	.0	.0	198.
TUNGS.	K	17.2	.0	.0	8.14	1.31	3.50	84.7
MANGAN	K	.952D+04	.335D+04	675.	112.	.0	.0	.231D+05
CHROM.	K	.181D+04	.118D+04	574.	.0	84.4	470.	.563D+04
GOLD		685.	.260D+04	57.3	112.	.0	.0	.366D+04
SILVER	K	4.07	.444	.114	2.87	.434	.210	25.7
MERC.	K	2.41	.0	.0	.760	1.43	3.33	12.7
VANAD.	K	24.8	5.48	.0	9.66	4.76	.0	45.0
PLATIN		50.4	98.5	.0	.491	.0	.0	155.
TITAN.	K	378.	.0	.0	390.	571.	.0	.344D+04
TIN	K	59.5	5.06	19.3	.0	7.61	1.53	504.
SILIC.	K	.151D+04	92.1	.0	893.	.110D+04	112.	.440D+04
FLOUR.	M	.399	.553	.0	.147	.190	.104	2.47
POTASH	M	14.6	.0	.974	2.50	12.2	.0	53.4
SD.ASH	M	15.0	.0	.0	13.5	11.3	1.10	53.5
BORATE	K	106.	.0	.0	427.	.0	177.	749.
PHOSPH	M	54.2	5.21	4.42	78.8	.0	.0	222.
SULFUR	M	19.7	.0	.0	16.9	6.31	1.70	87.5
CHLOR.	M	7.55	.137	.726D-01	17.8	7.84	.503	51.9
MAGNES	M	2.11	.325D-02	.0	1.35	.792	1.80	10.8

Note: The countries included in each of the sixteen regions are listed in appendix H.

Table 7–35
Cumulative Output up to 1990

	AAF	ASC	ASL	CAN	EEM	JAP	LAL	LAM	OCH	OIL
MOLYB. K	.0	36.7	.0	260.	.0	5.00	13.2	136.	.0	.0
TUNGS. K	.0	203.	90.4	29.1	.0	8.40	57.7	17.7	23.0	.0
MANGAN K	.121D+04	.803D+04	.692D+04	.0	.124D+04	447.	.0	.215D+05	.136D+05	.217D+05
CHROM. K	.0	.0	.453D+04	.0	.469D+04	.0	.0	186.	.0	.230D+04
GOLD	.0	.0	.0	.115D+04	.0	.0	.0	.0	464.	.0
SILVER K	.519	.0	2.36	38.4	3.65	1.50	36.9	29.9	25.1	.0
MERC. K	.0	5.94	.477	3.69	6.63	1.58	1.17	11.1	.0	6.31
VANAD. K	.0	2.04	.0	.0	.0	.0	.0	.0	.0	.0
PLATIN	.0	.0	.0	41.9	.0	.0	11.0	.0	.0	.0
TITAN. K	.0	.0	.182D+04	.670D+04	.0	204.	.0	.0	.780D+04	.0
TIN K	.0	415.	.174D+04	.0	16.0	.0	507.	93.1	172.	242.
SILIC. K	.0	.0	232.	.150D+04	187.	.291D+04	.0	397.	152.	.0
FLOUR. M	.193	1.36	1.56	.682	.536	.0	.0	4.62	.0	.0
POTASH M	16.8	12.6	.0	108.	43.7	.0	.0	.998	.0	.0
SD.ASH M	.0	.0	9.69	.0	42.1	42.7	.0	.720D-01	.0	.0
BORATE K	.0	111.	.0	.0	.0	.0	.0	196.	.0	.0
PHOSPH M	484.	81.8	.0	.0	.0	.0	.0	.0	32.8	.0
SULFUR M	.0	39.3	.0	73.7	49.1	45.6	.0	63.0	.0	63.3
CHLOR. M	.974	11.5	5.07	14.6	14.3	70.9	2.42	7.19	4.02	9.46
MAGNES M	.0	22.7	2.23	1.72	7.70	.598	.0	.0	.0	.0

	RUH	SAF	TAF	USA	WEH	WEM	WD TOTAL
MOLYB. K	189.	.0	.0	958.	.0	.0	.160D+04
TUNGS. K	136.	.0	.0	71.4	10.2	27.1	674.
MANGAN K	.737D+05	.271D+05	.534D+04	994.	.0	.0	.182D+06
CHROM. K	.144D+05	.960D+04	.461D+04	.0	658.	.379D+04	.448D+05
GOLD	.544D+04	.205D+05	474.	961.	.0	.0	.290D+05
SILVER K	33.3	3.50	.981	24.8	4.34	2.10	207.
MERC. K	18.5	.0	.0	6.55	11.2	26.5	99.7
VANAD. K	193.	44.8	.0	85.1	37.1	.0	362.
PLATIN	397.	775.	.0	4.25	.0	.0	.123D+04
TITAN. K	.295D+04	.0	.0	.335D+04	.449D+04	.0	.273D+05
TIN K	450.	41.3	150.	.0	58.0	12.3	.389D+04
SILIC. K	.115D+05	693.	.0	.790D+04	.841D+04	853.	.348D+05
FLOUR. M	3.14	3.91	.0	1.29	1.78	.834	19.9
POTASH M	119.	.0	7.13	25.0	101.	.0	434.
SD.ASH M	114.	.0	.0	117.	86.7	8.39	420.
BORATE K	813.	.0	.0	.339D+04	.0	.132D+04	.584D+04
PHOSPH M	440.	38.4	32.9	636.	.0	.0	.175D+04
SULFUR M	168.	.0	.0	144.	54.3	14.0	714.
CHLOR. M	61.1	1.07	.628	154.	63.3	4.02	425.
MAGNES M	15.4	.262D-01	.0	12.0	5.87	13.0	81.2

Note: The countries included in each of the sixteen regions are listed in appendix H.

Table 7–36
Consumption 1990

	AAF	ASC	ASL	CAN	EEM	JAP	LAL	LAM	OCH	OIL
MOLYB. K	.764	2.41	2.60	3.24	12.4	34.4	4.21	5.34	1.27	29.6
TUNGS. K	.862D-01	8.41	.285	2.19	15.6	5.80	.292	.952	.374	2.88
MANGAN K	22.0	.122D+04	631.	210.	.252D+04	.355D+04	164.	214.	202.	.105D+04
CHROM. K	7.68	93.9	33.3	47.1	658.	.111D+04	44.4	72.4	9.97	236.
GOLD	16.1	203.	120.	46.1	395.	133.	85.1	163.	35.4	327.
SILVER K	.785D-01	.980	.573	.595	1.79	3.01	.399	.783	.222	1.60
MERC. K	.201D-01	1.69	.903D-01	.844D-01	2.76	.925	.753D-01	.196	.605D-01	.329
VANAD. K	.166D-01	.253	.625D-01	1.49	.666	6.94	.103	.183	.403	.709
PLATIN	.302	3.71	1.80	8.30	12.0	50.5	1.09	2.64	1.02	8.31
TITAN. K	9.18	141.	49.8	107.	278.	263.	50.5	116.	156.	120.
TIN K	.548	33.5	11.5	12.1	46.3	84.2	10.6	11.1	14.9	21.2
SILIC. K	8.96	109.	32.2	111.	274.	528.	61.3	83.6	29.0	411.
FLOUR. M	.250D-02	.263D-01	.808D-02	.198	.481D-01	.254	.129D-01	.221D-01	.409D-01	.493D-01
POTASH M	1.48	6.46	3.46	.413	2.82	4.27	.719	4.22	.177	4.22
SD.ASH M	.811D-01	1.15	.425	.769	2.36	5.70	.484	.927	.592	2.27
BORATE K	5.93	10.3	16.3	13.9	40.3	127.	9.62	27.0	3.44	45.8
PHOSPH M	6.45	8.88	15.1	9.69	6.59	11.8	3.22	18.6	2.48	18.5
SULFUR M	2.19	2.88	5.19	6.09	2.59	6.44	1.17	6.43	.301	6.79
CHLOR. M	.122	1.38	.592	1.77	1.77	9.03	.339	.918	.507	1.57
MAGNES M	.605D-01	.588	.268	.220	1.42	1.08	.382	.492	.126	.709

	RUH	SAF	TAF	USA	WEH	WEM	WD TOTAL
MOLYB. K	25.2	.228	.117	37.8	37.0	1.60	198.
TUNGS. K	23.0	.572D-01	.283D-01	13.7	10.5	.442	84.7
MANGAN K	.791D+04	931.	16.1	.224D+04	.215D+04	79.4	.231D+05
CHROM. K	753.	251.	13.3	963.	.131D+04	34.5	.563D+04
GOLD	183.	7.70	29.0	388.	.144D+04	90.9	.366D+04
SILVER K	4.06	.170	.139	4.66	6.21	.417	25.7
MERC. K	2.40	.492D-01	.896D-02	1.51	2.35	.113	12.7
VANAD. K	16.4	3.37	.148D-01	12.2	2.12	.771D-01	45.0
PLATIN	15.2	.252	.166	24.1	24.6	1.43	155.
TITAN. K	40.9	8.66	4.40	776.	.125D+04	67.2	.344D+04
TIN K	59.5	3.86	4.75	53.1	126.	10.3	504.
SILIC. K	938.	3.64	5.19	894.	877.	31.0	.440D+04
FLOUR. M	.795	.676D-01	.895D-02	.734	.190	.874D-02	2.47
POTASH M	8.51	.680	.391D-01	7.60	7.39	.915	53.4
SD.ASH M	15.6	.140	.589D-01	12.3	10.0	.514	53.5
BORATE K	107.	3.52	.742	159.	169.	10.1	749.
PHOSPH M	38.9	1.33	.174	48.3	27.6	4.07	222.
SULFUR M	19.0	.683	.106	15.3	10.8	1.46	87.5
CHLOR. M	7.55	.137	.726D-01	17.8	7.84	.503	51.9
MAGNES M	1.59	.226D-01	.240D-01	1.37	2.14	.300	10.8

Note: The countries included in each of the sixteen regions are listed in appendix H.

Table 7–37
Net Trade 1990

		AAF	ASC	ASL	CAN	EEM	JAP	LAL	LAM	OCH	OIL
MOLYB.	K	-.764	2.17	-2.60	29.7	-12.4	-33.9	-2.72	12.3	-1.27	-29.6
TUNGS.	K	-.862D-01	17.3	11.3	1.38	-15.6	-4.96	7.17	1.41	2.60	-2.88
MANGAN	K	133.	-223.	154.	-210.	-.236D+04	-.351D+04	-164.	.251D+04	.153D+04	.181D+04
CHROM.	K	-7.68	-93.9	526.	-47.1	-58.3	-.111D+04	-44.4	-47.3	-9.97	98.5
GOLD		-16.1	-203.	-120.	95.8	-395.	-133.	-85.1	-163.	24.1	-327.
SILVER	K	-.190D-01	-.980	-.337	4.46	-1.42	-3.01	4.60	2.62	3.20	-1.60
MERC.	K	-.201D-01	-.965	-.360D-01	.375	-1.90	-.716	.816D-01	1.21	-.605D-01	.546
VANAD.	K	-.166D-01	.0	-.625D-01	-1.49	-.666	-6.94	-.103	-.183	-.403	-.709
PLATIN		-.302	-3.71	-1.80	-3.40	-12.0	-50.5	.0	-2.64	-1.02	-8.31
TITAN.	K	-9.18	-141.	174.	744.	-278.	-236.	-50.5	-116.	839.	-120.
TIN	K	-.548	18.3	209.	-12.1	-44.7	-84.2	55.6	1.54	7.90	14.6
SILIC.	K	-8.96	-109.	-5.97	75.7	-249.	-147.	-61.3	-28.9	-8.64	-411.
FLOUR.	M	.215D-01	.102	.184	-.108	.195D-01	-.254	-.129D-01	.551	-.409D-01	-.493D-01
POTASH	M	1.06	-5.01	-3.46	13.8	1.96	-4.27	-.719	-4.10	-.177	-4.22
SD.ASH	M	-.811D-01	-1.15	.819	-.769	3.29	-.400D-01	-.484	-.918	-.592	-2.27
BORATE	K	-5.93	3.25	-16.3	-13.9	-40.3	-127.	-9.62	-2.06	-3.44	-45.8
PHOSPH	M	58.6	.958	-15.1	-9.69	-6.59	-11.8	-3.22	-18.6	1.63	-18.5
SULFUR	M	-2.19	1.96	-5.19	1.85	3.29	.0	-1.17	1.41	-.301	3.20
CHLOR.	M	.0	.0	.0	.0	.0	.0	.0	.0	.0	.0
MAGNES	M	-.605D-01	2.55	.184D-01	-.600D-02	-.406	-.995	-.382	-.492	-.126	-.709

		RUH	SAF	TAF	USA	WEH	WEM	WD TOTAL
MOLYB.	K	-1.64	-.228	-.117	79.7	-37.0	-1.60	.0
TUNGS.	K	-5.74	-.572D-01	-.283D-01	-5.58	-9.23	3.06	.0
MANGAN	K	.160D+04	.242D+04	659.	-.213D+04	-.215D+04	-79.4	.0
CHROM.	K	.105D+04	928.	561.	-963.	-.122D+04	436.	.0
GOLD		502.	.259D+04	28.3	-276.	-.144D+04	-90.9	.0
SILVER	K	.786D-02	.275	-.254D-01	-1.79	-5.78	-.207	.0
MERC.	K	.396D-02	-.492D-01	-.896D-02	-.754	-.917	3.22	.0
VANAD.	K	8.46	2.10	-.148D-01	-2.55	2.64	-.771D-01	.0
PLATIN		35.2	98.2	-.166	-23.6	-24.6	-1.43	.0
TITAN.	K	337.	-8.66	-4.40	-386.	-678.	-67.2	.0
TIN	K	.0	1.20	14.5	-53.1	-119.	-8.81	.0
SILIC.	K	569.	88.5	-5.19	-1.16	222.	80.7	.0
FLOUR.	M	-.396	.485	-.895D-02	-.587	.0	.948D-01	.0
POTASH	M	6.05	-.680	.935	-5.10	4.83	-.915	.0
SD.ASH	M	-.663	-.140	-.589D-01	1.16	1.31	.588	.0
BORATE	K	-.336D-01	-3.52	-.742	268.	-169.	167.	.0
PHOSPH	M	15.3	3.88	4.24	30.6	-27.6	-4.07	.0
SULFUR	M	.623	-.683	-.106	1.57	-4.51	.242	.172D-04
CHLOR.	M	.0	.0	.0	.0	.0	.0	.0
MAGNES	M	.515	-.193D-01	-.240D-01	-.134D-01	-1.35	1.50	.0

Note: The countries included in each of the sixteen regions are listed in appendix H.

Table 7–38
Production 2000

	AAF	ASC	ASL	CAN	EEM	JAP	LAL	LAM	OCH	OIL
MOLYB. K	.0	7.61	.0	43.7	.0	.500	2.35	28.2	.0	.0
TUNGS. K	.0	41.0	16.1	4.66	.0	.840	10.5	4.42	3.96	.0
MANGAN K	211.	.186D+04	.118D+04	.0	235.	.0	.0	.386D+04	.227D+04	.410D+04
CHROM. K	.0	.0	748.	.0	869.	.0	.0	57.4	.0	534.
GOLD	.0	.0	.0	56.9	.0	.0	.0	.0	151.	.0
SILVER K	.851D-01	.0	.236	8.03	.365	.0	2.94	5.23	5.71	.0
MERC. K	.0	1.41	.875D-01	.673	1.25	.293	.269	2.25	.0	1.43
VANAD. K	.0	.480	.0	.0	.0	.0	.0	.0	.0	.0
PLATIN	.0	.0	.0	5.74	.0	.0	2.72	.0	.0	.0
TITAN. K	.0	.0	337.	.125D+04	.0	37.0	.0	.0	.144D+04	.0
TIN K	.0	86.7	291.	.0	1.60	.0	91.5	28.9	26.6	56.6
SILIC. K	.0	.0	40.8	250.	35.8	520.	.0	126.	21.9	.0
FLOUR. M	.335D-01	.193	.273	.113	.101	.0	.0	.618	.0	.0
POTASH M	3.57	2.70	.0	22.0	5.97	.0	.0	.167	.0	.0
SD.ASH M	.0	.0	2.13	.0	9.18	7.82	.0	.211D-01	.0	.0
BORATE K	.0	23.6	.0	.0	.0	.0	.0	44.2	.0	.0
PHOSPH M	96.7	19.6	.0	.0	.0	.0	.0	.0	6.31	.0
SULFUR M	.0	7.94	.0	9.51	7.18	10.5	.0	10.9	.0	16.3
CHLOR. M	.174	2.50	.929	2.38	2.45	13.2	.690	1.95	.654	3.15
MAGNES M	.0	5.09	.546	.234	1.50	.112	.0	.0	.0	.0

	RUH	SAF	TAF	USA	WEH	WEM	WD TOTAL
MOLYB. K	34.4	.0	.0	155.	.0	.0	272.
TUNGS. K	26.0	.0	.0	10.4	1.54	5.11	125.
MANGAN K	.141D+05	.673D+04	908.	142.	.0	.0	.356D+05
CHROM. K	.254D+04	.197D+04	756.	.0	102.	635.	.822D+04
GOLD	.104D+04	.389D+04	159.	145.	.0	.0	.544D+04
SILVER K	10.0	.784	.180	3.71	.434	.210	37.9
MERC. K	3.57	.0	.0	.996	1.82	5.04	19.1
VANAD. K	36.6	15.5	.0	12.3	6.07	.0	71.0
PLATIN	72.8	139.	.0	.491	.0	.0	221.
TITAN. K	559.	.0	.0	506.	742.	.0	.487D+04
TIN K	90.7	15.0	26.5	.0	9.04	2.77	727.
SILIC. K	.230D+04	151.	.0	.115D+04	.139D+04	184.	.617D+04
FLOUR. M	.624	1.08	.0	.193	.237	.148	3.62
POTASH M	20.1	.0	1.70	2.50	15.5	.0	74.2
SD.ASH M	23.2	.0	.0	17.9	14.8	1.92	77.0
BORATE K	158.	.0	.0	568.	.0	245.	.104D+04
PHOSPH M	76.6	8.59	7.51	110.	.0	.0	325.
SULFUR M	25.4	.0	.0	20.9	7.01	2.39	118.
CHLOR. M	11.5	.372	.121	23.5	10.2	.977	74.7
MAGNES M	3.33	.103D-01	.0	1.75	.971	2.83	16.4

Note: The countries included in each of the sixteen regions are listed in appendix H.

Table 7–39
Cumulative Output up to 2000

	AAF	ASC	ASL	CAN	EEM	JAP	LAL	LAM	OCH	OIL
MOLYB. K	.0	97.7	.0	643.	.0	10.0	32.4	364.	.0	.0
TUNGS. K	.0	536.	229.	70.3	.0	16.8	148.	51.6	57.7	.0
MANGAN K	.304D+04	.223D+05	.168D+05	.0	.322D+04	670.	.0	.544D+05	.336D+05	.565D+05
CHROM. K	.0	.0	.111D+05	.0	.120D+05	.0	.0	599.	.0	.664D+04
GOLD	.0	.0	.C	.215D+04	.0	.0	.0	.0	.151D+04	.0
SILVER K	1.24	.0	4.72	104.	7.30	.0	76.6	73.1	70.8	.0
MERC. K	.0	16.6	1.19	9.36	17.2	4.08	3.30	29.3	.0	17.8
VANAD. K	.0	5.71	.0	.0	.0	.0	.0	.0	.0	.0
PLATIN	.0	.0	.0	95.1	.0	.0	30.0	.0	.0	.0
TITAN. K	.0	.0	.463D+04	.172D+05	.0	524.	.0	.0	.200D+05	.0
TIN K	.0	.111D+04	.429D+04	.0	32.0	.0	.130D+04	301.	420.	705.
SILIC. K	.0	.0	567.	.369D+04	488.	.742D+04	.0	.130D+04	363.	.0
FLOUR. M	.480	2.97	3.89	1.70	1.38	.0	.0	10.6	.0	.0
POTASH M	47.3	33.3	.0	289.	97.5	.0	.0	2.44	.0	.0
SD.ASH M	.0	.0	26.6	.0	116.	110.	.0	.225	.0	.0
BORATE K	.0	296.	.0	.0	.0	.0	.0	541.	.0	.0
PHOSPH M	.129D+04	229.	.0	.0	.0	.0	.0	.0	84.9	.0
SULFUR M	.0	103.	.0	161.	114.	130.	.0	157.	.0	195.
CHLOR. M	2.45	30.9	12.7	35.4	35.4	182.	7.56	21.5	9.82	33.0
MAGNES M	.0	63.9	6.39	3.96	20.2	1.57	.0	.0	.0	.0

	RUH	SAF	TAF	USA	WEH	WEM	WD TOTAL
MOLYB. K	479.	.0	.0	.232D+04	.0	.0	.395D+04
TUNGS. K	352.	.0	.0	164.	24.5	70.1	.172D+04
MANGAN K	.192D+06	.775D+05	.133D+05	.226D+04	.0	.0	.475D+06
CHROM. K	.361D+05	.254D+05	.113D+05	.0	.159D+04	.931D+04	.114D+06
GOLD	.141D+05	.530D+05	.156D+04	.225D+04	.0	.0	.745D+05
SILVER K	104.	9.65	2.45	57.6	8.68	4.20	524.
MERC. K	48.4	.0	.0	15.3	27.4	68.4	258.
VANAD. K	500.	150.	.0	195.	91.3	.0	942.
PLATIN	.101D+04	.196D+04	.0	9.16	.0	.0	.311D+04
TITAN. K	.763D+04	.0	.0	.783D+04	.111D+05	.0	.688D+05
TIN K	.120D+04	141.	379.	.0	141.	33.8	.100D+05
SILIC. K	.306D+05	.191D+04	.0	.181D+05	.209D+05	.233D+04	.876D+05
FLOUR. M	8.25	12.1	.0	2.99	3.91	2.09	50.3
POTASH M	292.	.0	20.5	50.0	239.	.0	.107D+04
SD.ASH M	305.	.0	.0	274.	217.	23.5	.107D+04
BORATE K	.213D+04	.0	.0	.837D+04	.0	.343D+04	.148D+05
PHOSPH M	.109D+04	107.	92.5	.158D+04	.0	.0	.448D+04
SULFUR M	393.	.0	.0	333.	121.	34.4	.174D+04
CHLOR. M	156.	3.62	1.60	360.	154.	11.4	.106D+04
MAGNES M	42.6	.939D-01	.0	27.5	14.7	36.1	217.

Note: The countries included in each of the sixteen regions are listed in appendix H.

Table 7-40
Consumption 2000

	AAF	ASC	ASL	CAN	EEM	JAP	LAL	LAM	OCH	OIL
MOLYB. K	1.04	4.70	3.39	3.80	17.9	46.3	6.61	11.7	1.37	42.8
TUNGS. K	.109	17.2	.543	2.76	22.6	7.67	.633	2.48	.386	4.39
MANGAN K	31.9	.227D+04	977.	246.	.368D+04	.478D+04	253.	488.	216.	.167D+04
CHROM. K	10.6	176.	55.3	57.7	954.	.150D+04	75.9	166.	11.0	405.
GOLD	23.2	385.	190.	56.9	572.	194.	144.	331.	43.3	654.
SILVER K	.112	1.83	.915	.741	2.56	4.38	.682	1.59	.273	3.12
MERC. K	.288D-01	3.29	.145	.111	4.05	1.30	.147	.439	.720D-01	.609
VANAD. K	.237D-01	.480	.109	1.78	.972	9.35	.175	.438	.429	1.20
PLATIN	.437	6.96	2.42	9.73	17.4	70.3	2.72	5.74	1.21	16.1
TITAN. K	13.9	261.	78.5	147.	401.	360.	95.7	245.	196.	238.
TIN K	.786	62.9	18.7	14.5	67.7	113.	19.1	26.9	16.3	37.5
SILIC. K	12.8	205.	50.1	132.	403.	721.	94.3	193.	31.2	640.
FLOUR. M	.336D-02	.502D-01	.156D-01	.250	.734D-01	.347	.225D-01	.499D-01	.449D-01	.878D-01
POTASH M	1.91	12.0	4.38	.470	2.90	7.30	1.11	5.84	.253	7.45
SD.ASH M	.123	2.10	.690	1.01	3.39	7.87	.897	2.04	.680	4.19
BORATE K	7.90	19.1	22.3	17.7	56.6	175.	16.8	47.8	4.00	82.1
PHOSPH M	9.17	18.2	21.1	12.3	7.59	21.9	5.50	28.4	3.87	36.1
SULFUR M	2.84	5.37	6.61	7.09	2.88	10.5	1.87	9.10	.409	12.1
CHLOR. M	.174	2.50	.929	2.38	2.45	13.2	.690	1.95	.654	3.15
MAGNES M	.988D-01	1.13	.518	.240	2.10	1.48	.742	1.17	.140	1.23

	RUH	SAF	TAF	USA	WEH	WEM	WD TOTAL
MOLYB. K	36.8	.690	.203	48.2	43.3	3.09	272.
TUNGS. K	34.6	.207	.438D-01	17.6	12.3	.903	125.
MANGAN K	.120D+05	.348D+04	24.6	.286D+04	.252D+04	140.	.356D+05
CHROM. K	.115D+04	752.	17.6	.124D+04	.158D+04	61.1	.822D+04
GOLD	290.	17.2	45.7	503.	.184D+04	142.	.544D+04
SILVER K	6.50	.376	.220	6.02	7.95	.657	37.9
MERC. K	3.56	.142	.147D-01	1.98	2.98	.210	19.1
VANAD. K	25.2	12.7	.225D-01	15.6	2.49	.142	71.0
PLATIN	23.2	.650	.269	31.4	29.7	2.57	221.
TITAN. K	59.8	22.4	7.08	.101D+04	.162D+04	119.	.487D+04
TIN K	90.7	13.4	7.60	68.3	150.	18.7	727.
SILIC. K	.142D+04	13.5	7.92	.115D+04	.105D+04	57.8	.617D+04
FLOUR. M	1.24	.203	.109D-01	.963	.237	.159D-01	3.62
POTASH M	10.6	1.30	.243	9.15	7.97	1.26	74.2
SD.ASH M	24.3	.391	.941D-01	15.9	12.5	.886	77.0
BORATE K	158.	8.28	1.80	201.	205.	15.6	.104D+04
PHOSPH M	53.7	2.78	1.17	64.1	33.1	6.20	325.
SULFUR M	24.6	1.32	.419	18.9	12.0	2.07	118.
CHLOR. M	11.5	.372	.121	23.5	10.2	.977	74.7
MAGNES M	2.53	.714D-01	.359D-01	1.76	2.62	.501	16.4

Note: The countries included in each of the sixteen regions are listed in appendix H.

Table 7-41
Net Trade 2000

	AAF	ASC	ASL	CAN	EEM	JAP	LAL	LAM	OCH	OIL
MOLYB. K	-1.04	2.91	-3.39	39.9	-17.9	-45.8	-4.26	16.5	-1.37	-42.8
TUNGS. K	-.109	23.8	15.6	1.90	-22.6	-6.83	9.86	1.94	3.58	-4.39
MANGAN K	179.	-415.	207.	-246.	-.344D+04	-.478D+04	-253.	.337D+04	.205D+04	.243D+04
CHROM. K	-10.6	-176.	692.	-57.7	-84.5	-.150D+04	-75.9	-108.	-11.0	130.
GOLD	-23.2	-385.	-190.	.0	-572.	-194.	-144.	-331.	107.	-654.
SILVER K	-.271D-01	-1.83	-.679	7.29	-2.20	-4.38	2.26	3.64	5.44	-3.12
MERC. K	-.288D-01	-1.88	-.580D-01	.563	-2.79	-1.01	.122	1.81	-.720D-01	.819
VANAD. K	-.237D-01	.0	-.109	-1.78	-.972	-9.35	-.175	-.438	-.429	-1.20
PLATIN	-.437	-6.96	-2.42	-3.99	-17.4	-70.3	.0	-5.74	-1.21	-16.1
TITAN. K	-13.9	-261.	259.	.110D+04	-401.	-323.	-95.7	-245.	.124D+04	-238.
TIN K	-.786	23.8	272.	-14.5	-66.1	-113.	72.5	2.01	10.3	19.1
SILIC. K	-12.8	-205.	-9.29	118.	-367.	-200.	-94.3	-66.8	-9.27	-640.
FLOUR. M	.301D-01	.143	.258	-.137	.272D-01	-.347	-.225D-01	.568	-.449D-01	-.878D-01
POTASH M	1.66	-9.34	-4.38	21.6	3.07	-7.30	-1.11	-5.68	-.253	-7.45
SD.ASH M	-.123	-2.10	1.44	-1.01	5.79	-.551D-01	-.897	-2.02	-.680	-4.19
BORATE K	-7.90	4.46	-22.3	-17.7	-56.6	-175.	-16.8	-3.66	-4.00	-82.1
PHOSPH M	87.6	1.43	-21.1	-12.3	-7.59	-21.9	-5.50	-28.4	2.44	-36.1
SULFUR M	-2.84	2.57	-6.61	2.42	4.30	.0	-1.87	1.84	-.409	4.18
CHLOR. M	.0	.0	.0	.0	.0	.0	.0	.0	.0	.0
MAGNES M	-.988D-01	3.96	.285D-01	-.654D-02	-.602	-1.36	-.742	-1.17	-.140	-1.23

	RUH	SAF	TAF	USA	WEH	WEM	WD TOTAL
MOLYB. K	-2.39	-.690	-.203	107.	-43.3	-3.09	.0
TUNGS. K	-8.66	-.207	-.438D-01	-7.14	-10.8	4.20	.0
MANGAN K	.215D+04	.325D+04	884.	-.272D+04	-.252D+04	-140.	.0
CHROM. K	.139D+04	.122D+04	739.	-.124D+04	-.148D+04	574.	.0
GOLD	748.	.387D+04	114.	-358.	-.184D+04	-142.	-.143D-03
SILVER K	3.52	.409	-.400D-01	-2.31	-7.52	-.447	.109D-04
MERC. K	.594D-02	-.142	-.147D-01	-.989	-1.16	4.83	.0
VANAD. K	11.5	2.85	-.225D-01	-3.25	3.58	-.142	.0
PLATIN	49.6	138.	-.269	-30.9	-29.7	-2.57	.0
TITAN. K	500.	-22.4	-7.08	-501.	-881.	-119.	.0
TIN K	.0	1.56	18.9	-68.3	-141.	-15.9	.0
SILIC. K	887.	138.	-7.92	-1.49	346.	126.	.0
FLOUR. M	-.619	.881	-.109D-01	-.770	.0	.133	.0
POTASH M	9.45	-1.30	1.46	-6.65	7.53	-1.26	.0
SD.ASH M	-1.03	-.391	-.941D-01	2.03	2.30	1.03	.0
BORATE K	-.498D-01	-8.28	-1.80	368.	-205.	229.	.0
PHOSPH M	22.9	5.80	6.34	45.7	-33.1	-6.20	.0
SULFUR M	.814	-1.32	-.419	2.05	-5.01	.315	.225D-04
CHLOR. M	.0	.0	.0	.0	.0	.0	.0
MAGNES M	.800	-.611D-01	-.359D-01	-.173D-01	-1.65	2.33	.0

Note: The countries included in each of the sixteen regions are listed in appendix H.

Table 7-42
Production 2010—Optimistic Scenario

		AAF	ASC	ASL	CAN	EEM	JAP	LAL	LAM	OCH	OIL
MOLYB.	K	.0	8.90	.0	58.1	.0	.500	3.90	36.4	.0	.0
TUNGS.	K	.0	45.7	19.6	6.40	.0	.840	13.1	5.21	4.91	.0
MANGAN	K	288.	.215D+04	.152D+04	.0	305.	.0	.0	.515D+04	.306D+04	.480D+04
CHROM.	K	.0	.0	.101D+04	.0	.113D+04	.0	.0	75.8	.0	593.
GOLD		.0	.0	.0	76.9	.0	.0	.0	.0	206.	.0
SILVER	K	.129	.0	.236	10.9	.365	.0	1.15	8.74	7.75	.0
MERC.	K	.0	1.77	.120	.877	1.62	.382	.403	2.94	.0	1.77
VANAD.	K	.0	.530	.0	.0	.0	.0	.0	.0	.0	.0
PLATIN		.0	.0	.0	8.31	.0	.0	4.56	.0	.0	.0
TITAN.	K	.0	.0	459.	.167D+04	.0	48.8	.0	.0	.193D+04	.0
TIN	K	.0	106.	392.	.0	1.60	.0	130.	36.6	36.8	56.4
SILIC.	K	.0	.0	51.2	338.	46.1	687.	.0	157.	30.6	.0
FLOUR.	M	.467D-01	.253	.366	.204	.131	.0	.0	.827	.0	.0
POTASH	M	3.92	2.57	.0	23.7	6.13	.0	.0	.191	.0	.0
SD.ASH	M	.0	.0	2.69	.0	11.6	10.3	.0	.283D-01	.0	.0
BORATE	K	.0	26.5	.0	.0	.0	.0	.0	55.0	.0	.0
PHOSPH	M	105.	19.0	.0	.0	.0	.0	.0	.0	5.47	.0
SULFUR	M	.0	8.01	.0	10.7	7.82	11.5	.0	12.5	.0	18.8
CHLOR.	M	.234	3.28	1.29	3.24	3.31	17.8	1.15	2.80	.879	4.34
MAGNES	M	.0	6.40	.646	.357	2.02	.144	.0	.0	.0	.0

		RUH	SAF	TAF	USA	WEH	WEM	WD TOTAL
MOLYB.	K	39.0	.0	.0	201.	.0	.0	348.
TUNGS.	K	30.9	.0	.0	13.8	1.98	6.15	149.
MANGAN	K	.177D+05	.814D+04	.121D+04	198.	.0	.0	.445D+05
CHROM.	K	.331D+04	.267D+04	.101D+04	.0	138.	851.	.108D+05
GOLD		.141D+04	.528D+04	212.	192.	.0	.0	.737D+04
SILVER	K	15.1	1.06	.226	4.90	.434	.210	51.2
MERC.	K	4.52	.0	.0	1.32	2.38	6.52	24.6
VANAD.	K	46.6	17.4	.0	17.1	8.11	.0	89.8
PLATIN		94.6	185.	.0	.491	.0	.0	293.
TITAN.	K	748.	.0	.0	676.	981.	.0	.652D+04
TIN	K	115.	17.4	31.2	.0	12.4	3.60	940.
SILIC.	K	.278D+04	177.	.0	.156D+04	.183D+04	223.	.788D+04
FLOUR.	M	.776	1.53	.0	.264	.318	.198	4.92
POTASH	M	19.9	.0	1.58	2.50	15.9	.0	76.4
SD.ASH	M	30.1	.0	.0	24.7	19.7	2.43	101.
BORATE	K	194.	.0	.0	733.	.0	302.	.131D+04
PHOSPH	M	75.2	8.85	6.97	122.	.0	.0	343.
SULFUR	M	25.0	.0	.0	24.6	7.42	1.87	128.
CHLOR.	M	15.1	.489	.138	31.1	13.8	1.28	100.
MAGNES	M	4.22	.112D-01	.0	2.53	1.31	3.54	21.2

Note: The countries included in each of the sixteen regions are listed in appendix H.

Table 7-43
Cumulative Output up to 2010—Optimistic Scenario

		AAF	ASC	ASL	CAN	EEM	JAP	LAL	LAM	OCH	OIL
MOLYB.	K	.0	180.	.0	.115D+04	.0	15.0	63.7	687.	.0	.0
TUNGS.	K	.0	970.	407.	126.	.0	25.2	265.	99.8	102.	.0
MANGAN	K	.553D+04	.423D+05	.303D+05	.0	.592D+04	670.	.0	.994D+05	.602D+05	.101D+06
CHROM.	K	.0	.0	.198D+05	.0	.220D+05	.0	.0	.127D+04	.0	.123D+05
GOLD		.0	.0	.0	.282D+04	.0	.0	.0	.0	.330D+04	.0
SILVER	K	2.31	.0	7.08	198.	10.9	1.50	97.0	143.	138.	.0
MERC.	K	.0	32.5	2.22	17.1	31.5	7.46	6.67	55.3	.0	33.8
VANAD.	K	.0	10.8	.0	.0	.0	.0	.0	.0	.0	.0
PLATIN		.0	.0	.0	165.	.0	.0	66.4	.0	.0	.0
TITAN.	K	.0	.0	.861D+04	.318D+05	.0	952.	.0	.0	.368D+05	.0
TIN	K	.0	.207D+04	.771D+04	.0	48.0	.0	.240D+04	628.	737.	.127D+04
SILIC.	K	.0	.0	.103D+04	.663D+04	897.	.135D+05	.0	.272D+04	626.	.0
FLOUR.	M	.881	5.20	7.08	3.28	2.54	.0	.0	17.8	.0	.0
POTASH	M	84.8	59.7	.0	517.	158.	.0	.0	4.23	.0	.0
SD.ASH	M	.0	.0	50.7	.0	220.	201.	.0	.472	.0	.0
BORATE	K	.0	546.	.0	.0	.0	.0	.0	.104D+04	.0	.0
PHOSPH	M	.230D+04	422.	.0	.0	.0	.0	.0	.0	144.	.0
SULFUR	M	.0	183.	.0	262.	189.	240.	.0	274.	.0	370.
CHLOR.	M	4.49	59.9	23.7	63.5	64.2	336.	16.8	45.2	17.5	70.5
MAGNES	M	.0	121.	12.4	6.92	37.8	2.85	.0	.0	.0	.0

		RUH	SAF	TAF	USA	WEH	WEM	WD TOTAL
MOLYB.	K	847.	.0	.0	.410D+04	.0	.0	.705D+04
TUNGS.	K	637.	.0	.0	285.	42.1	126.	.309D+04
MANGAN	K	.351D+06	.152D+06	.238D+05	.396D+04	.0	.0	.876D+06
CHROM.	K	.654D+05	.486D+05	.201D+05	.0	.279D+04	.167D+05	.209D+06
GOLD		.263D+05	.988D+05	.342D+04	.393D+04	.0	.0	.139D+06
SILVER	K	229.	18.9	4.48	101.	13.0	6.30	971.
MERC.	K	88.9	.0	.0	26.9	48.4	126.	477.
VANAD.	K	916.	314.	.0	342.	162.	.0	.175D+04
PLATIN		.185D+04	.358D+04	.0	14.1	.0	.0	.568D+04
TITAN.	K	.142D+05	.0	.0	.137D+05	.197D+05	.0	.126D+06
TIN	K	.223D+04	303.	667.	.0	249.	65.6	.184D+05
SILIC.	K	.560D+05	.355D+04	.0	.316D+05	.370D+05	.436D+04	.158D+06
FLOUR.	M	15.3	25.2	.0	5.27	6.68	3.83	93.0
POTASH	M	493.	.0	36.9	75.0	396.	.0	.182D+04
SD.ASH	M	572.	.0	.0	487.	389.	45.3	.197D+04
BORATE	K	.390D+04	.0	.0	.149D+05	.0	.617D+04	.265D+05
PHOSPH	M	.185D+04	194.	165.	.274D+04	.0	.0	.782D+04
SULFUR	M	645.	.0	.0	561.	193.	55.7	.297D+04
CHLOR.	M	290.	7.92	2.89	633.	273.	22.7	.193D+04
MAGNES	M	80.3	.201	.0	48.8	26.1	68.0	405.

Note: The countries included in each of the sixteen regions are listed in appendix H.

Table 7-44
Consumption 2010—Optimistic Scenario

	AAF	ASC	ASL	CAN	EEM	JAP	LAL	LAM	OCH	OIL
MOLYB. K	1.50	5.11	3.80	6.16	22.4	60.0	11.0	14.9	1.80	56.9
TUNGS. K	.199	16.7	.646	4.08	27.5	9.86	1.03	2.85	.549	3.43
MANGAN K	47.0	.263D+04	.124D+04	375.	.478D+04	.631D+04	430.	606.	302.	.153D+04
CHROM. K	17.8	212.	74.2	92.9	.124D+04	.197D+04	125.	219.	15.7	418.
GOLD	34.7	510.	271.	76.9	775.	263.	244.	480.	60.1	827.
SILVER K	.169	2.37	1.30	1.01	3.46	5.98	1.15	2.28	.381	3.87
MERC. K	.415D-01	4.15	.199	.150	5.22	1.70	.245	.603	.966D-01	.710
VANAD. K	.365D-01	.530	.142	2.62	1.25	12.4	.297	.523	.612	.821
PLATIN	.607	8.94	3.00	14.1	23.0	93.0	4.56	8.01	1.59	22.1
TITAN. K	19.6	364.	112.	195.	527.	475.	161.	356.	260.	286.
TIN K	1.21	73.9	24.7	21.0	88.5	149.	32.3	33.9	22.8	30.6
SILIC. K	19.0	233.	62.9	200.	519.	952.	160.	240.	43.5	631.
FLOUR. M	.640D-02	.624D-01	.209D-01	.450	.949D-01	.453	.354D-01	.678D-01	.657D-01	.959D-01
POTASH M	2.14	11.5	5.05	.485	2.84	7.58	1.50	6.66	.174	8.66
SD.ASH M	.167	2.77	.947	1.39	4.55	10.4	1.52	2.74	.940	3.77
BORATE K	9.21	21.0	27.2	22.8	73.4	223.	27.1	59.6	5.17	73.6
PHOSPH M	10.3	17.4	24.3	12.9	7.62	23.0	7.49	32.4	2.82	42.1
SULFUR M	3.19	5.18	7.65	8.08	3.08	11.5	2.60	10.5	.345	14.2
CHLOR. M	.234	3.28	1.29	3.24	3.31	17.8	1.15	2.80	.879	4.34
MAGNES M	.121	1.45	.610	.367	2.84	1.90	1.16	1.49	.196	.997

	RUH	SAF	TAF	USA	WEH	WEM	WD TOTAL
MOLYB. K	41.7	.784	.122	62.3	56.0	3.59	348.
TUNGS. K	41.2	.232	.150D-01	23.2	15.9	1.03	149.
MANGAN K	.148D+05	.377D+04	16.3	.398D+04	.349D+04	186.	.445D+05
CHROM. K	.145D+04	.103D+04	15.4	.170D+04	.214D+04	78.2	.108D+05
GOLD	394.	23.2	58.0	665.	.251D+04	183.	.737D+04
SILVER K	8.80	.508	.277	7.95	10.8	.843	51.2
MERC. K	4.52	.178	.156D-01	2.63	3.91	.267	24.6
VANAD. K	31.7	13.7	.145D-01	21.6	3.43	.184	89.8
PLATIN	28.7	.825	.314	41.3	39.3	3.27	293.
TITAN. K	77.8	29.0	8.21	.134D+04	.215D+04	157.	.652D+04
TIN K	115.	15.3	5.68	95.1	206.	24.3	940.
SILIC. K	.174D+04	14.8	5.27	.156D+04	.142D+04	75.1	.788D+04
FLOUR. M	1.55	.354	.919D-02	1.32	.318	.208D-01	4.92
POTASH M	9.79	1.17	.131D-01	10.2	7.80	.837	76.4
SD.ASH M	31.4	.460	.968D-01	22.2	16.9	1.18	101.
BORATE K	195.	8.76	.941	278.	268.	17.9	.131D+04
PHOSPH M	50.3	2.54	.773D-01	72.5	33.2	4.21	343.
SULFUR M	24.1	1.24	.765D-01	22.3	12.7	1.52	128.
CHLOR. M	15.1	.489	.138	31.1	13.8	1.28	100.
MAGNES M	3.22	.777D-01	.338D-01	2.56	3.54	.631	21.2

Note: The countries included in each of the sixteen regions are listed in appendix H.

Table 7–45
Net Trade 2010—Optimistic Scenario

		AAF	ASC	ASL	CAN	EEM	JAP	LAL	LAM	OCH	OIL
MOLYB.	K	-1.50	3.79	-3.80	51.9	-22.4	-59.5	-7.10	21.4	-1.80	-56.9
TUNGS.	K	-.199	29.0	19.0	2.32	-27.5	-9.02	12.0	2.37	4.36	-3.43
MANGAN	K	241.	-480.	278.	-375.	-.448D+04	-.631D+04	-430.	.454D+04	.276D+04	.327D+04
CHROM.	K	-17.8	-212.	933.	-92.9	-110.	-.197D+04	-125.	-143.	-15.7	175.
GOLD		-34.7	-510.	-271.	.0	-775.	-263.	-244.	-480.	146.	-827.
SILVER	K	-.410D-01	-2.37	-1.07	9.88	-3.09	-5.98	.0	6.46	7.36	-3.87
MERC.	K	-.415D-01	-2.37	-.793D-01	.728	-3.60	-1.32	.158	2.34	-.966D-01	1.06
VANAD.	K	-.365D-01	.0	-.142	-2.62	-1.25	-12.4	-.297	-.523	-.612	-.821
PLATIN		-.607	-8.94	-3.00	-5.78	-23.0	-93.0	.0	-8.01	-1.59	-22.1
TITAN.	K	-19.6	-364.	347.	.148D+04	-527.	-426.	-161.	-356.	.167D+04	-286.
TIN	K	-1.21	32.2	368.	-21.0	-86.9	-149.	98.0	2.71	13.9	25.8
SILIC.	K	-19.0	-233.	-11.6	139.	-473.	-264.	-160.	-83.2	-13.0	-631.
FLOUR.	M	.403D-01	.191	.345	-.246	.364D-01	-.453	-.354D-01	.760	-.657D-01	-.959D-01
POTASH	M	1.78	-8.90	-5.05	23.2	3.30	-7.58	-1.50	-6.46	-.174	-8.66
SD.ASH	M	-.167	-2.77	1.74	-1.39	7.00	-.728D-01	-1.52	-2.72	-.940	-3.77
BORATE	K	-9.21	5.52	-27.2	-22.8	-73.4	-223.	-27.1	-4.56	-5.17	-73.6
PHOSPH	M	95.2	1.56	-24.3	-12.9	-7.62	-23.0	-7.49	-32.4	2.65	-42.1
SULFUR	M	-3.19	2.83	-7.65	2.67	4.75	.0	-2.60	2.03	-.345	4.61
CHLOR.	M	.0	.0	.0	.0	.0	.0	.0	.0	.0	.0
MAGNES	M	-.121	4.95	.357D-01	-.999D-02	-.814	-1.76	-1.16	-1.49	-.196	-.997

		RUH	SAF	TAF	USA	WEH	WEM	WD TOTAL
MOLYB.	K	-2.71	-.784	-.122	139.	-56.0	-3.59	.0
TUNGS.	K	-10.3	-.232	-.150D-01	-9.45	-13.9	5.12	.0
MANGAN	K	.289D+04	.437D+04	.119D+04	-.379D+04	-.349D+04	-186.	.0
CHROM.	K	.187D+04	.165D+04	996.	-.170D+04	-.200D+04	773.	.0
GOLD		.102D+04	.525D+04	154.	-474.	-.251D+04	-183.	-.210D-03
SILVER	K	6.30	.554	-.504D-01	-3.05	-10.4	-.633	.166D-04
MERC.	K	.768D-02	-.178	-.156D-01	-1.31	-1.53	6.25	.0
VANAD.	K	15.0	3.73	-.145D-01	-4.51	4.68	-.184	.0
PLATIN		65.9	184.	-.314	-40.8	-39.3	-3.27	.0
TITAN.	K	670.	-29.0	-8.21	-668.	-.117D+04	-157.	.0
TIN	K	.0	2.11	25.5	-95.1	-194.	-20.7	.0
SILIC.	K	.104D+04	162.	-5.27	-2.03	406.	148.	.0
FLOUR.	M	-.770	1.18	-.919D-02	-1.05	.0	.177	.0
POTASH	M	10.2	-1.17	1.57	-7.74	8.10	-.837	.0
SD.ASH	M	-1.33	-.460	-.968D-01	2.46	2.78	1.25	.0
BORATE	K	-.614D-01	-8.76	-.941	455.	-268.	284.	.0
PHOSPH	M	24.9	6.31	6.89	49.7	-33.2	-4.21	.0
SULFUR	M	.898	-1.24	-.765D-01	2.26	-5.30	.348	.249D-04
CHLOR.	M	.0	.0	.0	.0	.0	.0	.0
MAGNES	M	.999	-.665D-01	-.338D-01	-.251D-01	-2.23	2.91	.0

Note: The countries included in each of the sixteen regions are listed in appendix H.

Table 7–46
Production 2010—Pessimistic Scenario

	AAF	ASC	ASL	CAN	EEM	JAP	LAL	LAM	OCH	OIL
MOLYB. K	.0	8.40	.0	51.4	.0	.500	3.73	31.9	.0	.0
TUNGS. K	.0	41.2	16.8	5.52	.0	.840	11.3	4.38	4.22	.0
MANGAN K	250.	.211D+04	.123D+04	.0	260.	.0	.0	.441D+04	.262D+04	.418D+04
CHROM. K	.0	.0	883.	.0	979.	.0	.0	65.7	.0	545.
GOLD	.0	.0	.0	68.7	.0	.0	.0	.0	188.	.0
SILVER K	.117	.0	.236	9.94	.365	.0	1.17	7.94	7.07	.0
MERC. K	.0	1.76	.107	.806	1.45	.344	.390	2.69	.0	1.66
VANAD. K	.0	.518	.0	.0	.0	.0	.0	.0	.0	.0
PLATIN	.0	.0	.0	7.54	.0	.0	4.45	.0	.0	.0
TITAN. K	.0	.0	423.	.156D+04	.0	44.4	.0	.0	.179D+04	.0
TIN K	.0	100.	336.	.0	1.60	.0	116.	31.0	31.2	50.5
SILIC. K	.0	.0	41.6	296.	39.7	593.	.0	133.	25.6	.0
FLOUR. M	.416D-01	.231	.324	.178	.116	.0	.0	.733	.0	.0
POTASH M	4.08	2.61	.0	23.1	5.87	.0	.0	.172	.0	.0
SD.ASH M	.0	.0	2.43	.0	10.4	8.93	.0	.247D-01	.0	.0
BORATE K	.0	25.9	.0	.0	.0	.0	.0	48.2	.0	.0
PHOSPH M	104.	19.2	.0	.0	.0	.0	.0	.0	6.07	.0
SULFUR M	.0	8.06	.0	10.4	7.55	10.7	.0	11.5	.0	19.6
CHLOR. M	.231	3.28	1.19	3.00	3.03	16.3	1.14	2.56	.809	4.22
MAGNES M	.0	5.69	.546	.287	1.69	.121	.0	.0	.0	.0

	RUH	SAF	TAF	USA	WEH	WEM	WD TOTAL
MOLYB. K	34.5	.0	.0	178.	.0	.0	309.
TUNGS. K	26.4	.0	.0	11.9	1.71	5.29	130.
MANGAN K	.151D+05	.693D+04	.104D+04	171.	.0	.0	.383D+05
CHROM. K	.290D+04	.235D+04	890.	.0	121.	748.	.948D+04
GOLD	.129D+04	.487D+04	203.	174.	.0	.0	.680D+04
SILVER K	13.6	.971	.237	4.44	.434	.210	46.8
MERC. K	4.05	.0	.0	1.19	2.17	6.00	22.6
VANAD. K	39.8	14.7	.0	14.8	6.93	.0	76.7
PLATIN	85.9	168.	.0	.491	.0	.0	266.
TITAN. K	694.	.0	.0	618.	905.	.0	.603D+04
TIN K	98.8	14.9	28.9	.0	10.7	3.10	824.
SILIC. K	.244D+04	158.	.0	.136D+04	.159D+04	197.	.687D+04
FLOUR. M	.683	1.36	.0	.234	.284	.176	4.36
POTASH M	19.1	.0	1.54	2.50	15.6	.0	74.5
SD.ASH M	25.9	.0	.0	21.8	17.3	2.20	88.9
BORATE K	168.	.0	.0	646.	.0	268.	.116D+04
PHOSPH M	71.8	8.67	6.80	116.	.0	.0	333.
SULFUR M	23.5	.0	.0	22.9	7.15	1.79	123.
CHLOR. M	13.8	.452	.152	28.6	12.7	1.19	92.7
MAGNES M	3.56	.945D-02	.0	2.14	1.10	3.05	18.2

Note: The countries included in each of the sixteen regions are listed in appendix H.

Table 7–47
Cumulative Output up to 2010—Pessimistic Scenario

	AAF	ASC	ASL	CAN	EEM	JAP	LAL	LAM	OCH	OIL
MOLYB. K	.0	178.	.0	.112D+04	.0	15.0	62.8	665.	.0	.0
TUNGS. K	.0	947.	393.	121.	.0	25.2	257.	95.7	98.6	.0
MANGAN K	.535D+04	.421D+05	.288D+05	.0	.570D+04	670.	.0	.958D+05	.581D+05	.979D+05
CHROM. K	.0	.0	.192D+05	.0	.213D+05	.0	.0	.121D+04	.0	.120D+05
GOLD	.0	.0	.0	.277D+04	.0	.0	.0	.0	.321D+04	.0
SILVER K	2.25	.0	7.08	194.	10.9	1.50	97.1	139.	135.	.0
MERC. K	.0	32.4	2.16	16.8	30.7	7.27	6.60	54.0	.0	33.3
VANAD. K	.0	10.7	.0	.0	.0	.0	.0	.0	.0	.0
PLATIN	.0	.0	.0	161.	.0	.0	65.9	.0	.0	.0
TITAN. K	.0	.0	.843D+04	.312D+05	.0	930.	.0	.0	.361D+05	.0
TIN K	.0	.204D+04	.743D+04	.0	48.0	.0	.233D+04	600.	709.	.124D+04
SILIC. K	.0	.0	980.	.642D+04	865.	.130D+05	.0	.260D+04	601.	.0
FLOUR. M	.856	5.09	6.87	3.15	2.46	.0	.0	17.3	.0	.0
POTASH M	85.6	59.9	.0	514.	157.	.0	.0	4.14	.0	.0
SD.ASH M	.0	.0	49.4	.0	214.	194.	.0	.454	.0	.0
BORATE K	.0	543.	.0	.0	.0	.0	.0	.100D+04	.0	.0
PHOSPH M	.230D+04	423.	.0	.0	.0	.0	.0	.0	147.	.0
SULFUR M	.0	183.	.0	261.	188.	236.	.0	269.	.0	374.
CHLOR. M	4.48	59.8	23.3	62.3	62.8	329.	16.7	44.0	17.1	69.9
MAGNES M	.0	118.	11.9	6.57	36.2	2.73	.0	.0	.0	.0

	RUH	SAF	TAF	USA	WEH	WEM	WD TOTAL
MOLYB. K	824.	.0	.0	.399D+04	.0	.0	.685D+04
TUNGS. K	614.	.0	.0	276.	40.7	122.	.299D+04
MANGAN K	.338D+06	.146D+06	.230D+05	.383D+04	.0	.0	.845D+06
CHROM. K	.633D+05	.470D+05	.195D+05	.0	.270D+04	.162D+05	.202D+06
GOLD	.257D+05	.968D+05	.337D+04	.384D+04	.0	.0	.136D+06
SILVER K	222.	18.4	4.53	98.4	13.0	6.30	949.
MERC. K	86.5	.0	.0	26.2	47.4	124.	467.
VANAD. K	882.	301.	.0	330.	156.	.0	.168D+04
PLATIN	.181D+04	.350D+04	.0	14.1	.0	.0	.554D+04
TITAN. K	.139D+05	.0	.0	.135D+05	.193D+05	.0	.123D+06
TIN K	.215D+04	291.	656.	.0	240.	63.1	.178D+05
SILIC. K	.543D+05	.346D+04	.0	.306D+05	.358D+05	.423D+04	.153D+06
FLOUR. M	14.8	24.3	.0	5.12	6.51	3.72	90.2
POTASH M	489.	.0	36.7	75.0	395.	.0	.182D+04
SD.ASH M	551.	.0	.0	472.	377.	44.1	.190D+04
BORATE K	.377D+04	.0	.0	.144D+05	.0	.600D+04	.258D+05
PHOSPH M	.184D+04	194.	164.	.271D+04	.0	.0	.777D+04
SULFUR M	638.	.0	.0	553.	192.	55.3	.295D+04
CHLOR. M	283.	7.74	2.96	621.	268.	22.3	.189D+04
MAGNES M	77.1	.193	.0	46.9	25.0	65.5	390.

Note: The countries included in each of the sixteen regions are listed in appendix H.

Table 7-48
Consumption 2010—Pessimistic Scenario

	AAF	ASC	ASL	CAN	EEM	JAP	LAL	LAM	OCH	OIL
MOLYB. K	1.43	5.05	3.36	5.49	19.4	52.3	10.5	13.0	1.58	51.4
TUNGS. K	.178	16.3	.467	3.52	23.6	8.44	1.01	2.34	.469	3.13
MANGAN K	43.4	.258D+04	991.	319.	.408D+04	.538D+04	419.	509.	250.	.137D+04
CHROM. K	16.2	210.	62.6	80.2	.107D+04	.172D+04	123.	190.	13.4	392.
GOLD	31.3	506.	242.	68.7	692.	237.	248.	431.	53.6	821.
SILVER K	.154	2.35	1.16	.902	3.07	5.38	1.17	2.04	.339	3.83
MERC. K	.393D-01	4.11	.177	.136	4.68	1.53	.244	.537	.871D-01	.686
VANAD. K	.334D-01	.518	.109	2.22	1.07	10.6	.293	.432	.507	.738
PLATIN	.586	8.90	2.84	12.8	20.6	83.8	4.45	7.20	1.44	20.6
TITAN. K	18.3	362.	100.	181.	481.	432.	161.	321.	236.	278.
TIN K	1.11	72.7	19.7	17.9	76.1	128.	32.0	28.7	19.2	28.3
SILIC. K	17.6	228.	51.1	172.	447.	821.	155.	203.	36.5	567.
FLOUR. M	.586D-02	.619D-01	.181D-01	.392	.833D-01	.404	.338D-01	.595D-01	.567D-01	.913D-01
POTASH M	2.34	11.6	4.94	.477	2.66	7.11	1.41	6.00	.222	9.25
SD.ASH M	.159	2.74	.825	1.25	3.92	8.99	1.50	2.39	.794	3.52
BORATE K	9.75	21.0	25.1	20.7	63.3	193.	26.7	52.2	4.49	72.0
PHOSPH M	11.2	17.7	23.8	12.6	7.14	21.6	7.08	29.2	3.48	44.9
SULFUR M	3.49	5.26	7.47	7.76	2.85	10.7	2.46	9.47	.392	15.0
CHLOR. M	.231	3.28	1.19	3.00	3.03	16.3	1.14	2.56	.809	4.22
MAGNES M	.109	1.41	.516	.295	2.37	1.59	1.12	1.25	.159	.891

	RUH	SAF	TAF	USA	WEH	WEM	WD TOTAL
MOLYB. K	36.9	.689	.143	55.1	49.2	3.15	309.
TUNGS. K	35.2	.197	.205D-01	20.1	13.7	.878	130.
MANGAN K	.127D+05	.317D+04	21.6	.343D+04	.297D+04	157.	.383D+05
CHROM. K	.126D+04	900.	15.2	.149D+04	.187D+04	68.9	.948D+04
GOLD	353.	21.3	60.5	602.	.226D+04	166.	.680D+04
SILVER K	7.87	.465	.290	7.20	9.77	.762	46.8
MERC. K	4.05	.161	.173D-01	2.37	3.56	.242	22.6
VANAD. K	27.0	11.5	.193D-01	18.7	2.93	.155	76.7
PLATIN	26.0	.753	.346	37.4	35.7	3.00	266.
TITAN. K	70.7	26.4	8.93	.123D+04	.198D+04	144.	.603D+04
TIN K	98.8	13.1	6.94	82.4	177.	20.9	824.
SILIC. K	.150D+04	12.5	6.89	.136D+04	.123D+04	63.9	.687D+04
FLOUR. M	1.36	.313	.830D-02	1.17	.284	.187D-01	4.36
POTASH M	9.25	1.16	.133D-01	9.57	7.66	.800	74.5
SD.ASH M	27.1	.402	.109	19.5	14.7	1.04	88.9
BORATE K	168.	7.95	1.06	242.	234.	15.7	.116D+04
PHOSPH M	47.5	2.51	.735D-01	67.8	32.4	4.02	333.
SULFUR M	22.6	1.22	.741D-01	20.6	12.3	1.44	123.
CHLOR. M	13.8	.452	.152	28.6	12.7	1.19	92.7
MAGNES M	2.70	.655D-01	.356D-01	2.16	2.96	.535	18.2

Note: The countries included in each of the sixteen regions are listed in appendix H.

Table 7-49
Net Trade 2010—Pessimistic Scenario

	AAF	ASC	ASL	CAN	EEM	JAP	LAL	LAM	OCH	OIL
MOLYB. K	-1.43	3.35	-3.36	45.9	-19.4	-51.8	-6.78	19.0	-1.58	-51.4
TUNGS. K	-.178	24.9	16.3	1.99	-23.6	-7.60	10.3	2.04	3.75	-3.13
MANGAN K	207.	-471.	239.	-319.	-.382D+04	-.538D+04	-419.	.390D+04	.237D+04	.281D+04
CHROM. K	-16.2	-210.	820.	-80.2	-95.1	-.172D+04	-123.	-124.	-13.4	154.
GOLD	-31.3	-506.	-242.	.0	-692.	-237.	-248.	-431.	135.	-821.
SILVER K	-.371D-01	-2.35	-.920	9.03	-2.71	-5.38	.0	5.91	6.73	-3.83
MERC. K	-.393D-01	-2.35	-.706D-01	.670	-3.23	-1.18	.146	2.15	-.871D-01	.976
VANAD. K	-.334D-01	.0	-.109	-2.22	-1.07	-10.6	-.293	-.432	-.507	-.738
PLATIN	-.586	-8.90	-2.84	-5.24	-20.6	-83.8	.0	-7.20	-1.44	-20.6
TITAN. K	-18.3	-362.	323.	.138D+04	-481.	-388.	-161.	-321.	.155D+04	-278.
TIN K	-1.11	27.7	316.	-17.9	-74.5	-128.	84.3	2.33	12.0	22.2
SILIC. K	-17.6	-228.	-9.46	124.	-407.	-228.	-155.	-70.2	-10.9	-567.
FLOUR. M	.357D-01	.169	.306	-.214	.323D-01	-.404	-.338D-01	.673	-.567D-01	-.913D-01
POTASH M	1.73	-9.03	-4.94	22.6	3.21	-7.11	-1.41	-5.83	-.222	-9.25
SD.ASH M	-.159	-2.74	1.61	-1.25	6.46	-.630D-01	-1.50	-2.37	-.794	-3.52
BORATE K	-9.75	4.91	-25.1	-20.7	-63.3	-193.	-26.7	-3.99	-4.49	-72.0
PHOSPH M	93.0	1.52	-23.8	-12.6	-7.14	-21.6	-7.08	-29.2	2.59	-44.9
SULFUR M	-3.49	2.80	-7.47	2.64	4.70	.0	-2.46	2.01	-.392	4.57
CHLOR. M	.0	.0	.0	.0	.0	.0	.0	.0	.0	.0
MAGNES M	-.109	4.27	.308D-01	-.803D-02	-.681	-1.47	-1.12	-1.25	-.159	-.891

	RUH	SAF	TAF	USA	WEH	WEM	WD TOTAL
MOLYB. K	-2.39	-.689	-.143	123.	-49.2	-3.15	.0
TUNGS. K	-8.81	-.197	-.205D-01	-8.18	-12.0	4.41	.0
MANGAN K	.249D+04	.376D+04	.102D+04	-.326D+04	-.297D+04	-157.	.0
CHROM. K	.164D+04	.145D+04	875.	-.149D+04	-.175D+04	680.	.0
GOLD	939.	.485D+04	143.	-429.	-.226D+04	-166.	-.194D-03
SILVER K	5.76	.506	-.528D-01	-2.76	-9.34	-.552	.152D-04
MERC. K	.707D-02	-.161	-.173D-01	-1.18	-1.39	5.76	.0
VANAD. K	12.8	3.19	-.193D-01	-3.90	4.01	-.155	.0
PLATIN	60.0	167.	-.346	-36.9	-35.7	-3.00	.0
TITAN. K	623.	-26.4	-8.93	-611.	-.108D+04	-144.	.0
TIN K	.0	1.82	22.0	-82.4	-167.	-17.8	.0
SILIC. K	935.	145.	-6.89	-1.77	365.	133.	.0
FLOUR. M	-.677	1.04	-.830D-02	-.932	.0	.157	.0
POTASH M	9.89	-1.16	1.53	-7.07	7.89	-.800	.0
SD.ASH M	-1.15	-.402	-.109	2.27	2.57	1.15	.0
BORATE K	-.532D-01	-7.95	-1.06	404.	-234.	252.	.0
PHOSPH M	24.3	6.16	6.73	48.5	-32.4	-4.02	.0
SULFUR M	.889	-1.22	-.741D-01	2.24	-5.11	.345	.246D-04
CHLOR. M	.0	.0	.0	.0	.0	.0	.0
MAGNES M	.862	-.561D-01	-.356D-01	-.212D-01	-1.87	2.51	.0

Note: The countries included in each of the sixteen regions are listed in appendix H.

Table 7-50
Production 2020—Optimistic Scenario

	AAF	ASC	ASL	CAN	EEM	JAP	LAL	LAM	OCH	OIL
MOLYB. K	.0	12.3	.0	82.2	.0	.500	4.94	51.8	.0	.0
TUNGS. K	.0	64.3	28.4	7.74	.0	.840	18.4	7.34	6.99	.0
MANGAN K	380.	.284D+04	.210D+04	.0	419.	.0	.0	.702D+04	.418D+04	.724D+04
CHROM. K	.0	.0	.139D+04	.0	.157D+04	.0	.0	106.	.0	997.
GOLD	.0	.0	.0	98.8	.0	.0	.0	.0	290.	.0
SILVER K	.165	.0	.236	15.5	.365	.0	1.50	12.6	11.1	.0
MERC. K	.0	2.41	.172	1.19	2.21	.527	.525	4.05	.0	2.69
VANAD. K	.0	.699	.0	.0	.0	.0	.0	.0	.0	.0
PLATIN	.0	.0	.0	10.7	.0	.0	5.97	.0	.0	.0
TITAN. K	.0	.0	642.	.231D+04	.0	65.1	.0	.0	.265D+04	.0
TIN K	.0	143.	540.	.0	1.60	.0	167.	50.6	50.4	95.0
SILIC. K	.0	.0	70.8	472.	63.9	948.	.0	217.	41.6	.0
FLOUR. M	.606D-01	.344	.500	.193	.184	.0	.0	1.12	.0	.0
POTASH M	5.11	3.07	.0	31.1	7.98	.0	.0	.259	.0	.0
SD.ASH M	.0	.0	3.84	.0	16.2	13.9	.0	.394D-01	.0	.0
BORATE K	.0	33.9	.0	.0	.0	.0	.0	75.7	.0	.0
PHOSPH M	139.	22.9	.0	.0	.0	.0	.0	.0	6.95	.0
SULFUR M	.0	9.75	.0	13.7	9.96	14.4	.0	16.8	.0	26.9
CHLOR. M	.313	4.40	1.80	4.37	4.52	24.3	1.55	3.99	1.19	6.93
MAGNES M	.0	8.74	.932	.585	2.69	.195	.0	.0	.0	.0

	RUH	SAF	TAF	USA	WEH	WEM	WD TOTAL
MOLYB. K	52.1	.0	.0	284.	.0	.0	488.
TUNGS. K	43.3	.0	.0	18.9	2.78	8.87	208.
MANGAN K	.237D+05	.108D+05	.164D+04	272.	.0	.0	.606D+05
CHROM. K	.454D+04	.345D+04	.138D+04	.0	191.	.117D+04	.148D+05
GOLD	.200D+04	.752D+04	299.	259.	.0	.0	.105D+05
SILVER K	21.3	1.50	.307	6.60	.434	.210	71.9
MERC. K	6.09	.0	.0	1.80	3.30	8.91	33.9
VANAD. K	64.0	22.4	.0	23.5	11.3	.0	122.
PLATIN	129.	255.	.0	.491	.0	.0	402.
TITAN. K	.103D+04	.0	.0	912.	.134D+04	.0	.896D+04
TIN K	155.	22.6	43.0	.0	17.1	4.96	.129D+04
SILIC. K	.390D+04	260.	.0	.213D+04	.257D+04	324.	.110D+05
FLOUR. M	1.09	1.93	.0	.366	.444	.271	6.51
POTASH M	25.3	.0	2.06	2.50	20.3	.0	97.6
SD.ASH M	39.5	.0	.0	34.1	27.3	3.42	138.
BORATE K	252.	.0	.0	.100D+04	.0	413.	.178D+04
PHOSPH M	94.2	11.5	9.07	162.	.0	.0	445.
SULFUR M	31.2	.0	.0	32.8	9.50	2.36	167.
CHLOR. M	20.3	.663	.178	42.0	19.0	1.76	137.
MAGNES M	5.55	.151D-01	.0	3.51	1.81	4.90	28.9

Note: The countries included in each of the sixteen regions are listed in appendix H.

Table 7-51
Cumulative Output up to 2020—Optimistic Scenario

		AAF	ASC	ASL	CAN	EEM	JAP	LAL	LAM	OCH	OIL
MOLYB.	K	.0	286.	.0	.185D+04	.0	20.0	108.	.113D+04	.0	.0
TUNGS.	K	.0	.152D+04	647.	196.	.0	33.6	422.	163.	161.	.0
MANGAN	K	.887D+04	.672D+05	.483D+05	.0	.955D+04	670.	.0	.160D+06	.965D+05	.161D+06
CHROM.	K	.0	.0	.318D+05	.0	.355D+05	.0	.0	.217D+04	.0	.202D+05
GOLD		.0	.0	.0	.369D+04	.0	.0	.0	.0	.578D+04	.0
SILVER	K	3.78	.0	9.44	331.	14.6	1.50	110.	250.	233.	.0
MERC.	K	.0	53.4	3.68	27.4	50.7	12.0	11.3	90.2	.0	56.1
VANAD.	K	.0	16.9	.0	.0	.0	.0	.0	.0	.0	.0
PLATIN		.0	.0	.0	260.	.0	.0	119.	.0	.0	.0
TITAN.	K	.0	.0	.141D+05	.517D+05	.0	.152D+04	.0	.0	.597D+05	.0
TIN	K	.0	.332D+04	.124D+05	.0	64.0	.0	.389D+04	.106D+04	.117D+04	.203D+04
SILIC.	K	.0	.0	.164D+04	.107D+05	.145D+04	.216D+05	.0	.459D+04	987.	.0
FLOUR.	M	1.42	8.19	11.4	5.27	4.11	.0	.0	27.6	.0	.0
POTASH	M	130.	87.8	.0	791.	228.	.0	.0	6.48	.0	.0
SD.ASH	M	.0	.0	83.3	.0	359.	322.	.0	.810	.0	.0
BORATE	K	.0	848.	.0	.0	.0	.0	.0	.169D+04	.0	.0
PHOSPH	M	.352D+04	631.	.0	.0	.0	.0	.0	.0	206.	.0
SULFUR	M	.0	272.	.0	384.	278.	370.	.0	421.	.0	599.
CHLOR.	M	7.23	98.3	39.2	102.	103.	547.	30.3	79.2	27.8	127.
MAGNES	M	.0	197.	20.2	11.6	61.4	4.54	.0	.0	.0	.0

		RUH	SAF	TAF	USA	WEH	WEM	WD TOTAL
MOLYB.	K	.130D+04	.0	.0	.653D+04	.0	.0	.112D+05
TUNGS.	K	.101D+04	.0	.0	449.	65.9	202.	.487D+04
MANGAN	K	.558D+06	.246D+06	.381D+05	.631D+04	.0	.0	.140D+07
CHROM.	K	.105D+06	.792D+05	.321D+05	.0	.444D+04	.268D+05	.337D+06
GOLD		.433D+05	.163D+06	.597D+04	.618D+04	.0	.0	.228D+06
SILVER	K	411.	31.7	7.14	158.	17.4	8.40	.159D+04
MERC.	K	142.	.0	.0	42.5	76.8	203.	769.
VANAD.	K	.147D+04	513.	.0	545.	259.	.0	.280D+04
PLATIN		.297D+04	.578D+04	.0	19.0	.0	.0	.915D+04
TITAN.	K	.230D+05	.0	.0	.217D+05	.313D+05	.0	.203D+06
TIN	K	.358D+04	503.	.104D+04	.0	396.	108.	.295D+05
SILIC.	K	.894D+05	.573D+04	.0	.501D+05	.590D+05	.710D+04	.252D+06
FLOUR.	M	24.6	42.5	.0	8.42	10.5	6.17	150.
POTASH	M	719.	.0	55.1	100.	578.	.0	.270D+04
SD.ASH	M	920.	.0	.0	781.	624.	74.5	.316D+04
BORATE	K	.613D+04	.0	.0	.236D+05	.0	.974D+04	.420D+05
PHOSPH	M	.270D+04	296.	245.	.416D+04	.0	.0	.118D+05
SULFUR	M	927.	.0	.0	848.	278.	76.8	.445D+04
CHLOR.	M	467.	13.7	4.47	999.	437.	37.9	.312D+04
MAGNES	M	129.	.333	.0	79.1	41.7	110.	655.

Note: The countries included in each of the sixteen regions are listed in appendix H.

Table 7-52
Consumption 2020—Optimistic Scenario

	AAF	ASC	ASL	CAN	EEM	JAP	LAL	LAM	OCH	OIL
MOLYB. K	1.72	6.82	4.78	7.51	31.2	83.4	13.9	20.9	2.33	92.4
TUNGS. K	.138	22.4	.951	4.39	39.0	13.9	.983	3.92	.688	7.82
MANGAN K	51.7	.347D+04	.172D+04	523.	.657D+04	.863D+04	451.	831.	419.	.278D+04
CHROM. K	18.4	288.	110.	108.	.172D+04	.273D+04	143.	305.	19.6	757.
GOLD	45.9	701.	394.	98.8	.109D+04	367.	320.	690.	82.4	.138D+04
SILVER K	.217	3.25	1.89	1.29	4.86	8.33	1.50	3.27	.522	6.54
MERC. K	.519D-01	5.63	.286	.193	7.13	2.34	.308	.859	.128	1.25
VANAD. K	.350D-01	.699	.204	3.55	1.73	17.0	.272	.712	.839	1.76
PLATIN	.790	12.0	3.90	18.1	31.3	127.	5.97	11.3	2.13	35.4
TITAN. K	27.3	496.	164.	268.	698.	634.	209.	510.	352.	472.
TIN K	1.28	99.0	36.0	28.7	121.	204.	33.1	46.9	31.3	59.6
SILIC. K	20.8	311.	86.9	266.	719.	.131D+04	173.	332.	59.2	.114D+04
FLOUR. M	.592D-02	.845D-01	.321D-01	.425	.134	.635	.389D-01	.930D-01	.707D-01	.176
POTASH M	2.77	13.7	6.11	.643	3.65	9.20	1.82	9.04	.211	12.7
SD.ASH M	.220	3.68	1.34	1.92	6.12	14.0	1.70	3.82	1.33	6.65
BORATE K	11.9	26.4	35.3	31.3	97.6	296.	30.1	82.0	7.23	125.
PHOSPH M	13.3	20.8	29.4	17.1	9.79	28.0	9.15	44.1	3.47	61.9
SULFUR M	4.13	6.23	9.30	10.4	4.05	14.4	3.17	14.3	.418	21.2
CHLOR. M	.313	4.40	1.80	4.37	4.52	24.3	1.55	3.99	1.19	6.93
MAGNES M	.163	1.91	.883	.601	3.78	2.57	1.27	2.03	.281	1.79

	RUH	SAF	TAF	USA	WEH	WEM	WD TOTAL
MOLYB. K	55.8	.971	.185	83.8	77.1	5.01	488.
TUNGS. K	57.7	.276	.318D-01	31.9	22.3	1.46	208.
MANGAN K	.198D+05	.480D+04	22.6	.546D+04	.479D+04	256.	.606D+05
CHROM. K	.198D+04	.120D+04	18.4	.233D+04	.296D+04	110.	.148D+05
GOLD	546.	32.0	78.6	898.	.350D+04	249.	.105D+05
SILVER K	12.3	.697	.375	10.7	15.1	1.14	71.9
MERC. K	6.08	.238	.206D-01	3.58	5.41	.371	33.9
VANAD. K	43.1	17.2	.202D-01	29.7	4.74	.254	122.
PLATIN	38.1	1.11	.410	55.7	53.8	4.44	402.
TITAN. K	102.	39.8	10.7	.181D+04	.294D+04	215.	.896D+04
TIN K	155.	19.7	8.01	131.	283.	33.5	.129D+04
SILIC. K	.235D+04	18.7	7.34	.213D+04	.197D+04	104.	.110D+05
FLOUR. M	2.17	.335	.986D-02	1.82	.444	.307D-01	6.51
POTASH M	11.9	1.47	-.257D-02	13.7	9.69	1.04	97.6
SD.ASH M	41.3	.627	.126	30.5	23.3	1.62	138.
BORATE K	252.	11.6	1.20	381.	365.	24.1	.178D+04
PHOSPH M	61.5	3.18	.773D-02	96.8	41.5	5.24	445.
SULFUR M	30.1	1.54	.608D-01	30.0	16.3	1.93	167.
CHLOR. M	20.3	.663	.178	42.0	19.0	1.76	137.
MAGNES M	4.18	.105	.460D-01	3.55	4.90	.888	28.9

Note: The countries included in each of the sixteen regions are listed in appendix H.

Table 7-53
Net Trade 2020—Optimistic Scenario

		AAF	ASC	ASL	CAN	EEM	JAP	LAL	LAM	OCH	OIL
MOLYB.	K	-1.72	5.45	-4.78	74.7	-31.2	-82.9	-8.98	30.9	-2.33	-92.4
TUNGS.	K	-.138	41.9	27.4	3.35	-39.0	-13.0	17.4	3.43	6.30	-7.82
MANGAN	K	328.	-634.	379.	-523.	-.615D+04	-.863D+04	-451.	.619D+04	.376D+04	.445D+04
CHROM.	K	-18.4	-288.	.128D+04	-108.	-153.	-.273D+04	-143.	-199.	-19.6	240.
GOLD		-45.9	-701.	-394.	.0	-.109D+04	-367.	-320.	-690.	208.	-.138D+04
SILVER	K	-.525D-01	-3.25	-1.65	14.3	-4.49	-8.33	.0	9.32	10.6	-6.54
MERC.	K	-.519D-01	-3.22	-.114	.994	-4.92	-1.81	.216	3.20	-.128	1.45
VANAD.	K	-.350D-01	.0	-.204	-3.55	-1.73	-17.0	-.272	-.712	-.839	-1.76
PLATIN		-.790	-12.0	-3.90	-7.45	-31.3	-127.	.0	-11.3	-2.13	-35.4
TITAN.	K	-27.3	-496.	479.	.204D+04	-698.	-569.	-209.	-510.	.230D+04	-472.
TIN	K	-1.28	44.2	504.	-28.7	-119.	-204.	134.	3.72	19.1	35.4
SILIC.	K	-20.8	-311.	-16.1	206.	-656.	-365.	-173.	-115.	-17.6	-.114D+04
FLOUR.	M	.547D-01	.259	.468	-.232	.494D-01	-.635	-.389D-01	1.03	-.707D-01	-.176
POTASH	M	2.34	-10.6	-6.11	30.4	4.33	-9.20	-1.82	-8.78	-.211	-12.7
SD.ASH	M	-.220	-3.68	2.50	-1.92	10.1	-.983D-01	-1.70	-3.78	-1.33	-6.65
BORATE	K	-11.9	7.56	-35.3	-31.3	-97.6	-296.	-30.1	-6.28	-7.23	-125.
PHOSPH	M	125.	2.05	-29.4	-17.1	-9.79	-28.0	-9.15	-44.1	3.48	-61.9
SULFUR	M	-4.13	3.53	-9.30	3.32	5.91	.0	-3.17	2.53	-.418	5.74
CHLOR.	M	.0	.0	.0	.0	.0	.0	.0	.0	.0	.0
MAGNES	M	-.163	6.83	.492D-01	-.164D-01	-1.08	-2.38	-1.27	-2.03	-.281	-1.79

		RUH	SAF	TAF	USA	WEH	WEM	WD TOTAL
MOLYB.	K	-3.62	-.971	-.185	200.	-77.1	-5.01	.0
TUNGS.	K	-14.4	-.276	-.318D-01	-13.0	-19.5	7.41	.0
MANGAN	K	.394D+04	.596D+04	.162D+04	-.519D+04	-.479D+04	-256.	.0
CHROM.	K	.256D+04	.226D+04	.136D+04	-.233D+04	-.277D+04	.106D+04	.0
GOLD		.145D+04	.749D+04	220.	-639.	-.350D+04	-249.	-.300D-03
SILVER	K	9.09	.799	-.684D-01	-4.11	-14.7	-.935	.242D-04
MERC.	K	.105D-01	-.238	-.206D-01	-1.78	-2.12	8.54	.0
VANAD.	K	20.9	5.20	-.202D-01	-6.19	6.52	-.254	.0
PLATIN		91.1	254.	-.410	-55.2	-53.8	-4.44	.0
TITAN.	K	924.	-39.8	-10.7	-902.	-.160D+04	-215.	.115D-04
TIN	K	.0	2.90	35.0	-131.	-266.	-28.6	.0
SILIC.	K	.155D+04	241.	-7.34	-2.77	604.	220.	.0
FLOUR.	M	-1.08	1.60	-.986D-02	-1.46	.0	.241	.0
POTASH	M	13.3	-1.47	2.06	-11.2	10.6	-1.04	.0
SD.ASH	M	-1.75	-.627	-.126	3.53	4.00	1.80	.0
BORATE	K	-.796D-01	-11.6	-1.20	623.	-365.	389.	.0
PHOSPH	M	32.7	8.30	9.06	65.3	-41.5	-5.24	.0
SULFUR	M	1.12	-1.54	-.608D-01	2.82	-6.79	.433	.310D-04
CHLOR.	M	.0	.0	.0	.0	.0	.0	.0
MAGNES	M	1.38	-.894D-01	-.460D-01	-.348D-01	-3.08	4.01	.0

Note: The countries included in each of the sixteen regions are listed in appendix H.

Table 7–54
Production 2020—Pessimistic Scenario

	AAF	ASC	ASL	CAN	EEM	JAP	LAL	LAM	OCH	OIL
MOLYB. K	.0	11.3	.0	68.0	.0	.500	4.55	42.7	.0	.0
TUNGS. K	.0	56.0	22.8	6.15	.0	.840	15.0	5.86	5.61	.0
MANGAN K	299.	.281D+04	.164D+04	.0	323.	.0	.0	.546D+04	.323D+04	.596D+04
CHROM. K	.0	.0	.111D+04	.0	.124D+04	.0	.0	85.9	.0	905.
GOLD	.0	.0	.0	80.4	.0	.0	.0	.0	247.	.0
SILVER K	.138	.0	.236	13.3	.365	.0	1.53	10.7	9.52	.0
MERC. K	.0	2.39	.144	1.03	1.83	.439	.495	3.49	.0	2.46
VANAD. K	.0	.692	.0	.0	.0	.0	.0	.0	.0	.0
PLATIN	.0	.0	.0	8.92	.0	.0	5.69	.0	.0	.0
TITAN. K	.0	.0	554.	.202D+04	.0	54.2	.0	.0	.231D+04	.0
TIN K	.0	132.	420.	.0	1.60	.0	137.	40.1	38.4	84.3
SILIC. K	.0	.0	56.5	384.	50.2	742.	.0	172.	31.2	.0
FLOUR. M	.494D-01	.294	.405	.152	.147	.0	.0	.911	.0	.0
POTASH M	5.22	3.16	.0	29.5	7.35	.0	.0	.220	.0	.0
SD.ASH M	.0	.0	3.32	.0	13.7	10.8	.0	.320D-01	.0	.0
BORATE K	.0	32.8	.0	.0	.0	.0	.0	62.0	.0	.0
PHOSPH M	134.	23.4	.0	.0	.0	.0	.0	.0	7.40	.0
SULFUR M	.0	9.81	.0	12.6	9.21	12.5	.0	14.6	.0	28.0
CHLOR. M	.295	4.37	1.55	3.77	3.78	20.6	1.53	3.40	1.00	6.69
MAGNES M	.0	7.29	.751	.414	1.97	.145	.0	.0	.0	.0

	RUH	SAF	TAF	USA	WEH	WEM	WD TOTAL
MOLYB. K	42.3	.0	.0	234.	.0	.0	403.
TUNGS. K	33.9	.0	.0	15.0	2.23	7.14	170.
MANGAN K	.182D+05	.826D+04	.128D+04	210.	.0	.0	.477D+05
CHROM. K	.360D+04	.277D+04	.111D+04	.0	152.	939.	.119D+05
GOLD	.171D+04	.655D+04	270.	214.	.0	.0	.907D+04
SILVER K	17.7	1.27	.307	5.46	.434	.210	61.2
MERC. K	4.99	.0	.0	1.48	2.80	7.72	29.3
VANAD. K	49.3	17.1	.0	18.2	8.76	.0	94.0
PLATIN	108.	216.	.0	.491	.0	.0	339.
TITAN. K	893.	.0	.0	763.	.116D+04	.0	.775D+04
TIN K	119.	17.5	35.4	.0	13.3	3.92	.104D+04
SILIC. K	.317D+04	222.	.0	.168D+04	.207D+04	271.	.884D+04
FLOUR. M	.865	1.57	.0	.293	.363	.220	5.27
POTASH M	23.5	.0	1.90	2.50	19.3	.0	92.6
SD.ASH M	30.1	.0	.0	27.2	21.9	2.91	110.
BORATE K	194.	.0	.0	810.	.0	339.	.144D+04
PHOSPH M	86.5	10.9	8.36	147.	.0	.0	417.
SULFUR M	27.7	.0	.0	28.4	8.69	2.17	154.
CHLOR. M	16.9	.569	.196	35.4	16.3	1.52	118.
MAGNES M	4.15	.115D-01	.0	2.63	1.36	3.88	22.6

Note: The countries included in each of the sixteen regions are listed in appendix H.

Table 7–55
Cumulative Output up to 2020—Pessimistic Scenario

		AAF	ASC	ASL	CAN	EEM	JAP	LAL	LAM	OCH	OIL
MOLYB.	K	.0	276.	.0	.172D+04	.0	20.0	104.	.104D+04	.0	.0
TUNGS.	K	.0	.143D+04	591.	180.	.0	33.6	388.	147.	148.	.0
MANGAN	K	.810D+04	.667D+05	.432D+05	.0	.862D+04	670.	.0	.145D+06	.873D+05	.149D+06
CHROM.	K	.0	.0	.292D+05	.0	.324D+05	.0	.0	.197D+04	.0	.193D+05
GOLD		.0	.0	.0	.352D+04	.0	.0	.0	.0	.539D+04	.0
SILVER	K	3.52	.0	9.44	310.	14.6	1.50	111.	232.	218.	.0
MERC.	K	.0	53.2	3.41	25.9	47.1	11.2	11.0	84.9	.0	53.9
VANAD.	K	.0	16.7	.0	.0	.0	.0	.0	.0	.0	.0
PLATIN		.0	.0	.0	244.	.0	.0	117.	.0	.0	.0
TITAN.	K	.0	.0	.133D+05	.491D+05	.0	.142D+04	.0	.0	.566D+05	.0
TIN	K	.0	.321D+04	.112D+05	.0	64.0	.0	.360D+04	956.	.106D+04	.191D+04
SILIC.	K	.0	.0	.147D+04	.982D+04	.131D+04	.197D+05	.0	.412D+04	885.	.0
FLOUR.	M	1.31	7.71	10.5	4.80	3.77	.0	.0	25.5	.0	.0
POTASH	M	132.	88.7	.0	777.	223.	.0	.0	6.10	.0	.0
SD.ASH	M	.0	.0	78.2	.0	334.	292.	.0	.737	.0	.0
BORATE	K	.0	837.	.0	.0	.0	.0	.0	.155D+04	.0	.0
PHOSPH	M	.349D+04	636.	.0	.0	.0	.0	.0	.0	214.	.0
SULFUR	M	.0	273.	.0	376.	272.	352.	.0	399.	.0	612.
CHLOR.	M	7.11	98.1	37.0	96.2	96.8	514.	30.1	73.8	26.2	124.
MAGNES	M	.0	183.	18.3	10.1	54.5	4.06	.0	.0	.0	.0

		RUH	SAF	TAF	USA	WEH	WEM	WD TOTAL
MOLYB.	K	.121D+04	.0	.0	.605D+04	.0	.0	.104D+05
TUNGS.	K	916.	.0	.0	411.	60.5	184.	.449D+04
MANGAN	K	.505D+06	.222D+06	.347D+05	.573D+04	.0	.0	.128D+07
CHROM.	K	.958D+05	.725D+05	.295D+05	.0	.406D+04	.247D+05	.309D+06
GOLD		.407D+05	.154D+06	.574D+04	.578D+04	.0	.0	.215D+06
SILVER	K	379.	29.6	7.25	148.	17.4	8.40	.149D+04
MERC.	K	132.	.0	.0	39.6	72.2	192.	726.
VANAD.	K	.133D+04	459.	.0	495.	235.	.0	.253D+04
PLATIN		.278D+04	.542D+04	.0	19.0	.0	.0	.857D+04
TITAN.	K	.218D+05	.0	.0	.204D+05	.296D+05	.0	.192D+06
TIN	K	.324D+04	453.	977.	.0	360.	98.2	.271D+05
SILIC.	K	.823D+05	.536D+04	.0	.458D+05	.541D+05	.657D+04	.231D+06
FLOUR.	M	22.5	38.9	.0	7.75	9.75	5.70	138.
POTASH	M	702.	.0	54.0	100.	569.	.0	.265D+04
SD.ASH	M	831.	.0	.0	717.	573.	69.6	.290D+04
BORATE	K	.558D+04	.0	.0	.217D+05	.0	.904D+04	.387D+05
PHOSPH	M	.263D+04	292.	240.	.402D+04	.0	.0	.115D+05
SULFUR	M	893.	.0	.0	809.	271.	75.1	.433D+04
CHLOR.	M	436.	12.8	4.70	941.	414.	35.8	.295D+04
MAGNES	M	116.	.297	.0	70.8	37.3	100.	594.

Note: The countries included in each of the sixteen regions are listed in appendix H.

Table 7-56
Consumption 2020—Pessimistic Scenario

	AAF	ASC	ASL	CAN	EEM	JAP	LAL	LAM	OCH	OIL
MOLYB. K	1.57	6.80	4.16	6.13	25.2	66.7	12.8	17.1	1.85	79.7
TUNGS. K	.113	22.3	.722	3.45	31.2	10.9	.961	3.10	.534	7.39
MANGAN K	44.6	.344D+04	.135D+04	396.	.506D+04	.661D+04	427.	655.	308.	.250D+04
CHROM. K	15.7	285.	88.2	83.6	.136D+04	.215D+04	139.	248.	15.0	713.
GOLD	38.2	695.	324.	80.4	884.	303.	326.	577.	66.4	.138D+04
SILVER K	.182	3.22	1.55	1.05	3.94	6.86	1.53	2.72	.419	6.50
MERC. K	.463D-01	5.59	.240	.163	5.89	1.95	.307	.715	.106	1.21
VANAD. K	.299D-01	.692	.156	2.69	1.35	13.1	.265	.555	.620	1.66
PLATIN	.722	11.9	3.48	15.1	25.7	106.	5.69	9.54	1.76	31.9
TITAN. K	23.7	492.	136.	233.	582.	528.	208.	429.	294.	458.
TIN K	1.09	98.1	28.0	21.9	94.2	158.	32.6	37.3	23.5	56.8
SILIC. K	18.1	308.	69.3	206.	565.	.103D+04	163.	263.	44.4	.102D+04
FLOUR. M	.514D-02	.839D-01	.262D-01	.336	.107	.516	.371D-01	.766D-01	.551D-01	.168
POTASH M	2.99	14.1	5.88	.597	3.24	8.15	1.76	7.69	.259	13.8
SD.ASH M	.196	3.63	1.10	1.58	4.71	10.9	1.66	3.10	1.00	6.30
BORATE K	12.3	26.6	31.4	26.3	75.4	228.	29.4	67.2	5.62	124.
PHOSPH M	14.4	21.5	28.3	15.8	8.67	24.8	8.85	37.5	4.08	66.8
SULFUR M	4.45	6.40	8.92	9.41	3.50	12.5	3.05	12.2	.450	22.5
CHLOR. M	.295	4.37	1.55	3.77	3.78	20.6	1.53	3.40	1.00	6.69
MAGNES M	.138	1.86	.712	.426	2.76	1.91	1.18	1.59	.202	1.62

	RUH	SAF	TAF	USA	WEH	WEM	WD TOTAL
MOLYB. K	45.2	.788	.181	67.8	62.8	4.09	403.
TUNGS. K	45.2	.215	.287D-01	25.2	17.9	1.17	170.
MANGAN K	.151D+05	.363D+04	22.6	.422D+04	.368D+04	198.	.477D+05
CHROM. K	.155D+04	955.	17.0	.185D+04	.236D+04	89.2	.119D+05
GOLD	445.	26.8	78.5	744.	.290D+04	207.	.907D+04
SILVER K	9.95	.585	.376	8.86	12.5	.950	61.2
MERC. K	4.98	.197	.224D-01	2.95	4.60	.313	29.3
VANAD. K	33.0	13.0	.202D-01	23.1	3.66	.197	94.0
PLATIN	31.4	.935	.447	46.0	45.2	3.79	339.
TITAN. K	84.7	33.1	11.4	.152D+04	.253D+04	184.	.775D+04
TIN K	119.	15.3	8.12	102.	221.	26.5	.104D+04
SILIC. K	.183D+04	14.2	7.35	.168D+04	.154D+04	81.8	.884D+04
FLOUR. M	1.72	.271	.842D-02	1.46	.363	.256D-01	5.27
POTASH M	10.8	1.41	-.584D-01	11.9	9.16	.955	92.6
SD.ASH M	31.4	.501	.133	24.0	18.3	1.31	110.
BORATE K	194.	9.75	.994	297.	289.	19.5	.144D+04
PHOSPH M	55.4	3.05	-.267	84.3	38.9	4.80	417.
SULFUR M	26.6	1.46	-.256D-01	25.7	14.9	1.75	154.
CHLOR. M	16.9	.569	.196	35.4	16.3	1.52	118.
MAGNES M	3.05	.796D-01	.437D-01	2.66	3.67	.693	22.6

Note: The countries included in each of the sixteen regions are listed in appendix H.

Table 7–57
Net Trade 2020—Pessimistic Scenario

		AAF	ASC	ASL	CAN	EEM	JAP	LAL	LAM	OCH	OIL
MOLYB.	K	-1.57	4.51	-4.16	61.9	-25.2	-66.2	-8.27	25.6	-1.85	-79.7
TUNGS.	K	-.113	33.7	22.1	2.70	-31.2	-10.1	14.0	2.76	5.08	-7.39
MANGAN	K	255.	-627.	295.	-396.	-.474D+04	-.661D+04	-427.	.481D+04	.292D+04	.346D+04
CHROM.	K	-15.7	-285.	.103D+04	-83.6	-120.	-.215D+04	-139.	-162.	-15.0	192.
GOLD		-38.2	-695.	-324.	.0	-884.	-303.	-326.	-577.	181.	-.138D+04
SILVER	K	-.439D-01	-3.22	-1.32	12.2	-3.58	-6.86	.0	7.98	9.10	-6.50
MERC.	K	-.463D-01	-3.20	-.957D-01	.863	-4.06	-1.51	.188	2.77	-.106	1.26
VANAD.	K	-.299D-01	.0	-.156	-2.69	-1.35	-13.1	-.265	-.555	-.620	-1.66
PLATIN		-.722	-11.9	-3.48	-6.20	-25.7	-106.	.0	-9.54	-1.76	-31.9
TITAN.	K	-23.7	-492.	419.	.178D+04	-582.	-474.	-208.	-429.	.201D+04	-458.
TIN	K	-1.09	34.3	392.	-21.9	-92.6	-158.	104.	2.89	14.8	27.5
SILIC.	K	-18.1	-308.	-12.8	178.	-515.	-285.	-163.	-91.0	-13.2	-.102D+04
FLOUR.	M	.443D-01	.210	.379	-.184	.400D-01	-.516	-.371D-01	.834	-.551D-01	-.168
POTASH	M	2.22	-11.0	-5.88	28.9	4.12	-8.15	-1.76	-7.47	-.259	-13.8
SD.ASH	M	-.196	-3.63	2.23	-1.58	8.95	-.760D-01	-1.66	-3.07	-1.00	-6.30
BORATE	K	-12.3	6.21	-31.4	-26.3	-75.4	-228.	-29.4	-5.14	-5.62	-124.
PHOSPH	M	119.	1.95	-28.3	-15.8	-8.67	-24.8	-8.85	-37.5	3.32	-66.8
SULFUR	M	-4.45	3.40	-8.92	3.20	5.71	.0	-3.05	2.44	-.450	5.54
CHLOR.	M	.0	.0	.0	.0	.0	.0	.0	.0	.0	.0
MAGNES	M	-.138	5.43	.391D-01	-.116D-01	-.793	-1.76	-1.18	-1.59	-.202	-1.62

		RUH	SAF	TAF	USA	WEH	WEM	WD TOTAL
MOLYB.	K	-2.94	-.788	-.181	166.	-62.8	-4.09	.0
TUNGS.	K	-11.3	-.215	-.287D-01	-10.3	-15.7	5.97	.0
MANGAN	K	.306D+04	.463D+04	.126D+04	-.401D+04	-.368D+04	-198.	.0
CHROM.	K	.205D+04	.181D+04	.110D+04	-.185D+04	-.220D+04	850.	.0
GOLD		.126D+04	.652D+04	192.	-529.	-.290D+04	-207.	-.261D-03
SILVER	K	7.78	.684	-.684D-01	-3.40	-12.0	-.740	.207D-04
MERC.	K	.911D-02	-.197	-.224D-01	-1.47	-1.80	7.41	.0
VANAD.	K	16.3	4.06	-.202D-01	-4.81	5.09	-.197	.0
PLATIN		77.0	215.	-.447	-45.6	-45.2	-3.79	.0
TITAN.	K	808.	-33.1	-11.4	-755.	-.138D+04	-184.	.101D-04
TIN	K	.0	2.26	27.2	-102.	-207.	-22.6	.0
SILIC.	K	.134D+04	208.	-7.35	-2.18	521.	190.	.0
FLOUR.	M	-.858	1.29	-.842D-02	-1.17	.0	.195	.0
POTASH	M	12.7	-1.41	1.96	-9.43	10.1	-.955	.0
SD.ASH	M	-1.33	-.501	-.133	3.14	3.56	1.60	.0
BORATE	K	-.613D-01	-9.75	-.994	512.	-289.	320.	.0
PHOSPH	M	31.1	7.90	8.63	62.2	-38.9	-4.80	.0
SULFUR	M	1.08	-1.46	.256D-01	2.72	-6.21	.418	.299D-04
CHLOR.	M	.0	.0	.0	.0	.0	.0	.0
MAGNES	M	1.10	-.681D-01	-.437D-01	-.261D-01	-2.31	3.19	.0

Note: The countries included in each of the sixteen regions are listed in appendix H.

Table 7–58
Production 2030—Optimistic Scenario

	AAF	ASC	ASL	CAN	EEM	JAP	LAL	LAM	OCH	OIL
MOLYB. K	.0	16.5	.0	112.	.0	.500	6.68	70.9	.0	.0
TUNGS. K	.0	86.6	38.1	11.0	.0	.840	24.7	9.99	9.42	.0
MANGAN K	520.	.381D+04	.277D+04	.0	568.	.0	.0	.952D+04	.566D+04	.990D+04
CHROM. K	.0	.0	.188D+04	.0	.213D+04	.0	.0	147.	.0	.141D+04
GOLD	.0	.0	.0	138.	.0	.0	.0	.0	401.	.0
SILVER K	.238	.0	.236	21.6	.365	.0	1.97	17.6	15.5	.0
MERC. K	.0	3.26	.240	1.61	2.96	.709	.699	5.50	.0	3.76
VANAD. K	.0	.940	.0	.0	.0	.0	.0	.0	.0	.0
PLATIN	.0	.0	.0	14.5	.0	.0	7.95	.0	.0	.0
TITAN. K	.0	.0	886.	.315D+04	.0	87.2	.0	.0	.361D+04	.0
TIN K	.0	194.	733.	.0	1.60	.0	224.	69.7	68.2	131.
SILIC. K	.0	.0	92.9	652.	86.6	.128D+04	.0	299.	56.8	.0
FLOUR. M	.827D-01	.465	.679	.266	.250	.0	.0	1.53	.0	.0
POTASH M	6.32	3.50	.0	38.9	9.88	.0	.0	.327	.0	.0
SD.ASH M	.0	.0	5.29	.0	22.1	18.9	.0	.544D-01	.0	.0
BORATE K	.0	42.9	.0	.0	.0	.0	.0	101.	.0	.0
PHOSPH M	174.	26.4	.0	.0	.0	.0	.0	.0	8.45	.0
SULFUR M	.0	11.4	.0	17.5	12.3	17.7	.0	21.3	.0	35.9
CHLOR. M	.417	5.86	2.49	5.97	6.13	33.0	2.06	5.56	1.61	10.4
MAGNES M	.0	11.8	1.26	.775	3.66	.266	.0	.0	.0	.0

	RUH	SAF	TAF	USA	WEH	WEM	WD TOTAL
MOLYB. K	67.9	.0	.0	384.	.0	.0	659.
TUNGS. K	57.4	.0	.0	25.3	3.74	11.9	279.
MANGAN K	.318D+05	.146D+05	.223D+04	368.	.0	.0	.817D+05
CHROM. K	.612D+04	.469D+04	.187D+04	.0	259.	.158D+04	.201D+05
GOLD	.275D+04	.104D+05	413.	347.	.0	.0	.145D+05
SILVER K	29.4	2.06	.423	8.84	.434	.210	98.8
MERC. K	8.08	.0	.0	2.42	4.47	12.0	45.7
VANAD. K	86.4	30.5	.0	31.7	15.3	.0	165.
PLATIN	174.	348.	.0	.491	.0	.0	545.
TITAN. K	.140D+04	.0	.0	.123D+04	.183D+04	.0	.122D+05
TIN K	208.	30.6	59.5	.0	23.1	6.69	.175D+04
SILIC. K	.526D+04	356.	.0	.287D+04	.349D+04	442.	.149D+05
FLOUR. M	1.46	2.62	.0	.494	.600	.368	8.81
POTASH M	30.5	.0	2.54	2.50	25.0	.0	119.
SD.ASH M	53.2	.0	.0	46.4	37.2	4.66	188.
BORATE K	334.	.0	.0	.135D+04	.0	553.	.238D+04
PHOSPH M	113.	14.2	11.2	207.	.0	.0	554.
SULFUR M	37.6	.0	.0	42.5	11.9	2.78	211.
CHLOR. M	27.2	.897	.241	56.6	25.9	2.40	187.
MAGNES M	7.50	.201D-01	.0	4.81	2.47	6.64	39.2

Note: The countries included in each of the sixteen regions are listed in appendix H.

Table 7-59
Cumulative Output up to 2030—Optimistic Scenario

	AAF	ASC	ASL	CAN	EEM	JAP	LAL	LAM	OCH	OIL
MOLYB. K	.0	430.	.0	.283D+04	.0	25.0	166.	.174D+04	.0	.0
TUNGS. K	.0	.227D+04	979.	290.	.0	42.0	637.	249.	244.	.0
MANGAN K	.134D+05	.100D+06	.727D+05	.0	.145D+05	670.	.0	.243D+06	.146D+06	.247D+06
CHROM. K	.0	.0	.482D+05	.0	.540D+05	.0	.0	.344D+04	.0	.322D+05
GOLD	.0	.0	.0	.488D+04	.0	.0	.0	.0	.923D+04	.0
SILVER K	5.79	.0	11.8	517.	18.2	1.50	128.	400.	366.	.0
MERC. K	.0	81.8	5.74	41.4	76.5	18.2	17.4	138.	.0	88.4
VANAD. K	.0	25.1	.0	.0	.0	.0	.0	.0	.0	.0
PLATIN	.0	.0	.0	387.	.0	.0	189.	.0	.0	.0
TITAN. K	.0	.0	.218D+05	.790D+05	.0	.228D+04	.0	.0	.911D+05	.0
TIN K	.0	.500D+04	.187D+05	.0	80.0	.0	.585D+04	.167D+04	.177D+04	.316D+04
SILIC. K	.0	.0	.246D+04	.163D+05	.220D+04	.328D+05	.0	.717D+04	.148D+04	.0
FLOUR. M	2.13	12.2	17.3	7.56	6.28	.0	.0	40.8	.0	.0
POTASH M	187.	121.	.0	.114D+04	318.	.0	.0	9.41	.0	.0
SD.ASH M	.0	.0	129.	.0	550.	486.	.0	1.28	.0	.0
BORATE K	.0	.123D+04	.0	.0	.0	.0	.0	.258D+04	.0	.0
PHOSPH M	.509D+04	877.	.0	.0	.0	.0	.0	.0	283.	.0
SULFUR M	.0	378.	.0	540.	390.	530.	.0	612.	.0	913.
CHLOR. M	10.9	150.	60.6	153.	157.	834.	48.3	127.	41.8	213.
MAGNES M	.0	300.	31.2	18.4	93.2	6.85	.0	.0	.0	.0

	RUH	SAF	TAF	USA	WEH	WEM	WD TOTAL
MOLYB. K	.190D+04	.0	.0	.987D+04	.0	.0	.170D+05
TUNGS. K	.151D+04	.0	.0	670.	98.5	305.	.730D+04
MANGAN K	.836D+06	.373D+06	.574D+05	.951D+04	.0	.0	.211D+07
CHROM. K	.158D+06	.120D+06	.484D+05	.0	.669D+04	.406D+05	.511D+06
GOLD	.671D+05	.252D+06	.953D+04	.921D+04	.0	.0	.352D+06
SILVER K	665.	49.5	10.8	235.	21.7	10.5	.244D+04
MERC. K	213.	.0	.0	63.6	116.	308.	.117D+04
VANAD. K	.222D+04	777.	.0	821.	392.	.0	.424D+04
PLATIN	.449D+04	.879D+04	.0	23.9	.0	.0	.139D+05
TITAN. K	.352D+05	.0	.0	.324D+05	.472D+05	.0	.309D+06
TIN K	.539D+04	769.	.155D+04	.0	597.	167.	.447D+05
SILIC. K	.135D+06	.881D+04	.0	.751D+05	.893D+05	.109D+05	.382D+06
FLOUR. M	37.3	65.3	.0	12.7	15.7	9.37	227.
POTASH M	998.	.0	78.1	125.	804.	.0	.378D+04
SD.ASH M	.138D+04	.0	.0	.118D+04	947.	115.	.479D+04
BORATE K	.906D+04	.0	.0	.353D+05	.0	.146D+05	.628D+05
PHOSPH M	.374D+04	424.	347.	.600D+04	.0	.0	.168D+05
SULFUR M	.127D+04	.0	.0	.122D+04	384.	103.	.634D+04
CHLOR. M	704.	21.5	6.56	.149D+04	661.	58.7	.474D+04
MAGNES M	194.	.509	.0	121.	63.1	168.	996.

Note: The countries included in each of the sixteen regions are listed in appendix H.

Table 7-60
Consumption 2030—Optimistic Scenario

	AAF	ASC	ASL	CAN	EEM	JAP	LAL	LAM	OCH	OIL
MOLYB. K	2.47	9.07	5.79	10.5	42.0	112.	18.8	28.8	3.19	131.
TUNGS. K	.265	30.3	1.24	6.51	52.4	18.6	1.33	5.39	.957	10.4
MANGAN K	76.4	.466D+04	.226D+04	718.	.890D+04	.117D+05	570.	.114D+04	568.	.388D+04
CHROM. K	27.6	390.	153.	149.	.234D+04	.368D+04	188.	424.	26.8	.108D+04
GOLD	65.7	957.	556.	138.	.148D+04	500.	420.	970.	114.	.205D+04
SILVER K	.314	4.42	2.66	1.80	6.65	11.4	1.97	4.58	.721	9.65
MERC. K	.734D-01	7.62	.398	.266	9.54	3.15	.408	1.20	.173	1.81
VANAD. K	.551D-01	.940	.272	4.94	2.35	23.1	.332	.975	1.15	2.39
PLATIN	1.08	16.1	5.07	24.6	42.2	172.	7.95	15.7	2.87	51.7
TITAN. K	37.4	674.	232.	366.	929.	849.	273.	714.	472.	700.
TIN K	1.95	134.	49.0	39.5	164.	276.	41.7	64.7	42.4	83.5
SILIC. K	30.8	420.	114.	370.	975.	.178D+04	223.	457.	80.8	.159D+04
FLOUR. M	.863D-02	.114	.450D-01	.587	.183	.852	.522D-01	.130	.977D-01	.255
POTASH M	3.39	15.6	7.28	.813	4.46	11.0	2.15	11.4	.243	17.2
SD.ASH M	.299	4.92	1.86	2.61	8.30	19.1	2.12	5.27	1.78	9.54
BORATE K	15.0	32.8	45.0	41.9	131.	396.	37.0	110.	9.61	174.
PHOSPH M	16.3	23.8	35.1	21.7	12.1	33.7	10.9	55.8	4.07	84.1
SULFUR M	5.06	7.16	11.1	13.4	5.13	17.7	3.82	18.2	.514	29.0
CHLOR. M	.417	5.86	2.49	5.97	6.13	33.0	2.06	5.56	1.61	10.4
MAGNES M	.224	2.56	1.20	.796	5.14	3.50	1.57	2.78	.377	2.50

	RUH	SAF	TAF	USA	WEH	WEM	WD TOTAL
MOLYB. K	72.6	1.32	.277	111.	103.	6.65	659.
TUNGS. K	76.6	.388	.552D-01	42.6	30.0	1.94	279.
MANGAN K	.265D+05	.651D+04	35.1	.740D+04	.650D+04	346.	.817D+05
CHROM. K	.266D+04	.163D+04	26.7	.315D+04	.401D+04	149.	.201D+05
GOLD	743.	43.7	108.	.120D+04	.478D+04	334.	.145D+05
SILVER K	16.7	.953	.517	14.3	20.7	1.53	98.8
MERC. K	8.06	.324	.284D-01	4.82	7.33	.502	45.7
VANAD. K	58.1	23.4	.316D-01	40.1	6.44	.343	165.
PLATIN	50.4	1.50	.559	74.6	72.7	5.96	545.
TITAN. K	136.	53.7	14.6	.245D+04	.400D+04	293.	.122D+05
TIN K	208.	26.7	12.0	177.	384.	45.2	.175D+04
SILIC. K	.314D+04	25.5	11.4	.288D+04	.267D+04	141.	.149D+05
FLOUR. M	2.91	.458	.144D-01	2.46	.600	.419D-01	8.81
POTASH M	13.9	1.71	-.391D-01	17.5	11.6	1.16	119.
SD.ASH M	55.6	.843	.176	41.6	31.8	2.20	188.
BORATE K	334.	14.9	1.60	516.	493.	31.8	.238D+04
PHOSPH M	72.0	3.72	-.160	125.	50.5	5.93	554.
SULFUR M	36.2	1.81	.349D-01	39.1	20.3	2.25	211.
CHLOR. M	27.2	.897	.241	56.6	25.9	2.40	187.
MAGNES M	5.63	.140	.649D-01	4.86	6.68	1.20	39.2

Note: The countries included in each of the sixteen regions are listed in appendix H.

Table 7-61
Net Trade 2030—Optimistic Scenario

	AAF	ASC	ASL	CAN	EEM	JAP	LAL	LAM	OCH	OIL
MOLYB. K	-2.47	7.43	-5.79	102.	-42.0	-112.	-12.1	42.1	-3.19	-131.
TUNGS. K	-.265	56.3	36.8	4.50	-52.4	-17.7	23.3	4.60	8.47	-10.4
MANGAN K	444.	-852.	513.	-718.	-.833D+04	-.117D+05	-570.	.837D+04	.509D+04	.603D+04
CHROM. K	-27.6	-390.	.173D+04	-149.	-207.	-.368D+04	-188.	-277.	-26.8	324.
GOLD	-65.7	-957.	-556.	.0	-.148D+04	-500.	-420.	-970.	288.	-.205D+04
SILVER K	-.758D-01	-4.42	-2.42	19.8	-6.29	-11.4	.0	13.0	14.8	-9.65
MERC. K	-.734D-01	-4.36	-.159	1.34	-6.58	-2.44	.291	4.30	-.173	1.95
VANAD. K	-.551D-01	.0	-.272	-4.94	-2.35	-23.1	-.332	-.975	-1.15	-2.39
PLATIN	-1.08	-16.1	-5.07	-10.1	-42.2	-172.	.0	-15.7	-2.87	-51.7
TITAN. K	-37.4	-674.	653.	.278D+04	-929.	-762.	-273.	-714.	.314D+04	-700.
TIN K	-1.95	59.9	684.	-39.5	-162.	-276.	182.	5.04	25.9	47.9
SILIC. K	-30.8	-420.	-21.1	283.	-888.	-494.	-223.	-158.	-24.1	-.159D+04
FLOUR. M	.741D-01	.351	.634	-.321	.670D-01	-.852	-.522D-01	1.40	-.977D-01	-.255
POTASH M	2.93	-12.1	-7.28	38.1	5.42	-11.0	-2.15	-11.1	-.243	-17.2
SD.ASH M	-.299	-4.92	3.43	-2.61	13.8	-.134	-2.12	-5.22	-1.78	-9.54
BORATE K	-15.0	10.1	-45.0	-41.9	-131.	-396.	-37.0	-8.39	-9.61	-174.
PHOSPH M	158.	2.57	-35.1	-21.7	-12.1	-33.7	-10.9	-55.8	4.38	-84.1
SULFUR M	-5.06	4.28	-11.1	4.03	7.18	.0	-3.82	3.07	-.514	6.97
CHLOR. M	.0	.0	.0	.0	.0	.0	.0	.0	.0	.0
MAGNES M	-.224	9.25	.666D-01	-.217D-01	-1.48	-3.24	-1.57	-2.78	-.377	-2.50

	RUH	SAF	TAF	USA	WEH	WEM	WD TOTAL
MOLYB. K	-4.71	-1.32	-.277	273.	-103.	-6.65	.0
TUNGS. K	-19.2	-.388	-.552D-01	-17.3	-26.2	9.96	.0
MANGAN K	.533D+04	.806D+04	.219D+04	-.703D+04	-.650D+04	-346.	.0
CHROM. K	.346D+04	.305D+04	.185D+04	-.315D+04	-.376D+04	.143D+04	-.119D-04
GOLD	.201D+04	.104D+05	305.	-857.	-.478D+04	-334.	-.415D-03
SILVER K	12.6	1.11	-.942D-01	-5.50	-20.2	-1.32	.337D-04
MERC. K	.141D-01	-.324	-.284D-01	-2.40	-2.87	11.5	.0
VANAD. K	28.4	7.06	-.316D-01	-8.38	8.86	-.343	.0
PLATIN	124.	346.	-.559	-74.1	-72.7	-5.96	.0
TITAN. K	.126D+04	-53.7	-14.6	-.122D+04	-.217D+04	-293.	.157D-04
TIN K	.0	3.93	47.5	-177.	-361.	-38.5	.0
SILIC. K	.212D+04	330.	-11.4	-3.73	828.	301.	.0
FLOUR. M	-1.45	2.17	-.144D-01	-1.97	.0	.326	.0
POTASH M	16.7	-1.71	2.58	-15.0	13.3	-1.16	.0
SD.ASH M	-2.35	-.843	-.176	4.84	5.48	2.46	.0
BORATE K	-.105	-14.9	-1.60	836.	-493.	522.	.101D-04
PHOSPH M	41.1	10.4	11.4	82.2	-50.5	-5.93	.0
SULFUR M	1.36	-1.81	-.349D-01	3.42	-8.47	.527	.376D-04
CHLOR. M	.0	.0	.0	.0	.0	.0	.0
MAGNES M	1.87	-.119	-.649D-01	-.477D-01	-4.20	5.44	.0

Note: The countries included in each of the sixteen regions are listed in appendix H.

Table 7-62
Production 2030—Pessimistic Scenario

		AAF	ASC	ASL	CAN	EEM	JAP	LAL	LAM	OCH	OIL
MOLYB.	K	.0	14.8	.0	86.2	.0	.500	5.93	54.5	.0	.0
TUNGS.	K	.0	71.9	28.3	8.07	.0	.840	18.6	7.44	6.97	.0
MANGAN	K	377.	.378D+04	.206D+04	.0	400.	.0	.0	.679D+04	.399D+04	.764D+04
CHROM.	K	.0	.0	.139D+04	.0	.153D+04	.0	.0	112.	.0	.124D+04
GOLD		.0	.0	.0	102.	.0	.0	.0	.0	319.	.0
SILVER	K	.184	.0	.236	17.3	.365	.0	2.01	14.0	12.4	.0
MERC.	K	.0	3.23	.190	1.30	2.25	.545	.643	4.45	.0	3.33
VANAD.	K	.0	.930	.0	.0	.0	.0	.0	.0	.0	.0
PLATIN		.0	.0	.0	11.1	.0	.0	7.42	.0	.0	.0
TITAN.	K	.0	.0	721.	.259D+04	.0	66.4	.0	.0	.295D+04	.0
TIN	K	.0	175.	521.	.0	1.60	.0	170.	51.9	47.2	113.
SILIC.	K	.0	.0	70.7	493.	62.3	917.	.0	223.	38.5	.0
FLOUR.	M	.620D-01	.373	.505	.192	.183	.0	.0	1.14	.0	.0
POTASH	M	6.31	3.61	.0	35.6	8.65	.0	.0	.260	.0	.0
SD.ASH	M	.0	.0	4.39	.0	17.7	13.3	.0	.414D-01	.0	.0
BORATE	K	.0	40.8	.0	.0	.0	.0	.0	77.5	.0	.0
PHOSPH	M	162.	26.9	.0	.0	.0	.0	.0	.0	8.53	.0
SULFUR	M	.0	11.4	.0	15.1	10.8	14.2	.0	17.3	.0	37.3
CHLOR.	M	.373	5.79	2.02	4.80	4.68	25.7	2.01	4.43	1.25	9.92
MAGNES	M	.0	9.34	.967	.485	2.42	.179	.0	.0	.0	.0

		RUH	SAF	TAF	USA	WEH	WEM	WD TOTAL
MOLYB.	K	50.3	.0	.0	292.	.0	.0	504.
TUNGS.	K	41.0	.0	.0	18.2	2.77	8.85	213.
MANGAN	K	.221D+05	.102D+05	.159D+04	257.	.0	.0	.592D+05
CHROM.	K	.443D+04	.344D+04	.138D+04	.0	188.	.117D+04	.149D+05
GOLD		.220D+04	.855D+04	350.	262.	.0	.0	.118D+05
SILVER	K	22.4	1.63	.389	6.66	.434	.210	78.2
MERC.	K	6.02	.0	.0	1.82	3.54	9.79	37.1
VANAD.	K	60.4	21.3	.0	22.4	10.9	.0	116.
PLATIN		135.	273.	.0	.491	.0	.0	427.
TITAN.	K	.114D+04	.0	.0	940.	.147D+04	.0	.989D+04
TIN	K	145.	21.7	44.1	.0	16.5	4.86	.131D+04
SILIC.	K	.396D+04	289.	.0	.205D+04	.260D+04	349.	.110D+05
FLOUR.	M	1.05	1.95	.0	.361	.452	.274	6.54
POTASH	M	27.1	.0	2.17	2.50	22.6	.0	109.
SD.ASH	M	36.6	.0	.0	33.8	27.7	3.76	137.
BORATE	K	232.	.0	.0	.100D+04	.0	423.	.178D+04
PHOSPH	M	98.6	13.0	9.60	175.	.0	.0	495.
SULFUR	M	31.0	.0	.0	33.9	10.2	2.41	184.
CHLOR.	M	20.6	.712	.257	43.8	20.8	1.93	149.
MAGNES	M	5.08	.140D-01	.0	3.25	1.69	4.91	28.3

Note: The countries included in each of the sixteen regions are listed in appendix H.

Table 7-63
Cumulative Output up to 2030—Pessimistic Scenario

		AAF	ASC	ASL	CAN	EEM	JAP	LAL	LAM	OCH	OIL
MOLYB.	K	.0	407.	.0	.249D+04	.0	25.0	157.	.152D+04	.0	.0
TUNGS.	K	.0	.207D+04	847.	251.	.0	42.0	556.	213.	211.	.0
MANGAN	K	.115D+05	.996D+05	.617D+05	.0	.122D+05	670.	.0	.206D+06	.123D+06	.217D+06
CHROM.	K	.0	.0	.417D+05	.0	.462D+05	.0	.0	.296D+04	.0	.300D+05
GOLD		.0	.0	.0	.443D+04	.0	.0	.0	.0	.822D+04	.0
SILVER	K	5.13	.0	11.8	462.	18.2	1.50	128.	356.	327.	.0
MERC.	K	.0	81.3	5.08	37.6	67.4	16.1	16.7	125.	.0	82.9
VANAD.	K	.0	24.9	.0	.0	.0	.0	.0	.0	.0	.0
PLATIN		.0	.0	.0	344.	.0	.0	182.	.0	.0	.0
TITAN.	K	.0	.0	.197D+05	.722D+05	.0	.203D+04	.0	.0	.829D+05	.0
TIN	K	.0	.474D+04	.159D+05	.0	80.0	.0	.514D+04	.142D+04	.148D+04	.290D+04
SILIC.	K	.0	.0	.211D+04	.142D+05	.188D+04	.279D+05	.0	.609D+04	.123D+04	.0
FLOUR.	M	1.87	11.0	15.1	6.52	5.43	.0	.0	35.8	.0	.0
POTASH	M	190.	123.	.0	.110D+04	303.	.0	.0	8.50	.0	.0
SD.ASH	M	.0	.0	117.	.0	491.	413.	.0	1.10	.0	.0
BORATE	K	.0	.120D+04	.0	.0	.0	.0	.0	.225D+04	.0	.0
PHOSPH	M	.497D+04	888.	.0	.0	.0	.0	.0	.0	294.	.0
SULFUR	M	.0	378.	.0	514.	372.	486.	.0	559.	.0	938.
CHLOR.	M	10.5	149.	54.9	139.	139.	745.	47.8	113.	37.5	208.
MAGNES	M	.0	266.	26.9	14.6	76.4	5.68	.0	.0	.0	.0

		RUH	SAF	TAF	USA	WEH	WEM	WD TOTAL
MOLYB.	K	.167D+04	.0	.0	.868D+04	.0	.0	.149D+05
TUNGS.	K	.129D+04	.0	.0	577.	85.5	264.	.641D+04
MANGAN	K	.706D+06	.314D+06	.490D+05	.806D+04	.0	.0	.181D+07
CHROM.	K	.136D+06	.104D+06	.420D+05	.0	.577D+04	.352D+05	.443D+06
GOLD		.602D+05	.229D+06	.884D+04	.816D+04	.0	.0	.319D+06
SILVER	K	579.	44.1	10.7	208.	21.7	10.5	.219D+04
MERC.	K	187.	.0	.0	56.1	104.	280.	.106D+04
VANAD.	K	.188D+04	651.	.0	698.	333.	.0	.358D+04
PLATIN		.400D+04	.786D+04	.0	23.9	.0	.0	.124D+05
TITAN.	K	.320D+05	.0	.0	.289D+05	.428D+05	.0	.280D+06
TIN	K	.456D+04	650.	.137D+04	.0	508.	142.	.389D+05
SILIC.	K	.118D+06	.791D+04	.0	.645D+05	.774D+05	.967D+04	.331D+06
FLOUR.	M	32.1	56.5	.0	11.0	13.8	8.17	197.
POTASH	M	954.	.0	74.3	125.	778.	.0	.366D+04
SD.ASH	M	.116D+04	.0	.0	.102D+04	821.	103.	.413D+04
BORATE	K	.771D+04	.0	.0	.308D+05	.0	.128D+05	.548D+05
PHOSPH	M	.355D+04	412.	330.	.563D+04	.0	.0	.161D+05
SULFUR	M	.119D+04	.0	.0	.112D+04	365.	97.9	.602D+04
CHLOR.	M	624.	19.3	6.97	.134D+04	600.	53.1	.428D+04
MAGNES	M	162.	.425	.0	100.	52.5	144.	848.

Note: The countries included in each of the sixteen regions are listed in appendix H.

Table 7-64
Consumption 2030—Pessimistic Scenario

		AAF	ASC	ASL	CAN	EEM	JAP	LAL	LAM	OCH	OIL
MOLYB.	K	2.12	9.04	4.81	7.84	31.1	82.0	16.7	22.1	2.31	107.
TUNGS.	K	.208	30.0	.895	4.72	38.7	13.4	1.28	4.01	.677	9.61
MANGAN	K	61.9	.462D+04	.170D+04	492.	.626D+04	.813D+04	527.	846.	377.	.337D+04
CHROM.	K	22.0	386.	115.	105.	.168D+04	.265D+04	180.	323.	18.6	.100D+04
GOLD		50.5	946.	425.	102.	.110D+04	378.	430.	757.	82.8	.204D+04
SILVER	K	.243	4.37	2.03	1.33	4.92	8.55	2.01	3.57	.524	9.57
MERC.	K	.616D-01	7.54	.316	.208	7.24	2.42	.404	.935	.131	1.74
VANAD.	K	.444D-01	.930	.200	3.40	1.67	16.1	.321	.716	.765	2.24
PLATIN		.931	15.9	4.28	18.9	31.8	131.	7.42	12.4	2.17	45.1
TITAN.	K	30.4	666.	182.	298.	708.	647.	271.	562.	362.	675.
TIN	K	1.56	133.	36.4	27.3	116.	195.	40.8	48.3	28.8	79.1
SILIC.	K	25.2	415.	86.8	261.	701.	.127D+04	204.	341.	54.9	.137D+04
FLOUR.	M	.702D-02	.112	.345D-01	.423	.134	.634	.487D-01	.998D-01	.693D-01	.241
POTASH	M	3.63	16.1	6.84	.709	3.68	9.01	2.10	9.07	.282	18.7
SD.ASH	M	.251	4.84	1.44	1.98	5.81	13.4	2.06	4.01	1.22	8.97
BURATE	K	15.1	33.0	38.3	32.5	91.9	276.	36.1	84.0	6.75	172.
PHOSPH	M	17.4	24.5	32.9	18.8	9.89	27.5	10.6	44.3	4.50	91.2
SULFUR	M	5.39	7.37	10.4	11.4	4.09	14.2	3.70	14.4	.509	30.8
CHLOR.	M	.373	5.79	2.02	4.80	4.68	25.7	2.01	4.43	1.25	9.92
MAGNES	M	.178	2.47	.917	.498	3.40	2.36	1.43	2.05	.243	2.23

		RUH	SAF	TAF	USA	WEH	WEM	WD TOTAL
MOLYB.	K	53.8	.990	.236	82.0	77.4	5.02	504.
TUNGS.	K	54.6	.279	.422D-01	30.7	22.2	1.44	213.
MANGAN	K	.183D+05	.448D+04	28.4	.517D+04	.455D+04	245.	.592D+05
CHROM.	K	.188D+04	.119D+04	22.3	.227D+04	.292D+04	111.	.149D+05
GOLD		548.	33.7	99.2	908.	.362D+04	253.	.118D+05
SILVER	K	12.3	.733	.476	10.8	15.6	1.16	78.2
MERC.	K	6.01	.246	.294D-01	3.63	5.80	.393	37.1
VANAD.	K	40.1	16.2	.255D-01	28.3	4.54	.244	116.
PLATIN		37.8	1.16	.584	56.3	56.6	4.72	427.
TITAN.	K	102.	41.1	14.6	.187D+04	.322D+04	234.	.989D+04
TIN	K	145.	18.9	10.4	125.	273.	32.8	.131D+04
SILIC.	K	.221D+04	17.8	9.34	.206D+04	.192D+04	102.	.110D+05
FLOUR.	M	2.10	.339	.114D-01	1.80	.452	.324D-01	6.54
POTASH	M	11.8	1.59	-.191	14.1	10.4	1.01	109.
SD.ASH	M	38.2	.619	.170	29.6	23.0	1.65	137.
BORATE	K	232.	11.6	.842	364.	359.	23.8	.178D+04
PHOSPH	M	60.8	3.43	-.899	99.8	44.8	5.13	495.
SULFUR	M	29.8	1.66	-.199	30.7	17.5	1.92	184.
CHLOR.	M	20.6	.712	.257	43.8	20.8	1.93	149.
MAGNES	M	3.69	.970D-01	.559D-01	3.28	4.56	.873	28.3

Note: The countries included in each of the sixteen regions are listed in appendix H.

Table 7-65
Net Trade 2030—Pessimistic Scenario

		AAF	ASC	ASL	CAN	EEM	JAP	LAL	LAM	OCH	OIL
MOLYB.	K	-2.12	5.72	-4.81	78.4	-31.1	-81.5	-10.8	32.4	-2.31	-107.
TUNGS.	K	-.208	41.9	27.4	3.35	-38.7	-12.6	17.4	3.42	6.30	-9.61
MANGAN	K	315.	-844.	364.	-492.	-.586D+04	-.813D+04	-527.	.594D+04	.361D+04	.427D+04
CHROM.	K	-22.0	-386.	.127D+04	-105.	-149.	-.265D+04	-180.	-211.	-18.6	239.
GOLD		-50.5	-946.	-425.	.0	-.110D+04	-378.	-430.	-757.	236.	-.204D+04
SILVER	K	-.587D-01	-4.37	-1.80	15.9	-4.55	-8.55	.0	10.4	11.9	-9.57
MERC.	K	-.616D-01	-4.31	-.126	1.09	-5.00	-1.87	.238	3.52	-.131	1.59
VANAD.	K	-.444D-01	.0	-.200	-3.40	-1.67	-16.1	-.321	-.716	-.765	-2.24
PLATIN		-.931	-15.9	-4.28	-7.74	-31.8	-131.	.0	-12.4	-2.17	-45.1
TITAN.	K	-30.4	-666.	539.	.230D+04	-708.	-581.	-271.	-562.	.259D+04	-675.
TIN	K	-1.56	42.4	484.	-27.3	-115.	-195.	129.	3.57	18.3	34.0
SILIC.	K	-25.2	-415.	-16.1	232.	-639.	-353.	-204.	-118.	-16.3	-.137D+04
FLOUR.	M	.550D-01	.260	.470	-.231	.497D-01	-.634	-.487D-01	1.04	-.693D-01	-.241
POTASH	M	2.68	-12.5	-6.84	34.9	4.97	-9.01	-2.10	-8.81	-.282	-18.7
SD.ASH	M	-.251	-4.84	2.95	-1.98	11.9	-.939D-01	-2.06	-3.97	-1.22	-8.97
BORATE	K	-15.1	7.76	-38.3	-32.5	-91.9	-276.	-36.1	-6.43	-6.75	-172.
PHOSPH	M	145.	2.37	-32.9	-18.8	-9.89	-27.5	-10.6	-44.3	4.04	-91.2
SULFUR	M	-5.39	3.99	-10.4	3.76	6.69	.0	-3.70	2.86	-.509	6.49
CHLOR.	M	.0	.0	.0	.0	.0	.0	.0	.0	.0	.0
MAGNES	M	-.178	6.86	.494D-01	-.136D-01	-.975	-2.18	-1.43	-2.05	-.243	-2.23

		RUH	SAF	TAF	USA	WEH	WEM	WD TOTAL
MOLYB.	K	-3.49	-.990	-.236	210.	-77.4	-5.02	.0
TUNGS.	K	-13.7	-.279	-.422D-01	-12.5	-19.4	7.40	.0
MANGAN	K	.378D+04	.572D+04	.156D+04	-.491D+04	-.455D+04	-245.	.0
CHROM.	K	.255D+04	.225D+04	.136D+04	-.227D+04	-.273D+04	.106D+04	.0
GOLD		.165D+04	.852D+04	250.	-647.	-.362D+04	-253.	-.341D-03
SILVER	K	10.1	.392	-.867D-01	-4.15	-15.1	-.952	.270D-04
MERC.	K	.115D-01	-.246	-.294D-01	-1.81	-2.27	9.40	.0
VANAD.	K	20.3	5.04	-.255D-01	-5.90	6.33	-.244	.0
PLATIN		97.4	272.	-.584	-55.9	-56.6	-4.72	.0
TITAN.	K	.104D+04	-41.1	-14.6	-930.	-.175D+04	-234.	.129D-04
TIN	K	.0	2.79	33.7	-125.	-257.	-28.0	.0
SILIC.	K	.174D+04	271.	-9.34	-2.67	679.	247.	.0
FLOUR.	M	-1.05	1.61	-.114D-01	-1.44	.0	.242	.0
POTASH	M	15.3	-1.59	2.37	-11.6	12.2	-1.01	.0
SD.ASH	M	-1.62	-.619	-.170	4.16	4.71	2.12	.0
BORATE	K	-.733D-01	-11.6	-.842	640.	-359.	399.	.0
PHOSPH	M	37.9	9.61	10.5	75.7	-44.8	-5.13	.0
SULFUR	M	1.27	-1.66	.199	3.19	-7.27	.490	.350D-04
CHLOR.	M	.0	.0	.0	.0	.0	.0	.0
MAGNES	M	1.39	-.830D-01	-.559D-01	-.322D-01	-2.87	4.03	.0

Note: The countries included in each of the sixteen regions are listed in appendix H.

Cumulative regional and world output levels beginning in 1980 are presented in separate tables; these cumulative output levels enable the model to track potential problems in resource exhaustion, either regionally or globally. The potential resource exhaustion problems are based on the assumed export shares, resource endowments, consumption coefficients, and economic growth rates. The problem of resource exhaustion is already confronted in some minerals by some regions beginning in the year 2010. Depletion intensifies across regions and minerals as the projections are extended to the model's terminal year, 2030.

Because of its fifty-year time span, this study does not attempt to reallocate production and export activities after 2000 from those regions which had exhausted their official resource estimates to more favorably endowed regions. In addition to lead and zinc, four other minerals—gold, silver, mercury, and tin—are projected to be exhausted on a global level even when assuming low economic growth for the first thirty years of the next century.

Regional resource exhaustion into the next century is indicated by a box drawn around the cumulative output level of the particular region and mineral, beginning with the year in which the resource is exhausted and continuing in subsequent decades until 2030. As in the case of lead and zinc, new resource discoveries, improvements in use efficiency, conservation, substitution, or increased recycling of these minerals are all possible solutions to the problem of potential resource depletion.

Comparing Results with Other Projections

The projections in the previous sections are compared with other estimates of future resource requirements for the U.S. and world economies in table 7-66. Other global projections of future minerals requirements to the year 2000 have been made by the Bureau of Mines [47], Wilfred Malenbaum [28], and William Watson and Ronald Ridker [38].

Table 7-67 presents indexes of projections of global minerals production for the alternative scenarios—high and low GDP growth—to the year 2030. The last column of table 7-67 provides an index of materials requirements as the world economy moves to a higher trajectory in terms of GDP growth. It compares the requirements, mineral by mineral, of the high-growth scenario (0) with those of the low-growth scenario (P).

Table 7-68 compares the projected average annual growth rates in minerals consumption for the United States generated by the solutions of the World Model with the rates of growth projected by the Bureau of Mines, W. Malenbaum, and R. Ridker and W. Watson.

Table 7-69 presents a set of indexes of U.S. minerals consumption generated from the World Model solution of the twenty-six minerals for

Table 7-66
World Consumption of Nonfuel Minerals: Comparison of Alternative Projected Annual Rates of Growth (1970–2000)
(percentage)

	Bureau of Mines	*W. Malenbaum*[a]	*Ridker and Watson*[b]	*Institute for Economic Analysis*
Iron	2.6	2.95	2.05	4.03
Molybdenum	4.5	—	2.40	4.15
Nickel	4.3	2.94	3.21	2.70
Tungsten	3.5	3.26	2.09	4.47
Manganese	2.7	3.36	0.55	4.90
Chromium	3.3	3.27	–.49	4.60
Copper	3.9	2.94	2.70	4.35
Lead	3.5	—	3.34	4.47[b]
Zinc	2.1	3.05	2.60	3.53
Gold	1.5	—	—	4.48
Silver	2.3	—	—	4.30
Aluminum	5.4	4.29	5.10	3.67
Mercury	2.3	—	—	1.76
Vanadium	3.6	—	4.26	4.70
Platinum	2.6	3.75	—	4.60
Titanium	3.8	—	2.12	3.50
Tin	0.9	2.05	2.93	4.00
Silicon	3.7	—	—	4.50
Fluorine	3.5	—	—	2.20
Potash	3.6	—	2.57	4.70
Soda ash	2.4	—	—	4.80
Boron	3.5	—	—	4.80
Phosphate rock	5.0	—	3.61	4.60
Sulfur	4.6	—	1.95	3.76
Chlorine	—	—	—	1.04
Magnesium	2.1	—	—	3.90

Sources: U.S. Bureau of Mines: *Bulletin 671,* 1980 edition; W. Malenbaum (1977): *The Global 2000 Report to the President,* vol. 2, 1980, pp. 206–207; R.G. Ridker and W.D. Watson: *To Choose a Future,* Johns Hopkins University Press, 1980, pp. 153.

Note: Rates of growth computed on the basis of physical units.

[a]1975–2000.

[b]Includes primary as well as secondary demand.

Table 7-67
Index of World Nonfuel Minerals Production

Minerals	*2000/1970*	*2030 Pessimistic/ 1970*	*2030 Optimistic/ 1970*	*2030 Optimistic/ 2030 Pessimistic*
Iron	3.2775	6.1387	8.1509	1.3278
Molybdenum	3.3915	6.2843	8.2170	1.3075
Nickel	2.2475	4.0817	5.4780	1.3421
Tungsten	3.7202	6.3393	8.3036	1.3099
Manganese	4.3204	7.1845	9.9150	1.3801
Chromium	3.8957	7.0616	9.5261	1.3490
Copper	3.6462	6.8154	9.0000	1.3205
Lead	3.6316	6.9737	8.8947	1.2755
Zinc	2.8868	5.4340	7.1698	1.3194
Gold	3.7260	8.0822	9.9315	1.2288
Silver	3.6095	7.4476	9.4095	1.2634
Aluminum	2.9752	5.6116	7.5207	1.3402
Mercury	1.6903	3.2832	4.0442	1.2318
Vanadium	3.9665	6.4804	9.2179	1.4224
Platinum	3.8704	7.4781	9.5447	1.2763
Titanium	2.8314	5.7500	7.1047	1.2356
Tin	3.2748	5.9009	7.8829	1.3359
Silicon	3.7853	6.7485	9.1411	1.3545
Fluorine	1.9462	3.5161	4.7366	1.3471
Potash	3.9679	5.8289	6.3636	1.0917
Soda ash	4.1622	7.4054	10.1622	1.3723
Boron	4.0945	7.0079	9.3701	1.3371
Phosphate rock	3.9442	6.0073	6.7233	1.1192
Sulfur	3.0334	4.7301	5.4242	1.1467
Chlorine	3.3981	7.0616	8.8626	1.2550
Magnesium	3.1783	5.4845	7.5969	1.3852

Table 7-68
U.S. Nonfuel Minerals Consumption: Comparison of Alternative Projected Annual Rates of Growth (1970-2000)
(percentage)

	Bureau of Mines	*W. Malenbaum*[a]	*Ridker and Watson*[b]	*Institute for Economic Analysis*
Iron	1.18	1.78	1.87	2.22
Molybdenum	4.19	—	2.35	2.63
Nickel	3.15	2.00	4.05	2.66
Tungsten	3.90	2.30	2.07	2.95
Manganese	1.36	2.70	2.05	2.90
Chromium	2.70	1.20	2.40	3.14
Copper	2.70	1.90	2.30	3.05
Lead	0.90	—	4.40	4.80[b]
Zinc	1.15	2.00	2.60	1.46
Gold	−0.20	—	—	3.34
Silver	3.80	—	—	3.30
Aluminum	4.48	4.00	4.30	2.25
Mercury	−0.47	—	—	0.19
Vanadium	3.23	—	2.54	3.01
Platinum	3.80	2.60	—	3.20
Titanium	2.60	—	2.26	2.40
Tin	−0.33	0.90	2.20	0.79
Silicon	3.26	—	—	2.94
Fluorine	1.83	—	—	1.90
Potash	2.89	—	1.80	2.56
Soda ash	1.16	—	—	3.23
Boron	4.00	—	—	3.12
Phosphate rock	3.20	—	2.50	3.24
Sulfur	4.38	—	2.40	2.76
Chlorine	—	—	—	3.31
Magnesium	4.60	—	—	1.74

Sources: U.S. Bureau of Mines: *Bulletin 671, Mineral Facts and Problems,* 1980 edition; W. Malenbaum (1977): *The Global 2000 Report to the President,* vol.2, 1980, pp. 206-207; R.G. Ridker and W.D. Watson: *To Choose a Future,* Johns Hopkins University Press, 1980. pp. 153.

Note: Rates of growth computed on the basis of physical units.

[a]1975-2000.

[b]Includes primary as well as secondary demand.

Table 7-69
Index of U.S. Nonfuel Minerals Consumption[a]

Minerals	*2000/1970*	*2030 Pessimistic/ 1970*	*2030 Optimistic/ 1970*	*2030 Optimistic/ 2030 Pessimistic*
Iron	1.9314	3.5597	5.1166	1.4374
Molybdenum	2.1810	3.7104	5.0226	1.3537
Nickel	2.2018	4.0401	5.5646	1.3774
Tungsten	2.3946	4.1769	5.7959	1.3876
Manganese	2.3636	4.2727	6.1157	1.4313
Chromium	2.5255	4.6232	6.4155	1.3877
Copper	2.4667	4.5333	6.3333	1.3971
Lead[b]	4.1667	7.1667	9.6667	1.3488
Zinc	1.5455	2.9091	4.1818	1.4375
Gold	2.6755	4.8298	6.3830	1.3216
Silver	2.6520	4.7577	6.2996	1.3241
Aluminum	1.9524	3.6190	5.1905	1.4342
Mercury	1.0588	1.9412	2.5775	1.3278
Vanadium	2.4337	4.4150	6.2559	1.4170
Platinum	2.5738	4.6148	6.1148	1.3250
Titanium	2.0322	3.7626	4.9296	1.3102
Tin	1.2672	2.3191	3.2839	1.4160
Silicon	2.3859	4.2739	5.9751	1.3981
Fluorine	1.7541	3.2787	4.4809	1.3667
Potash	2.1329	3.2867	4.0793	1.2411
Soda ash	2.5938	4.8287	6.7863	1.4054
Boron	2.5125	4.5500	6.4500	1.4176
Phosphate rock	2.6057	4.0569	5.0813	1.2525
Sulfur	2.2608	3.6722	4.6770	1.2736
Chlorine	2.6524	4.9436	6.3883	1.2922
Magnesium	1.6762	3.1238	4.6286	1.4817

[a]Based on World Model Projections.

[b]The consumption indices reflect projected total (primary and secondary) consumption as a ratio of primary consumption in the 1970 base-year.

Table 7-70
Index of U.S. Nonfuel Minerals Production

Minerals	*2000/1970*	*2030 Pessimistic/ 1970*	*2030 Optimistic/ 1970*	*2030 Optimistic/ 2030 Pessimistic*
Iron	1.9201	5.2951	5.2951	1.0000
Molybdenum	3.0754	5.7937	7.6190	1.3151
Nickel	2.0839	3.9930	3.9930	1.0000
Tungsten	2.3853	4.1743	5.8028	1.3901
Manganese	2.3706	4.2905	6.1436	1.4319
Chromium	N.A.	N.A.	N.A.	N.A.
Copper	2.6429	5.9286	8.1429	1.3735
Lead	1.8000	8.6000	11.6000	1.3488
Zinc	3.2500	8.0000	11.5000	1.4375
Gold	2.6753	4.8339	6.4022	1.3244
Silver	2.6500	4.7571	6.3143	1.3273
Aluminum	3.5000	5.5000	5.5000	1.0000
Mercury	1.0596	1.9362	2.5745	1.3297
Vanadium	2.4260	4.4181	6.2525	1.4152
Platinum	1.9719	1.9719	1.9719	1.0000
Titanium	2.0240	3.7600	4.9200	1.3085
Tin	N.A.	N.A.	N.A.	N.A.
Silicon	2.3909	4.2620	5.9667	1.4000
Fluorine	1.7545	3.2818	4.4909	1.3684
Potash	1.0081	1.0081	1.0081	1.0000
Soda ash	2.7925	5.2730	7.2387	1.3728
Boron	3.5723	6.2893	8.4906	1.3500
Phosphate rock	3.1339	4.9858	5.8974	1.1829
Sulfur	2.4106	3.9100	4.9020	1.2537
Chlorine	2.6554	4.9492	6.3955	1.2922
Magnesium	1.6990	3.1553	4.6699	1.4800

Note: Based on World Model projections.

2000/1970 and for the alternative scenarios to the year 2030, high GDP growth (0) and low GDP growth (P) as well as a comparison of the two scenarios in 2030. Table 7-70 includes a set of U.S. production indexes through 2030 for the twenty-six minerals using the World Model solutions.

Notes

1. For the country and region aggregation, see appendix H.

2. For the sectoral classification, see appendix H.

3. Formerly, the region in the World Model designated North America also included the Canal Zone, Greenland, Puerto Rico, and the Virgin Islands. For this study, no attempt was made to incorporate these geographical areas into the restructured World Model.

4. The original World Model study, in addition to three fossil fuels, included six nonfuel minerals—iron, nickel, copper, lead, zinc, and aluminum.

5. The required modifications for assimilating the Canadian matrix into the World Model, in addition to the values of the exogenous macro-and other variables, are published in appendix M of Leontief, Koo, Nasar and Sohn [25a].

6. A technical description of the methodology and equations used to incorporate these twenty new minerals into the World Model appears in appendix P of Leontief, Koo, Nasar, and Sohn [25a].

7. Appendix D includes the technical definitions of these terms.

8. With the exception of lead, all minerals production and consumption levels for the sixteen World Model regions to the year 2030 are projections of primary metal. Beginning in 1980, consumption and trade figures for lead were available from various Bureau of Mines reports. Output of those regions which are *net importers,* such as the United States, are projected in terms of primary lead output. However, the projections of all regional consumption and the production of all regions that are self-sufficient or net exporters of lead are calculated as if secondary production were zero and all lead requirements are met by primary production. Because recycled lead currently contributes about 45 percent of total U.S. lead requirements, the effect of treating lead consumption as total (primary and secondary) requirements overstates imports of the importing regions and production and exports of the exporting and self-sufficient regions. In addition, this treatment results in an acceleration of the date of regional and global exhaustion of lead resources. When the computations were executed, only data for total lead consumption on a regional basis were available. As the data for primary lead consumption become available, the lead projections will be revised downward to be consistent with the other minerals.

9. The detailed assumptions used in these projections are presented in appendix N of Leontief, Koo, Nasar, and Sohn [25a].

10. See the discussion of lead, note 8.

8 Summaries and Conclusions

To pull together all the threads of this study, it would be appropriate to summarize the alternative projections that emerge from the various sets of assumptions. The full range of projections allows us to consider concrete answers to questions regarding the adequacy of the existing world stock of nonfuel mineral resources to meet the necessary levels for sustained economic growth in the U.S. and the world economies through 2030.

Demand for each of the twenty-six minerals is summarized first from the point of view of the specific requirements of the U.S. economy through 2000 and then, more generally, in terms of the global requirements through 2030.

The general format of the summary for each mineral is as follows. First, the base-year demand and supply patterns of the United States are discussed, and the projections of the levels of future U.S. demand and supply computed from the different scenarios are presented. In the second part of the summary of each mineral, the base-year global demand-and-supply patterns are described, followed by the projected future global requirements through 2030 under alternative assumptions. Tentative conclusions regarding future regional (and in some cases, global) resource exhaustion concludes each summary.

Iron

United States

Base-Year Demand Pattern. In 1972, the United States consumed 87.65 MMT of primary iron contained in ore, iron oxide, and iron and steel, as well as 37.8 MMT of scrap iron and steel.

Transportation represented the largest end use of iron, at 30 percent of U.S. consumption. In the automobile sector, iron and steel comprised structural members of the frame, body coverings, and mechanical elements of the engine and drive train. Most railroad equipment is built of steel, and alloy steels are used for aircraft bodies and parts of engines. Construction was the second largest end use of iron; 29 percent of total iron consumed was used in frames for buildings, bridges, rails, and highway structures. Steel is used as the frame and reinforcement for concrete in buildings, bridges, and docks. It is also used by contractors in structural shapes, street piling, pipe, rod, and bars. Machinery represented 21 percent of iron use in 1972; nearly all industrial mining and agricultural equipment is made of steel. The frames of commercial and consumer electrical equipment and appliances are normally made of steel; steel containers represented 7 percent of total iron use in 1972.

Capital investment represented about 50 percent of U.S. direct plus indirect iron requirements. Personal consumption expenditures represented 35 percent of direct plus indirect iron requirements. Capital expenditures—on a per dollar basis—were 5.7 times as iron-intensive as personal consumption expenditures. Altogether, military spending only accounted for 6 percent of total iron requirements; however, on a per dollar basis, military spending was about twice as iron-intensive as personal consumption expenditures and one-third as iron-intensive as capital expenditures.

Base-Year Supply Pattern. In 1972, 70 percent of a total iron demand of 87.7 MMT was satisfied from primary sources. These supplies came from domestic mine production (53 percent), net imports (39 percent), and inventory reductions (7 percent). By-product output of iron was negligible. Nearly 30 percent (37.8 MMT) of U.S. total iron demand was met by recycling, in equal shares, prompt industrial and obsolete scrap. The sources of obsolete scrap iron range from obsolete structures to scrapped automobiles and ships; a small but growing share of iron scrap comes from municipal resource-recovery operations.

Projected Demand. U.S. iron demand will reach 140.8 to 196 MMT in 2000, depending on the rate of economic growth and technological change. Under the scenario that assumes technological change but no change in recycling rates, primary iron demand will increase to 162.3 MMT by 2000. A low projection of 92.4 MMT is attained only if the recycling rate can be doubled by 2000. Except in primary aluminum processing, the use of iron is assumed to decline per dollar of output in nearly all sectors as a result of substitution by other materials including plastics and aluminum, and increased use of alloy steels; downsizing, downweighting, and miniaturization are further causes of this trend. In motor vehicles, for example, the smaller size of vehicles and the increased use of plastics, aluminum, and high-strength steels are expected to reduce direct requirements for iron by 20 percent. Similarly, in aircraft, where fuel efficiency and speed are major concerns, some reduction of iron use per dollar of output is anticipated. Iron used in construction—either in structural applications or for pipes, valves, and plumbing—is also expected to decline because of greater use of substitute materials and the changing composition of output. Declining requirements per dollar of output in machinery sectors reflect similar trends.

Some shifts in end-use markets for iron will occur. Transportation, machinery, and appliances markets will expand, while construction markets will decline as a share of the total market. The market for cars and containers will stay roughly the same as in 1972.

Projected Supply. Under the technological-change scenario, domestic output of primary iron in 2000 is projected at 93.7 MMT. However, if the ratio of imports to primary demand increases or decreases to its feasible limits, then domestic production of primary iron can be expected to range between 68.3 to 128.9 MMT by the year 2000.

World

Base-Year Demand. The developed regions—the OECD and the East Bloc countries—consumed 90 percent of world iron production in 1970. Of the 425 MMT of iron consumed in 1970, Western Europe (high-income) accounted for 23 percent, the USSR for 20 percent, the United States for 17 percent, and Japan for 14 percent. Japan, which imported 57.6 MMT of iron in 1970, was the world's largest importing region, followed by Western Europe (high-income) at 47.1 MMT.

Base-Year Supply. In the 1970 base year, the USSR was the largest producer and net exporter of iron, followed by Western Europe (high-income) and the United States. Major exporters of iron included the USSR, Tropical Africa, Latin America (medium-income), Oceania, and Asia (low-income).

Projected Demand and Supply. World iron consumption is projected to increase from 425 MMT in 1970 to 1,389 MMT in 2000, an annual rate of growth of 4 percent. By 2000, the share of world consumption represented by the developed regions will drop from 90 percent in 1970 to 82 percent. Major importing regions include Japan, Eastern Europe, Western Europe (high-income), and the United States.

The major producing regions in 2000 are projected to be the USSR (432 MMT), the Latin America (medium-income) region (172 MMT), the Western Europe (high-income) region (140 MMT), and China (100 MMT). The leading exporters include the same regions, except China and Western Europe (high-income).

With respect to the fifty-year projections, global primary iron requirements in the year 2030 are projected to range between 2,609 and 3,463 MMT, depending on the future rate of growth of the world economy. Current estimates suggest that world iron resources are sufficient to meet the expected cumulative world iron requirements to the year 2030, even under the high-growth scenario.

Molybdenum

United States

Base-Year Demand Pattern. In 1972, the United States consumed 23.5 KMT of primary molybdenum, principally in the form of molybdic oxide and ferromolybdenum; in addition, an undisclosed amount of molybdenum was recycled from molybdenum metal, alloys, and industrial wastes.

More than 90 percent of domestic molybdenum consumption in 1972 was in metallurgical applications, largely for the production of molybdenum-containing steels. Major uses included engines and turbines (17 percent of total consumption), industrial machinery (17 percent), transportation vehicles and equipment (12 percent), plumbing and heating equipment (12 percent), and electrical machinery and components (8 percent). Nineteen percent of domestic consumption was attributed to various metal-processing sectors because of molybdenum's role as an alloy constituent in steels and superalloys as well as a refractory metal in electric furnaces. A small percentage of molybdenum was used in the chemicals sectors, chiefly for lubricants, catalysts, and pigments.

With regard to molybdenum requirements by final demand component, personal consumption expenditures represented 46 percent of direct plus indirect use of molybdenum in the U.S. economy in 1972. Capital expenditures accounted for 42 percent of molybdenum requirements, and defense expenditures accounted for approximately 8 percent. Per dollar of final demand expenditure, investment goods were 3.6 times and military goods 1.8 times as molybdenum-intensive as personal consumption goods.

Base-Year Supply Pattern. Nearly all reported molybdenum supplies in 1972 consisted of primary molybdenum. The base-year supply of molybdenum came chiefly from domestic production (including 42 percent of total domestic mine output produced as a by-product of copper mining). The United States is both the world's largest producer and exporter of molybdenum.

Projected Demand. Molybdenum consumption is projected to range from 48.2 to 64.1 KMT by 2000, depending on the rate of economic growth and technological change. The low projection assumes a lower rate of GDP growth and no technological change. Under the technological-change scenario, molybdenum demand will increase by 173 percent to 64.1 KMT by 2000. Eighty-six percent of the increase in molybdenum consumption is the result of GDP growth, and 14 percent the result of projected changes in technology. Per dollar of industry output, molybdenum requirements are expected to increase by 10 to 20 percent in all sectors, except chemicals where demand is expected to remain at 1972 levels. The use of molybdenum-bearing steels and high-temperature superalloys is expected to increase significantly in the engine, turbine, and transportation sectors. The use of molybdenum-bearing stainless steels, alloy steels, and cast irons in industrial machinery and in high-speed and other tool steels will also increase substantially.

Changes in the composition of final demand as well as technological changes will increase the importance of some end-use markets. Demand for molybdenum in the industrial and electrical machinery markets will increase from 17 and 8 percent in 1972 to 25 and 12 percent by the year 2000.

Projected Supply. All projected domestic consumption of molybdenum will continue to come from domestic mine production through 2000. Under the technological-change scenario, cumulative production by 2000 will represent 38 percent of currently identified U.S. reserves.

World

Base-Year Demand. The developed regions, both market-oriented and socialist bloc countries, consumed more than 81 percent of the world's molybdenum production in 1970. Of the 80.2 KMT consumed globally in 1970, the United States consumed 28 percent, Western Europe (high-income), 20 percent; Japan, 14 percent; and the USSR, 11 percent. The main importing regions were Western Europe (high-income), Japan, and the Middle-East oil-producing countries.

Base-Year Supply. The United States was the largest producer of molybdenum, supplying 63 percent of world production in 1970. Other significant producers were Canada, the USSR, and Latin America (medium-income, including Chile). The United States, Canada, and Latin America (medium-income) were the world's leading exporters of molybdenum.

Projected Demand. World molybdenum consumption is projected to increase from 80.2 KMT in 1970 to 272 KMT by 2000. The share of world consumption represented by the developed regions will drop from 81 percent in 1970 to 74 percent by 2000. No significant changes in the structure of world production or exports are expected by 2000.

World consumption of molybdenum is expected to range between 504 and 659 KMT by 2030, depending on rate of growth of the world economy.

Currently identified global resources of molybdenum will be adequate to meet projected demand through 2030, even under the optimistic scenario. However, the leading producing regions—the United States, Canada, and the USSR—will deplete their molybdenum resources unless new discoveries are made or a lower rate of production prevails.

Nickel

United States

Base-Year Demand Pattern. In 1972, the United States consumed 152 KMT of primary nickel in the form of ore, nickel oxide, and ferronickel and 64.4 KMT of secondary nickel recovered chiefly by recycling nonferrous nickel alloys and stainless steels.

As an alloying element, nickel imparts strength as well as heat and corrosion resistance. About 90 percent of nickel used is in the form of metal: stainless steels, alloy steels, superalloys, and alloys of nickel and copper. Chemical applications of nickel include batteries, dies, pigments, catalysts, and insecticides. Of the total nickel consumed in 1972, 34 percent was consumed in fabricated metal and machinery, 20 percent in transportation, 15 percent in oil and gas exploration and development, 12 percent in chemicals, and 9 percent in electrical machinery, equipment, and supplies.

Personal consumption expenditures represented 38 percent of direct plus indirect nickel requirements, investment about 48 percent, and military spending about 7 percent. Capital expenditures were roughly five times as nickel-intensive as personal consumption expenditures. Military spending was about twice as nickel-intensive as personal consumption expenditures.

Base-Year Supply Pattern. In 1972, 70 percent of total (primary plus secondary) U.S. nickel requirements came from primary sources, chiefly from imports. More than half of U.S. nickel imports came from Canada. The remainder came from a number of sources, including the Dominican Republic and Norway. Recycled nickel-based alloys were the largest source of secondary nickel, while nickel stainless steels were the next largest. Some secondary nickel was also recovered by recycling superalloys.

Projected Demand. Domestic primary nickel consumption is projected to range between 319 and 356 KMT in 2000, depending on the set of income-growth and technological-change assumptions used. Nearly the entire projected increase in nickel consumption in 2000 is the result of GDP growth; only 1.5 percent of the projected increase between 1972 and 2000 can be attributed to changes in technology.

Some end-use markets will grow more rapidly than others, thus increasing their share of total nickel consumption by 2000. For example, machinery industries' share of total nickel consumption will increase to 40 percent; the electrical machinery sector will also increase its share of the total slightly, to 10 percent of nickel consumption. The oil and gas industry and the chemical industry will represent a smaller share of total nickel consumption in 2000 than in 1972.

Per dollar of output, direct nickel requirements by industries are projected to increase in some sectors—aircraft, machinery, and electrical components—and decrease in others—automobiles, fabricated metal products, and household appliances.

Projected Supply. Currently estimated U.S. nickel reserves will be depleted during the 1980s, and continued domestic production of primary nickel even at the present low level depends largely on whether low-grade domestic resources are developed. Although small amounts of nickel will be produced as a by-product of copper, under the technological-change scenario, the United States is expected to import 328 KMT of nickel and to recycle 144 KMT out of total (primary and secondary) nickel requirements of 500 KMT by 2000.

World

Base-Year Demand. In 1970, eight developed regions—the OECD and the Eastern Bloc countries—consumed approximately 85 percent of world nickel production. Of the 741.8 KMT consumed in 1970, the United States accounted for 19.5 percent; Western Europe (high-income), 15.2 percent; the USSR, 4.8 percent; and Japan, 13.9 percent. Western Europe, the United States, and Japan were the leading nickel-importing regions.

Base-Year Supply. Oceania, Canada, the USSR, and Asia (low-income, including Indonesia) rank as the world's largest nickel-producing regions. Among the major nickel-consuming regions, the USSR is the only region that is self-sufficient. Oceania accounted for 25 percent of 1980 world nickel production, the Soviet Union for 23 percent, and Canada for 21 percent. Oceania, Canada, and Asia (low-income) were the world's leading nickel-exporting regions.

Projected Demand and Supply. World nickel consumption is projected to increase to 1,667 KMT by 2000. The share of world production consumed by the developed regions will increase to 90 percent. Production of nickel in 2000 is expected to be concentrated in the same four major nickel-producing regions as in 1970. However, additional supply by the year 2000 is expected to come from mining deep-sea nodules on the ocean floor.

The consumption of nickel in 2030 is projected to range between 3,028 and 4,063 KMT, depending on the assumed rate of growth in the world economy. World resources of nickel—not including the nickel deposits on the ocean floor—appear to be sufficient to meet cumulative world demand through 2030, even under the high GDP-growth assumption.

Tungsten

United States

Base-Year Demand Pattern. In 1972, the United States consumed 6.45 KMT of primary tungsten in the form of tungsten ores and concentrates, carbide powder, and ferrotungsten; it also consumed 0.35 KMT of secondary tungsten recovered by recycling tungsten-carbide products and other tungsten-bearing scrap.

Metalworking machinery, including high-speed machine tools, represented the single largest end use of tungsten—55 percent of U.S. consumption in 1972; construction and mining machinery accounted for 19 percent; transportation sectors, including the aerospace industry, 12 percent; electrical uses—wire filaments, electric lamps, and cathodes for electrical tubes—another 11 percent.

Capital-investment expenditures represented 60 percent of direct plus indirect U.S. tungsten consumption in 1972. Personal consumption and military expenditures represented 21 percent and 5 percent of total tungsten consumption, respectively. For the same year, per dollar, investment goods were eleven times as tungsten-intensive as personal consumption goods and 2.5 times as tungsten-intensive as military goods.

Base-Year Supply Pattern. In 1972, more than 95 percent of tungsten supplies came from primary sources. Domestic mine production (including by-product output) contributed about 50 percent to total primary demand, and net imports accounted for 42 percent of primary demand. Imports in 1972 came chiefly from Canada and Bolivia.

Superalloy scrap, tungsten carbide manufactures, and tungsten-wire mill wrappings—mostly prompt industrial scrap—constitute major sources of secondary tungsten supplies.

Projected Demand. At the assumed level and composition of GDP, U.S. tungsten demand is projected to reach 16.1 to 17.6 KMT by 2000, depending on the rate of economic growth and technological change. Under the standard scenario (which assumes technological change but no change in the recycling rate from that of the base-year), primary tungsten demand will increase 170 percent to 17.4 KMT by 2000.

GDP growth and changes in GDP composition account for about 90 percent of the growth in tungsten consumption, even though projected relative shifts away from capital and military spending dampen the contribution of GDP effects slightly. About 10 percent of the growth in tungsten consumption is the result of forecasted technological changes. Although tungsten use per dollar of output in the automobile and electric-lighting sectors is expected to decline by 20 to 35 percent, respectively, large increases of tungsten use in metalworking machinery, aircraft, electrical uses, and chemicals more than offset these projected declines. The result of these changes, as well as shifts in the composition of final demand, result in a larger market share for metalworking machinery by the year 2000.

Projected Supply. If the recycling rate and the import share of tungsten are both held at their 1972 base-year values, domestic mine output of tungsten (including by-product output) can be expected to reach 9.3 KMT by 2000. However, changes in either the recycling rate or the import share in the future will affect the level of projected domestic mine output.

World

Base-Year Demand. The developed regions—the OECD and Eastern Bloc countries—consumed 88 percent of the world's primary tungsten production in 1970. The USSR consumed 26 percent, the United States, 22 percent; Eastern Europe, 16 percent; Western Europe (high-income), 14.5 percent; and Japan, 6 percent. Eastern Europe, Western Europe (high-income), the United States, and the USSR were the world's leading importers of tungsten.

Base-Year Supply. Of the developed regions, only the USSR and the United States were major producers as well as consumers in 1970. Mainland China accounted for 28 percent of world production in 1970, and other important producing regions were Asia (low-income) and Latin America (low-income, including Bolivia). Major exporters were China, Asia (low-income, including Korea and Thailand), and Latin America (low-income, including Bolivia).

Projected Demand and Supply. World tungsten consumption in 2000 is projected to increase to 125 KMT. By 2000, the share of world consumption represented by the developed regions is projected to decline to 78 percent.

No significant changes are expected in the pattern of regional production, consumption, imports, and exports from the base-year structure.

By 2030, world tungsten consumption is expected to range between 213 to 279 KMT, depending on the rate of economic growth. Even though currently estimated global resources appear to be sufficient to meet projected cumulative world tungsten requirements under the optimistic scenario, some regional endowments will be depleted, including those of current major exporting regions such as Asia (low-income) and Latin America (low-income).

Manganese

United States

Base-Year Demand Pattern. In 1972, the United States consumed 1,236.8 KMT of manganese, an essential constituent of nearly all steels as well as an important input in cast iron and aluminum production. Manganese is consumed chiefly as ferromanganese, silicomanganese, and electrolytic metal.

The end-use distribution of manganese parallels that of iron because of manganese's role in steel making. In 1972, 27 percent of manganese consumption was attributed to the transportation sectors, 27 percent to construction, and 20 percent to machinery. Cans and containers represented 7 percent of domestic consumption; appliances, oil and gas, and chemical processing each represented 5 percent.

In 1972, 47 percent of direct plus indirect use of manganese was for capital-investment expenditures, as compared with 37 percent for personal-consumption expenditures. Military spending represented about 5 percent of total manganese use. As was the case with iron, U.S. capital expenditures were 4.9 times as manganese-intensive as personal-consumption expenditures. Military spending was 1.5 times as manganese-intensive as personal-consumption expenditures.

Base-Year Supply Pattern. In 1972 there was no domestic mine production of manganese. Thus, all new supply except reductions in inventories was imported. South Africa and Gabon were the largest sources of imports to the United States. An unknown quantity of manganese was recovered by recycling steel slags and high-manganese steels.

Projected Demand. U.S. manganese demand will reach 2,390 to 2,860 KMT by 2000, depending on the rate of economic growth and technological change. Under the technological-change scenario, manganese demand will increase by 93 percent over the base-year level to 2,390 KMT by 2000. The high projection assumes no technological change, whereas the low projection assumes increases in technical efficiency in the use of steels and some substitution for steels by plastics, aluminum, and other materials.

Without technological improvements, manganese consumption will increase by 131 percent by 2000 as a result of GDP growth, even with shifts in the composition of GDP from military spending and capital investment to personal-consumption expenditures. With changes in technology, except in chemicals and batteries—where per dollar of output requirements for manganese are expected to increase by 45 percent and 10 percent, respectively—manganese inputs in most sectors will decline by around 20 percent per dollar of output. By and large, these declines reflect more efficient use of steel; substitution by aluminum, plastics, and nonferrous alloys; and design changes resulting in smaller components. Some shifts in the relative importance of end-use markets will occur by 2000. For the 1972–2000 period, manganese requirements for construction are likely to decline substantially from 27 to 17 percent of manganese consumption; machinery use is expected to grow from 20 percent to 26 percent.

Projected Supply. Manganese will continue to be imported, chiefly from South Africa and the USSR, because the United States possesses only relatively small low-grade resources of this metal.

World

Base-Year Demand Pattern. The developed regions consumed 79 percent of the world's manganese production in 1970. Of the 8,240 KMT world consumption in 1970, the USSR accounted for 30 percent, the United States, 15 percent; Japan, 12 percent; and Western Europe, 10 percent. The largest importing regions of manganese in 1970 were the United States, Japan, Eastern Europe, and Western Europe (high-income).

Base-Year Supply Pattern. Of the developed regions, only the USSR is a major producer and exporter. In 1970, the Soviet Union accounted for 37 percent of world production. The second largest manganese producer was South Africa, supplying 18 percent of the world total in 1970. Other important producers included Latin America (medium-income) and the Middle East region (including Gabon). Major exporters of manganese included Latin America (medium-income), South Africa, and the Middle East.

Projected Demand. World manganese consumption is projected to increase to 35,600 KMT in 2000; by 2030, world manganese demand can be expected to range from 59,200 to 81,700 KMT, depending on rates of economic growth. By 2000, the share of world consumption represented by the developed regions is projected to drop to 74 percent of total world consumption.

Projected Supply. In 2000, the leading producing regions are projected to be the USSR, South Africa, the Middle East region, and Latin America (medium-income). In addition, significant amounts of manganese are expected to be recovered from nodules on the ocean floor.

Given current resource estimates, manganese resources are likely to be depleted in some regions after the turn of the century. By 2010, Eastern Europe and Asia (centrally planned countries) are projected, on the basis of current estimates, to exhaust their manganese resources; by 2020, Asia (low-income), Latin American (medium-income), the Middle East countries, and Tropical Africa also are expected to deplete their deposits. Arid Africa will exhaust its resources by 2030. However, the rates of regional depletion of manganese resources are subject to great uncertainty because of prospects for mining the ocean floor.

Chromium

United States

Base-Year Demand Pattern. In 1972, the United States consumed 461 KMT of primary chromium in the form of chromite, ferrochromium, and chromium-based chemicals and 87 KMT of secondary chromium, recovered by recycling stainless steel and other chromium-bearing scrap.

Metal-processing was the largest single end use of chromium, representing 26 percent of U.S. primary consumption in 1972. Chromium was consumed in the iron and steel industry and in the nonferrous metals industry as a constituent of chromium refractories and as an alloying element in cast irons and nonferrous alloys. Fabricated metals—which encompass materials used in construction, pipelines, equipment, and utensils—represented the second largest end use of chromium with 21 percent of the U.S. primary consumption. Machinery accounted for another 18 percent of U.S. consumption. Most of the chromium in these two end uses is consumed in the form of stainless steels. Transportation sectors—automobiles, aircraft, and shipbuilding—accounted for 7 percent of the U.S. consumption. The chemical industry consumed 10 percent of U.S. primary chromium, primarily to produce pigments, but also for leather tanning and such critical uses as metal plating.

Personal-consumption expenditures represented 40 percent of the direct plus indirect use of chromium in the U.S. economy in 1972. More importantly, 46 percent of direct plus indirect use of chromium was used to produce goods delivered to the investment component of the GDP. Per dollar of expenditure, investment goods were 4.5 times as chromium-intensive as personal-consumption goods. Military spending—which is nearly as chromium-intensive as is capital expenditures—represented about 7 percent of direct plus indirect chromium consumption.

Base-Year Supply Pattern. In 1972, 84 percent (461 KMT) of chromium supplies came from primary chromium, of which 80 percent came from imports and 20 percent from inventory reductions. There was no domestic mine production or by-product production of chromium.

The largest regional sources of chromium imports were Tropical Africa (including Zimbabwe) and South Africa. These regions supplied chromium largely in the forms of high-iron chromite and ferrochromium. The USSR and Turkey were the leading suppliers of high-chromium chromite; the Philippines was the main supplier of refractory or high-aluminum chromite.

The remaining 16 percent (97.1 KMT) of U.S. chromium supplies came from prompt industrial and obsolete scrap. Stainless steel scrap was the main source of this scrap, but small amounts involved superalloys, structural AISI steels, cast heat and corrosion-resistant steels, and cast irons.

Projected Demand. At the projected levels and composition of GDP, chromium consumption in the United States is expected to range between 1,080 and 1,240 KMT in 2000. Under the scenario that assumes technological change but no change in the recycling rate, primary chromium demand will increase to 1,150 KMT by 2000. A low projection of 851 KMT is attained only if the recycling rate can be doubled by 2000.

Growth and changes in the composition of GDP account for nearly all the growth in chromium consumption between 1972 and 2000. Indeed, such compositional changes as lower military and investment spending as a proportion of GDP in the year 2000 as compared with 1972 dampen the contribution slightly. About 6 percent of the growth can be attributed to forecasted technological changes. In most sectors, chromium consumption is assumed to increase per dollar of output. However, refractories' consumption of chromium is expected to be halved by 2000 because of the adoption of furnaces that do not use chromite refractories and because of substitution by magnesite brick.

A combination of increased use of stainless steel per vehicle (for example, catalytic converter shells in automobiles) and shifts within transportation sectors toward leisure vehicles and military aircraft are assumed to result in increased per dollar of output direct requirements for chromium.

In construction uses, chromium inputs per dollar are expected to increase because of an increase in the construction of nuclear and synthetic fuel facilities. The production of more sophisticated machinery will require additional amounts of chromium-based alloys for their corrosion and heat-resistant properties.

Changes in the composition of final demand as well as changes in technology between 1972 and 2000 account for slight shifts in the importance of different end-use markets. Metal production, fabricated metals, and machinery account for even larger shares of chromium consumption in 2000 than in the base-year, whereas chemicals represent a smaller share.

Projected Supply. Under the technological-change scenario, the United States will continue to import all of its primary chromium, 1,150 KMT in the year 2000. The only domestic source of production will continue to be from recycling, 232 KMT. No prospects are held open for domestic chromite mining, except under emergency conditions. On the other hand, under the high recycling scenarios, as much as 509 KMT or 37 percent of total demand in 2000 could come from scrap and industrial wastes.

World

Base-Year Demand. Eight developed regions—the OECD and the Eastern Bloc countries—consumed more than 86 percent the world's chromium production in 1970. Of the 2,110 KMT of world consumption, Western Europe (high-income) accounted for 26 percent; the United States, 23 percent; Japan, 16 percent; and the Soviet Union, 12 percent. The United States, Western Europe (high-income), and Japan were the major importing regions.

Base-Year Supply. Of the major consuming regions, only the USSR is a major producer and exporter. South Africa and Zimbabwe accounted for 37 percent of world chromium production in 1970; other important producers included Asia (low-income, including the Philippines), Eastern Europe (medium-income, including Albania), and Western Europe (medium-income, including Turkey and Yugoslavia). In addition to the USSR, the major exporters were South Africa, Zimbabwe, Asia (low-income), and Western Europe (medium-income).

Projected Demand. World chromium consumption is projected to increase to 8,220 KMT in 2000. The share of world consumption of chromium absorbed by the developed regions is expected to drop from 86 percent in 1970 to 80 percent in 2000.

Chromium demand in 2030 is estimated to range between 14,900 and 20,100 KMT, depending on the rate of world economic growth.

Projected Supply. Based on the current regional patterns of production and exports, several regions can be expected to deplete their currently estimated resource deposits by 2000: Eastern Europe (medium-income), Western Europe (medium-income), and the Middle East. Consequently, South Africa and Tropical Africa (including Zimbabwe and Madagascar) can be expected to become the two leading exporting regions of chromium by the turn of the century, as the current exporters, principally the USSR, begin to reduce their share in world exports to retard resource depletion.

By 2030, only South Africa and Tropical Africa are expected to export chromium on the basis of current regional resource endowments.

Copper

United States

Base-Year Demand Pattern. In 1972, apparent U.S. consumption of primary copper reached 1.77 MMT, more than half of which was purchased by the electrical machinery and equipment sector. Of the 920,000 MT of copper consumed by the electrical-related sectors in 1972, two subsectors—electrical transmission and distribution equipment and radio, television, and communication equipment—accounted for more than 60 percent. Significant quantities of copper and its derivatives, such as bronze and brass, were also used to produce architectural and ornamental metal work for buildings and private houses. The construction industry absorbed 19 percent of the base year's primary copper consumption, approximately 317,000 MT.

In 1972, 221,000 MT of primary copper were consumed by different industrial and commercial machinery and equipment producers, mainly because of the metal's excellent heat-exchange and high corrosive-resistance properties.

The use of copper and its alloys in the transportation equipment industry was mainly concentrated in automobile radiators and some electrical applications. The transportation sector accounted for about 9 percent of the primary copper end-use market in 1972, with about 115,000 MT directly used for motor vehicles and parts.

Other uses of copper in areas such as ordnance parts and equipment, coinage, chemical products, and other miscellaneous uses totaled 145,000 MT in the base-year and represented approximately 8 percent of the end-use demand market.

In 1972, given the U.S. end-use consumption pattern and the mix of goods and services purchased by the private sector, nearly 45 percent (788,000 MT) of total primary copper was used, directly and indirectly, to satisfy private business investment. In 1972, about half a million metric tons of primary copper were absorbed in the production of goods and services purchased by consumers. For government defense users, 14 percent (250,000 MT of total demand for primary copper) was directly and indirectly consumed in 1972. The combined requirements for nondefense government uses totaled 193,000 MT.

Demand for secondary copper (both new and old scrap) was approximately 1.18 MMT in 1972.

Base-Year Supply Pattern. In 1972, total U.S. production of primary copper from domestically mined ores reached a level of 1.5 MMT, of which about 1 percent was produced as a by-product from lead, zinc, silver, and gold mining. The United States had net imports of 197,000 MT of primary copper, in ore or refined form, which was approximately 11 percent of U.S. domestic requirements. The United States imported copper mainly from Canada, in addition to small quantities from Chile, Zambia, and Peru. In 1972, the copper scrap generated was nearly 0.2 MMT and the amount of copper scrap imported was negligible.

Projected Demand. The U.S. demand for primary copper in 2000 is projected to range between 3.7 and 4.3 MMT, depending on the different underlying assumptions regarding income growth and materials substitution.

Gradual substitution of copper by aluminum is likely to continue for applications in the electrical construction, machinery, transportation, and ordnance sectors. By 2000, about 60 percent of all copper consumption will be for electrical uses. According to the projections, the construction sector will cease to be the second largest consumer of copper by 2000 as copper or copper-related products are replaced by less-expensive materials, such as plastics. The construction sector's share of the end-use market in 2000 is expected to be 10 percent. The relative share of primary copper for other industries, such as machinery, transportation, and ordnance, is expected to remain the same.

Although the recycling rate for secondary copper in the base-year is about 40 percent, the current highest technically attainable rate is estimated at 72 percent. Given the assumptions regarding anticipated technological change, economic growth, and the assumed technically maximum rates of recycling, projected demand for primary copper in 2000 could be only 2.1 MMT.

Projected Supply. U.S. copper reserves, which can be economically extracted at today's prices, are estimated at 92 MMT in 1980 by the Bureau of Mines. The vast quantity of U.S. reserves is likely to accommodate even the highest of the projected levels of domestic production until the end of the century. Given the base-year's import share and rate of recycling, domestic copper mining activities will have to increase significantly and the output of primary copper required from domestic mines could reach 4.13 MMT by 2000, more than 2.5 times the base-year's output level. However, the alternative policy—that is, increasing the secondary supply of copper by increasing the recycling rate to 72 percent in 2000—could significantly reduce the projected domestic mining output of 4.13 MMT to about 2.1 MMT. Foreign-trade activities for copper are not likely to increase significantly from 1980 through the year 2000. However, if the United States reduces its import dependence, self-sufficiency is feasible and domestic mine output for primary copper will exceed 5 MMT by the year 2000. By-product output of copper will continue to contribute minimal amounts to the total domestic supply of copper.

World

Base-Year Demand Pattern. In 1970, world consumption of primary copper totaled 6.5 MMT. Besides the United States, the largest copper-consuming countries included Japan, the USSR, and the other European industrial nations. In 1970, the Western European (high-income) region consumed an estimated 28 percent of the world's copper production, followed by the United States with 23 percent; Japan and the Soviet Union each required 12 percent of total world copper production. Although the United States and the Soviet Union consumed mostly their own domestic production of copper, Japan and the developed European countries relied heavily on imports.

Base-Year Supply Pattern. In 1970, out of the 6.5 MMT of total world output of primary copper, 22 percent was mined in the United States, 17 percent in Tropical Africa (mainly Zaire and Zambia), 14 percent in the USSR, and 12 percent in Latin America (medium-income, mainly Chile and Mexico). For the same year, an estimated 42 percent of the world's copper exports, based on the regional classification used in the World Model, came from Tropical Africa, 15 percent from Latin America (medium-income), 15 percent from Canada, and 8 percent from Latin America (low-income, mainly Peru).

Projected Demand and Supply. World consumption of primary copper in 2000 is projected to increase by almost 150 percent to 24 MMT. Western Europe (high-income) is expected to account for 22 percent, the United States, 15 percent; the USSR, 15 percent; and Japan, 14 percent of total world copper consumption.

The share of exports from these exporting regions is not expected to change significantly by the end of this century. Similarly, Japan and the industrialized Western European countries are expected to continue to rely on imports of primary copper to satisfy their domestic needs.

By 2030, long-range projections of world primary demand range from 44 MMT under the low-economic growth scenario to about 58 MMT under the high-growth scenario. As of 1980, estimates of world copper resources totaled 1,627 MMT, and exhaustion of these global resources is not expected through the year 2030.

Lead

United States

Base-Year Demand Pattern. In 1972, total U.S. primary demand for lead amounted to 861,000 MT. The largest end-use sector, consuming over 45 percent of the year's domestic demand for primary lead, was the transportation industry, where the metal was mainly used in storage batteries. In 1972, close to one-fifth of all primary lead was consumed by the U.S. petroleum refining industry, mainly as gasoline additives. Primary lead used in cables or other electrical applications constituted another 9 percent of the end-use primary-demand market in 1972. As a high corrosion-resistant coating, lead paint was applied to a wide variety of metal products; demand for primary lead by the paint industry accounted for about 6 percent of the year's primary consumption. Other users of lead included the construction, ammunition, printing, and packaging industries. In the base-year, demand for both old and new lead scrap amounted to 560,000 MT.

Although 30 percent of the 1972 demand for lead was directly and indirectly used to produce capital-investment goods and services, over 50 percent went to satisfy consumer needs. In addition, another 14 percent was directly and indirectly consumed for government military uses.

Base-Year Supply Pattern. In 1972, over 60 percent of the base-year total U.S. demand for primary lead was supplied by domestic mines, of which about 12 percent was by-product or co-product output. Imports of primary lead, principally from Canada, Australia, and Peru, provided another 36 percent of total primary supply of the mineral. The remaining 3 percent was contributed by government stockpile releases.

Projected Demand. Based on the assumed low and high rates of growth for the U.S. economy, projected domestic demand for primary lead can range between 1.53 and 1.94 MMT by the year 2000. However, by incorporating additional technological-change assumptions regarding materials substitution, a nearly across-the-board reduction in lead use by its traditional consuming sectors results in a U.S. demand projection for primary lead of only 1.34 MMT by 2000.

Under this scenario, further reduction of lead content in gasoline is anticipated; demand for primary lead by the petroleum-refining industry is also expected to decrease significantly to only 4 percent of the total U.S. primary demand. As a result, even though uses of lead are not expected to increase per dollar of output, the relative end-use shares of the transportation and electrical sectors are projected to reach about 57 percent and 14 percent, respectively, by 2000. At the same time, the demand for secondary lead is projected at about 871,000 MT, representing nearly 40 percent of total domestic demand.

Projected Supply. Given the assumed changes in materials substitution and the different assumptions regarding import dependence and recycling activities, lead output from domestic mines can be expected to range between 600,000 and 960,000 MT by 2000. Under the scenario in which the United States is assumed to maintain its present import dependence and recycling activities, domestic output of lead is projected to reach 760,000 MT. By increasing the recycling rate to its technically maximum level, the United States can lower its mine outputs to 640,000 MT and still satisfy its domestic requirements.

World

Base-Year Demand. In 1970, global demand for primary lead amounted to 3.8 MMT, of which about 34 percent was consumed by the Western European (high-income) nations, 16 percent by the United States, and 13 percent by the USSR. Except for the USSR, Western Europe (high-income) and the United States were the world's major lead importers. Nearly 85 percent of Western European (high-income) total consumption of primary lead was satisfied by imports; similarly, the United States relied on imported primary lead for about 30 percent of its domestic requirements.

Base-Year Supply. World output of primary lead in the base-year was concentrated in five regions: Canada, the United States, and the USSR each contributed about 13 percent of global output; Australia contributed an additional 11 percent, and Western Europe (high-income) accounted for 8 percent.

In 1970, exports of primary lead from Canada totaled about 400,000 MT, followed by Australia, with approximately 300,000 MT, and Latin America (low-income, mainly Peru), with about 200,000 MT.

Projected Demand and Supply. Projected total (primary and secondary) demand for lead on a global basis is expected to reach 13.7 MMT by the year 2000, while demand for primary lead depending on the recycling regime in force is estimated to range from 6.2-8.2 MMT.[1] Western Europe (high-income), the United States, and the USSR are expected to remain the major consumers of lead, with shares of the world's total demand projected at 22 percent, 18 percent, and 16 percent, respectively. However, because of the relatively high rates of economic growth assumed for developing regions, the demand for lead by the oil-producing Middle East nations as well as by the middle-income Eastern European and Latin American countries is also expected to become increasingly important in the future global consumption pattern. At the same time, the United States, the Middle East, and Japan are expected to be major importers of lead.

In 2000, major contributions in meeting the world's global lead requirements are projected to be provided by Western Europe (high-income), the USSR, Australia, and the United States. Canada, the Eastern Europe middle-income countries, and two Latin American regions are also expected to produce significant amounts of lead. Oceania, the low-income Latin American countries, and Canada are likely to remain the major exporting regions.

Turning to the fifty-year long-range projections, primary and secondary global requirements for lead under the pessimistic and optimistic scenarios are expected to range between 26.4 and 33.8 MMT by 2030. Based on the estimates of global lead resources, cumulative world production is expected to deplete the estimated resource base after 2020. Presumably, as in the case of silver, zinc, and gold, additional resources will be located, increased recycling of secondary lead will occur, or substitution of other materials will provide a solution to this problem of global exhaustion by 2030.

Zinc

United States

Base-Year Demand Pattern. In 1970, the United States consumed 1.31 MMT of primary zinc, 85 percent of which was consumed as metal and the remaining 15 percent for nonmetallic purposes. The largest consuming sector of primary zinc was the construction industry, which absorbed 37 percent of the mineral's end-use demand. Zinc die castings were widely used in passenger cars and trucks; primary zinc used in these applications represented 27 percent of the year's primary demand. Electrical and machinery industries were also important consumers of zinc concentrates and metals; the purchases of these two industries amounted to approximately 13 percent and 8 percent, respectively, of the 1972 primary zinc demand.

As for the nonmetallic uses of zinc, about 9 percent of total primary demand was consumed in the manufacture of rubber products. Small amounts of zinc industrial compounds were also used in the paint and chemical manufacturing industries, which together accounted for 6 percent of the base-year's total primary zinc demand.

In 1972, demand for secondary zinc amounted to 0.35 MMT, approximately 21 percent of total zinc demand in the United States.

In analyzing the base-year data, nearly 50 percent of the primary zinc in 1972 was consumed by industries, either directly or indirectly, to produce output that was delivered to the investment component of the GDP. At the same time, 31 percent was required to produce the 1972 bundle of consumer goods and services. In addition, close to 20 percent of primary zinc was consumed directly and indirectly to satisfy government purchases, of which one-third was for military purposes.

Base Year-Supply Pattern. In 1972, the United States produced a total of 0.44 MMT of zinc (in metal content), of which 0.11 MMT was the result of by-product and co-product output. The balance of private-industry stock adjustment and government stockpile releases together added another 0.16 MMT to the total 1972 domestic supply. However, the major source of the U.S. supply of primary zinc was provided by foreign imports. With approximately two-thirds in the form of metal, primary zinc net imports amounted to 0.72 MMT and represented 55 percent of the year's primary zinc requirements.

Zinc recovered from both old and new scrap was 0.35 MMT in the base-year, primarily generated from galvanized pipe and sheet in steel mills and die-cast skimmings in foundries.

Projected Demand. By 2000, total U.S. demand for primary zinc is expected to range between 1.7 and 2.8 MMT, depending on the assumed rate of growth of the U.S. economy. However, if the potential of zinc use in housing construction, automobiles, die castings, and corrosion-protection coatings is realized, the materials-substitution trend accompanied by high-income growth assumptions could increase total U.S. primary demand for zinc to about 3.5 MMT by 2000. Under this scenario, consumption of primary zinc metal and concentrate by the construction and electrical industries is likely to account for almost half of total end-use primary demand. Zinc required by the transportation sector will likely capture about 30 percent of the end-use market in 2000, and the machinery manufacturing industry is expected to consume another 11 percent.

Referring to other nonmetal uses of primary zinc, the relative shares of rubber products, chemicals, and paints in 2000 are expected to remain essentially the same as those of the base-year.

After stepping up the zinc recycling rate to achieve its technically possible maximum by the year 2000, U.S. consumption of primary zinc is expected to increase to 2.9 MMT as compared with the higher projection under the high-GDP and materials-substitution scenarios.

Projected Supply. Maintaining both the 1972 share of imports to domestic demand and the base-year recycling rate of zinc, income growth in conjunction with the anticipated trend of materials substitution is expected to require domestic mines to produce about 1.34 MMT of zinc in the year 2000. However, projected results indicated that even under the high-import dependence and low-domestic production scenario, the U.S. cumulative output of the mineral is expected to exceed its currently estimated reserve base.

World

Base-Year Demand. World primary consumption of zinc amounted to 5.3 MMT in 1970, with four industrialized regions consuming close to 75 percent of the total: the Western Europe (high-income) region (28 percent), the United States (21 percent), Japan (15 percent), and the USSR (9 percent). The three market-oriented economies all imported most of their zinc requirements.

Base-Year Supply. In 1970, the world's leading zinc producer was Canada, whose output amounted to about 1.3 MMT or approximately one-quarter of the world's total output. The three other important producing regions were Oceania, the USSR, and Western Europe (high-income). In addition, the United States and Latin America (low-income, mainly Peru) each supplied about 8 percent of the world's output of primary zinc.

As a result, in the base-year, Canada was the major exporting region followed by Oceania and Latin America (low-income).

Projected Demand and Supply. By 2000, the world economy is expected to require about 15.4 MMT of zinc. Along with the major developed regions, several developing regions, such as middle-income Latin American countries, oil-producing Middle east nations, and low-income Asian countries, are also expected to become increasingly important consumers of zinc. Moreover, Western Europe (high-income) and Japan will continue to be the major importers throughout this century.

Almost one-third of global zinc output in 2000 is expected to be produced by the world's two largest producers: the USSR and Canada. However, significant quantities are also projected to come from the middle- and low-income Latin American countries, Australia, and the United States. As a result of this regional demand-and-supply distribution, the principal zinc exporters are projected to be Canada, Australia, and the low-income Latin American countries.

Long-range projections indicate that global requirements of zinc may range between 28.7 and 38.1 MMT by 2030, depending on the underlying assumptions about world economic growth. Projected cumulative zinc requirements indicate that, on a global basis, current resource estimates of zinc will likely be depleted before the year 2010. Presumably, as in the case of lead, silver, and gold, additional resources will be located, higher recycling of secondary zinc will occur, or substitution from zinc to other materials will provide a solution to this problem of global exhaustion.

Gold

United States

Base-Year Demand Pattern. In 1972, U.S. consumption of fabricated gold from primary sources was 206 MT. Nonfinancial uses for gold were concentrated in three sectors: jewelry and arts (60 percent), electronics (30 percent), and dental supplies (10 percent). As a result, about two-thirds of the total use of gold could be attributed to satisfying the household bill of goods. Another 20 percent of primary gold consumption was required to produce the investment bill of goods.

Base-Year Supply Pattern. Domestic supply of primary gold accounted for only about 20 percent of total primary consumption for domestic use. In the base-year, by-product output of gold—mainly from copper mining—accounted for almost half of the domestic output. The balance of primary gold consumption—over 80 percent—was supplied by imports. U.S. imports of gold originate mainly in Canada, South Africa, and the USSR, though gold from the latter two travels through the entrepôt of Switzerland. Recycling of secondary gold equaled 22.9 MT, about 10 percent of primary consumption for domestic use.

Projected Demand. U.S. primary consumption of gold for nonfinancial uses for 2000 is projected to range between 500 and 585 MT, on the basis of the assumptions regarding income growth, materials substitution, and recycling.

Projected Supply. U.S. domestic production of gold in 2000 is projected to range between 80 and 147 MT, based on two extreme scenarios that combine assumptions regarding income growth, technological change, and import dependency. Holding all other parameters constant at their 1972 values except income growth, U.S. domestic output of gold is projected to be 126.9 MT in the year 2000.

World

Base-Year Demand and Supply. World demand for gold in the 1970 base-year was approximately 1,460 MT, with Western Europe (high-income) consuming about 45 percent; the United States, about 13 percent; and Eastern Europe, about 10 percent. Gold production was dominated by two countries: South Africa, which supplied about 70 percent of the total, and the USSR, which supplied about 20 percent of the total primary demand. The two leading producing regions accounted for 95 percent of world exports of gold. Canada, the United States, Oceania and Tropical Africa all contributed marginal amounts to world production.

Projected Demand and Supply. Because the demand for investment bullion and gold coins is greater than the demand for fabricated gold, long-term projections of total demand for gold are difficult, if not impossible, to make. Political and economic uncertainties contribute to these difficulties. Both the USSR and South Africa indirectly settle their international accounts by the sale of gold for hard currency. Because the World Model is a general equilibrium framework, projections of the future demand for gold—including both fabricated and fiduciary gold—is balanced with supply by assuming that the proportions of the two uses will not change significantly from the base-year. As a result, total demand for fabricated and investment gold is projected at 5,440 MT by 2000. However, the demand for fabricated gold is expected to be about half that figure. No significant changes are expected in the relative positions of the gold-producing and -exporting regions by 2000; the consuming and importing regions are also expected to remain the same.

Projected total demand for fabricated and monetary gold can be expected to range between 11,800 and 14,500 MT, depending on the rate of world economic growth. However, even after disregarding entirely the demand for gold for financial purposes, on the basis of the projected demand for fabricated gold, the world's currently estimated stock of gold resources will be insufficient by 2030 to meet projected world demand, even under relatively slow world economic growth.

Silver

United States

Base-Year Demand Pattern. In 1972, the U.S. economy consumed 3.8 KMT of primary silver and an additional 1.95 KMT of secondary, recycled silver.

About 60 percent of all silver is consumed in two broad sectors: silverware, coins, medallions, jewelry, and arts (34 percent); and photography (26 percent). Other significant amounts of silver were consumed in the electrical equipment sector (11 percent), electrical components (8 percent), appliances (7 percent), and refrigeration (7 percent).

As a result of these silver consumption patterns, about 60 percent of the total amount of silver consumed in the economy could be predictably traced, directly and indirectly, to satisfying the personal-consumption expenditures of households.

Base-Year Supply Pattern. In 1972, silver production from domestic mines accounted for 1.2 KMT, approximately 30 percent of total U.S. consumption of primary silver for domestic use. Over 70 percent of the domestic mine output of silver in 1972 was obtained as a by-product of copper, lead, and zinc. Net imports of silver in the base-year were 10 percent greater than the level of domestic output, implying that almost 34 percent of total consumption of primary silver was obtained from foreign sources, largely from Canada, Mexico, and Peru.

Projected Demand. U.S. demand for silver in 2000 is expected to be distributed in the same sectors as in the base-year in practically the same proportions. Projected demand for primary silver resulting from income growth is expected to range between 6 and 10.5 KMT by 2000. If the prospective technological-change assumptions are realized, domestic consumption could increase to more than 11.8 KMT by 2000.

Projected Supply. If the recycling rate and the ratio of imports to domestic output remain unchanged, projected U.S. mine output can be expected to reach over 5 KMT by 2000. However, because of the continuing depletion of U.S. reserves, it is likely that imports will supply a major and even increasing share of U.S. silver requirements in 2000. If that proves to be the case, current U.S. reserves should just suffice to 2000.

World

Base-Year Demand. In 1970, world demand for silver reached a level of 10.5 KMT, 90 percent of which was consumed in the developed market-oriented and socialist countries. The major importing regions were Western Europe, the United States, Japan, and Eastern Europe.

Base-Year Supply. Of the 10.5 KMT of silver produced in 1970, world output was largely concentrated in six regions: the two Latin American regions (including Mexico and Peru), the USSR, Canada, the United States, and Oceania. The leading exporters include all of the major producers except the United States.

Projected Demand and Supply. World consumption of primary silver in 2000 is projected at a level of 37.9 KMT, given the projected GDP growth rates used in the computations. The same five regions can be expected to provide the required exports needed to satisfy global requirements. By 2030, world silver consumption is projected to range between 78.2 and 98.8 KMT. At the currently estimated level of world silver resources, the major producing regions can be expected to deplete their endowments by 2030. Presumably, as in the case of lead, zinc, and gold, additional resources will be located, higher recycling of secondary silver will occur, or substitution by other materials will provide a solution to this problem of global exhaustion.

Aluminum

United States

Base-Year Demand. In the United States, 1972 apparent consumption of primary aluminum amounted to 4.8 MMT. The construction sector remained the leading consuming sector of primary aluminum, capturing 26 percent of the end-use market in part because of the increasing use of aluminum (substituting mainly for wood and steel products) in residential sidings, mobile homes, windows, guard rails, bridges, and highways. The transportation sectors consumed 876,000 MT of primary aluminum in 1972, representing an end-use market share of 18 percent. Automobiles continued to be the dominant user of aluminum in those sectors. However, its light weight and high-strength characteristics account for its wide use in the aircraft and shipbuilding sectors. In 1972, these two sectors accounted for about 30 percent of aluminum consumed by the transportation sectors.

Consumption of primary aluminum by the electrical industry constituted 13 percent of the total end-use market in 1972, the third largest user. Similarly, properties such as corrosion resistance and good malleability also contributed to the broad use of aluminum in industries producing metal cans, flexible packaging, large containers, household appliances, and machinery. The end-use shares in 1972 for cans and containers, appliances, and equipment, and machinery were 10 percent, 9 percent and 6 percent, respectively. The share for miscellaneous use of aluminum has gradually declined since 1970 and dropped to 6 percent in 1972, reflecting mainly reduced demand during the 1970s for the output of the arms and ammunition sector, in which aluminum is a major material input. Different nonmetallic uses such as alumina-fused refractories, alumina catalyst, and aluminum oxide together accounted for 11 percent of primary aluminum demand in 1972.

In light of the heavy use of aluminum in the construction, transportation, electrical, and machinery manufacturing industries, almost 44 percent of the 4.8 MMT of primary aluminum consumed in 1972 was directly and indirectly consumed to satisfy $185 billion in private-business capital investment (in which construction accounted for over 50 percent). Although personal-consumption expenditures for 1972 amounted to $738 billion, four times as much as capital investment, only 32 percent of the metal was consumed for delivering the various goods and services to satisfy consumer needs.

Industrial use of secondary aluminum, both old and new scrap, amounted to an estimated 857,000 MT in 1972, almost 15 percent of the year's total aluminum requirements.

Base-Year Supply Pattern. In 1972, total U.S. gross imports of bauxite, alumina, and intermediate aluminum products reached 4.8 MMT in aluminum content. For the same year, aluminum produced from domestically mined bauxite amounted to 424,000 metric tons. In 1972, secondary supply of aluminum from both old and new scrap met 15 percent of the year's total aluminum demand.

Projected Demand. Depending on the rate of growth assumed for the U.S. economy from the year 1972 to 2000, all other parameters held unchanged, the demand for primary aluminum is projected to range from 8.2 to 10.2 MMT at the end of this century. Because the competitive positions of such materials as wood, plastics, steel, and magnesium is also expected to shift with respect to aluminum, the changing industrial composition of materials consumption when accompanied by relatively high economic growth could increase the demand for primary aluminum to 14.3 MMT by 2000. The most noticeable increase in aluminum use will take place in the automobile sector. In 1971, a typical passenger car contained an average of 72 pounds of aluminum. To achieve higher fuel efficiency through weight reduction, automakers now use aluminum in place of, wherever possible, heavy metals such as carbon steel, in most cars. In other applications, substitution will continue between aluminum and other materials. More extensive use of aluminum in the packaging, machinery, and consumer-durable industries is also expected. Given the anticipated economic growth and materials substitution, the transportation sector will replace the construction industry as the leading consumer of primary aluminum by 2000. About 4.4 MMT of the metal (or 31 percent of total demand) will be used to produce various transportation equipment. In the construction industry, much of the substitution for steel and wood products by aluminum has already taken place. A future wave of substitution will probably make plastics more attractive than other substances. The construction sector will account for about 9 percent of primary aluminum requirements in 2000. Use of aluminum in electrical machinery and equipment is expected to grow at an average annual rate of 3.3 percent from 1972 to 2000. As industries that manufacture metal cans and containers, home appliances, and machinery and equipment continue to purchase increasing amounts of aluminum per dollar of their outputs, the share of the mineral for nonmetalic uses in refractories and chemical plants is not likely to change significantly by 2000.

Another variable that will significantly affect the future demand for primary aluminum is the supply of secondary aluminum. If the recycling rate for aluminum can be increased from 15 percent in 1972 to 37 percent in 2000 (the highest projected rate at that time), the level of total primary aluminum consumption will increase to only 11.2 MMT by 2000 instead of the 14.3 MMT.

Projected Supply. Given the income-growth and technical-change assumptions, while holding the base-year recycling rate and proportion of imports to primary demand constant, production of aluminum from domestically mined bauxite is projected at 1.4 MMT in 2000. The 1980 estimate of recoverable aluminum from domestic bauxite reserves is 9.1 MMT. Without increased primary aluminum prices and without improvements in technology, the cumulative domestic production of aluminum from 1980 to 2000 will probably exceed the 14.3 MMT estimate of recoverable domestic reserves. If imports of bauxite, alumina, and aluminum to the United States increase gradually from 92 percent in 1972 to 97 percent in 2000, aluminum production in 2000 from domestically mined bauxite will remain at the same level as in the 1972. Cumulative production from 1980 to 2000 under this scenario would exceed the estimated aluminum reserve by 21 percent. A low- import policy for aluminum will no doubt hasten the depletion of currently estimated reserves. Projected results imply that the United States is likely to become almost entirely dependent on imports to satisfy its aluminum requirements by the year 2000.

World

Base-Year Demand and Supply. Except for the USSR, regional supply of aluminum-based ores is very different from regional demand for aluminum. For example, bauxitic ores are principally mined in Latin America (low-income, mainly Guyana, Jamaica, and Surinam) and Oceania (mainly Australia), but the major consuming countries are the United States, Japan, and the Western European industrial countries. In 1970, 45 percent and 16 percent of the world's output of aluminum originated in Latin America (low-income) and Oceania, respectively. At the same time, the United States consumed 35 percent of the world's bauxite production (in aluminum content), with Western Europe (high-income) requiring 26 percent, the Soviet Union, 9 percent, and Japan, 9 percent. In 1970, the world's exports, based on the World Model's regional classification, originated from the Latin American (low-income) region (64 percent), Oceania (21 percent), and the Western Europe (medium-income) region (8 percent), respectively.

Projected Demand and Supply. Estimates of world consumption of bauxite (in aluminum content) in 2000 is 35.9 MMT. Latin America (low-income) and Oceania will continue to be major producers of bauxite, and the export pattern is not expected to change significantly by 2000. However, the U.S. and the Western European (high-income) shares of consumption will likely decline, mainly because of the increasing demand for the mineral in faster-growing developing regions.

Long-term fifty-year projections estimate that global aluminum consumption could range between 67.8 and 91.0 MMT by 2030, depending on the rate of economic growth assumed in the World Model. In 1980, the world's bauxite reserves, in recoverable aluminum equivalent, were estimated at 4,716 MMT. With this sizable reserve base, the potential world supply of aluminum should be sufficient to meet world demand through the next fifty years.

Mercury

United States

Base-Year Demand Pattern. Although more than twenty other minerals are known to contain mercury, primary mercury is recovered almost exclusively from red sulfide and cinnabar deposits. In 1972, 1,405 MT (equivalent to about 40,765 units of 76-pound flasks) of primary mercury were consumed by various industries in the United States. The largest share, approximately 29 percent of the year's primary demand, was consumed by the electrical sector. The second largest end-use of mercury was in caustic soda and chlorine production. In 1972, about 306 MT, about 20 percent of the total primary demand, were consumed for this purpose. Other significant uses of mercury were in dental amalgams for fillings and various industrial and control instruments. These uses accounted for about 19 percent of 1972 consumption.

The paint-producing industry, consuming about 267 MT of primary mercury in the base-year, represented another 19 percent of total primary demand. Mercury use in some plastics and synthetic products, together with its use in pharmaceutical products, represented a share of about 3 percent. Finally, other chemical and miscellaneous uses of primary mercury accounted for roughly 9 percent of 1972 primary demand.

In the same year, demand for secondary mercury (new and old scrap) amounted to about 418 MT (equivalent to approximately 12,130 flasks). As a result, 23 percent of the total U.S. demand for mercury in 1972 was met from secondary sources of supply.

On the basis of these end-use patterns, it was estimated that close to 50 percent of the total base-year's demand for primary mercury was directly and indirectly used to produce goods and services purchased by households. Another 30 percent was used directly and indirectly to satisfy private capital-investment requirements.

Base-Year Supply Pattern. In 1972, 70 percent (980 MT) of the U.S. supply of primary mercury came from foreign sources. The largest supplier was Canada, where the United States obtained almost 40 percent of its imports. Algeria and Spain were also important sources; in 1972, these two countries together provided approximately another 40 percent of U.S. imports. The remaining U.S. imports were procured mainly from Yugoslavia, Mexico, and Italy.

Domestic mine production contributed about 18 percent (253 MT) to the country's primary mercury supply in 1972. Private industry inventory reductions in 1972 amounted to 154 MT, constituting nearly 12 percent of the year's primary supply.

Projected Demand. Projected U.S. demand for mercury in 2000, based on income growth and possible technological changes, is expected to increase to 1,680 MT (about 48,700 flasks). Under this scenario, it is expected that mercury use in caustic-chlorine production will be entirely displaced by the use of diaphragm cells by 2000. Similarly, mercury in paint products is expected to be replaced, primarily by plastic and copper paints. However, domestic consumption of primary mercury by the electrical industries may amount to 940 MT by 2000, despite the penetration of some nonmercury batteries such as graphite dry cells. Similarly, allowing for limited substitution of porcelain and plastics for dental amalgams, the dental supplies and professional instruments industries are expected to consume 350 MT of primary mercury by 2000. Because few substitutions of mercury are anticipated in the pharmaceuticals, plastics and synthetic products industries, or the chemical industries, their consumption of primary mercury in 2000 is projected at about 100 and 280 MT, respectively.

Because the recycling rate for secondary mercury in 2000 is not likely to exceed the present one, projected demand for this secondary source of mercury is a moderate 500 MT.

Projected Supply. On the basis of the rates and patterns of income growth, technological change, and materials substitution that are assumed to take place in the U.S. economy, the computations indicate that domestic mine production of mercury can be expected to range between 351 and 1,110 MT in 2000. Because the recycling rate of mercury has already achieved its peak and it is likely that it will be maintained at that high level at least until the end of the century, this forecasting range is mainly the result of different assumptions concerning U.S. import policies. When the forecast is compared to the 1980 estimate of the total U.S. mercury reserve base of 12,060 MT, it becomes clear that any supply policies intended to reduce import dependence by encouraging domestic production at a rate assumed in the high projections will likely exhaust domestic reserves as early as 1997. Conversely, if the United States increases its reliance on foreign production, the low projected results show that only 57 percent of the reserve base will be depleted by 2000.

World

Base-Year Demand and Supply. World production of primary mercury totaled 11,300 MT in 1970 and decreased to an estimated 7,270 MT in 1980. The shares of global production in 1970 was 27 percent by Western Europe (medium-income, mainly Spain, Turkey, and Yugoslavia), 17 percent by the Soviet Union, 12 percent by Western Europe (high-income, mainly the Federal Republic of Germany and Italy), 11 percent by Latin America (medium-income, mainly Mexico), 8 percent by the United States, and 6 percent by Eastern Europe (medium-income, mainly Czechoslovakia). Major mercury-consuming regions were Western Europe (high-income) with 20 percent, Eastern Europe (medium-income) with 20 percent, the

Soviet Union with 17 percent, the United States with 17 percent, and China with 14 percent.

As a result of this regional supply-and-demand pattern, close to 60 percent of the world's mercury exports in 1970 originated from Western Europe (medium-income), 22 percent from Latin America (medium-income), and 16 percent from the Middle East (mainly Algeria).

Projected Demand and Supply. Assuming no technological change and materials substitution in the use of mercury, global consumption of primary mercury may be as high as 19,000 MT by 2000. Under the optimistic scenario, world production of mercury necessary to satisfy demand in 2030 is estimated at 45,700 MT; under the pessimistic scenario, the projection is that about 37,000 MT will be required. Given the world's identified mercury resources of 570,000 MT, it is likely that without major economies in the future use of mercury, global mercury requirements will exceed global mercury endowments by 2020. Moreover, projected results indicate that by 2010, three regions (Canada, low-income Latin America, and the Middle East countries) that now export mercury are likely to deplete their regional resources. In the optimistic scenario, the same situation is likely to occur by 2020 in the USSR and by 2030 in Latin America (low-income) and Western Europe (medium-income). As a result, limitations in supply will probably enhance the conservation and substitution efforts in the use of mercury.

Vanadium

United States

Base-Year Demand Pattern. In 1972, the United States consumed 6.55 KMT of vanadium in the form of ferrovanadium, vanadium master alloys, and vanadium-based chemical compounds. An unknown quantity of secondary vanadium recycled from spent catalysts and tool steels was also consumed. The major use of vanadium is as an alloying element in steels. Major end-use sectors of vanadium were transportation (37 percent), construction (24 percent), metal-working machinery (17 percent), construction machinery (15 percent), and chemicals (7 percent).

Personal-consumption expenditures represented less than one-third of direct plus indirect vanadium requirements in 1972, as compared to 54 percent for capital-investment expenditures; military spending accounted for about 7 percent of total vanadium requirements. Investment spending was 7.6 times as vanadium-intensive as personal-consumption expenditures.

Base-Year Supply Pattern. Vanadium is typically produced as a by-product of uranium, phosphorous, iron (outside the United States), or crude oil and rarely by itself; only one U.S. mine produced vanadium from vanadium ores. In 1972, 4.8 KMT or 73 percent of vanadium demand was satisfied from domestic vanadium, including two-thirds as by-product of phosphorous and uranium production. One-quarter of 1972 demand was met by imports originating primarily in South Africa. A small but unknown quantity of vanadium was recycled from tool steels and spent catalysts.

Projected Demand. U.S. domestic consumption of vanadium is projected to range between 14.3 and 15.9 KMT in 2000, depending on the assumed rate of GDP growth and technological change. Over 92 percent of the growth in vanadium consumption can be attributed to GDP growth; the remaining 8 percent is the result of changes in technology. The transportation and construction machinery sectors increase in relative importance as end-use markets for vanadium, while end-use markets in construction and chemicals shrink slightly.

Per dollar of output, the increase in vanadium use by the construction sectors is the most dramatic—about 40 percent higher than in 1972.

Projected Supply. Under the technological-change scenario, the United States will produce 4.25 KMT from domestic vanadium deposits and 8.5 KMT as a by-product, totaling about 80 percent of total U.S. primary demand in 2000.

World

Base-Year Demand. In 1970 the developed regions—the OECD and Eastern Bloc countries—consumed about 87 percent of the world's vanadium production. Of the 17.9 KMT of world vanadium consumption in 1970, the United States accounted for 36 percent, the USSR for 29 percent, Japan for 12 percent, and Western Europe (high-income) for 5 percent.

Base-Year Supply. The major producing regions in 1970 were the USSR (46 percent), the United States (28 percent), and South Africa (16 percent). About 10 percent of world output was produced in Western Europe (high-income). These producing regions, except the United States, were also major exporters.

Projected Demand. World vanadium consumption is projected to increase to 71 KMT in 2000. The developed regions' share of world consumption is projected to decline to about 80 percent in 2000. No significant changes are expected in the patterns of world vanadium production or exports.

In the long-range fifty-year projections, world vanadium consumption is expected to range between 116 and 165 KMT, depending on the rate of world economic growth, by 2030. No long-term resource problem for vanadium is expected globally or regionally, even under the optimistic high-growth scenario.

Platinum

United States

Base-Year Demand Pattern. In 1972, U.S. consumption of primary platinum was 14.6 MT. Over 45 percent of the total was consumed in the chemical sector. Almost another 40 percent of total consumption was used about equally in the petroleum-refining sector and in electrical uses. The remaining 16 percent of total consumption was used in three other sectors: ceramics and glass, dental supplies, and jewelry and art.

Platinum's role in the component parts of final demand is clear; almost 60 percent of its use was needed to satisfy households expenditures; another 20 percent could be traced directly and indirectly to the investment component of the GDP.

Base-Year Supply Pattern. Less than 10 percent of total U.S. primary platinum requirements in 1972 came from domestic sources. As a result, over 80 percent of U.S. platinum requirements were satisfied from imports. The residual base-year supply came from inventory reductions. U.S. imports were procured mainly from the Soviet Union, Colombia, and South Africa.

Projected Demand. The use of platinum over the next twenty years is expected to be limited to the same six consuming sectors in more or less the same proportions as in 1972. However, a reduction in the use of platinum as an emission control catalyst and in petroleum refining is projected for 2000. Glass fibers are expected to continue to displace textile fibers, and, as a result, more platinum is expected to be used per dollar of glass output.

By 2000, expected U.S. consumption of platinum is projected to range between 31.4 and 33.6 MT, about 2.3 times the 1972 level, depending on the rate of growth of GDP and technological-change assumptions used in the projections.

Projected Supply. Given its current reserve estimates, the United States can be expected to continue its near-total dependence on foreign sources of supply through 2000 and beyond.

World

Base-Year Demand. In the 1970 base-year, world consumption of primary platinum was 57.1 MT. The largest consuming regions were Japan, the United States, and Western Europe (high-income), together accounting for over 65 percent of global consumption. Approximately 16 percent of world demand for platinum came from the socialist bloc (USSR and Eastern Europe). With the exception of the USSR, all the other major consuming regions imported most, if not all, of their platinum requirements.

Base-Year Supply. Over 60 percent of 1970 world platinum supplies were produced in South Africa and over 30 percent of global output of platinum was produced in the USSR. The only other region producing any significant amount of platinum was Canada, with about 5 percent of global output. World exports in the base-year were concentrated in the two largest producing regions.

Projected Demand and Supply. World consumption of platinum in 2000 is projected to reach 221 MT, an annual growth rate of 4.5 percent. World platinum requirements are expected to continue to be satisfied by South Africa and the USSR.

The fifty-year long-range projections of global demand for platinum estimate a range between 427 and 545 MT, depending on the rate of growth of the world economy. Current global resource estimates are sufficient to meet cumulative world output requirements by 2030. However, without major new discoveries of platinum deposits in the Soviet Union, it is likely that the Soviet Union's role in world exports will diminish after 2000 to delay the date of exhaustion of its resource base of platinum.

Titanium

United States

Base-Year Demand Pattern. Rutile and ilmenite are the principal mineral sources of titanium. Both minerals can be used to make titanium dioxide pigment; rutile is essential for use in making sponge metal. Rutile and sponge metal are included in the national stockpile list of strategic materials.

The United States consumed 550,220 MT of titanium in 1972; nonmetal uses accounted for about 98 percent of total consumption. Over 90 percent of the domestic demand for titanium concentrates were used to produce titanium dioxide pigment. This pigment, because of its special characteristics, is widely used in surface coatings of paints, paper, ceramics, and fiberglass. In 1972, titanium used in paints, varnishes, and lacquers accounted for 50 percent (275,000 MT) of total domestic demand. To improve printability, titanium pigment was also extensively used in paper coatings, and this market represented 20 percent (112,000 MT) of the base-year's total titanium use. Given its stable chemical structure, the pigment has widespread applications in plastics and rubber products, such as polyethylene, polyvinyl chloride, and rubber tires. The titanium consumed in manufacturing these products accounted for about 10 percent (58,000 MT) of total 1972 demand. In addition, approximately 2 percent (12,650 MT) of titanium concentrates were consumed in producing porcelain, enamels, ceramic capacitors, and glass fibers. Use of titanium in welding fluxes and titanium carbide cutting tools, along with its uses in catalysts and water repellants, accounted for the remaining 16 percent of the nonmetallic use of titanium.

In 1972, consumption of titanium for metal use represented close to 2 percent of the total titanium demand, about 9,000 MT. The consumption of titanium metal has been closely related to the growth of the aerospace industry. In 1972, about 60 percent of titanium mill products were consumed in aerospace applications. Utilization of titanium metal in nonaerospace applications, such as chemical process equipment, marine parts, and desalination equipment, accounted for the remaining 40 percent.

In 1972, over 57 percent (298,000 MT) of total titanium demand was directly and indirectly used to produce goods and services delivered to households. At the same time, about 25 percent (134,000 MT) was consumed directly and indirectly to satisfy private capital investment requirements.

Base-Year Supply Pattern. U.S. supplies of titanium primarily came from both domestic mine production and foreign imports. Although the United States continued to produce ilmenite at a level that satisfied demand, most titanium slag (used to produce titanium pigment) and nearly all rutile (used to make both titanium pigment and sponge metal) were imported from Canada and Australia, respectively. In 1972, U.S. mines produced 207,000 MT of titanium and accounted for almost 37 percent of the year's primary demand. Net imports in titanium content, on the other hand, amounted to 258,000 MT and represented 47 percent of total demand. The remaining 15 percent of titanium demand in 1972 was satisfied through industry stock reductions. Because of both technical and environmental problems, the contribution of secondary titanium to total domestic demand was still insignificant in 1972.

Future Demand. Depending on the alternative GDP growth rates assumed in the study, the use of titanium in the United States by 2000 can be expected to range between 1.0 and 1.2 MMT. However, by further incorporating the anticipated technological changes leading to proportionally larger usage of titanium in paints, paper products, ceramics and glass, aerospace equipment, and other industrial equipment, the probable demand for titanium will be even higher and can reach 1.4 MMT in 2000, about 2.5 times higher than the base-year demand. Under scenarios incorporating the technological-change assumptions, sectors producing paints and allied products will continue to be the leading users of titanium dioxide (TiO_2). However, the use of titanium dioxide as surface coatings for paper products, particularly for printing and packaging, is likely to accelerate. The unique characteristics of TiO_2 will continue allowing the pigment to be used broadly in plastics and synthetic products; titanium usage in these products will likely consume about 9 percent of total demand. Use of titanium as fluxes, catalysts, and water repellants is expected to account for the other 13 percent of nonmetal titanium requirements.

Use of titanium metal in aerospace and nonaerospace applications will probably require 20,000 MT by 2000. The principal consuming sector, aerospace manufacturing, may account for about 60 percent. Recently announced plans to build 100 B-1 bombers and proposals to purchase 50 C-52s, the world's largest cargo planes, for military use would also increase the demand for titanium metal above the 20,000–MT level.[1]

Future Supply. If the United States maintains the current ratio of titanium imports to primary demand through 2000, domestic production is expected to range between 506,000 and 632,000 MT by 2000, based on alternative GDP growth rates and the anticipated changing structure of the U.S. economy. Cumulative production from 1980 to 2000 under the most ambitious domestic production schedule will amount to 11 MMT. Even though 1980 domestic resources for titanium are estimated at about 16 MMT, pollution control regulations, land-use disputes, and low-grade deposits will prohibit a large increase in domestic production above 11 MMT. A more probable scenario is that the United States will continue to import slightly more than half its requirements from Canada (titanium slag) and Australia (rutile). In this case, cumulative domestic production of titanium (mainly ilmenite) from 1980 to 2000 is likely to reach about 9.5 MMT.

World

Base-Year Demand and Supply. World demand for titanium concentrates (about 90 percent ilmenite and 10 percent rutile-based) in 1970 amounted to 1.72 MMT. In 1970, ilmenite was mostly mined in Australia (27 percent),

Canada (24 percent), high-income Western European countries (16 percent), the United States (15 percent), and the Soviet Union (10 percent). Production of rutile is concentrated almost exclusively in Australia. In 1970, the Western European industrial countries together absorbed an estimated 36 percent of the world's total titanium production, followed by the United States with 29 percent, Eastern European countries with 8 percent, and Japan with 5 percent.

In light of the world supply and demand patterns, and based on the World Model's regional breakdown, in 1970 Australia and Canada together accounted for an estimated 76 percent of world exports of titanium. Australia exported 40 percent and Canada 36 percent, with the remaining amounts contributed mainly by the USSR, India, and Malaysia. This export pattern is not expected to change significantly through 2000.

In 1970, major titanium importers were Western Europe, the United States, Eastern Europe, and Japan.

Future Demand and Supply. By 2000, Western Europe and the United States will consume more than half the world's output of titanium. The Eastern European countries are expected to consume another 9 percent, and Japan will consume an additional 7 percent. Australia is expected to export 1.2 MMT of titanium and Canada is likely to contribute another 1 MMT to world exports by 2000.

To satisfy that demand, world titanium production is projected to reach a level of 4.9 MMT by 2000. Depending on the growth of the world economy, global requirements of titanium may range between 10 and 12 MMT by 2030.

Global titanium resources, as estimated in 1980, are sufficient to sustain world production through 2030, even though such countries as Japan and Australia are likely to deplete their current resources by 2010. The Soviet Union is expected to exhaust its regional endowment by 2020. However, it is more likely that some shifts in the export policies of Australia and the Soviet Union before 2020 will be required to accommodate their regional needs.

Tin

United States

Base-Year Demand Pattern. In 1972, U.S. primary tin consumption totaled almost 50 KMT in contained metal. The cans and containers industry continued to be the dominant end-use sector for primary tin, particularly tin in the form of tinplate. Although materials such as tin-free steel (TFS) and aluminum are slowly penetrating into this industry, the cans and containers sector still consumed 16,000 MT of primary tin, which represented 40 percent of the domestic primary demand in 1972. For the same year, the second largest end-user of primary tin was the electrical industry. Either in pure form or alloyed with other materials, such as solder or alloy coatings, approximately 8,500 MT of primary tin was consumed in electrical applications in 1972, representing 17 percent of the year's total primary demand. As an element in various alloys, tin was also employed extensively in construction-related products, such as pumps, valves, and pipe unions (about 15 percent of the total 1972 U.S. primary tin demand). The transportation sector ranked fourth in tin usage, with tin alloys used mostly in automobile radiators, pistons, gasoline tanks, and bearings for motorcar and marine engines. In the base-year, more than 6,700 metric tons of primary tin was consumed by the transportation industries. Tin can often be found in a variety of machinery products as an alloying component in castings, stampings, bearings, and fittings, as well as a coating material for machinery parts. The amount of primary tin reported to be used in the machinery sector was about 5,500 MT in 1972. Tin was also an essential substance in a number of chemical compounds. Chemical requirements for tin totaled about 4,200 MT in 1972, nearly 8 percent of the year's primary tin demand.

Demand for secondary tin has been hampered by the high cost of scrap collection and cleaning; the base-year consumption of old and new scrap in contained tin was estimated at about 15,000 MT.

In light of these demand patterns, almost half the total demand was allocated to satisfy consumer needs. At the same time, about one-third of the primary tin was consumed to make possible the level and mix of the year's private business capital investment.

Base-Year Supply Pattern. The United States relies almost exclusively on imports of tin metal and concentrate to meet its domestic needs of primary tin. Although domestic reserves were estimated at only 50,000 MT in 1980, the amount of tin recycled has not been substantial because of the relatively high cost of recycling. In 1972, the United States imported 57,700 MT of contained tin, of which about 8 percent was sent to domestic smelters in the form of tin concentrate. Major sources of U.S. tin imports are Malaysia, Thailand, Bolivia, and Indonesia.

The U.S. government, which classifies tin as a strategic and critical mineral, has instituted a tin stockpile policy. In 1972, the government released a small quantity of primary tin (237 MT). In the same year, about 23 percent of the total U.S. demand for tin (65,000 MT) was satisfied by recycled secondary tin scrap.

Projected Demand. Given the average annual rates of growth, structural changes, and materials substitution assumed to take place in the U.S. economy by 2000, total primary tin demand is projected to range between 68,300 and 113,000 MT. A gradual but significant displacement of tin used in the cans and containers sector is likely to continue through the end of the century. Recent developments of plastic materials in food packaging and continuing conversion of beverage containers to tin-free steel and aluminum signify a clear movement away from the use of tin in the cans and containers industry.

If a policy of increasing the recycling rate of tin is adopted, structural limitations as well as higher tin prices will likely pull down domestic demand for primary tin to about 68,000 MT by 2000, along with an increase of secondary tin consumption to nearly 58,000 MT.

Projected Supply. With limited domestic economic reserves, the United States is expected to continue to rely on foreign imports for nearly all its requirements for primary tin. A handful of developing Asian countries, as well as Bolivia, will remain the major sources of supply. In 1978, Malaysia provided more than half of U.S. imports of primary tin metal and concentrate, while Thailand, Indonesia, and Bolivia each contributed about 10 percent.

The current domestic recycling rate for tin is about 23 percent. However, if conditions are favorable to increase recycling, the technically maximum rate can be as high as 55 percent.

World

Base-Year Demand and Supply. In 1970, world demand for primary tin amounted to 222,000 MT, of which 28 percent was consumed by the Western European industrial countries, 24 percent by the United States, 13 percent by Japan, 8 percent by the USSR, 6 percent by the Eastern European countries, and the remaining 20 percent by the rest of the world. Major producers of tin were concentrated in two regions: Asia (low-income, especially Malaysia, Indonesia, and Thailand), produced nearly 53 percent of the world's output, and Latin America (low-income, mainly Bolivia) contributed another 14 percent. These two regions were the major exporters to the world economy, contributing about 140,000 MT.

Projected Demand and Supply. Given the relatively high future rates of economic growth in the major consuming regions in the World Model, world demand for primary tin may reach as high as 727,000 MT by 2000. Based on the World Model's regional classification, Asia (low-income) and Latin America (low-income) are expected to supply 65 percent and 17 percent of world exports, respectively, in 2000. Modest amounts of exports can also be expected from Australia and China.

In 2030, world production levels necessary to satisfy demand requirements are projected to range between 1,310 to 1,750 KMT. Currently estimated world resource levels suggest that cumulative global tin requirements under the high economic growth assumption through 2030 will exceed world endowments.

Silicon

United States

Base-Year Demand Pattern. In 1972, the United States consumed 509.5 KMT of silicon and silicon metal. Its chief use is for deoxidizing and as a strengthening alloy in the production of iron, steel, and nonferrous alloys, principally aluminum. In metallurgical uses, silicon is used in the form of ferrosilicon and silicon metal. Silicon is also used in various nonmetallurgical applications, including silanes, silicones, and semiconductor devices.

The end-use distribution for silicon—except for chemicals—reflects the end use of cast iron, steel, and aluminum. The most important end uses of silicon in 1972 were transportation, machinery, and construction. These three sectors accounted for 36, 22, and 18 percent of total demand, respectively. The primary use of silicon in the transportation sector is in the production of cast-iron parts for the automotive industry. Cast iron in machinery is a second major user of silicon, while cast-iron pipe is the main use of silicon in construction. Chemical uses of silicon are represented chiefly by silanes and silicone.

Personal-consumption expenditures represented about 37 percent of the direct plus indirect uses of silicon in the U.S. economy in 1972. Capital expenditures were more significant, accounting for about 47 percent of total silicon consumption; military spending accounted for about 7 percent. Per dollar, investment goods were 5.5 times as silicon-intensive as personal-consumption goods; military goods were twice as silicon-intensive as personal-consumption expenditures.

Base-Year Supply Pattern. Because silicon alone cannot be recycled, all supplies come from primary sources. Silicon reserves consist of nearly limitless supplies of quartzite, quartz sand, and gravel. However, largely because of lower smelting costs abroad, the United States imports some silicon from foreign sources, including Canada, Norway, Yugoslavia, and France. In 1972, about 4 percent of total domestic demand was imported.

Projected Demand. U.S. silicon demand is expected to reach 1,150 to 1,244 KMT in 2000, depending on the rate of economic growth and technical change. Under the technological-change scenario (scenario 8), silicon demand will increase 144 percent to 1,242 KMT in 2000.

Because of the projected changes in the structure of final demand in 2000, transportation, machinery, and construction are expected to consume 41 percent, 29 percent, and 11 percent, respectively, of all silicon used in the economy.

Under the technological-change scenario, more than 98 percent of the growth in projected silicon consumption is attributable to changes in the level and composition of GDP, even though the silicon-intensive components of GDP—capital investment and military spending—comprised slightly smaller shares of total projected GDP in 2000 than in 1972. In aircraft, ships, and machinery, silicon input per dollar of output is expected to remain nearly stable. Increases in silicon requirements (again per dollar of output) of between 5 and 20 percent are projected for motor vehicles and electrical uses, because of the higher silicon content in cast irons and increased use of electrical steels. Declines in silicon use in construction, oil and gas, and appliance sectors reflect, in part, substitution of other substances for iron and steel.

Projected Supply. Under scenario 8, the United States will import 68 KMT of silicon in 2000, more than 5 percent of total supply. The balance of supply will come from domestic sources. However, it is likely that imports will be nearer to 232 KMT or about 20 percent of projected demand under the high-import scenario.

World

Silicon reserves and resources are widespread throughout the world and are, for all practical purposes, limitless. In 1970, the United States, the USSR, Western Europe (high-income), and Japan accounted for a total of 78.6 percent of world silicon demand. These same regions together produced 88.7 percent of the world's silicon supply. Of these regions, only the USSR and Western Europe (high-income) are net exporters.

World consumption is projected to increase from 1,630 KMT in 1970 to 6,170 KMT in 2000, an increase of about 4.5 percent per year. By 2030, world silicon demand may range from 11,000 KMT to 14,900 KMT, depending on the assumed rates of economic growth for the different regions of the world.

Fluorine

United States

Base-Year Demand Pattern. The iron and steel sector was the single largest user of fluorine in 1972, absorbing about 41 percent of the nearly 600,000

MT of fluorine consumed in pickling stainless steel. Approximately 210 KMT of fluorine (35 percent of total consumption) was delivered to the chemical sector for the production of hydrofluoric acid, which is used to manufacture fluorocarbons, primarily for use in aerosol sprays and refrigerants. A third relatively large user of fluorine was the primary aluminum sector, consuming about 23 percent of total U.S. fluorine requirements in 1972, about 138 KMT.

Almost half the fluorine required in the U.S. economy in 1972 was used to satisfy personal-consumption expenditures, reflecting its use in chemical products and household durable goods using steel and aluminum. About 40 percent of total fluorine consumption was used directly and indirectly for producing the investment component of the 1972 final demand, undoubtedly reflecting the use of metals by the construction sector.

Base-Year Supply Pattern. In the base-year, domestic fluorine production contributed only about 16 percent to total U.S. requirements. Mexico was the major source of U.S. imports in 1972, shipping approximately half a million metric tons of fluorine.

Projected Demand. The major users of fluorine in 2000 are expected to be the traditional consuming sectors—chemicals, iron and steel, and aluminum. However, recent legislation, which restricts the use of fluorocarbons as aerosol propellants because of alleged environmental hazards, is expected to moderate the demand for fluorine by the chemical sectors. The projected levels of U.S. fluorine consumption in 2000, incorporating the technological-change and income-growth assumptions, are expected to range from 0.96 to 1.6 MMT, 1.6 to 2.7 times the 1972 level.

Projected Supply. To meet the expected demand for fluorine, the domestic share of total supply is not expected to register a significant increase. Under the high-import scenarios, the United States can be expected to turn to imports to satisfy all but 5 percent of its fluorine requirements in 2000.

World

Base-Year Demand Pattern. In 1970, world fluorine consumption was 1.86 MMT; over 95 percent consumed in the developed OECD and Eastern European regions. The main importers included Japan, the United States, and the USSR.

Base-Year Supply Pattern. The leading producing and exporting regions—in descending order of their share of total production and exports—were Latin America (medium-income, including Mexico), Western Europe (high-income), China, Western Europe (medium-income), Asia (low-income), and South Africa.

Projected Demand and Supply. Under the regional income-growth assumptions used in these projections, world demand for fluorine can be expected to reach 3.6 MMT, or about twice the 1970 base year level. This level represents a global annual rate of growth of about 2.2 percent over the 1970–2000 period. By 2000, South Africa can be expected to be the leading exporter of fluorine as some of the other exporting regions reduce or divert exports to satisfy their own regional requirements.

Global fluorine requirements in 2030 under the pessimistic scenario are projected to be 6.54 MMT, an annual rate of growth of about 2 percent. Under the optimistic high income-growth scenario, world production is expected to be around 8.81 MMT, an annual rate of growth of approximately 2.6 percent. Even though global resources of fluorine in fluorspar are sufficient to satisfy global requirements only through 2000, presumably the total resource base of fluorine can be extended to include the fluorine contained in phosphate rock. In this case, current total world fluorine resources will suffice to meet global requirements to the year 2030, even under the high-income growth scenario.

Potash

United States

Base-Year Demand Pattern. In 1972, 95 percent of the 4.37 MMT of the potash used in the United States was consumed in the agricultural sectors, as fertilizer. The remaining 250,000 MT of potash was used to manufacture chemicals. However, all potash is first channeled through the chemical sector for processing. Not surprisingly, more than 95 percent of the total potash consumed in the economy could directly and indirectly be traced to personal-consumption expenditures and the export bill of goods. Both GDP components are heavily weighted by food items.

Base-Year Supply Pattern. During the base-year, domestic potash production satisfied about 55 percent of total U.S. consumption. The remaining 45 percent was obtained from imports, 95 percent of which came from Canada.

Projected Demand. Projected use of potash by 2000 is expected to be limited to the same two sectors—agriculture and chemicals. The projected levels of U.S. consumption, based on the assumed GDP growth rates, range between 8.75 and 10.79 MMT, approximately 2 to 2.5 times the 1972 base-year level.

Projected Supply. To meet the projected requirements for potash in 2000, the share of imports to total consumption is expected to range from the 1972 ratio of 52.8 percent to as much as 91.7 percent. As a result, domestic output of potash in 2000 can be expected to range between 1 MMT—if the domestic potash sector continues to contract as it encounters lower grade ores—and a maximum domestic output of 6.66 MMT, if the current ratio of consumption to output is maintained by 2000.

World

Base-Year Demand and Supply. Of the total world potash consumption of 18.7 MMT in the 1970 base-year, almost 80 percent was consumed in the developed countries—both market-based and socialist economies. The main importers, in physical levels, were the United States, Japan, China, and the other developing Asian countries. Canada, the USSR, and high-income Western Europe were the leading producers and exporters.

Projected Demand and Supply. With the expected increase in future world population, which in turn will require an increased use of fertilizers for food production, world demand for potash in 2000 is projected at 74.2 MMT, a 4.7 percent annual rate of growth.

At the end of the fifty-year forecast period, global potash requirements under the pessimistic scenario—low-income growth in the developed regions and high population growth in the developing regions—are projected to be 109 MMT. Under the optimistic scenario—high-income growth in the developed regions and low population growth in the developing regions—global requirements are projected at 119 MMT. Even under the optimistic scenario, which requires more potash on a global scale, current estimates of the world's potash resources appear likely to meet the world's projected cumulative demand for potash through 2030.

Soda Ash

United States

Base-Year Demand Pattern. In 1972, over one-half the 6.4 MMT of soda ash consumed in the United States was used in the glass sector. The next largest user of soda ash was the chemical sector, consuming about 1.7

MMT. The remainder was consumed by the pulp-and-paper, water-treatment, and soap-and-detergents sectors. As a result of this pattern of intermediate use of the mineral, almost 70 percent of the total consumption of soda ash could be traced to personal-consumption expenditures.

Base-Year Supply Pattern. The U.S. domestic output of soda ash in 1972 was sufficient to satisfy all domestic consumption requirements as well as exports of approximately 500,000 MT.

Projected Demand. Projected use of soda ash in 2000 is expected to be concentrated in the same end-use sectors as in the base-year. The projected levels of U.S. consumption of soda ash in 2000, based on alternative rates of growth in GDP, are estimated to range between 14.1 and 15.9 MMT, an annual rate of growth in demand from 2.7 to 3.1 percent.

Projected Supply. The United States can be expected to continue to satisfy its requirements of soda ash to 2000 by relying on domestic sources of supply. Using current reserve estimates of soda ash, there appear to be sufficient quantities to meet cumulative demand to 2000.

World

Base-Year Demand and Supply. In 1970, world consumption of soda ash was 18.5 MMT. The United States consumed about one-third, followed by the USSR and Western Europe (high-income). The industrialized countries—both market-oriented and socialist economies—collectively consumed over 90 percent of total world demand for soda ash.

The major producing and exporting regions in 1970 were the United States and the three European regions.

Projected Demand and Supply. In 2000, world demand for soda ash is projected at 77 MMT, an annual rate of growth of 4.8 percent, given the rates of growth in GDP used in the projections. No significant changes are expected in import-export patterns in 2000.

By 2030, depending on the growth path taken by the world economy, world soda ash requirements are estimated to range between 137 MMT (under the low GDP-growth assumption) and 188 MMT (under the high GDP-growth assumption). Although no apparent global shortage of soda ash is likely, some traditional exporting regions are likely to deplete their estimated resource bases. As a result, they can be expected to reduce the rate of growth of their exports and output levels to extend the life of their resource base into the twenty-first century.

Boron

United States

Base-Year Demand Pattern. In 1972, five sectors accounted for all 83 KMT of boron consumed in the U.S. economy. Over half was used by the ceramic-and-glass sectors; another 20 percent by the soap-and-detergent sector; 13 percent by the agricultural sector in herbicides and fertilizers; and the residual consumed as fluxes in metallurgical uses. As a result, much of the output of the sectors in which boron is used is delivered to personal-consumption expenditures. Consequently, almost 70 percent of the boron consumed in the United States in 1972 was used, directly and indirectly, to satisfy household demand.

Base-Year Supply Pattern. In 1972, more than half of U.S. boron production was exported, indicating that the United States is self-sufficient in meeting its boron requirements.

Projected Demand. Projected future use of boron is expected to be limited to the same five sectors as in the base-year. However, because of technological change, increased use of boron, per dollar of output, in the ceramic-and-glass and fabricated-metal sectors is expected by 2000. At the same time, the agricultural sectors are expected to decrease their use of boron, per unit of output. The projected levels of U.S. boron consumption in 2000, based on the technological-change and income-growth assumptions used, range from 201 to 276 KMT.

The United States is expected to remain a leading exporter of boron by 2000 and later; therefore, requirements will continue to be satisfied by domestic production.

World

Base-Year Demand and Supply. Over 90 percent of the total 1970 world demand for boron of 254 KMT was consumed in the developed, market-oriented, and socialist economies. In addition to the United States, Western Europe (medium-income, including Turkey) was the other significant producing and exporting region in 1970. Western Europe (high-income), Japan, and Eastern Europe, were, as expected, the main importing regions.

Projected Demand and Supply. World boron requirements are projected to reach 1,040 KMT in 2000, an annual rate of growth of 4.8 percent. No significant changes from the base-year are projected for the patterns of production, consumption, exports, and imports.

Over the long-range, fifty-year forecast horizon, under the pessimistic set of assumptions, world boron requirements are estimated at approximately 1,780 KMT. Under the optimistic scenario, world demand for boron in 2030 is projected to be 2,380 KMT. Current world estimates of boron resources appear sufficient to meet the world's cumulative demand for boron through 2030, even under the high GDP-growth assumptions. However, as a result of the possible depletion of the currently estimated resource base in certain regions, such as the United States, other producing regions will be expected to assume a larger share of the future world export market of boron.

Phosphate Rock

United States

Base-Year Demand Pattern. In 1972, about 88 percent of the 26.8 MMT of phosphates consumed in the United States was used in the agricultural sectors for fertilizers. However, before phosphates enter the agricultural sector, they are first shipped to the chemical sector for processing. The remaining uses of phosphates were in detergents, animal feeds, and food products.

Turning to the GDP components, as in the case of potash, practically all phosphate use was directed to satisfying personal-consumption expenditures and the export bill of goods; both components are heavily weighted by food items.

Base-Year Supply Pattern. All U.S. phosphate requirements in the base-year were met by domestic production. In addition, about 35 percent of domestic output was exported.

Projected Demand and Supply. Projected use of phosphates is expected to be concentrated in the same few sectors, again primarily in the agricultural sectors. The projected levels of U.S. consumption of phosphates in 2000 under the various GDP growth rates assumed in the study range between 61.5 and 66.5 MMT, approximately 2.3 and 2.5 times the 1972 level of consumption. The United States is expected to continue to be self-sufficient in meeting its phosphates requirements as well as to export phosphates to the world economy. Projected output is expected to range between 110 and 127 MMT in 2000.

World

Base-Year Demand and Supply. Of the total world phosphate production of 82.4 MMT in the World Model's base-year, 1970, over 85 percent was consumed in the developed regions.

The leading exporting regions in 1970 were Arid Africa (including Morocco, Jordan, Tunisia, and Israel), the United States, and the USSR. Residual amounts were also shipped from China, South Africa, and Tropical Africa. The main importers were Western Europe, Eastern Europe, Japan, Canada, Asia (low-income), and South America.

Projected Demand and Supply. With the expected increase in future world population requiring increased use of fertilizers for food production, world consumption in 2000 is projected to increase to 325 MMT, an annual rate of growth of 4.7 percent from the base-year.

By 2030, under the pessimistic scenario—low GDP growth in the developed regions and high population growth in the developing regions—world phosphate demand is projected to be 495 MMT. Under the optimistic scenario—high GDP growth in the developed regions and low population growth in the developing regions—world consumption of phosphate is projected at 554 MMT. The estimated resource base in all the regions currently producing phosphates appears to be sufficient to satisfy cumulative world output by 2030, even under the high-growth optimistic scenario.

Sulfur

United States

Base-Year Demand Pattern. In 1972, over one-half the 10 MMT of sulfur consumed in the United States was used in the form of sulfuric acid to manufacture fertilizers used in the agricultural sectors. The other major consumers of sulfur, either as sulfuric acid, sulfur dioxide, or carbon disulfide, are the plastics and synthetic-products sectors, metals-mining and -processing sectors, paints and inorganic pigments-producing sectors, paper-products sector, petroleum-refining sector, and the iron-and-steel sector. As a result of this demand pattern, over 70 percent of total sulfur consumption in the base year could be traced to personal-consumption expenditures made by households.

Base-Year Supply Pattern. Output of the domestic sulfur sector in 1972 was 7.2 MMT. Legislation enacted to curb sulfur emissions by the chemical, materials, and energy-related sectors increased supplies of sulfur. By-product sulfur contributed more than 30 percent to total 1972 U.S. requirements. Net exports of sulfur from the United States in 1972 amounted to approximately 680 KMT.

Projected Demand. Sulfur demand in the United States by 2000 is expected to be concentrated in the same seven sectors as in the base-year. Because of possible problems of oversupply as by-product sulfur supplies increase, new nonconventional uses of sulfur are being sought. The first pilot use of sulfur as a substitute for regular asphalt in road-paving is underway. Other potential new uses for sulfur currently under consideration are in sulfur concretes in place of regular cement in certain environments.

The projected levels of sulfur demand for the U.S. economy in 2000 range from 18.9 to 24.3 MMT, depending on the mix of assumptions regarding technological change and income growth in the future.

Projected Supply. The United States can be expected to satisfy its requirements for sulfur from domestic sources at least until 2000. If the trend of by-product sulfur production continues, small amounts of sulfur may be exported even though the currently estimated stock of reserves is expected to be exhausted by that year.

World

Base-Year Demand and Supply. In 1970, total world consumption of sulfur was approximately 39 MMT. About 90 percent was consumed in the developed regions of the world economy. The principal producers were the United States, Canada, the USSR, Western Europe, Eastern Europe, and Japan. Japan, Eastern Europe, the Middle East, and Africa (Centrally Planned) were the chief exporting regions in the base-year.

Projected Demand and Supply. World consumption in 2000 is projected to reach 118 MMT, an annual rate of growth of 3.8 percent from 1970.

World sulfur consumption by 2030 is projected to range between 184 and 211 MMT, depending on the rate of growth of the world economy.

Even though regional resource estimates indicate that sulfur resources may be depleted by 2030 in many regions, the sulfur recovered from the control of sulfur emissions can be expected to meet global requirements.

Chlorine

United States

Base-Year Demand and Supply. In 1972, consumption of chlorine in the United States reached 8.97 MMT. The four consuming sectors, in descending order, were: chemicals (43 percent), plastics and synthetic products (36 percent), paper products (14 percent), and water treatment and chemicals (7 percent). About 65 percent of total chlorine consumption was used directly and indirectly to satisfy the household bill of goods.

Because chlorine is produced from salt, the United States will continue to remain self-sufficient in meeting its chlorine requirements.

Projected Demand. Projected U.S. demand for chlorine for 2000 is expected to range from 20.2 to 23.5 MMT, about 2.5 times the 1972 base-year level. No significant changes from the 1972 base-year are projected in the distribution of chlorine use.

World

World demand for chlorine in 1970 was 21.1 MMT, nearly of all of which was produced and consumed in the developed regions. Because of the difficulty of transporting chlorine, the movement of the chemical is restricted to short distances. Consequently, there is little or no trade in chlorine so far as the regional breakdown of countries in this model is concerned. Projected world demand in 2000 is estimated at 74.7 MMT.

Projected demand for chlorine in 2030 is estimated to range between 149 and 187 MMT, depending on the selected rate of growth of world output. No long-term resource-supply problem is anticipated for chlorine.

Magnesium

United States

Base-Year Demand Pattern. Magnesium is consumed in the United States in the form of both metals and chemical compounds. However, of the 1 MMT of magnesium consumed in 1972, only 12 percent was in metallic form. The rest was used in its chemical form. Over 80 percent of total magnesium consumption was used in refractories. The next single largest magnesium user was the chemical sector, consuming about 9 percent of total U.S. requirements. On the other hand, over 70 percent of the 110 KMT of magnesium metal was absorbed by two sectors—transportation (primarily in aircraft) and machinery (for use in aluminum and magnesium alloys). Residual metallic uses were in the chemicals and ferrous and nonferrous metals-processing sectors.

Judging by its end-use markets, 80 percent of the magnesium consumed in the economy was used to satisfy, directly and indirectly, household demand and private business investment. However, eight times more magnesium was used per dollar of expenditures on business investment than on personal consumption.

Base-Year Supply Pattern. The United States was self-sufficient in 1972, satisfying its requirements of magnesium from domestic sources of production.

Projected Demand and Supply. Projected use of magnesium is expected to be restricted to the same consuming sectors as in the base-year in more or less the same proportions. In 2000, U.S. consumption of magnesium—both as a compound and in its metallic form—is projected to range between 1.3 and 2.3 MMT, depending on the selected set of technological-change and recycling assumptions.

The United States is expected to continue to be self-sufficient by 2000 or to rely on imports only marginally.

World

World consumption of magnesium in 1970 was 5.2 MMT. Almost 90 percent of this amount was used in the developed countries. The main exporting regions were Asia Centrally Planned (including China and North Korea), the USSR, and Western Europe (medium-income, including Greece). In 2000, world consumption requirements are projected to be 16.4 MMT, based on the rates of growth in world economy implicit in the model.

By 2030, global requirements of magnesium are projected to range between 28.3 and 39.2 MMT, depending on whether the low or high GDP projection is used. Because most of the magnesium consumed in the world is derived from dolomite, sea water, and well and lake brines, magnesium is a nearly inexhaustible resource. Therefore, even if the world's limited magnesite resources are depleted, no physical problems are expected in securing the world's cumulative requirements of magnesium to the year 2030.

Notes

1. See note 8, chapter 7, for the treatment of primary and secondary lead in the World Model.
2. R. Haloran, "Pentagon Urges 50 Cargo Planes for Rapid Forces," *The New York Times,* January 21, 1982, Section A, p. 1.

Appendix A: Sector Classification for IEA/USMIN

In the U.S. minerals input-output model, IEA/USMIN, the economy is divided into 106 producing and consuming sectors. The first 106 rows and columns of the coefficients matrix—the interindustry submatrix A in figure 3–1—represent a complete, commodity-by-commodity input-output matrix. The remaining 209 rows and columns provide coordinated information on the production and consumption of twenty-six nonfuel minerals, imports, energy requirements, and other production-related variables.

Table A–1 lists row and column names for the model and the units in which each variable is measured (such as billions of 1972 dollars, millions of metric tons, millions of BTU). Table A–1 also maps the rows and columns of the 496-section BEA table into the IEA classification.

The input-output sectors (submatrix A_c, rows and columns 1 to 106 in figure 3–1) are defined as they are in the BEA 85-order classification, except when a more detailed representation of individual sectors seemed appropriate. Those sectors were defined at a BEA 496-order level: mineral mining and metals processing, and sectors that either produce important metals substitutes (hydraulic cement) or else are major end-users (nonclay refractories and fertilizers). Table A–2 maps the SIC code into BEA sectors for the 85- and 496-order tables.

The production and consumption of individual minerals were treated by creating a row and column for each mineral. In the coefficients matrix, these appear as submatrix B_c (domestically mined minerals) and as rows 254, 255, and 256 in submatrix M_c (minerals wholly imported and not produced domestically). Minerals are defined in terms of their metal or elemental content; no distinction is made between the different product forms in which minerals are actually used.

For example, the row for magnesium includes both magnesium contained in magnesite, which is mined in the stone and clay mining industry and used chiefly to produce refractory brick, and magnesium contained in magnesium metal, which is produced from sea water in the industrial chemicals industry and used in transportation equipment and machinery. Input requirements for minerals mining and metal processing are represented in the interindustry part of the model (rows and columns 1 to 106) at the BEA 496-order level, not at the level of individual minerals.

Table A-1
Row and Column Variables

		IEA/USMIN	*Corresponding BEA Sector*
AGR 1	$	Livestock and livestrock products ($)	01.0100-01.0302
AGR 2	$	Oil crops ($)	02.0100, 02.0600-02.0702
AGR 3	$	Grains ($)	02.0201-02.0203
AGR 4	$	Residual agriculture ($)	02.0300-02.0503,03.0000
AGR 5	$	Agricultural services ($)	04.0000
EXT 1	$	Iron ($)	05.0000
EXT 2	$	Copper ($)	06.0100
EXT 3	$	Nonferrous minerals ($)	06.0200
EXT 4	$	Stone and clay minerals ($)	09.0000
EXT 5	$	Chemical and fertilizer minerals ($)	10.0000
EXT 6	$	Coal mining ($)	07.0000
EXT 7	$	Oil and gas ($)	08.0000
CON 1	$	New construction ($)	11.0101-11.0508
CON 2	$	Maintenance and repair ($)	12.0100-12.0216
MAN 1	$	Arms and ammunition ($)	13.0100-13.0700
MAN 2	$	Food and kindred products ($)	14.0101-14.3200
MAN 3	$	Tobacco manufactures ($)	15.0101-15.0200
MAN 4	$	Broad and narrow fabrics ($)	16.0100-16.0400
MAN 5	$	Miscellaneous fabrics ($)	17.0100-17.1002
MAN 6	$	Apparel ($)	18.0100-18.0400
MAN 7	$	Miscellaneous fabricated textile products ($)	19.0100-19.0306
MAN 8	$	Lumber and wood products (except containers) ($)	20.0100-20.0903
MAN 9	$	Wood containers ($)	21.000
MAN10	$	Household furniture ($)	22.0101-22.0400
MAN11	$	Other furniture ($)	23.0100-23.0700
MAN12	$	Paper and allied products, except containers and boxes ($)	24.0100-24.0706
MAN13	$	Paperboard containers and boxes ($)	25.000
MAN14	$	Printing and publishing ($)	26.0100-26.0805
CHE 1	$	Industrial chemicals ($)	27.0100
CHE 2	$	Fertilizers ($)	27.0201-27.0202
CHE 3	$	Other chemicals ($)	27.0310-27.0406
MAN15	$	Plastics and synthetic materials ($)	28.0100-28.0400
MAN16	$	Drugs, cleaning and toilet preparations ($)	29.0100-29.0300
MAN17	$	Paints and allied products ($)	30.0000
MAN18	$	Petroleum refining and related products ($)	31.0100-31.0300
MAN19	$	Rubber and miscellaneous plastics products ($)	32.0100-32.0500

MAN20	$	Leather tanning and finishing ($)	33.0001
MAN21	$	Footwear and other leather products ($)	34.0100–34.0305
MAN22	$	Glass and glass products ($)	35.0100–35.0200
MAN23	$	Hydraulic cement ($)	36.0100
MAN24	$	Nonclay refractories ($)	36.2100
MAN25	$	Other stone and clay products ($)	36.0200–36.2000, 36.2200
MET 1	$	Blast furnaces and steel mills ($)	37.0101
MET 2	$	Electrometallurgical products ($)	37.0102
MET 3	$	Steel wire and related products ($)	37.0103
MET 4	$	Cold finishing of steel shapes ($)	37.0104
MET 5	$	Steel pipe and tubes ($)	37.0105
MET 6	$	Iron and steel foundries ($)	37.0200
MET 7	$	Iron and steel forgings ($)	37.0300
MET 8	$	Metal heat treating ($)	37.0401
MET 9	$	Primary metal products, not elsewhere classified ($)	37.0402
MET10	$	Primary copper processing ($)	38.0100
MET11	$	Primary lead processing ($)	38.0200
MET12	$	Primary zinc processing ($)	38.0300
MET13	$	Primary aluminum processing ($)	38.0400
MET14	$	Primary nonferrous metals processing, n.e.c.[4] ($)	38.0500
MET15	$	Secondary nonferrous metals processing ($)	38.0600
MET16	$	Copper rolling and drawing ($)	38.0700
MET17	$	Aluminum rolling and drawing ($)	38.0800
MET18	$	Nonferrous rolling and drawing ($)	38.0900
MET19	$	Nonferrous wire drawing and insulating ($)	38.1000
MET20	$	Aluminum castings ($)	38.1100
MET21	$	Brass, bronze, and copper castings ($)	38.1200
MET22	$	Nonferrous castings, not elsewhere classified ($)	38.1300
MET23	$	Nonferrous forgings ($)	38.1400
MAN26	$	Metal containers ($)	39.0100–39.0200
MAN27	$	Heating, plumbing, and fabricated metal products ($)	40.0100–40.0902
MAN28	$	Screw machine parts and stampings ($)	41.0100–41.0203
MAN29	$	Other fabricated metal products ($)	42.0100–42.1100
MAN30	$	Engines and turbines ($)	43.0100–43.0200
MAN31	$	Farm and garden machinery ($)	44.0001–44.0002
MAN32	$	Construction and mining machinery ($)	45.0100–45.0300
MAN33	$	Materials handling machinery and equipment ($)	46.0100–46.0400
MAN34	$	Metalworking machinery and equipment ($)	47.0100–47.0403
MAN35	$	Special handling machinery and equipment ($)	48.0100–48.0600
MAN36	$	General industrial machinery and equipment ($)	49.0100–49.0700

Table A-1 continued

		IEA/USMIN	*Corresponding BEA Sector*
MAN37	$	Machine shop products ($)	50.0001-50.0002
MAN38	$	Office, computing, and accounting machinery ($)	51.0101-51.0400
MAN39	$	Service industry machinery ($)	52.0100-52.0500
MAN40	$	Electric transmission and distribution ($)	53.0100-53.0800
MAN41	$	Household appliances ($)	54.0100-54.0700
MAN42	$	Electric lighting and wiring equipment ($)	55.0100-55.0300
MAN43	$	Radio, TV, and communication equipment ($)	56.0100-56.0400
MAN44	$	Electronic components and accessories ($)	57.0100-57.0300
MAN45	$	Miscellaneous electric Machinery ($)	58.0100-58.0500
MAN46	$	Motor vehicles and equipment ($)	59.0100-59.0302
MAN47	$	Aircraft and parts ($)	60.0100-60.0040
MAN48	$	Other transportation equipment ($)	61.0100-61.0700
MAN49	$	Professional scientific instruments and supplies ($)	62.0100-62.0700
MAN50	$	Optical, ophthalmic and photo equipment ($)	63.0100-63.0300
MAN51	$	Miscellaneous manufacturing ($)	64.0100-64.1200
SRV 1	$	Transportation and warehousing ($)	65.0100-65.0700
SRV 2	$	Communications (except radio and TV) ($)	66.0000
SRV 3	$	Radio and TV broadcasting ($)	67.0000
SRV 4	$	Electric, gas, water, and sanitary services ($)	68.0100-68.0300
SRV 5	$	Wholesale and retail trade ($)	69.0100-69.0200
SRV 6	$	Finance and insurance ($)	70.0100-70.0500
SRV 7	$	Real estate and rental services ($)	71.0100-71.0200
SRV 8	$	Hotels, personal and repair services, and eating and drinking places ($)	72.0100-72.0300, 74.0000
SRV 9	$	Business services ($)	73.0100-73.0300
SRV10	$	Automobile repairs and services ($)	75.0000
SRV11	$	Amusements ($)	76.0100-76.0200
SRV12	$	Health, educational, and social services ($)	77.0100-77.0900
SRV13	$	Federal government enterprises ($)	78.0100-78.0400
SRV14	$	State and local government ($)	79.0100-79.0300
SCR27	$	Other scrap, used in second hand goods ($)	81.0000
MIN 1	T	Iron (MMT)	
MIN 2	T	Molybdenum (KMT)	
MIN 3	T	Nickel (KMT)	
MIN 4	T	Tungsten (KMT)	
MIN 7	T	Copper (MMT)	
MIN 8	T	Lead (MMT)	

MIN 9	T	Zinc (MMT)
MIN10	T	Gold (MT)
MIN11	T	Silver (KMT)
MIN12	T	Aluminum (MMT)
MIN13	T	Mercury (KMT)
MIN14	T	Vanadium (KMT)
MIN15	T	Platinum (MT)
MIN16	T	Titanium (KMT)
MIN21	T	Silicon (KMT)
MIN23	T	Fluorine (MMT)
MIN24	T	Potash (MMT)
MIN25	T	Soda ash (MMT)
MIN26	T	Boron (KMT)
MIN27	T	Phosphate rock (MMT)
MIN28	T	Sulfur (MMT)
CHM 1	T	Chlorine (MMT)
CHM 2	T	Magnesium (MMT)
SCR 1	T	Iron and steel scrap (MMT)
SCR 2	T	Copper scrap (MMT)
SCR 3	T	Lead scrap (MMT)
SCR 4	T	Zinc scrap (MMT)
SCR 5	T	Aluminum scrap (MMT)
SCR 6	T	Nickel scrap (KMT)
SCR 7	T	Chromium scrap (KMT)
SCR 8	T	Gold scrap (MT)
SCR 9	T	Silver scrap (KMT)
SCR10	T	Tungsten scrap (KMT)
SCR11	T	Mercury scrap (KMT)
SCR12	T	Tin scrap (KMT)
SCR13	T	Magnesium scrap (MMT)
IMA 1	$	Livestock and livestock products ($)
IMA 2	$	Oil crops ($)
IMA 3	$	Grains ($)
IMA 4	$	Residual agriculture ($)
IMA 5	$	Agricultural services ($)
IME 1	$	Iron ($)
IME 2	$	Copper ($)
IME 3	$	Nonferrous minerals ($)
IME 4	$	Stone and clay minerals ($)
IME 4	$	Chemical and fertilizer minerals ($)
IME 6	$	Coal mining ($)
IME 7	$	Oil and gas ($)
IMC 1	$	New construction ($)

Table A-1 continued

		IEA/USMIN	*Corresponding BEA Sector*
IMC 2	$	Maintenance and repair ($)	
IMX 1	$	Arms and ammunition ($)	
IMX 2	$	Food and kindred products ($)	
IMX 3	$	Tobacco manufactures ($)	
IMX 4	$	Broad and narrow fabrics ($)	
IMX 5	$	Miscellaneous fabrics ($)	
IMX 6	$	Apparel ($)	
IMX 7	$	Miscellaneous fabricated textile products ($)	
IMX 8	$	Lumber and wood products (except containers) ($)	
IMX 9	$	Wood containers ($)	
IMX10	$	Household furniture ($)	
IMX11	$	Other furniture ($)	
IMX12	$	Paper and allied products, excluding containers and boxes ($)	
IMX13	$	Paperboard containers and boxes ($)	
IMX14	$	Printing and publishing ($)	
IMH 1	$	Industrial chemicals ($)	
IMH 2	$	Fertilizers ($)	
IMH 3	$	Other chemicals ($)	
IMX15	$	Plastics and synthetic materials ($)	
IMX16	$	Drugs, cleaning, and toilet preparations ($)	
IMX17	$	Paints and allied products ($)	
IMX18	$	Petroleum refining and related products ($)	
IMX19	$	Rubber and miscellaneous plastics products ($)	
IMX20	$	Leather tanning and finishing ($)	
IMX21	$	Footwear and other leather products ($)	
IMX22	$	Glass and glass products ($)	
IMX23	$	Hydraulic cement ($)	
IMX24	$	Nonclay refractories ($)	
IMX25	$	Other stone and clay products ($)	
IMT 1	$	Blast furnaces and steel mills ($)	
IMT 2	$	Electrometallurgical products ($)	
IMT 3	$	Steel wire and related products ($)	
IMT 4	$	Cold finishing of steel shapes ($)	
IMT 5	$	Steel pipe and tubes ($)	
IMT 6	$	Iron and steel foundries ($)	
IMT 7	$	Iron and steel forgings ($)	
IMT 8	$	Metal heat treating ($)	

IMT 9	$	Primary metal products, not elsewhere classified ($)
IMT10	$	Primary copper processing ($)
IMT11	$	Primary lead processing ($)
IMT12	$	Primary zinc processing ($)
IMT13	$	Primary aluminum processing ($)
IMT14	$	Primary nonferrous metals processing, not elsewhere classified ($)
IMT15	$	Secondary nonferrous metals processing ($)
IMT16	$	Copper rolling and drawing ($)
IMT17	$	Aluminum rolling and drawing ($)
IMT18	$	Nonferrous rolling and drawing ($)
IMT19	$	Nonferrous wire drawing and insulating ($)
IMT20	$	Aluminum castings ($)
IMT21	$	Brass, bronze, and copper castings ($)
IMT22	$	Nonferrous castings, not elsewhere classified ($)
IMT23	$	Nonferrous forgings ($)
IMX26	$	Metal containers ($)
IMX27	$	Heating, plumbing, and fabricated metal products ($)
IMX28	$	Screw machine parts and stampings ($)
IMX29	$	Other fabricated metal products ($)
IMX30	$	Engines and turbines ($)
IMX31	$	Farm and garden machinery ($)
IMX32	$	Construction and mining machinery ($)
IMX33	$	Materials handling machinery and equipment ($)
IMX34	$	Metalworking machinery and equipment ($)
IMX35	$	Special industry machinery and equipment ($)
IMX36	$	General industrial machinery and equipment ($)
IMX37	$	Machine shop products ($)
IMX38	$	Office, computing, and accounting machinery ($)
IMX39	$	Service industry machinery ($)
IMX40	$	Electric transmission and distribution ($)
IMX41	$	Household appliances ($)
IMX42	$	Electric lighting and wiring equipment ($)
IMX43	$	Radio, TV, and communication equipment ($)
IMX44	$	Electronic components and accessories ($)
IMX45	$	Miscellaneous electric machinery ($)
IMX46	$	Motor vehicles and equipment ($)
IMX47	$	Aircraft and parts ($)
IMX48	$	Other transportation equipment ($)

Table A-1 continued

		IEA/USMIN	*Corresponding BEA Sector*
IMX49	$	Professional scientific instruments and supplies ($)	
IMX50	$	Optical, ophthalmic and photographic equipment ($)	
IMX51	$	Miscellaneous manufacturing ($)	
IMS 1	$	Transportation and warehousing ($)	
IMS 2	$	Communications (except radio and TV) ($)	
IMS 3	$	Radio and TV broadcasting ($)	
IMS 4	$	Electric, gas, water, and sanitary services ($)	
IMS 5	$	Wholesale and retail trade ($)	
IMS 6	$	Finance and insurance ($)	
IMS 7	$	Real estate and rental services ($)	
IMS 8	$	Hotel, personal and repair services, eating and drinking places ($)	
IMS 9	$	Business services ($)	
IMS10	$	Automobile repairs and services ($)	
IMS11	$	Amusements ($)	
IMS12	$	Health, educational, and social services ($)	
IMS13	$	Federal government enterprises ($)	
IMS14	$	State and local government ($)	
IMR27	$	Other scrap, used in second hand goods ($)	
IMM 1	T	Iron (MMT)	
IMM 2	T	Molybdenum (KMT)	
IMM 3	T	Nickel (KMT)	
IMM 4	T	Tungsten (KMT)	
IMM 5	T	Manganese (KMT)	
IMM 6	T	Chromium (KMT)	
IMM 7	T	Copper (MMT)	
IMM 8	T	Lead (MMT)	
IMM 9	T	Zinc (MMT)	
IMM10	T	Gold (MT)	
IMM11	T	Silver (KMT)	
IMM12	T	Aluminum (MMT)	
IMM13	T	Mercury (KMT)	
IMM14	T	Vanadium (KMT)	
IMM15	T	Platinum (MT)	
IMM16	T	Titanium (KMT)	
IMM17	T	Tin (KMT)	
IMM21	T	Silicon (KMT)	
IMM23	T	Fluorine (MMT)	
IMM24	T	Potash (MMT)	

IMM25	T	Soda Ash (MMT)	
IMM26	T	Boron (KMT)	
IMM27	T	Phosphate rock (MMT)	
IMM28	T	Sulfur (MMT)	
IMCH1	T	Chlorine (MMT)	
IMCH2	T	Magnesium (MMT)	
IMP 1	T	Iron and steel scrap (MMT)	
IMP 2	T	Copper scrap (MMT)	
IMP 3	T	Lead scrap (MMT)	
IMP 4	T	Zinc scrap (MMT)	
IMP 5	T	Aluminum scrap (MMT)	
IMP 6	T	Nickel scrap (KMT)	
IMP 7	T	Chromium scrap (KMT)	
IMP 8	T	Gold scrap (MT)	
IMP 9	T	Silver scrap (KMT)	
IMP10	T	Tungsten scrap (KMT)	
IMP11	T	Mercury scrap (KMT)	
IMP12	T	Tin scrap (KMT)	
IMP13	T	Magnesium scrap (MMT)	
NCI 1	$	Noncomparable imports ($)	80.0000
DUM	$	Dummy sector ($)	82.0000–85.0000
VA	$	Value added ($)	87.0000
LAB		Thousands of employees	
NRG 1	BTU	Coal mining (BTU)	
NRG 2	BTU	Crude petroleum and natural gas (BTU)	
NRG 3	BTU	Petroleum refining (BTU)	
NRG 4	BTU	Gas utilities (BTU)	
NRG 5	BTU	Electrical utilities (BTU)	
NSC 1	T	New iron and steel scrap (MMT)	
NSC 2	T	New copper scrap (MMT)	
NSC 3	T	New lead scrap (MMT)	
NSC 4	T	New zinc scrap (MMT)	
NSC 5	T	New aluminum scrap (MMT)	
NSC 6	T	New nickel scrap (KMT)	
NSC 7	T	New chromium scrap (KMT)	
NSC 8	T	New gold scrap (MT)	
NSC 9	T	New silver scrap (KMT)	
NSC10	T	New tungsten scrap (KMT)	
NSC11	T	New mercury scrap (KMT)	
NSC12	T	New tin scrap (KMT)	
NSC13	T	New magnesium scrap (MMT)	
POL 1	T	5-day biochemical oxygen demand (MLB)	
POL 2	T	Suspended solids (MLB)	
POL 3	T	Oil and greses (MLB)	

Table A-1 continued

		IEA/USMIN	Corresponding BEA Sector
POL 4	T	Total dissolved solids (MLB)	
POL 5	T	Total phosphates (MLB)	
POL 6	T	Heavy metals (MLB)	
POL 7	T	Waste stream (MLB)	
POL 8	T	Particulates (MLT)	
POL 9	T	Sulfur oxide (MLT)	
POL10	T	Carbon monoxide (MLT)	
POL11	T	Hydrocarbons (MLT)	
POL12	T	Nitrogen oxide (MLT)	
		Additional Columns in IEA/USMIN	*Corresponding BEA Sector*
PCE	$	Personal-consumption expenditures ($)	91.0000
CAP	$	Gross private fixed investments ($)	92.0000
INV	$	Change in business inventories ($)	93.0000
EX	$	Exports ($)	94.0000
IM	$	Imports ($)	95.0000
NDG	$	Nondefense, government purchases ($)	97.0000–99.3000
DEF	$	Defense, government purchases ($)	96.0000

Table A-2
BEA Industry Classification of the 1972 Input-Output Table

Industry Number and Title	*Related Census-SIC codes (1972 edition)*
Agriculture, Forestry, and Fisheries	
1. Livestock and livestock products	
1.0100 Dairy farm products	0241, pt. 0191, pt. 0259, pt. 0291
1.0200 Poultry and eggs	025 (excl. 0254 and pt. 0259), pt. 0191, pt 0219, pt. 0291
1.0301 Meat animals	021 (excl. pt. 0219), pt. 0191, pt. 0259, pt. 0291
1.0302 Miscellaneous livestock	027, pt. 0191, pt. 0219, pt. 0259, pt. 0291
2. Other agricultural products	
2.0100 Cotton	0131, pt. 0191, pt. 0219, pt. 0259, pt. 0291
2.0201 Food grains	pt. 011, pt. 0191, pt. 0219, pt. 0259, pt. 0291

2.0202 Feed grains	pt. 011, pt. 0139, pt. 0191, pt. 0219, pt. 0259, pt. 0291
2.0203 Grass seeds	pt. 0139, pt. 0191, pt. 0219, pt. 0259, pt. 0291
2.0300 Tobacco	0132, pt. 0191, pt. 0219, pt. 0259, pt. 0291
2.0401 Fruits	pt. 017, pt. 0191, pt. 0219, pt. 0259, pt. 0291
2.0402 Tree nuts	0173, pt. 0179,pt. 0191, pt. 0219, pt.0259, pt. 0291
2.0501 Vegetables	0134, 0161, pt. 0119, pt. 0139, pt. 0191, pt. 0219, pt. 0259, pt. 0291
2.0502 Sugar crops	0133, pt. 0191, pt. 0219, pt. 0259, pt. 0291
2.0503 Miscellaneous crops	pt. 0191, pt. 0139, pt. 0191, pt. 0219, pt. 0259, pt. 0291
2.0600 Oil bearing crops	0116, pt. 0119, pt. 013, pt. 0173, pt. 0219, pt. 0259, pt. 0291
2.0701 Forest products	pt. 018, pt. 0191, pt. 0219, pt. 0259, pt. 0291
2.0702 Greenhouse and nursery products	pt. 018, pt. 0191, pt. 0219, pt. 0291
3. Forestry and fishery products	
3.0000 Forestry and fishery products	081-4, 091, 097
4. Agricultural, forestry, and fishery services	
4.0000 Agricultural, forestry, and fishery services	0254, 07 (excl. 074), 085, 092
Mining	
5. Iron and ferroalloy ores mining	
5.0000 Iron and ferroalloy ores mining	101, 106
6. Nonferrous metal ores mining	
6.0100 Copper ore mining	102
6.0200 Nonferrous metal ores mining, except copper	103-5, pt. 108, 109
7. Coal mining	
7.0000 Coal mining	1111, pt. 1112, 1211, pt. 1213
8. Crude petroleum and natural gas	
8.0000 Crude petroleum and natural gas	131, 132, pt. 138
9. Stone and clay mining and quarrying	
9.0000 Stone and clay mining and quarrying	141-5, pt. 148, 149
10. Chemical and fertilizer mineral mining	
10.0000 Chemical and fertilizer mineral mining	147

Table A-2 continued

Industry Number and Title		*Related Census-SIC codes (1972 edition)*
Construction		
	11. New construction	
11.0101	New residential 1-unit structures, nonfarm	pt. 15, pt. 17
11.0102	New residential 2-4 unit structures, nonfarm	pt. 15, pt. 17
11.0103	New residential garden apartments	pt. 15–17
11.0104	New residential high-rise apartments	pt. 15–17
11.0105	New residential additions and alterations, nonfarm	pt. 15, pt. 17
11.0106	New hotels and motels	pt. 15–17
11.0107	New dormitories	pt. 15, pt. 17
11.0201	New industrial buildings	pt. 15–17
11.0202	New office buildings	pt. 15, pt. 17
11.0203	New warehouses	pt. 15, pt. 17
11.0204	New garages and service stations	pt. 15, pt. 17
11.0205	New stores and restaurants	pt. 15, pt. 17
11.0206	New religious buildings	pt. 15, pt. 17
11.0207	New educational buildings	pt. 15, pt. 17
11.0208	New hospital and institutional buildings	pt. 15, pt. 17
11.0209	New other nonfarm buildings	pt. 15, pt. 17
11.0301	New telephone and telegraph facilities	pt. 16, pt. 17
11.0302	New railroads	pt. 16, pt. 17
11.0303	New electric utility facilities	pt. 16, pt. 17
11.0304	New gas utility facilities	pt. 16, pt. 17
11.0305	New petroleum pipelines	pt. 16, pt. 17
11.0306	New water supply facilities	pt. 16, pt. 17
11.0307	New sewer system facilities	pt. 16, pt. 17
11.0308	New local transit facilities	pt. 16, pt. 17
11.0400	New highways and streets	pt. 16, pt. 17
11.0501	New farm housing units and additions and alterations	pt. 15, pt. 17
11.0502	New farm service facilities	pt. 15, pt. 17
11.0503	New petroleum and natural gas well drilling	pt. 138
11.0504	New petroleum, natural gas, and solid mineral exploration	pt. 108, pt. 1112, pt. 1213, pt. 138, pt. 148
11.0505	New military facilities	pt. 15–17
11.0506	New conservation and development facilities	pt. 15–17
11.0507	Other new nonbuilding facilities	pt. 15–17
11.0508	New access structures for solid mineral development	pt. 108, pt. 1112, pt. 1213, pt. 148

12. Maintenance and repair construction

12.0100	Maintenance and repair, residential	pt. 15, pt. 17
12.0201	Maintenance and repair of other nonfarm buildings	pt. 15, pt. 17
12.0202	Maintenance and repair of farm residential buildings	pt. 15, pt. 17
12.0203	Maintenance and repair of farm service facilities	pt. 15, pt. 17
12.0204	Maintenance and repair of telephone and telegraph facilities	pt. 16, pt. 17
12.0205	Maintenance and repair of railroads	pt. 16, pt. 17
12.0206	Maintenance and repair of electric utility facilities	pt. 16, pt. 17
12.0207	Maintenance and repair of gas utility facilities	pt. 16, pt. 17
12.0208	Maintenance and repair of petroleum pipelines	pt. 16, pt. 17
12.0209	Maintenance and repair of water supply facilities	pt. 16, pt. 17
12.0210	Maintenance and repair of sewer facilities	pt. 16, pt. 17
12.0211	Maintenance and repair of local transit facilities	pt. 16, pt. 17
12.0212	Maintenance and repair of military facilities	pt. 15–17
12.0213	Maintenance and repair of conservation and development facilities	pt. 15–17
12.0214	Maintenance and repair of highways and streets	pt. 16, pt. 17
12.0215	Maintenance and repair of petroleum and natural gas wells	pt. 138
12.0216	Maintenance and repair of other nonbuilding facilities	pt. 15–17

Manufacturing

13. Ordnance and accessories

13.0100	Complete guided missiles	3761
13.0200	Ammunition, except for small arms, n.e.c.[4]	3483
13.0300	Tanks and tank components	3795
13.0500	Small arms	3484
13.0600	Small arms ammunition	3482
13.0700	Other ordnance and accessories	3489

14. Food and kindred products

14.0101	Meat packing plants	2011
14.0102	Sausages and other prepared meats	2013
14.0103	Poultry dressing plants	2016
14.0104	Poultry and egg processing	2017
14.0200	Creamery butter	2021
14.0300	Cheese, natural and processed	2022
14.0400	Condensed and evaporated milk	2023
14.0500	Ice cream and frozen desserts	2024
14.0600	Fluid milk	2026

Table A-2 continued

Industry Number and Title	*Related Census-SIC codes (1972 edition)*
14.0700 Canned and cured sea foods	2091
14.0800 Canned specialties	2032
14.0900 Canned fruits and vegetables	2033
14.1000 Dehydrated food products	2034
14.1100 Pickles, sauces, and salad dressings	2035
14.1200 Fresh or frozen packaged fish	2092
14.1300 Frozen fruits and vegetables	2037-8
14.1401 Flour and other grain mill products	2041
14.1402 Cereal preparations	2043
14.1403 Blended and prepared flour	2045
14.1501 Dog, cat, and other pet foods	2047
14.1502 Prepared feeds, n.e.c.[4]	2048
14.1600 Rice milling	2044
14.1700 Wet corn milling	2046
14.1801 Bread, cake, and related products	2051
14.1802 Cookies and crackers	2052
14.1900 Sugar	2061-3
14.2001 Confectionary products	2065
14.2002 Chocolate and cocoa products	2066
14.2003 Chewing gum	2067
14.2101 Malt liquors	2082
14.2102 Malt	2083
14.2103 Wines, brandy, and brandy spirits	2084
14.2104 Distilled liquor, except brandy	2085
14.2200 Bottled and canned soft drinks	2086
14.2300 Flavoring extracts and sirups, n.e.c.[4]	2087
14.2400 Cottonseed oil mills	2074
14.2500 Soybean oil mills	2075
14.2600 Vegetable oil mills, n.e.c.[4]	2076
14.2700 Animal and marine fats and oil	2077
14.2800 Roasted coffee	2095
14.2900 Shortening and cooking oils	2079
14.3000 Manufactured ice	2097
14.3100 Macaroni and spaghetti	2098
14.3200 Food preparations, n.e.c.[4]	2099
15. Tobacco manufactures	
15.0101 Cigarettes	211
15.0102 Cigars	212

15.0103	Chewing and smoking tobacco	213
15.0200	Tobacco stemming and redrying	214
	16. Broad and narrow fabrics, yarn and thread mills	
16.0100	Broadwoven fabric mills and fabric finishing plants	221-3, 2261-2
16.0200	Narrow fabric mills	224
16.0300	Yarn mills and finishing of textiles, n.e.c.[4]	2269, 2281-3
16.0400	Thread mills	2284
	17. Miscellaneous textile goods and floor coverings	
17.0100	Floor coverings	227
17.0200	Felt goods, n.e.c.[4]	2291
17.0300	Lace goods	2292
17.0400	Padding and upholstery filling	2293
17.0500	Processed textile waste	2294
17.0600	Coated fabrics, not rubberized	2295
17.0700	Tire and cord fabric	2296
17.0900	Cordage and twine	2298
17.1001	Nonwoven fabrics	2297
17.1002	Textile goods, n.e.c.[4]	2299
	18. Apparel	
18.0101	Women's hosiery, except socks	2251
18.0102	Hosiery, n.e.c.[4]	2252
18.0201	Knit outerwear mills	2253
18.0202	Knit underwear mills	2254
18.0203	Knitting mills, n.e.c.[4]	2259
18.0300	Knit fabric mills	2257-8
18.0400	Apparel made from purchased materials	231-8, 39996
	19. Miscellaneous fabricated textile products	
19.0100	Curtains and draperies	2391
19.0200	Housefurnishings, n.e.c.[4]	2392
19.0301	Textile bags	2393
19.0302	Canvas products	2394
19.0303	Pleating and stitching	2395
19.0304	Automotive and apparel trimmings	2396
19.0305	Schiffli machinery embroideries	2397
19.0306	Fabricated textile products, n.e.c.[4]	2399
	20. Lumber and wood product, except containers	
20.0100	Logging camps and logging contractors	2411
20.0200	Sawmills and planing mills, general	2421
20.0300	Hardwood dimension and flooring mills	2426
20.0400	Special product sawmills, n.e.c.[4]	2429

Table A-2 continued

Industry Number and Title	*Related Census-SIC codes (1972 edition)*
20.0501 Millwork	2431
20.0502 Wood kitchen cabinets	2434
20.0600 Veneer and plywood	2435–6
20.0701 Structural wood members, n.e.c.[4]	2439
20.0702 Prefabricated wood buildings	2452
20.0800 Wood preserving	2491
20.0901 Wood pallets and skids	2448
20.0902 Particleboard	2492
20.0903 Wood products, n.e.c.[4]	2499
21. Wood containers	
21.0000 Wood containers	2441, 2449
22. Household furniture	
22.0101 Wood household furniture	2511
22.0102 Household furniture, n.e.c.[4]	2519
22.0103 Wood TV and radio cabinets	2517
22.0200 Upholstered household furniture	2512
22.0300 Metal household furniture	2514
22.0400 Mattresses and bedsprings	2515
23. Other furniture and fixtures	
23.0100 Wood office furniture	2521
23.0200 Metal office furniture	2522
23.0300 Public building furniture	2531
23.0400 Wood partitions and fixtures	2541
23.0500 Metal partitions and fixtures	2542
23.0600 Blinds, shades, and drapery hardware	2591
23.0700 Furniture and fixtures, n.e.c.[4]	2599
24. Paper and allied products, except containers and boxes	
24.0100 Pulp mills	261
24.0200 Paper mills, except building paper	262
24.0300 Paperboard mills	263
24.0400 Envelopes	2642
24.0500 Sanitary paper products	2647
24.0602 Building paper and board mills	266
24.0701 Paper coating and glazing	2641
24.0702 Bags, except textile	2643
24.0703 Die-cut paper and board	2645
24.0704 Pressed and molded pulp goods	2646

24.0705 Stationary products	2648
24.0706 Converted paper products, n.e.c.[4]	2649
25. Paperboard containers and boxes	
25.0000 Paperboard containers and boxes	265
26. Printing and publishing	
26.0100 Newspapers	271
26.0200 Periodicals	272
26.0301 Book publishing	2731
26.0302 Book printing	2732
26.0400 Miscellaneous publishing	274
26.0501 Commercial printing	2751-2, 2754
26.0502 Lithographic platemaking and services	2795
26.0601 Manifold business forms	276
26.0602 Blankbooks and looseleaf binders	2782
26.0700 Greeting card publishing	277
26.0801 Engraving and plate printing	2753
26.0802 Bookbinding and related work	2789
26.0803 Typesetting	2791
26.0804 Photoengraving	2793
26.0805 Eletrotyping and stereotyping	2794
27. Chemicals and selected chemical products	
27.0100 Industrial inorganic and organic chemicals	281 (excl. 28195), 2865, 2869
27.0201 Nigrogenous and phosphatic fertilizers	2873-4
27.0202 Fertilizers, mixing only	2875
27.0300 Agricultural chemicals, n.e.c.[4]	2879
27.0401 Gum and wood chemicals	2861
27.0402 Adhesives and sealants	2891
27.0403 Explosives	2892
27.0404 Printing ink	2893
27.0405 Carbon black	2895
27.0406 Chemical preparations, n.e.c.[4]	2899
28. Plastics and synthetic materials	
28.0101 Plastics materials and resins	2821
28.0200 Synthetic rubber	2822
28.0300 Cellulosic man-made fibers	2823
28.0400 Organic fibers, noncellulosic	2824
29. Drugs, cleaning and toilet preparations	
29.0100 Drugs	283
29.0201 Soap and other detergents	2841
29.0202 Polishes and sanitation goods	2842

Table A-2 continued

Industry Number and Title	*Related Census-SIC codes (1972 edition)*
29.0203 Surface active agents	2843
29.0300 Toilet preparations	2844
30. Paints and allied products	
30.0000 Paints and allied products	285
31. Petroleum refining and related industries	
31.0100 Petroleum refining and miscellaneous products of petroleum and coal	291, 299
31.0200 Paving mixtures and blocks	2951
31.0300 Asphalt felts and coatings	2952
32. Rubber and miscellaneous plastics products	
32.0100 Tires and inner tubes	301
32.0200 Rubber and plastics footwear	302
32.0301 Reclaimed rubber	303
32.0302 Fabricated rubber products, n.e.c.[4]	306
32.0400 Miscellaneous plastics products	307
32.0500 Rubber and plastics hose and belting	304
33. Leather tanning and finishing	
33.0001 Leather tanning and finishing	311
34. Footwear and other leather products	
34.0100 Footwear cut stock	313
34.0201 Shoes, except rubber	3143-4, 3149
34.0202 House slippers	3142
34.0301 Leather gloves and mittens	315
34.0302 Luggage	316
34.0303 Women's handbags and purses	3171
34.0304 Personal leather goods	3172
34.0305 Leather goods, n.e.c.[4]	319
35. Glass and glass products	
35.0100 Glass and glass products, except containers	321, 3229, 323
35.0200 Glass containers	3221
36. Stone and clay products	
36.0100 Cement, hydraulic	324
36.0200 Brick and structural clay tile	3251
36.0300 Ceramic wall and floor tile	3253
36.0400 Clay refractories	3255

36.0500	Structural clay products, n.e.c.[4]	3259
36.0600	Vitreous plumbing fixtures	3261
36.0701	Vitreous china food utensils	3262
36.0702	Fine earthenware food utensils	3263
36.0800	Porcelain electrical supplies	3264
36.0900	Pottery products, n.e.c.[4]	3269
36.1000	Concrete block and brick	3271
36.1100	Concrete products, n.e.c.[4]	3272
36.1200	Ready-mix concrete	3273
36.1300	Lime	3274
36.1400	Gypsum products	3275
36.1500	Cut stone and stone products	328
36.1600	Abrasive products	3291
36.1700	Asbestos products	3292
36.1800	Gaskets, packing and sealing devices	3293
36.1900	Minerals, ground or treated	3295
36.2000	Mineral wool	3296
36.2100	Nonclay refractories	3297
36.2200	Nometallic mineral products, n.e.c.[4]	3299
	37. Primary iron and steel manufacturing	
37.0101	Blast furnaces and steel mills	3312
37.0102	Electrometallurgical products	3313
37.0103	Steel wire and related products	3315
37.0104	Cold finishing of steel shapes	3316
37.0105	Steel pipe and tubes	3317
37.0200	Iron and steel foundries	332
37.0300	Iron and steel forgings	3462
37.0401	Metal heat treating	3398
37.0402	Primary metal products, n.e.c.[4]	3399
	38. Primary nonferrous metals manufacturing	
38.0100	Primary copper	3331
38.0200	Primary lead	3332
38.0300	Primary zinc	3333
38.0400	Primary aluminum	3334, 28195
38.0500	Primary nonferrous metals, n.e.c.[4]	3339
38.0600	Secondary nonferrous metals	334
38.0700	Copper rolling and drawing	3351
38.0800	Aluminum rolling and drawing	3353–5
38.0900	Nonferrous rolling and drawing, n.e.c.[4]	3356
38.1000	Nonferrous wire drawing and insulating	3357
38.1100	Aluminum castings	3361
38.1200	Brass, bronze, and copper castings	3362

Table A-2 continued

Industry Number and Title	*Related Census-SIC codes (1972 edition)*
38.1300 Nonferrous castings, n.e.c.[4]	3369
38.1400 Nonferrous forgings	3463
39. Metal containers	
39.0100 Metal cans	3411
39.0200 Metal barrels, drums, and pails	3412
40. Heating, plumbing, and fabricated structural metal products	
40.0100 Metal sanitary ware	3431
40.0200 Plumbing fixture fittings and trim	3432
40.0300 Heating equipment, except electric	3433
40.0400 Fabricated structural metal	3441
40.0500 Metal doors, sash, and trim	3442
40.0600 Fabricated plate work (boiler shops)	3443
40.0700 Sheet metal work	3444
40.0800 Architectural metal work	3446
40.0901 Prefabricated metal buildings	3448
40.0902 Miscellaneous metal work	3449
41. Screw machine products and stampings	
41.0100 Screw machine products and bolts, nuts, rivets, and washers	345
41.0201 Automotive stampings	3465
41.0202 Crowns and closures	3466
41.0203 Metal stampings, n.e.c.[4]	3469
42. Other fabricated metal products	
42.0100 Cutlary	3421
42.0201 Hand and edge tools, n.e.c.[4]	3423
42.0202 Hand saws and saw blades	3425
42.0300 Hardware, n.e.c.[4]	3429
42.0401 Plating and polishing	3471
42.0402 Metal coating and allied services	3479
42.0500 Miscellaneous fabricated wire products	3495–6
42.0700 Steel springs, except wire	3493
42.0800 Pipe, valves, and pipe fittings	3494, 3498
42.1000 Metal foil and leaf	3497
42.1100 Fabricated metal products, n.e.c.[4]	3499
43. Engines and turbines	
43.0100 Steam engines and turbines	3511

43.0200	Internal combustion engines, n.e.c.[4]	3519
	44. Farm and garden machinery	
44.0001	Farm machinery and equipment	3523
44.0002	Lawn and garden equipment	3524
	45. Construction and mining machinery	
45.0100	Construction machinery and equipment	3531
45.0200	Mining machinery, except oilfield	3532
45.0300	Oilfield machinery	3533
	46. Materials handling machinery and equipment	
46.0100	Elevators and moving stairways	3534
46.0200	Conveyors and conveying equipment	3535
46.0300	Hoists, cranes, and monorails	3536
46.0400	Industrial trucks and tractors	3537
	47. Metalworking machinery and equipment	
47.0100	Machine tools, metal cutting types	3541
47.0200	Machine tools, metal forming types	3542
47.0300	Special dies and tools and machine tool accessories	3544–5
47.0401	Power driven hand tools	3546
47.0402	Rolling mill machinery	3547
47.0403	Metalworking machinery, n.e.c.[4]	3549
	48. Special industry machinery and equipment	
48.0100	Food products machinery	3551
48.0200	Textile machinery	3552
48.0300	Woodworking machinery	3553
48.0400	Paper industries machinery	3554
48.0500	Printing trades machinery	3555
48.0600	Special industry machinery, n.e.c.[4]	3559
	49. General industrial machinery and equipment	
49.0100	Pumps and compressors	3561, 3563
49.0200	Ball and roller bearings	3562
49.0300	Blowers and fans	3564
49.0400	Industrial patterns	3565
49.0500	Power transmission equipment	3566,3568
49.0600	Industrial furnaces and ovens	3567
49.0700	General industrial machinery	3569
	50. Miscellaneous machinery, except electrical	
50.0001	Carburetors, pistons, rings, valves	3592
50.0002	Machinery, except electrical, n.e.c.[4]	3599

Table A-2 continued

Industry Number and Title	*Related Census-SIC codes (1972 edition)*
51. Office, computing, and accounting machines	
51.0101 Electronic computing equipment	3573
51.0102 Calculating and accounting machines	3574
51.0200 Typewriters	3572
51.0300 Scales and balances	3576
51.0400 Office machines, n.e.c.[4]	3579
52. Service industry machines	
52.0100 Automatic merchandising machines	3581
52.0200 Commercial laundry equipment	3582
52.0300 Refrigeration and heating equipment	3585
52.0400 Measuring and dispensing pumps	3586
52.0500 Service industry machines, n.e.c.[4]	3589
53. Electric transmission and distribution equipment and industrial apparatus	
53.0100 Instruments to measure electricity	3825
53.0200 Transformers	3612
53.0300 Switchgear and switchboard apparatus	3613
53.0400 Motors and generators	3621
53.0500 Industrial controls	3622
53.0600 Welding apparatus, electric	3623
53.0700 Carbon and graphite products	3624
53.0800 Electrical industrial apparatus, n.e.c.[4]	3629
54. Household appliances	
54.0100 Household cooking equipment	3631
54.0200 Household refrigerators and freezers	3632
54.0300 Household laundry equipment	3633
54.0400 Electric housewares and fans	3634
54.0500 Household vacuum cleaners	3635
54.0600 Sewing machines	3636
54.0700 Household appliances, n.e.c.[4]	3639
55. Electric lighting and wiring equipment	
55.0100 Electric lamps	3641
55.0200 Lighting fixtures and equipment	3645-8
55.0300 Wiring devices	3643-4
56. Radio, TV, and communication equipment	
56.0100 Radio and TV receiving sets	3651
56.0200 Phonograph records and tape	3652

56.0300 Telephone and telegraph apparatus	3661
56.0400 Radio and TV communication equipment	3662
57. Electronic components and accessories	
57.0100 Electron tubes	3671-3
57.0200 Semiconductors and related devices	3674
57.0300 Electronic components, n.e.c.[4]	3675-9
58. Miscellaneous electric machinery, equipment, and supplies	
58.0100 Storage batteries	3691
58.0200 Primary batteries, dry and wet	3692
58.0300 X-ray apparatus and tubes	3693
58.0400 Engine electrical equipment	3694
58.0500 Electrical equipment, n.e.c.[4]	3699
59. Motor vehicles and equipment	
59.0100 Truck and bus bodies	3713
59.0200 Truck trailers	3715
59.0301 Motor vehicles	3711
59.0302 Motor vehicle parts and accessories	3714
60. Aircraft and parts	
60.0100 Aircraft	3721
60.0200 Aircraft and missile engines and engine parts	3724, 3764
60.0400 Aircraft and missile equipment, n.e.c.[4]	3728, 3769
61. Other transportation equipment	
61.0100 Ship building and repairing	3731
61.0200 Boat building and repairing	3732
61.0300 Railroad equipment	374
61.0500 Motorcycles, bicycles, and parts	375
61.0601 Travel trailers and campers	3792
61.0602 Mobile homes	2451
61.0700 Transportation equipment, n.e.c.[4]	3799
62. Professional, scientific, and controlling instruments and supplies	
62.0100 Engineering and scientific instruments	3811
62.0200 Mechanical measuring devices	3823-4, 3829
62.0300 Automatic temperature controls	3822
62.0400 Surgical and medical instruments	3841
62.0500 Surgical appliances and supplies	3842
62.0600 Dental equipment and supplies	3843
62.0700 Watches, clocks, and parts	387
63. Optical, ophthalmic, and photographic equipment and supplies	
63.0100 Optical instruments and lenses	383

Table A-2 continued

Industry Number and Title	*Related Census-SIC codes (1972 edition)*
63.0200 Ophthalmic goods	385
63.0300 Photographic equipment and supplies	386
64. Miscellaneous manufacturing	
64.0101 Jewelry, precious metal	3911
64.0102 Jewelers materials and lapidary work	3915
64.0104 Silverware and plated ware	3914
64.0105 Costume jewelry	3961
64.0200 Musical instruments	393
64.0301 Games, toys, and children's vehicles	3944
64.0302 Dolls	3942
64.0400 Sporting and athletic goods, n.e.c.[4]	3949
64.0501 Pens and mechanical pencils	3951
64.0502 Lead pencils and art goods	3952
64.0503 Marking devices	3953
64.0504 Carbon paper and inked ribbons	3955
64.0600 Artificial trees and flowers	3962
64.0701 Buttons	3963
64.0702 Needles, pins, and fasteners	3964
64.0800 Brooms and brushes	3991
64.0900 Hard surface floor coverings	3996
64.1000 Burial caskets and vaults	3995
64.1100 Signs and advertising displays	3993
64.1200 Manufacturing industries, n.e.c.[4]	3999 (excl. 39996)
Transportation, Communication, and Utilities	
65. Transportation and warehousing	
65.0100 Railroads and related services	40, 474, pt. 4789
65.0200 Local, suburban, and interurban highway passenger transportation	41
65.0300 Motor freight transportation and warehousing	42, pt. 4789
65.0400 Water transportation	44
65.0500 Air transportation	46
65.0700 Transportation services	47 (excl. 474 and pt. 4789)
66. Communications, except radio and TV	
66.0000 Communications, except radio and TV	48 (excl. 483)
67. Radio and TV broadcasting	
67.0000 Radio and TV broadcasting	483

68. Electric, gas, water, and sanitary services[2]	
68.0100 Electric services (utilities)	491, pt. 493
68.0200 Gas production and distribution (utilities)	492, pt. 493
68.0300 Water supply and sanitary services	497-7, pt. 493
Wholesale and Retail Trade	
69. Wholesale and retail trade	
69.0100 Wholesale trade	50, 51 (excl. manufacturer's sales offices)
69.0200 Retail trade	52-7, 59, 7396, 8042
Finance, Insurance, and Real Estate	
70. Finance and insurance[2]	
70.0100 Banking	60
70.0200 Credit agencies	61 (excl. pt. 613), 67
70.0300 Security and commodity brokers	62
70.0400 Insurance carriers	63
70.0500 Insurance agents and brokers	64
71. Real estate and rental	
71.0100 Owner-occupied dwellings	not applicable
71.0200 Real estate	65-6, pt. 1531
Services	
72. Hotels and lodging, personal and repair services (except auto)	
72.0100 Hotels and lodging places	70 (excl. dining)
72.0200 Personal and repair services, except auto repair and beauty and barber shops	72 (excl. 723-4), 762-4, pt. 7699
72.0300 Beauty and barber shops	723-4
73. Business services	
73.0100 Miscellaneous business services	732-9 (excl. 7396), 7692, 7694, pt. 7699
73.0200 Advertising	731
73.0300 Miscellaneous professional services	81, 89 (excl. 8922)
74. Eating and drinking places	
74.0000 Eating and drinking places	58, pt. 70
75. Automobile repair and services	
75.0000 Automobile repair and services	75
76. Amusements	
76.0100 Motion pictures	78
76,0200 Amusement and recreation services	79

Table A-2 continued

Industry Number and Title	*Related Census-SIC codes (1972 edition)*
77. Health, educational, and social services and nonprofit organizations	
77.0100 Doctors and dentists	801-3, 8041
77.0200 Hospitals	806
77.0300 Other medical and health services	074, 8049, 805, 807-9
77.0400 Educational services	82
77.0500 Nonprofit organizations	84, 86, 8922
77.0600 Job training and related services	8331
77.0700 Child day care services	8351
77.0800 Residential care	8361
77.0900 Social services, n.e.c.	8321, 8399
Government Enterprises	
78. Federal Government enterprises	
78.0100 U.S. Postal Service	4311
78.0200 Federal electric utilities	pt. 491
78.0300 Commodity Credit Corporation	pt. 613
78.0400 Other Federal Government enterprises	several[3]
79. State and local government enterprises	
79.0100 Local government passenger transit	pt. 41
79.0200 State and local electric utilities	pt. 491
79.0300 Other State and local government enterprises	several[3]
Dummy and Special Industries	
80. Noncomparable imports	
80.0000 Noncomparable imports	
81. Scrap, used, and secondhand goods	
81.0000 Scrap, used, and secondhand goods	
82. Government industry	
82.0000 Government industry	
83. Rest of the world industry	
83.0000 Rest of the world industry	
84. Household industry	
84.0000 Household industry	
85. Inventory valuation adjustment	
85.0000 Inventory valuation adjustment	

Value Added and Final Demand

V.A. Value added, total

88 Employee compensation

89 Indirect business taxes

90 Property-type income

91. Personal consumption expenditures

91.0000 Personal consumption expenditures

92. Gross private domestic fixed investment

92.0000 Gross private domestic fixed investment

93. Change in business inventories

93.0000 Change in business inventories

94. Exports

94.0000 Exports

95. Imports

95.0000 Imports

96. Federal Government purchases, national defense

96.0000 Federal Government purchases, national defense

97. Federal Government purchases, nondefense

97.0000 Federal Government purchases, nondefense

98. State and local government purchases, education

98.0000 State and local government purchases, education

99. State and local government purchases, other

99.1000 State and local government purchases, health, welfare, and sanitation

99.2000 State and local government purchases, safety

99.3000 State and local government purchases, other general government

---

Note: The titles in italics represent the groupings of industries used for the summary version of the 1972 tables.

1. The industry classification is usually identical with that for the commodity which is the primary product of the industry. However, for some industries the primary product, or a component thereof, is the same as the primary product of another industry. In such cases, commodity output is included with the industry most definitively associated with the commodity, usually the largest producer.

2. Excluding government enterprises.

3. In the 1972 SIC, government enterprise activities are generally classified with the similar private activity. In I-0, activities of enterprises are classified in groups 78 and 79 and the corresponding SIC's are shown except for 78.0400 and 79.0300, each of which includes a number of SIC's and several activities for which no comparable SIC exists.

4. n.e.c. = not elsewhere classified.

Columns 107 to 142, which correspond to the mineral rows (primary and scrap), have only two nonzero entries for by-product and import-to-output ratios.

The choice of classification was based on several considerations:

1. The need to represent the production and consumption of individual minerals that are treated only as aggregates even in the BEA 496-order input-output table.
2. The desire for a level of sectoral detail appropriate to the analysis of materials substitution and technological change in the extraction and minerals processing sectors.
3. Availability of data, particularly data to compute input-output coefficients for commodities defined at a greater level of detail than the 496-order BEA input-output table.
4. Restrictions, imposed by computational methods, on the range of absolute magnitudes of entries in the coefficient matrix.

The first and second considerations sometimes conflicted with the third and fourth—for example, when we attempted to incorporate individual vectors of input-output coefficients for the mining of each mineral.[1] The relevant Bureau of Mines data were organized at a 4-digit SIC level and grouped together in aggregate sectors for nearly half the minerals being studied, even though the level of detail was substantially greater than at a BEA 496-order level. Creating subvectors that corresponded to individual minerals by a rule-of-thumb proved unsuccessful because the resulting system of equations in the model proved to be colinear. However, a second problem arose from the same effort to disaggregate the BEA 496-order mining sectors. In a disaggregated matrix, mineral input-output coefficients were measured in metric units of input per $1 billion of output. By-product coefficients were measured in metric units of by-product output per metric unit of own-industry output. The range in the absolute magnitudes of entries in the matrix was enormous, the ratio of the largest to smallest value being 10^{14}. Solutions starting from a matrix with this range of values proved unreliable, using the computational technique appropriate for solving a model of this size.

Note

1. Economic Analysis Division, Bureau of Mines, unpublished worksheets with 1972 input-output data; tables 2B and summary tables for 1972.

Appendix B: Base-Year 1972 Input-Output Coefficients in IEA/USMIN

This appendix describes the data sources, aggregation, and estimation procedures used in calculating the base-year input-output coefficients for IEA/USMIN. The description covers both the interindustry coefficients and the mineral and scrap coefficients.

Interindustry Coefficients

IEA rows 1–106 and columns 1–106 comprise an interindustry coefficients matrix that depicts the 1972 structure of the U.S. economy in terms of inputs and outputs of 106 commodities. The coefficients, expressed as dollars of input *i* required per unit of output *j*, were calculated from a 496-order commodity-by-commodity flow matrix for 1972. To obtain the flow matrix, commodity-by-commodity direct requirements coefficients were multiplied by a vector of commodity outputs.[1] The flow matrix was aggregated into 106 sectors (see table A–1 of appendix A for the BEA into IEA aggregation scheme). Each column of the 106-sector aggregated flow matrix was divided by appropriate commodity-output totals to derive coefficients.

Mineral Coefficients

The 1972 input-output coefficients for twenty-six nonfuel minerals and thirteen types of metal scrap appear in rows 107 to 142 and columns 1 to 106 (submatrix B_c in figure 3–1) and rows 254, 255, and 265, columns 1 to 106. The coefficients are expressed in metric units of mineral input per $1 billion of sector output (1972 dollars) and are derived from sectoral data on mineral consumption and sector output.

Control totals for 1972 interindustry mineral consumption were calculated from a supply-demand balance for each mineral and type of scrap. Mineral supply is defined as the sum of domestic mine output (including by-product output from other sectors), imports, and inventory reductions:

$$\text{Supply} = X + BX + M + \text{INV} + \text{ERROR}$$

where X the mining sector's output of its principal commodity (for example, the output of gold by the gold-mining sector); BX is the by-product output produced in other sectors (such as the output of by-product gold produced in the copper mining sector); M is imports; INV is releases from private or government inventories; and ERROR is an apparent deficit in supply.

Mineral demand is defined as the sum of product deliveries to individual consuming sectors and to final demand:

$$\text{Demand} = C + \text{EX} + \text{INV} + \text{ERROR}$$

where C is consumption by industries; EX is exports; INV is increases in private or government inventories; and ERROR is apparent surplus in supply.

Failures of supply and demand to balance—resulting in either excess supply or excess demand—were generally treated as errors charged to the inventory-change data.

After establishing control totals, interindustry consumption was distributed to consuming sectors on an end-use basis. End use refers to the last statistically identifiable consumption of a mineral, either in its original form (as ore or concentrate) or embodied in a finished product of another sector (for example, lead in an automobile battery). Thus, in the coefficients matrix the automobile sector is shown as a consumer of such minerals as iron, nickel, and chromium, even though actual purchases by the automobile sector are in the form of rolled sheet, die castings, wire, and other products.[2]

Notes

1. These data sets were taken from a computer tape, *496-Industry Input-Output Tables for the United States, 1972 (BEA IED 79*-005), A Transactions and Direct Requirements Coefficients, obtained from the Interindustry Economics Division, Bureau of Economic Analysis, U.S. Department of Commerce, Tower Building, Washington, DC 20230. Detailed documentaion of these data is contained in Bureau of Economic Analysis, *Definitions and Conventions of the 1972 Input-Output Study* (BEA-SP 80-034), U.S. Department of Commerce, Bureau of Economic Analysis, July 1980.

2. Tables B-1 to B-26 of appendix B in [25A] provide the names of each of the twenty-six minerals covered by the study. The paper also lists every

major reference or data source consulted for each metal. The producing and consuming input-output sectors are identified and the units of measurement for metal flows are given. In all cases in which data used to calculate coefficients were not taken directly from published sources, the method for establishing 1972 interindustry flows is described.

References

U.S. Department of Commerce, Bureau of Economic Analysis, *496–Industry Input-Output Tables for the United States, 1972 (BEA IED 79–005), A Transactions and Direct Requirements Coefficients,* Washington, D.C., 1979.

U.S. Department of Commerce, Bureau of Economic Analysis, *Definitions and Conventions of the 1972 Input-Output Study,* Washington, D.C., 1980.

Appendix C: Technological Change in Minerals Use: Updating Base-Year Minerals Coefficients for 1980, 1990, and 2000 in IEA/USMIN

This appendix describes procedures used to update interindustry and mineral coefficients to reflect possible future substitution, process improvements, and changes in product mix of end-use sectors.

The method used to update base-year coefficients included estimating updaters by which individual 1972 input-output coefficients were multiplied to obtain the projected values for 1980, 1990, and 2000. For example, an individual updater u may assume the following values:

1. u_{ij} = if a_{ij} is expected to remain unchanged from the base-year

2. u_{ij} ≤ if a_{ij} is expected to decrease relative to the base-year

3. $u_{ij} \geq 1$ if a_{ij} is expected to increase relative to the base-year.

Updaters were first estimated for the year 2000 with 1980 and 1990 values obtained by linear interpolation.

Updaters for Interindustry and Mineral Coefficients

Selected base-year coefficients in IEA/USMIN rows 1 to 106 and columns 1 to 106 were multiplied by updaters to calculate projected coefficient values for 1980, 1990, and 2000. Estimates were based largely on forecasts prepared by Shapanka, Ayres, and Noble (1974) and Shapanka (1978). Shapanka's projections involved across-the-row or cell-by-cell materials substitutions. For the most recent projections (1978), Shapanka concentrated on such sectors as construction, primary iron and steel, primary nonferrous metals, appliances, and automobiles, in which technological change was expected to be the most extensive.

Nearly all base-year coefficients in IEA/USMIN rows 107 to 142 and columns 1 to 106 were multiplied by updaters.[1]

Updaters for all minerals and scrap coefficients were estimated by Professor Michael B. Bever of the Department of Materials Science and Engineering at the Massachusetts Institute of Technology. A variety of sources concerning changing materials requirements were consulted including contingency forecasts by Bureau of Mines experts (1980), Shapanka (1978), and extrapolations of time series data prepared by Jack Faucett Associates, Inc. (1977).

The projections were made by estimating the value of an updater (the ratio of the estimated 2000 coefficient to the actual 1972 coefficient) by which ore and scrap coefficients were multiplied to update metal input requirements. The recycling rate was not assumed to improve between 1972 and 2000, except in a high-recycling scenario. Furthermore, the updaters for some end uses of iron, copper, aluminum, lead, and zinc were applied to selected cells in the interindustry matrix.

No changes in input requirements for the extraction industries were assumed, as was done in the original study. Although ore grades and related factors are expected to increase mining costs, these increases should be offset by improvements in mining technologies and new discoveries. Energy inputs into metals processing were projected according to judgments by Shapanka; additional changes, stemming from shifts in manufacturing to more energy-efficient processes and techniques, were also assumed to affect metals processing. Estimates for the degree of change were taken from the CONAES study (1978).

Labor requirements were taken from Bureau of the Census (1976) and Bureau of Labor Statistics (1975a, 1975b) publications. Energy inputs coefficients were calculated independently based on data prepared by Peter Penner (1978).

Note

1. The updaters for each mineral and selected interindustry sector for 2000 are presented in appendix C of [25a].

References

Committee on Nuclear and Alternative Energy Systems (CONAES), "United States Energy Demand: Some Low Energy Futures," *Science,* vol. 200, April 14, 1978.

Jack Faucett Associates, Inc., *Development of Input-Output Tables 1958–72: Final Report,* Chevy Chase, Maryland, 1977.

Penner, P.S., *Direct Energy Transaction Matrix for 1971,* technical memorandum 98, Center for Advanced Computation, University of Illinois, Urbana-Champaign, Illinois, 1978.

Shapanka, A., "Long Range Technological Forecasts for Use in Studying the Resource and Environmental Consequences of U.S. Population and Economic Growth: 1975–2025," discussion paper D-31, Division of Renewable Resources, Resources for the Future, May 1978.

———, R.U. Ayres, and S. Noble, "The Use of Explicit Technological Forecasts in Long Range Input-Output Models—Application to a 42-Sector Model," IRT-362-R, International Research and Technology Corporation, Washington, D.C., 1974.

U.S. Department of Commerce, Bureau of the Census, *1972 Census of Manufacturers,* vol. 2, *Industry Statistics,* part 3, Washington, D.C., 1976.

U.S. Department of Labor, Bureau of Labor Statistics, The Structure of the United States Economy in 1980 and 1985, bulletin 1831, Washington, D.C., 1975a.

———, *The U.S. Economy in 1985,* bulletin 1809, Washington, D.C., 1975b.

Appendix D: Economic Reserves and Subeconomic Resources for the United States

Table D–1 of this appendix gives current estimates of identified U.S. economic reserves and subeconomic resources for twenty-three nonfuel minerals. Estimates are not presented for silicon and chlorine—manufactured from sand and sea water, respectively—because these minerals are virtually inexhaustible. Information for platinum reserves and resources is not available.

Table D–1
U.S. Reserves and Resources Estimates for 26 Nonfuel Minerals (1980)

IEA Code	*Mineral (Units)*	*Reserves*	*Resources* [a]
MIN 1	Iron (MMT)	3,628.0	17,777.0
MIN 2	Molybdenum (KMT)	4,983.0	8,652
MIN 3	Nickel (KMT)[b]	326.5	13,877.1
MIN 4	Tungsten (KMT)	124.5	450.7
IMM 5	Manganese (KMT)[b]	0.0	66,755.2
IMM 6	Chromium (KMT)	0.0	2,294.7
MIN 7	Copper (MMT)[b]	92.0	382.0
MIN 8	Lead (MMT)	27.0	74.0
MIN 9	Zinc (MMT)	15.0	65.0
MIN 10	Gold (MT)	1,399.5	7,464.0
MIN 11	Silver (KMT)	47.0	177.6
MIN 12	Aluminum (MMT)	9.1	45.4
MIN 13	Mercury (KMT)	12.1	27.6
MIN 14	Vanadium (KMT)	104.3	9,101.0
MIN 15	Platinum (MT)	(1)	(1)
MIN 16	Titanium (KMT)	16,326.0	94,872.2
IMM 17	Tin (KMT)	50.0	200.0
MIN 21	Silicon (KMT)	(2)	(2)
MIN 23	Fluorine (MMT)	12.4	43.3

Table D–1 continued

IEA Code	*Mineral (Units)*	*Reserves*	*Resources*
MIN 24	Potash (MMT)	300.0	6,000.0
MIN 25	Soda ash (MMT)	26,937.0	(2)
MIN 26	Boron (KMT)	18,100.0	(1)
MIN 27	Phosphate rock (MMT)	1,800.0	9,250.0
MIN 28	Sulfur (MMT)	175.0	330.0
CHM 1	Chlorine (MMT)	(2)	(2)
CHM 2	Magnesium (MMT)	9.1	(2)

Source: *Mineral Facts and Problems,* 1980 ed., Bureau of Mines, United States Department of the Interior, Washington, D.C., 1980.

[a]Includes reserves.

[b]Excludes deposits in seabed nodules.

(1) Not available.

(2) Virtually unlimited.

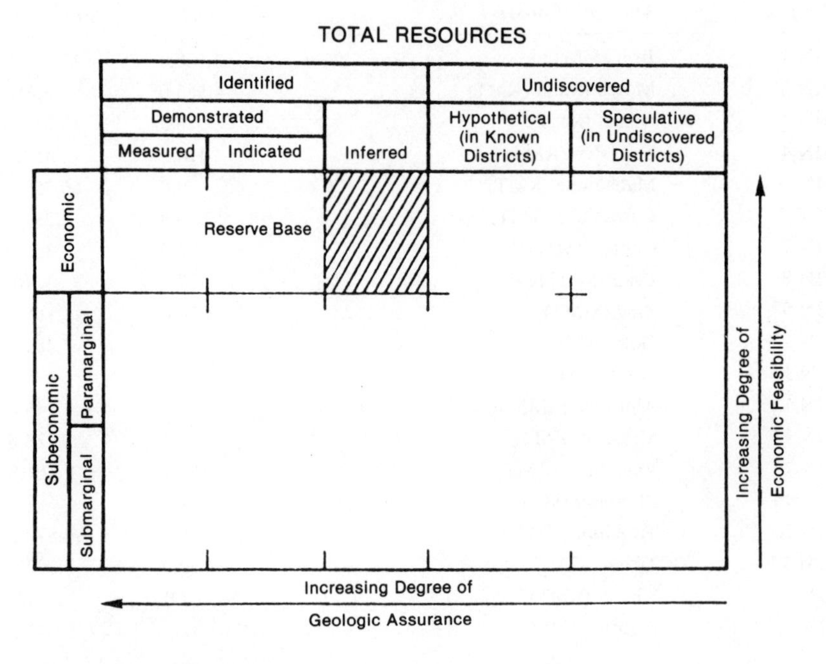

Source: Bureau of Mines.

Figure D–1. Classification of Resources

The Bureau of Mines classifies total resources into four categories (see figure D–1) defined as follows:

Resources: A concentration of naturally occurring solid, liquid, or gaseous materials in or on the Earth's crust in such form that economic extraction of a commodity is currently or potentially feasible.

Identified Resources: Specific bodies of mineral-bearing material whose location, quality, and quantity are known from geologic evidence supported by engineering measurements.

Undiscovered Resources: Unspecified bodies of mineral-bearing material surmised to exist on the basis of broad geologic knowledge and theory.

Reserve: That portion of the identified resource from which a usable mineral and energy commodity can be economically and legally extracted at the time of determination.[1]

Note

1. *Mineral Facts and Problems* (Washington, D.C.: 1976), pp.16–17. 1975 edition, Bureau of Mines, United States Department of the Interior.

References

U.S. Department of the Interior, Bureau of Mines, *Mineral Facts and Problems,* bulletin 667, 1975 ed. Washington, D.C., 1976.

———, *Mineral Facts and Problems,* bulletin 671, 1980 ed., Washington, D.C., 1980.

Appendix E Final Demand Projections for 1980, 1990, and 2000 in IEA/USMIN

This appendix describes the sources of final-demand data and projections, and the adaptation of BLS projections to IEA/USMIN. The adaptation of BLS final demand projections involved two operations:

1. Reclassification of the BLS 1980 and 1990 projections to conform to the IEA/USMIN sector scheme.
2. Extension of the BLS projections for 1990 to 2000.

Data Sources

Base-year final demand data for 1972 were taken from the 496-sector U.S. input-output table (1972).

BLS final demand projections for 1980 and 1990 appear in *The U.S. Economy in 1985* (1975). The projections were developed for the BLS U.S. input-output model. For each year, a 154 × 210 matrix of final demands is given. The 154 rows correspond to the producing sectors that deliver output to final demand. The 210 columns represent the set of final-demand activities for the economy. By summing together these detailed columns of final demands, a bill of goods which corresponds to the National Income and Product Accounts (NIPA) can be derived: consumption, investment, imports, exports, and both defense and nondefense government spending.

For 2000, BLS made available to IEA an unpublished BLS document giving projections of aggregate demand by the NIPA categories. These projections were generated by the BLS macroeconomic model.

A sector-correspondence between IEA/USMIN and the BLS sectors was developed using sectors in terms of SIC definitions given in the detailed U.S. input-output model (1975) and *The U.S. Economy in 1985* (1975). See tables E-1 and E-2.

Table E-1
Correspondence Between BLS Input-Output Sectors and SIC Codes

BLS Number	*Name*	*BEA Sector*	*SIC (1967)*
Agriculture, Forestry, and Fishery			
1.	Dairy and poultry products	1.01– 1.02	
2.	Meat animals and livestock	1.03	01
3.	Cotton	2.01	
4.	Food and feed grains	2.02	
5.	Other agricultural products	2.03– 2.07	
6.	Forestry, and fishery products	3.00	074, 08, 091
7.	Agricultural, forestry and fishery services	4.00	071, 0723, 073, PT. 0729, 085, 098
Mining			
8.	Iron and ferroalloy ores mining	5.00	1011, 106
9.	Copper ore mining	6.01	102
10.	Other nonferrous ore mining	6.02	103–109 (Except 106)
11.	Coal mining	7.00	11, 12
12.	Crude petroleum and natural gas	8.00	1311, 1321
13.	Stone and clay mining and quarrying	9.00	14 (Except 147)
14.	Chemical and fertilizer mineral mining	10.00	147
Construction			
15.	New residential building construction	11.01	
16.	New nonresidential building construction	11.02	PT. 15, PT. 16, PT. 17, PT. 6561
17.	New public utility construction	11.03	
18.	New highway construction	11.04	PT. 16, PT. 17
19.	All other new construction	11.05	PT. 15, PT. 16, PT. 17

20. Oil and gas well drilling and exploration	11.0503–11.0504	PT. 138
21. Maintenance and repair construction	12.01–12.02	PT. 15, PT. 16, PT. 17, 138
Manufacturing		
22. Ordnance	13.02–13.07	19 (Except 1925)
23. Complete guided missiles	13.01	1925
24. Meat products	14.01	201
25. Dairy products	14.02–14.06	202
26. Canned and frozen foods	14.07–14.13	203
27. Grain mill products	14.14–14.17	204
28. Bakery products	14.18	205
29. Sugar	14.19	206
30. Confectionery products	14.20	207
31. Alcoholic beverages	14.21	2082–2085
32. Soft drinks and flavorings	14.22–14.23	2086–2087
33. Miscellaneous food products	14.24–14.32	209
34. Tobacco manufacturing	15.01–15.02	21
35. Fabrics, yarn, and thread mills	16.01–16.04	2211, 2221, 2231, 2241, 226, 228
36. Floor coverings	17.01	227
37. Miscellaneous textile goods	17.02–17.10	229
38. Hosiery and knit goods	18.01–18.03	225
39. Apparel	18.04	23 (Except 239), 399996
40. Miscellaneous fabricated textile products	19.01–19.03	239
41. Logging	20.01	2411
42. Sawmills and planing mills	20.02–20.04	242
43. Millwork, plywood, and other wood products	20.05–20.09	243, 249
44. Wooden containers	21.00	244
45. Household furniture	22.01–22.04	251

Table E-1 continued

BLS Number	Name	BEA Sector	SIC (1967)
46.	Other furniture and fixtures	23.01–23.07	252–259
47.	Paper products	24.01–24.07	26 (Except 265)
48.	Paperboard	25.00	265
49.	Newspaper printing and publishing	26.01	2711
50.	Periodicals and book printing, publishing	26.02–26.04	272–274
51.	Miscellaneous printing and publishing	26.05–26.08	275–279
52.	Industrial inorganic and organic chemicals	27.01	281 (Except 28195)
53.	Agricultural chemicals	27.02–27.03	287
54.	Miscellaneous chemical products	27.04	2861, 289
55.	Plastic materials and synthetic rubber	28.01–28.02	2821–2822
56.	Synthetic fibers	28.03–28.04	2823–2824
57.	Drugs	29.01	283
58.	Cleaning and toilet preparations	29.02–29.03	284
59.	Paints and allied products	30.00	2851
60.	Petroleum refining and related products	31.01–31.03	29
61.	Tires and inner tubes	32.01	3011
62.	Miscellaneous rubber products	32.02–32.03	3021, 3031, 3069
63.	Plastic products	32.04	3079
64.	Leather tanning and industrial leather	33.00	3111, 3121
65.	Footwear and other leather products	34.01–34.03	3131, 314, 3151, 3161, 317, 3199
66.	Glass	35.01–35.02	3211, 322, 3231
67.	Cement and concrete products	36.01 and 36.10–36.14	3241, 327
68.	Structural clay products	36.02–36.05	325
69.	Pottery and related products	36.06–36.09	326

70. Miscellaneous stone and clay products	36.15–36.22	3281, 329
71. Blast furnaces and basic steel products	37.01	331
72. Iron and steel foundries and forgings	37.02–37.04	332, 3391, 3399
73. Primary copper and copper products	38.01, 38.07, 38.10, 38.12	3331, 3351, 3357, 3362
74. Primary aluminum and aluminum products	38.04, 38.08, 38.11	3334, 28195, 3352, 3361
75. Other primary nonferrous products	38.02, 38.03, 38.05, 38.06, 38.09, 38.13, 38.14	3332, 3333, 3339, 3341, 3356, 3369, 3392
76. Metal containers	39.01–39.02	3411, 3491
77. Heating apparatus and plumbing fixtures	40.01–40.03	343
78. Fabricated structural metal	40.04–40.09	344
79. Screw machine products	41.01	345
80. Metal stampings	41.02	3461
81. Cutlery, handtools, and general hardware	42.01–42.03	342
82. Other fabricated metal products	42.04–42.11	347–349 (Except 3491)
83. Engine, turbines, and generators	43.01–43.02	351
84. Farm machinery	44.00	3522
85. Construction, mining, and oilfield machinery	45.01–45.03	3531–3533
86. Material handling equipment	46.01–46.04	3534–3537
87. Metal working machines	47.01–47.04	354
88. Special industry machinery	48.01–48.06	355
89. General industrial machinery	49.01–49.07	356
90. Machine shop products	50.00	359
91. Computers and peripheral equipment	51.01	3573–3574
92. Typewriters and other office equipment	52.02–51.04	357 (Except 3573 and 3574)
93. Service industry machines	52.01–52.05	358
94. Electric transmission equipment	53.01–53.03	361

Table E–1 continued

BLS Number	*Name*	*BEA Sector*	*SIC (1967)*
95.	Electrical industrial apparatus	53.04–53.08	362
96.	Household appliances	54.01–54.07	363
97.	Electric lighting and wiring	55.01–55.03	364
98.	Radio and TV receiving sets	56.01–56.02	365
99.	Telephone and telegraph apparatus	56.03	3661
100.	Radio and communication equipment	56.04	3662
101.	Electronic components	57.01–57.03	367
102.	Miscellaneous electrical products	58.01–58.05	369
103.	Motor vehicles	59.01–59.03	371
104.	Aircraft	60.01–60.04	372
105.	Ship and boat building and repair	61.01–61.02	373
106.	Railroad equipment	61.03–61.04	374
107.	Motorcycles, bicycles, and parts	61.05	3751
108.	Other transportation equipment	61.06–61.07	379
109.	Scientific and controlling instruments	62.01–62.03	3811, 382
110.	Medical and dental instruments	62.04–62.06	384
111.	Optical and ophthalmic equipment	63.01–63.02	3831, 3851
112.	Photographic equipment and supplies	63.03	3861
113.	Watches, clocks, and clock operated devices	62.07	387
114.	Jewelry and silverware	64.01	391, 3961
115.	Musical instruments and sporting goods	64.02–64.04	393, 394
116.	Other miscellaneous manufactured products	64.05–64.12	395, 396, 399 (Except 39996)

Transportation		
117. Railroad transportation	65.01	40, 474
118. Local transit, intercity buses	65.02	41
119. Truck transportation	65.03	42, 473
120. Water transportation	65.04	44
121. Air transportation	65.05	45
122. Pipeline transportation	65.06	46
123. Transportation services	65.07	47 (Except 473 and 474)
Communication		
124. Communication, except radio and TV	66.00	48 (Except 483)
125. Radio and TV broadcasting	67.00	483
Public Utilities		
126. Electric utilities	68.01	491, PT. 493
127. Gas utilities	68.02	492, PT. 493
128. Water and sanitary services	68.03	494–497, PT. 493
Wholesale and Retail Trade		
129. Wholesale trade	69.01	50
130. Retail trade	69.02	52–59, 7396, PT. 8099
Finance, Insurance, and Real Estate		
131. Banking	70.01	60
132. Credit agencies and financial brokers	70.02–70.03	61, 62, 67
133. Insurance	70.04–70.05	63–64
134. Owner-occupied real estate	71.01	NA
135. Real estate	71.02	65 (Except PT. 6561), 66
136. Hotels and lodging places	72.01	70
137. Personal and repair services	72.02	72 (Except 723, 724), 76 (Except 7692, 7694 and PT. 7699)

Table E–1 continued

BLS Number	*Name*	*BEA Sector*	*SIC (1967)*
138.	Barber and beauty shops	72.03	723, 724
139.	Miscellaneous business services	73.01	73 (Except 731, 7396), 7692, 7694, PT. 7699
140.	Advertising	73.02	731
141.	Miscellaneous professional services	73.03	81, 89 (Except 8921)
142.	Automobile repair	75.00	75
143.	Motion pictures	76.01	78
144.	Amusements and recreation services	76.02	79
145.	Doctors and dentists services	77.01	801–804
146.	Hospitals	77.02	8061
147.	Other medical services	77.03	0722, 807, 809 (Except PT. 8099)
148.	Educational services	77.04	82
149.	Nonprofit organizations	77.05	84, 86, 8921
Government Enterprises			
150.	Post Office	78.01	NA
151.	Commodity Credit Corporation	78.03	NA
152.	Other federal enterprises	78.02, 78.04	NA
153.	Local government passenger transit	79.01	NA
154.	Other state and local government	79.02, 79.03	NA

Table E–2
Correspondence Between IEA and BLS Input-Output Sectors

IEA	*BLS*	*IEA*	*BLS*
AGR 1	1, 2	MAN 21	65
AGR 2	3, part 5	MAN 22	66
AGR 3	4	MAN 23	Part 67
AGR 4	6, part 5	MAN 24	Part 70
AGR 5	7	MAN 25	68, 69, Part 67, Part 70
EXT 1	8		
EXT 2	9	MAN 26	76
EXT 3	10	MAN 27	77, 78
EXT 4	13	MAN 28	87
EXT 5	14	MAN 29	88
EXT 6	11	MAN 30	89
EXT 7	12	MAN 31	90
CON 1	15–20	MAN 32	91, 92
CON 2	21	MAN 34	87
MAN 1	22, 23	MAN 35	88
MAN 2	24–33	MAN 36	89
MAN 3	34	MAN 37	90
MAN 4	35	MAN 38	91, 92
MAN 5	36, 37	MAN 39	93
MAN 6	38, 39	MAN 40	94, 95
MAN 7	40	MAN 42	97
MAN 8	41, 42, 43	MAN 43	98, 99, 100
MAN 9	44	MAN 45	102
MAN 10	45	MAN 46	103
MAN 11	46	MAN 47	104
MAN 12	47	MAN 48	105–108
MAN 13	48	MAN 49	109, 110, 113
MAN 14	49–51	MAN 50	111, 112
		MAN 51	114–116
CHM 1	52		
		SRV 1	117–123
CHM 2	53	SRV 2	124
CHM 3	54	SRV 3	125
		SRV 4	126–128
MAN 15	55, 56	SRV 5	129, Part 130
MAN 16	57,58	SRV 6	131–133
MAN 17	59	SRV 7	134, 135
MAN 18	60	SRV 8	136–138, Part 130
MAN 19	61–63		
MAN 20	64	SRV 9	139–141

IEA	BLS	IEA	BLS
SRV 10	142	MET 10, 16,	Part 73
SRV 11	143, 144	MET 19, 21	Part 73
SRV 12	145–149	MET 13, 17, 20	Part 74
SRV 13	150–152	MET 11, 12, 14	Part 75
SRV 14	153, 154		
		MET 15, 18, 22	Part 75
MET 1–5	Part 71	MET 23	Part 75
MET 6–9	Part 72		

Data Development

Reclassification of the 1980 and 1990 Projections:

The 210 BLS final demand columns were aggregated according to the following scheme:

IEA Final Demand Column Number and Name	*BLS Column Numbers*
322 Personal Consumption Expenditures	166–247
323 Gross Private Fixed Investment	248–342
325 Exports	344
326 Imports	345
327 Government Purchases, Nondefense	346–349, 353–372
328 Government Purchases, Defense	350–352

The rows were aggregated according to the correspondence given in table E–2, which shows the constituent BLS sectors for each IEA producing sector. This correspondence was derived by matching the underlying SIC codes for the IEA and BLS models. Twenty-eight IEA sectors shown in table E–2 contain partial BLS sectors. In those cases, it was necessary to partition the BLS sectors before performing the aggregation.

The partitioning was accomplished by deriving base-year (1972) final demands for both the IEA and BLS sectors, for each final-demand category. Using these quantities, a ratio (IEA-sector final demand divided by BLS-sector final demand) was calculated for each of the twenty-eight split BLS sectors in each final-demand category. The BLS final-demand projections for the twenty-eight split sectors were then multiplied by these scaling coefficients before the row aggregation was performed.

Table E-3
Projected Final Demand for IEA/USMIN for 2000
(millions of 1972 dollars)

GDP Account	*Projected 2000 Level*	*Constituent BLS Input-Output Columns*
Personal consumption expenditure—durables	341.4	166-176
Personal consumption expenditure—nondurables	673.1	177-197
Personal consumption expenditure—services	853.1	198-247
Business investment—equipment	221.8	248-324
Business investment—structures	89.1	325-341
Residential housing	93.7	342
Exports	183.6	344
Imports	-146.7	345
Federal, state, and local government—nondefense	117.4	346-349, 353-372
Federal government—defense	53.8	350-352

Note: Figures excluded dummy and special industries entries. For more detailed information, see chapter 3, table 3-1.

Extension of BLS Projections to the Year 2000:

The BLS macroeconomic projections of aggregate demand by GDP account for 2000 were used to develop final demand projections for IEA/USMIN for 2000.

Table E-3 gives these projections, in millions of 1972 dollars, along with the numbers of the corresponding final-demand columns of the 154 × 210 BLS 1990 input-output model final-demand projections.

The final-demand projections for 2000 were derived by first, aggregating the BLS 1990 final-demand projections to correspond to the GDP accounts given in table E-3. Next, the resulting eleven columns were converted to coefficients by dividing the elements of each column by its respective column total. Each of these columns of coefficients provides the percentage distribution of one category of final demand over the 154 producing sectors of the BLS model. To extend the 1990 projection to 2000, these percentage distributions were multiplied by the relevant corresponding GNP-account levels for 2000. The resulting projected final demands for 2000 were then aggregated to the IEA scheme, using the same correspondences.

References

U.S. Department of Commerce, Bureau of Economic Analysis, *Detailed Input-Output Structure of the U.S. Economy 1972,* vol. 1, Washington, D.C., 1979.

U.S. Department of Labor, Bureau of Labor Statistics, *The U.S. Economy in 1985,* bulletin 1809, Washington, D.C., 1975.

Appendix F: Assumptions and Data for Recycling and Trade Scenarios in IEA/USMIN

This appendix describes the major assumptions and data used to specify the zero- and high-recycling scenarios and the high- and low-import scenarios.

Recycling Scenarios—Data

For the high-recycling scenario, estimates of potential recovery for iron, copper, aluminum, nickel, and tungsten were taken from a report by the Office of Technology Assessment (1979). Estimates for potential recovery for lead and zinc were taken from an earlier investigation by Battelle Columbus Laboratories (1972). Estimates for chromium, tin, and magnesium were based on National Materials Advisory Board studies. The current recycling levels for gold and silver were maintained through 2000 under the high-recycling scenario, because there is little expectation that higher recycling rates are technically feasible.

Large exports of iron and steel scrap and failure to recycle old ferrous scrap in containers, construction materials, castings, and discarded automobiles offer opportunities for increased domestic recycling. Nearly 100 percent of prompt industrial copper scrap is recycled and very little is exported, but additional opportunities to recycle exist for recycling old scrap, obsolete brass products, copper wire and tube, magnet wire, and cartridge brass. For lead, nearly all new scrap and industrial wastes are recycled and little is exported; some opportunities exist to recycle additional batteries, lead cable sheathing, and solder. Zinc flue dusts and new galvanized clippings offer additional opportunities for recycling new scrap; other opportunities include old zinc die castings in automobiles, home appliances, farm machinery, old galvanized zinc, and zinc in old rubber products.

About 90 percent of aluminum prompt industrial scrap, including industrial wastes, are recycled, but only a small fraction of obsolete scrap. Packaging materials, including cans, and aluminum components of automobiles and appliances constitute the bulk of unrecovered obsolete scrap. Nearly all new nickel-base obsolete scrap is recycled, although downgrading is a serious problem in some alloy classes. Exports constitute more than 10 percent of nickel-base scrap consumption. Nickel-bearing industrial wastes, containing 13,000 tons of nickel annually, are not currently recovered. Only one-quarter of old nickel base scrap is recycled, the remainder being primarily a constituent of low-nickel alloys (obsolete stainless steels, low-alloy steels, cast irons, and copper-base alloys).

For chromium, most new stainless steel scrap is recycled; a possible exception is straight-chromium grades. Substantial amounts—60 KMT per year in 1976—of recoverable chromium is discarded in industrial wastes ranging from stainless steel fine dusts to spent catalysts. As in the case of nickel, only about one-quarter of chromium-bearing obsolete scrap (stainless steels, alloy steels, and cast irons) is currently recovered.

Additional opportunities exist for recycling tin; less than 40 percent of the recoverable tin in obsolete brass and bronze, solder, type metal, babbit, and antimonial lead is currently recovered. In the case of magnesium, only a quarter of magnesium-based scrap considered recoverable is currently recycled.

For tungsten, new scrap constitutes most of the recycled scrap. Tungsten carbides from machine tools is currently being reclaimed and will, in the future, represent a larger fraction of total scrap supply. Furthermore, tungsten carbide sludges, discarded tungsten-bearing components, and tools not now recycled constitute important future sources of scrap supply.

See tables F-1 and F-2.

Trade Scenarios—Assumptions and Data

This section presents the method used to estimate U.S. import requirements for twenty-six nonfuel minerals for the 1990–2000 period. In addition, the import coefficients, both for the high and low import-dependence scenarios, are given in tables F-3 and F-4.

There are alternative sources of supply for all resources, both domestic and foreign. Sources of metallic minerals can also be replenished through secondary production from recycled materials. In projecting the minerals requirements of the U.S. economy to 2000, these alternative sources of supply—domestic mine output, by-product output, secondary production, and imports—must be considered simultaneously and the likely role that each of these sources of supply will play in satisfying the demand for minerals. Without knowing the likely share of each source to the total for 1990 and 2000, alternative feasible assumptions were considered for the possible future contribution of imports and recycled metals to the total supply of minerals.

Although the United States is already heavily dependent on foreign sources of supply for most nonfuel minerals (see figure 4-30), this depen-

Table F-1
Recycling Rates Assumed for Three Alternative Scenarios[a]

	Iron	*Copper*	*Aluminum*	*Lead*	*Zinc*	*Nickel*	*Chromium*	*Gold*	*Silver*	*Tungsten*	*Mercury*	*Tin*	*Magnesium*
							Base Recycling Scenario						
1972, 1980, 1990, 2000	0.3	0.4	0.15	0.39	0.21	0.3	0.16	0.10	0.34	0.04	0.23	0.23	0.014
							Zero Recycling Scenario						
1972, 1980,1990, 2000	0	0	0	0	0	0	0	0	0	0	0	0	0
							High Recycling Scenario[b]						
1972	0.3	0.4	0.15	0.39	0.21	0.3	0.16	0.10	0.34	0.04	0.23	0.23	0.014
1980	0.3	0.4	0.15	0.39	0.21	0.3	0.16	0.10	0.34	0.04	0.23	0.23	0.014
1990	0.45	0.56	0.26	0.46	0.28	0.42	0.27	0.10	0.34	0.28	0.23	0.40	0.164
2000	0.6	0.72	0.37	0.54	0.36	0.54	0.37	0.10	0.34	0.52	0.23	0.56	0.314

[a]The recycling rate of a metal is defined as the ratio of metal contained in purchased scrap (prompt industrial plus old) to total metal consumption.

[b]The maximum potential recycling rate was calculated by taking the ratio of available scrap (old plus prompt industrial plus exports) to total metal consumption.

Table F-2
Scrap Coefficient Updaters under High Recycling Scenario

Type of Scrap	*1990 Coefficient Updater*	*2000 Coefficient Updater*
Iron and steel	1.50	2.00
Copper	1.40	1.80
Lead	1.20	1.40
Zinc	1.35	1.70
Aluminum	1.75	2.50
Nickel	1.40	1.80
Chromium	1.65	2.30
Gold	1.00	1.00
Silver	1.00	1.00
Tungsten	6.88	12.75
Mercury	1.00	1.00
Tin	1.70	2.40
Magnesium	11.70	22.40

dency is likely to increase for several reasons: continued depletion of domestic reserves in the absence of increased prices for nonfuel minerals, continued increases in the cost of mining and processing minerals because of increases in the price of energy, and increased costs associated with compliance with environmental regulations.

As a result, three alternative shares of imports as a percentage of domestic mine output were selected for this study, reflecting the base-year (1972) share and the highest and lowest projected share of imports (out of domestic production) to 2000.

The various levels of future import shares (tables F-3 and F-4) are based on forecasts made by Bureau of Mines commodity experts, reported in *Mineral Facts and Problems,* bulletin 667 (1975 ed.). In deriving the shares, the changing share of imports at both the mining and primary metals level were taken into account. For example, in 1972, although only approximately 12.5 percent of total finished steel shipments were imported (in net terms), just under 30 percent of U.S. iron ore requirements was imported. Therefore the 1972 U.S. import share for iron was 0.44. Bureau of Mines forecasts have predicted that the share of foreign finished-steel imports is likely to equal 25 percent by 2000, in addition to an increasing foreign share of total iron ore requirements, yielding an import share of 60 percent. After accounting for the exogenously determined export level, the adjusted net import share is estimated at 58 percent. On the other hand, the low import-dependency scenario, while it is unlikely, shows that the United

Table F-3
Derivation of Net Import to Primary Demand Coefficients of 26 Nonfuel Minerals for 1990[a]

Mineral Name	Estimated Primary Demand	Low-End Estimate of Domestic Output	High-End Estimate of Domestic Output	Import to Primary Demand Coefficient[b]	
				High Import Dependence	*Low Import Dependence*
Iron and steel	—[c]	—[c]	—[c]	.497	.283
Molybdenum (MLB)	102	154	202.6	0	0
Nickel (TST)	260	24.1	100	.899	.606
Tungsten (MLB)	26.7	4.65	15.19	.807	.371
Manganese (MST)	1.68	0	negl[d]	NCI[d]	NCI[d]
Chromium (MST)	0.7	0	0	NCI[d]	NCI[d]
Copper (TST)	2,700	2,100	2,500	.054	0
Lead (TST)	1,200	772	872	.344	.259
Zinc (TST)	2,120	750	1,200	.642	.428
Gold (MTOZ)	9.3	1.4635	2.2135	.818	.722
Silver (MTOZ)	190	38.9	44.4	.731	.693
Aluminum (TST)	11,100	497.5	1,279	.945	.852
Mercury (FKS)	53,000	8,000	25,000	.845	.517
Vanadium (ST)	15,500	6,534	13,184	.550	.091
Platinum (TOZ)	805	2	5	.994	.986

Table F–3 continued

Mineral Name	Estimated Primary Demand	Low-End Estimate of Domestic Output	High-End Estimate of Domestic Output	Import to Primary Demand Coefficient[b]	
				High Import Dependence	Low Import Dependence
Titanium (TST)	906	325	453	.631	.486
Tin (LT)	58,000	0	negl[e]	NCI[d]	NCI[d]
Silicon (TST)	755	753	over 755	.002	.002
Fluorine (TST)	1,570	90.5	325.5	.942	.776
Potash (MST)	9	1.776	3.5	.735	.478
Soda ash (TST)	10.18	8.99	11.0	0	0
Boron (TST)	185	275	350	0	0
Phosphate rock (MST)	44.53	62.84	68.34	0	0
Sulfur (MLT)	14.5	12.8	16	0	0
Chlorine (MST)	18.9	15.284	19.0	.184	0
Magnesium (TST)	1,740	over 1,740	over 1,740	0	0

Source: U.S. Department of the Interior, Bureau of Mines, *Mineral Facts and Problems,* bulletin 667, Washington, D.C., 1975.

[a]Bureau of Mines Projections for 1985 are used to derive 1990 import coefficients.

[b]Included exogenously determined exported amount.

[c]See note b, table F–4.

[d]Noncompetitive imports.

[e]Negligible amount produced as by-product.

Unit: Molybdenum and tungsten are in million pounds (MLB); nickel, copper, lead, zinc, aluminum, titanium, silicon, fluorine, soda ash, boron, and magnesium are in thousand short tons (TST); manganese, chromium, potash, and chlorine are in million short tons (MST); gold and silver are in million troy ounces (MTOZ); mercury is in flasks (FKS); vanadium is in short tons (ST); platinum is in troy ounces (TOZ); tin is in million long tons (MLT); sulfur is in million long tons (MLT).

Table F–4
Derivation of Net Import to Primary Demand Coefficients of 26 Nonfuel Minerals for 2000

				Import to Primary Demand Coefficient[a]	
Mineral Name	*Estimated Primary Demand*	*Low-End Estimate of Domestic Output*	*High-End Estimate of Domestic Output*	*High Import Dependence*	*Low Import Dependence*
Iron and steel	—[b]	—[b]	—[b]	.579	.210
Molybdenum (MLB)	193	208	383.4	0	0
Nickel (TST)	385	32.6	270	.912	.287
Tungsten (MLB)	49.4	2.03	23	.956	.501
Manganese (MST)	2.13	0	negl[c]	NCI[d]	NCI[d]
Chromium (MST)	1.1	0	0	NCI[d]	NCI[d]
Copper (TST)	4,200	3,300	4,200	.051	0
Lead (TST)	1,530	880	1,080	.411	.277
Zinc (TST)	3,050	1,000	1,900	.669	.371
Gold (MTOZ)	15.3	1.8	3.3	.862	.747
Silver (MTOZ)	230	44	55	.749	.699
Aluminum (TST)	20,960	537	2,100	.970	.880
Mercury (FKS)	47,000	15,000	30,000	.670	.341
Vanadium (ST)	33,000	8,700	21,000	.718	.319
Platinum (TOZ)	1,225	0	6	NCI[d]	.989

Table F–4 continued

Mineral Name	Estimated Primary Demand	Low-End Estimate of Domestic Output	High-End Estimate of Domestic Output	Import to Primary Demand Coefficient[a]	
				High Import Dependence	Low Import Dependence
Titanium (TST)	1,425	444	710	.680	.488
Tin (LT)	64,000	0	negl[c]	NCI[d]	NCI[d]
Silicon (TST)	1,200	981	over 1,200	.159	0
Fluorine (TST)	1,940	80	550	.958	.716
Potash (MST)	12	1	4.1	.886	.531
Soda ash (TST)	15.6	10.96	17.0	.195	0
Boron (TST)	340	380	500	0	0
Phosphate rock (MST)	69	80	91	0	0
Sulfur (MLT)	23	16.2	over 23	.030	0
Chlorine (MST)	39.5	over 39.5	over 39.5	.449	0
Magnesium (TST)	2,915	over 2,915	over 2,915	0	0

Source: U.S. Department of the Interior, Bureau of Mines, *Mineral Facts and Problems,* bulletin 667, Washington, D.C., 1975.

[a]Includes exogenously determined exported amount.

[b]Bureau of Mines' estimates suggest that by 2000, imports could constitute 60 percent of total U.S. supply (either as ore or as finished steel). It is possible, though unlikely, that the U.S. will become self-sufficient in steel-making by 2000, as it was prior to 1959, but continuing to import about 25 percent of iron ore requirements. After adjusting for exports, the net import to primary demand coefficient for 2000 is 58 percent under high import dependence and is 21 percent under low import dependence. The 1990 import coefficients are interpolated from the base year, 1972 value and the coefficients obtained for 2000. See also table F–5.

[c]Negligible amount produced as by-product.

[d]Noncompetitive imports.

Unit: Molybdenum and tungsten are in million pounds (MLB); nickel, copper, lead, zinc, aluminum, titanium, silicon, fluorine, soda ash, boron, and magnesium are in thousand short tons (TST); manganese, chromium, potash, and chlorine are in million short tons (MST); gold and silver are in million troy ounces (MTOZ); mercury is in flasks (FKS); vanadium is in short tons (ST); platinum is in troy ounces (TOZ); tin is in thousand long tons (TLT); sulfur is in million long tons (MLT).

States could be self-sufficient in steel-making capacity by 2000, according to Bureu of Mines officials, although continuing to import 25 percent (or adjusted net import share of 21 percent) of its iron ore requirements.

A similar methodology was used to derive the shares for the other nonfuel minerals included in this study based on the high and low projections by the Bureau of Mines. The 1972 shares are used for the 1980 projections and the 1990 shares are interpolations of the 1972 and 2000 shares.

Tables F-5 and F-6 incorporate the information in tables F-3 and F-4 with respect to the relevant extraction sectors valued in dollars (EXT 1-2) and the metals-processing sectors (MET 1-3 and MET 10-13). As a result, the metal content of imports at the mining and processing stages of production is consistent with the shares presented in tables F-3 and F-4.

EXT3 includes all nonferrous metals except copper. To determine the import to primary demand coefficient for this sector would require calculating weights for each of the metals involved.

As a result, only the metals-processing sector for the major nonferrous metals—lead, zinc, and aluminum; MET11, MET12, and MET13, respectively—are considered. The import coefficients for these three sectors are summarized in table F-6.

The 1980 import coefficients are figures obtained from BLS projections used in the model. The coefficients for (low) high import dependence for both years 1990 and 2000 are calculated according to the following method:

$$\text{Coefficient} = \frac{\dfrac{\text{1972 imported metal}}{\text{1972 imported metal} + \text{ores}} \times \begin{array}{l}\text{Import requirement under} \\ \text{assumption of (low)} \\ \text{high domestic production}\end{array}}{\text{Estimated primary demand}}$$

Table F–5
Derivation of Import to Domestic Own Industry Mine Production for EXT1, MET1, MET2, MET3, MIN1, EXT2, MET10, and MIN7 for 1980, 1990, and 2000

			1990[b]		2000[c]	
IEA Sectors	*1972 Baseline*	*1980*[a]	*Low Import Dependence*	*High Import Dependence*	*Low Import Dependence*	*High Import Dependence*
EXT1	.478	.478	.320	.350	.260	.250
MET1[d]	.123	.123	.210	.250	.070	0
MET2[d]	.269	.269	.210	.250	.070	0
MET3[d]	.301	.301	.210	.250	.070	0
MIN1[e]	.825	.825	.990	1.13	.620	.468
EXT2	.053	.053	.140	.214	0	0
MET10[f]	.072	.072	.080	0	.074	0
MIN7[e]	.243	.243	.289	.275	.081	0

Source: U.S. Department of Interior, Bureau of Mines, *Mineral Facts and Problems,* bulletin 667, Washington, D.C., 1975.

[a]Import coefficients for the base-year apply to 1980.

[b]Coefficient for year 1990 are derived by interpolating the base-year's and 2000 estimated imports.

[c]Year 2000 import coefficients for EXT1, MET1, MET2, MET3, and MIN1 are estimated from information contained in *Mineral Facts and Problems,* 1975 ed., bulletin 667, Bureau of Mines, U.S. Department of the Interior, 1976, pp. 553, 574–575. For EXT2, MET10, and MIN7, import coefficients are estimated from information contained in the same source, pp. 308–309.

[d]MET1, MET2, and MET3 are the relevant metal-processing sectors with respect to EXT1 (see appendix A for sector names).

[e]See also tables F–3 and F–4 for import to primary demand coefficients.

[f]MET10 is the relevant metal-processing sector for EXT2 (see appendix A for sector names).

Table F-6
Derivation of Import to Primary Demand Coefficients for the Extraction and Metal-Processing Industries of Nonferrous Metals Other than Copper for 1980, 1990, and 2000

		1990		*2000*	
	1980	*High Import Dependence*	*Low Import Dependence*	*High Import Dependence*	*Low Import Dependence*
MET11	.081	.246	.163	.315	.184
MET12	.273	.436	.222	.461	.267
MET13	.068	.150	.055	.200	.060

Source: Figures were obtained from *Mineral Facts and Problems,* for 1972 imported metals and ores. Estimated high import requirements and primary demand are taken from tables F-3 and F-4.

References

Battelle Memorial Institute, Columbus Laboratories, *A Study to Identify Opportunities for Increased Solid Waste Utilization,* vols. 2-7, Columbus, Ohio, 1972, pp. 28a, 213, 249.

National Academy of Sciences, National Materials Advisory Board, *Contingency Plans for Chromium Utilization,* Washington, D.C., 1978.

———, *Trends in the Use of Magnesium,* Washington, D.C., 1975, pp. 67-68.

———, *Trends in the Use of Tin,* Washington, D.C., 1970, pp. 7-9.

U.S. Congress, Office of Technology Assessment, *Technical Options for Conservation of Scarce Metals,* Washington, D.C., September, 1979.

U.S. Department of the Interior, Bureau of Mines, *Minerals in the U.S. Economy: Ten Year Supply-Demand Profiles for Non-fuel Mineral Commodities* (1968-1977), Washington, D.C., 1978.

Appendix G Total Requirements Matrix of Minerals for 1972 and 2000 Based on IEA/USMIN

The two sets of tables reproduced in this appendix present the minerals portion of the 1972 and 2000 inverse matrices. These matrices were used to compute the minerals requirements by final demand component in chapter 6.

Each element in these matrices represents the total (direct and indirect) mineral requirements needed to deliver to final demand a billion dollars of its column's output.

The minerals are abbreviated in the two tables as follows:

MIN 1 T	Iron	MIN 28 T	Sulfur
MIN 2 T	Molybdenum	CHM 1 T	Chlorine
MIN 3 T	Nickel	CHM 2 T	Magnesium
MIN 4 T	Tungsten	SCR 1 T	Iron and steel scrap
MIN 7 T	Copper	SCR 2 T	Copper scrap
MIN 8 T	Lead	SCR 3 T	Lead scrap
MIN 9 T	Zinc	SCR 4 T	Zinc scrap
MIN 10 T	Gold	SCR 5 T	Aluminum scrap
MIN 11 T	Silver	SCR 6 T	Nickel scrap
MIN 12 T	Aluminum	SCR 7 T	Chromium scrap
MIN 13 T	Mercury	SCR 8 K	Gold scrap
MIN 14 T	Vanadium	SCR 9 T	Silver scrap
MIN 15 K	Platinum	SCR 10 T	Tungsten scrap
MIN 16 T	Titanium	SCR 11 T	Mercury scrap
MIN 21 T	Silicon	SCR 12 T	Tin scrap
MIN 23 T	Fluorine	SCR 13 T	Magnesium scrap
MIN 24 T	Potash	IMM 5 T	Manganese
MIN 25 T	Soda ash	IMM 6 T	Chromium
MIN 26 T	Boron	IMM 17 T	Tin
MIN 27 T	Phosphate rock		

Table G-1
Inverse Matrix ("Minerals Portion"):1972

	AGR 1 $	AGR 2 $	AGR 3 $	AGR 4 $	AGR 5 $	EXT 1 $	EXT 2 $	EXT 3 $	EXT 4 $	EXT 5 $
MIN 1 T	.224D-01	.150D-01	.173D-01	.212D-01	.202D-01	.287D-01	.191D-01	.318D-01	.374D-01	.290D-01
MIN 2 T	.111D-01	.167D-01	.186D-01	.166D-01	.148D-01	.134D-01	.993D-02	.125D-01	.141D-01	.112D-01
MIN 3 T	.386D-01	.359D-01	.415D-01	.452D-01	.544D-01	.106	.771D-01	.108	.119	.973
MIN 4 T	.860D-03	.106D-02	.997D-03	.104D-02	.138D-02	.541D-02	.498D-02	.769D-02	.866D-02	.622D-02
MIN 7 T	.266D-03	.417D-03	.337D-03	.294D-03	.404D-03	.513D-03	.350D-03	.580D-03	.547D-03	.465D-03
MIN 8 T	.214D-03	.226D-03	.280D-03	.245D-03	.297D-03	.291D-03	.161D-03	.190D-03	.322D-03	.233D-03
MIN 9 T	.240D-03	.406D-03	.300D-03	.279D-03	.330D-03	.392D-03	.329D-03	.241D-03	.276D-03	.191D-03
MIN10 T	.159D-01	.127D-01	.127D-01	.230D-01	.364D-01	.420D-01	.146D-01	.911D-01	.306D-01	.808D-01
MIN11 T	.523D-03	.778D-03	.610D-03	.609D-03	.105D-02	.829D-03	.419D-03	.134D-02	.663D-03	.115D-02
MIN12 T	.181D-02	.241D-02	.181D-02	.192D-02	.183D-02	.179D-02	.192D-02	.156D-02	.157D-02	.130D-02
MIN13 T	.835D-03	.157D-02	.133D-02	.119D-02	.107D-02	.939D-03	.124D-02	.124D-02	.679D-03	.866D-03
MIN14 T	.134D-02	.288D-02	.283D-02	.195D-02	.223D-02	.344D-02	.244D-02	.329D-02	.385D-02	.238D-02
MIN15 K	.167D-01	.489D-01	.287D-01	.237D-01	.261D-01	.182D-01	.140D-01	.177D-01	.179D-01	.132D-01
MIN16 T	.328	.436	.343	.377	.471	.364	.268	.293	.299	.262
MIN21 T	.149	.375	.216	.188	.228	.258	.171	.264	.292	.216
MIN23 T	.551D-03	.125D-02	.852D-03	.742D-03	.830D-03	.864D-03	.613D-03	.870D-03	.830D-03	.717D-03
MIN24 T	.366D-01	.536D-01	.985D-01	.105	.391D-01	.298D-02	.433D-02	.191D-02	.371D-02	.101D-02
MIN25 T	.724D-02	.124D-01	.833D-02	.690D-02	.867D-02	.590D-02	.445D-02	.561D-02	.503D-02	.382D-02
MIN26 T	.130	.141	.257	.276	.119	.272D-01	.267D-01	.221D-01	.233D-01	.163D-01
MIN27 T	.210	.306	.564	.603	.224	.164D-01	.243D-01	.104D-01	.207D-01	.539D-02
MIN28 T	.636D-01	.936D-01	.169	.180	.689D-01	.739D-02	.935D-02	.524D-02	.860D-02	.356D-02
CHM 1 T	.126D-01	.304D-01	.204D-01	.168D-01	.205D-01	.142D-01	.110D-01	.130D-01	.124D-01	.984D-02
CHM 2 T	.376D-03	.745D-03	.553D-03	.434D-03	.605D-03	.733D-03	.917D-03	.402D-03	.359D-02	.427D-03
SCR 1 T	.967D-02	.648D-02	.747D-02	.916D-02	.872D-02	.124D-01	.822D-02	.137D-01	.161D-01	.125D-01
SCR 2 T	.133D-03	.151D-03	.143D-03	.131D-03	.195D-03	.295D-03	.198D-03	.344D-03	.325D-03	.282D-03
SCR 3 T	.139D-03	.147D-03	.182D-03	.159D-03	.193D-03	.189D-03	.194D-03	.123D-03	.209D-03	.151D-03
SCR 4 T	.149D-03	.350D-03	.237D-03	.199D-03	.230D-03	.198D-03	.155D-03	.154D-03	.157D-03	.111D-03
SCR 5 T	.215D-03	.122D-03	.120D-03	.183D-03	.160D-03	.152D-03	.782D-04	.112D-03	.133D-03	.105D-03
SCR 6 T	.160D-01	.146D-01	.166D-01	.184D-01	.227D-01	.448D-01	.325D-01	.454D-01	.502D-01	.234
SCR 7 T	.228D-01	.172D-01	.194D-01	.232D-01	.275D-01	.525D-01	.408D-01	.624D-01	.613D-01	.128
SCR 8 K	.177D-02	.141D-02	.141D-02	.255D-02	.404D-02	.467D-02	.162D-02	.101D-01	.340D-02	.898D-02
SCR 9 T	.269D-03	.400D-03	.314D-03	.313D-03	.543D-03	.427D-03	.216D-03	.700D-03	.341D-03	.593D-03
SCR10 T	.372D-04	.459D-04	.431D-04	.448D-04	.599D-04	.234D-03	.215D-03	.333D-03	.374D-03	.269D-03
SCR11 T	.249D-03	.467D-03	.413D-03	.353D-03	.319D-03	.280D-03	.368D-03	.369D-03	.202D-03	.258D-03
SCR12 T	.117D-01	.911D-02	.745D-02	.105D-01	.771D-02	.651D-02	.507D-02	.639D-02	.685D-02	.515D-02
SCR13 T	.531D-05	.106D-04	.783D-05	.613D-05	.856D-05	.104D-04	.129D-04	.113D-04	.511D-04	.605D-05
IMM 5 T	.443	.603	.494	.482	.497	.537	.380	.486	.596	.455
IMM 6 T	.157	.169	.187	.186	.166	.324	.325	.396	.335	.691
IMM17 T	.369D-01	.288D-01	.235D-01	.332D-01	.243D-01	.206D-01	.150D-01	.202D-01	.216D-01	.163D-01

	EXT 6 $	EXT 7 $	CON 1 $	CON 2 $	MAN 1 $	MAN 2 $	MAN 3 $	MAN 4 $	MAN 5 $	MAN 6 $
MIN 1 T	.304D-01	.313	.221	.246D-01	.379D-01	.603D-01	.773D-02	.182D-01	.181D-01	.137D-01
MIN 2 T	.101D-01	.216	.141D-01	.136D-01	.249D-01	.899D-02	.579D-02	.996D-02	.100D-01	.588D-02
MIN 3 T	.178	.389	.313	.127	.345	.468D-01	.259D-01	.420D-01	.417D-01	.364D-01
MIN 4 T	.112D-03	.210D-02	.351D-02	.334D-02	.623D-02	.100D-02	.503D-03	.966D-03	.105D-02	.613D-03
MIN 7 T	.529D-03	.362D-03	.329D-02	.902D-03	.102D-01	.220D-03	.130D-03	.324D-03	.322D-03	.236D-03
MIN 8 T	.174D-03	.105D-03	.882D-03	.790D-03	.812D-02	.192D-03	.117D-03	.157D-03	.174D-03	.115D-03
MIN 9 T	.212D-03	.969D-04	.406D-02	.770D-03	.568D-03	.254D-03	.186D-03	.277D-03	.429D-03	.179D-03
MIN10 T	.263D-01	.143D-01	.578D-01	.188D-01	.257	.187D-01	.149D-01	.292D-01	.331D-01	.256
MIN11 T	.491D-03	.341D-03	.119D-02	.188D-02	.224D-02	.518D-03	.382D-03	.777D-03	.933D-03	.303D-02
MIN12 T	.108D-02	.104D-02	.124D-01	.522D-02	.431D-01	.454D-02	.619D-03	.145D-02	.143D-02	.776D-03
MIN13 T	.611D-03	.491D-03	.167D-02	.362D-02	.815D-03	.763D-03	.413D-03	.330D-02	.321D-02	.151D-02
MIN14 T	.406D-02	.106D-02	.139D-01	.124D-02	.452D-02	.122D-02	.702D-03	.208D-02	.214D-02	.103D-02
MIN15 K	.140D-01	.489D-02	.138D-01	.130D-01	.166D-01	.130D-01	.703D-02	.180D-01	.182D-01	.968D-02
MIN16 T	.261	.225	.858	2.61	.311	.502	.287	1.31	1.38	.706
MIN21 T	.235	1.71	.888	.188	.299	.114	.653D-01	.173	.172	.916D-01
MIN23 T	.687D-03	.287D-03	.109D-02	.663D-03	.146D-02	.637D-03	.246D-03	.609D-03	.623D-03	.314D-03
MIN24 T	.609D-02	.423D-03	.201D-02	.198D-02	.940D-03	.194D-01	.230D-01	.757D-02	.486D-02	.332D-02
MIN25 T	.410D-02	.163D-02	.480D-02	.436D-02	.410D-02	.168D-01	.260D-02	.954D-02	.995D-02	.419D-02
MIN26 T	.284D-01	.134D-01	.731D-01	.100	.267D-01	.210	.672D-01	.763D-01	.725D-01	.344D-01
MIN27 T	.344D-01	.270D-02	.113D-01	.596D-02	.514D-02	.121	.132	.433D-01	.282D-01	.192D-01
MIN28 T	.161D-01	.127D-02	.643D-02	.488D-02	.522D-02	.359D-01	.404D-01	.305D-01	.252D-01	.138D-01
CAM 1 T	.163D-01	.342D-02	.799D-02	.738D-02	.856D-02	.114D-01	.715D-02	.577D-01	.552D-01	.259D-01
CMM 2 T	.515D-03	.384D-03	.461D-02	.444D-02	.418D-03	.319D-03	.168D-03	.420D-03	.421D-03	.243D-03
SCR 1 T	.131D-01	.135	.953D-01	.123D-01	.163D-01	.260D-01	.333D-02	.785D-02	.780D-02	.593D-02
SCR 2 T	.322D-03	.235D-03	.722D-02	.593D-03	.692D-02	.115D-03	.683D-04	.166D-03	.164D-03	.137D-03
SCR 3 T	.113D-03	.684D-04	.572D-03	.512D-03	.527D-02	.125D-03	.761D-04	.102D-03	.113D-03	.748D-04
SCR 4 T	.126D-03	.472D-04	.100D-02	.225D-03	.193D-03	.128D-03	.830D-04	.170D-03	.207D-03	.884D-04
SCR 5 T	.776D-04	.725D-04	.219D-02	.845D-03	.822D-02	.767D-03	.610D-04	.126D-03	.120D-03	.780D-04
SCR 6 T	.750D-01	.164	.132	.538D-01	.146	.196D-01	.108D-01	.175D-01	.174D-01	.152D-01
SCR 7 T	.142	.132	.146	.630D-01	.256	.426D-01	.126D-01	.246D-01	.248D-01	.245D-01
SCR 8 K	.248D-02	.203D-02	.619D-02	.877D-02	.297D-01	.208D-02	.166D-02	.324D-02	.368D-02	.285D-01
SCR 9 T	.253D-03	.175D-03	.614D-03	.967D-03	.116D-02	.256D-03	.197D-03	.400D-03	.480D-03	.156D-02
SCR10 T	.483D-03	.908D-04	.152D-03	.144D-03	.269D-03	.434D-04	.217D-04	.417D-04	.454D-04	.265D-04
SCR11 T	.182D-03	.146D-03	.497D-03	.108D-02	.243D-03	.227D-03	.123D-03	.984D-03	.956D-03	.449D-03
SCR12 T	.694D-02	.256D-02	.245D-01	.825D-02	.875D-02	.446D-01	.323D-02	.613D-02	.611D-02	.331D-02
SCR13 T	.729D-05	.544D-05	.653D-04	.629D-04	.585D-05	.446D-05	.237D-05	.591D-05	.591D-05	.340D-05
IMM 5 T	.517	4.07	2.98	.496	.527	.893	.159	.401	.399	.218
IMM 6 T	.682	.439	.828	.373	1.14	.275	.847D-01	.637	.497	.389
IMM17 T	.219D-01	.809D-02	.774D-01	.261D-01	.276D-01	.141	.102D-01	.194D-01	.193D-01	.104D-01

Table G–1 continued

	MAN 7 $	MAN 8 $	MAN 9 $	MAN10 $	MAN11 $	MAN12 $	MAN13 $	MAN14 $	CHM 1 $	CHM 2 $
MIN 1 T	.144D-01	.177D-01	.137D-01	.132D-01	.139D-01	.156D-01	.146D-01	.936D-02	.451D-01	.374D-01
MIN 2 T	.766D-02	.166D-01	.109D-01	.143D-01	.150D-01	.127D-01	.153D-01	.795D-02	.228D-01	.111
MIN 3 T	.382D-01	.131	.103	.294	.352	.591D-01	.514D-01	.324D-01	.898D-01	.102
MIN 4 T	.861D-03	.174D-02	.146D-02	.160D-02	.265D-02	.129D-02	.234D-02	.827D-03	.149D-02	.135D-02
MIN 7 T	.257D-03	.335D-03	.245D-03	.275D-03	.323D-03	.310D-03	.381D-03	.311D-03	.352D-03	.258D-03
MIN 8 T	.137D-03	.342D-03	.228D-03	.319D-03	.277D-03	.231D-03	.197D-03	.244D-02	.243D-03	.191D-03
MIN 9 T	.344D-03	.345D-03	.197D-03	.591D-03	.553D-03	.397D-03	.386D-03	.306D-03	.244D-03	.171D-03
MIN10 T	.898D-01	.317D-01	.364D-01	.662D-01	.548D-01	.280D-01	.336D-01	.678D-01	.289D-01	.240D-01
MIN11 T	.138D-02	.695D-03	.404D-03	.100D-02	.814D-03	.700D-03	.904D-03	.247D-02	.609D-03	.493D-03
MIN12 T	.111D-02	.151D-02	.888D-03	.100D-02	.101D-02	.156D-02	.200D-02	.139D-02	.183D-02	.205D-02
MIN13 T	.191D-02	.137D-02	.753D-03	.183D-02	.148D-02	.174D-02	.146D-02	.939D-03	.222D-01	.274D-02
MIN14 T	.149D-02	.207D-02	.137D-02	.139D-02	.163D-02	.211D-02	.307D-02	.187D-02	.214D-02	.151D-02
MIN15 K	.135D-01	.193D-01	.103D-01	.113D-01	.858D-02	.231D-01	.323D-01	.219D-01	.134D-01	.883D-02
MIN16 T	2.72	.855	.444	1.40	1.13	7.60	3.36	1.49	.466	.361
MIN21 T	.127	.175	.114	.108	.108	.183	.231	.152	.257	.148
MIN23 T	.492D-03	.773D-03	.718D-03	.939D-03	.183D-02	.797D-03	.121D-02	.787D-03	.682D-03	.511D-03
MIN24 T	.422D-02	.140D-01	.606D-02	.336D-02	.193D-02	.427D-02	.303D-02	.186D-02	.115D-01	1.78
MIN25 T	[illegible]	[illegible]	[illegible]	[illegible]	.872D-02	[illegible]	[illegible]	[illegible]	[illegible]	.309D-02
MIN26 T	.490D-01	.870D-01	.456D-01	.995D-01	.106	.313D-01	.198D-01	.143D-01	.484D-01	4.52
MIN27 T	.241D-01	.797D-01	.346D-01	.191D-01	.109D-01	.239D-01	.162D-01	.988D-02	.657D-01	10.2
MIN28 T	.171D-01	.263D-01	.121D-01	.102D-01	.717D-02	.396D-01	.207D-01	.100D-01	.226D-01	3.01
CHM 1 T	.337D-01	.144D-01	.731D-02	.146D-01	.992D-02	.103	.612D-01	.310D-01	.112D-01	.843D-02
CHM 2 T	.327D-03	.800D-03	.440D-03	.557D-03	.459D-03	.626D-03	.682D-03	.471D-03	.481D-03	.359D-03
SCR 1 T	.620D-02	.761D-02	.589D-02	.568D-02	.601D-02	.672D-02	.630D-02	.403D-02	.194D-01	.161D-01
SCR 2 T	.135D-03	.174D-03	.141D-03	.158D-03	.199D-03	.141D-03	.155D-03	.141D-03	.205D-03	.155D-03
SCR 3 T	.890D-04	.222D-03	.148D-03	.207D-03	.180D-03	.150D-03	.128D-03	.158D-02	.157D-03	.124D-03
SCR 4 T	.157D-03	.185D-03	.963D-04	.196D-03	.171D-03	.227D-03	.291D-03	.208D-03	.124D-03	.817D-04
SCR 5 T	.100D-03	.129D-03	.862D-04	.986D-04	.880D-04	.109D-03	.107D-03	.789D-04	.224D-03	.315D-03
SCR 6 T	.160D-01	.555D-01	.433D-01	.124	.149	.248D-01	.216D-01	.137D-01	.360D-01	.332D-01
SCR 7 T	.232D-01	.585D-01	.471D-01	.167	.196	.288D-01	.257D-01	.188D-01	.522D-01	.397D-01
SCR 8 K	.998D-02	.352D-02	.404D-02	.735D-02	.609D-02	.311D-02	.373D-02	.753D-02	.321D-02	.266D-02
SCR 9 T	.710D-03	.358D-03	.311D-03	.515D-03	.419D-03	.360D-03	.465D-03	.127D-02	.313D-03	.254D-03
SCR10 T	.372D-04	.768D-04	.630D-04	.693D-04	.114D-03	.558D-04	.101D-03	.358D-04	.643D-04	.583D-04
SCR11 T	.569D-03	.407D-03	.224D-03	.546D-03	.440D-03	.518D-03	.433D-03	.280D-03	.660D-02	.816D-03
SCR12 T	.471D-02	.660D-02	.371D-02	.576D-02	.503D-02	.623D-02	.747D-02	.510D-02	.106D-01	.167D-01
SCR13 T	.460D-05	.113D-04	.619D-05	.777D-05	.628D-05	.883D-05	.961D-05	.663D-05	.651D-05	.497D-05
IMM 5 T	.299	.387	.245	.247	.242	.409	.499	.327	.695	.552
IMM 6 T	.435	.599	.424	.895	1.17	.305	.227	.142	3.54	.546
IMM17 T	.149D-01	.208D-01	.117D-01	.182D-01	.159D-01	.197D-01	.236D-01	.161D-01	.335D-01	.526D-01

	CHM 3 $	MAN15 $	MAN16 $	MAN17 $	MAN18 $	MAN19 $	MAN20 $	MAN21 $	MAN22 $	MAN23 $
MIN 1 T	.533D-01	.288D-01	.454D-01	.970D-01	.175	.170D-01	.361D-01	.176D-01	.145D-01	.205D-01
MIN 2 T	.229	.159D-01	.102D-01	.150D-01	.159	.158D-01	.112D-01	.930D-02	.105D-01	.114D-01
MIN 3 T	.167	.579D-01	.523D-01	.735D-01	.217	.682D-01	.394D-01	.634D-01	.418D-01	.606D-01
MIN 4 T	.156D-01	.147D-02	.119D-02	.140D-02	.542D-02	.320D-02	.999D-03	.105D-02	.240D-02	.279D-02
MIN 7 T	.839D-02	.357D-03	.290D-03	.314D-03	.316D-03	.353D-03	.299D-03	.306D-03	.288D-03	.345D-03
MIN 8 T	.336D-03	.226D-03	.190D-03	.153D-01	.597D-02	.166D-03	.178D-03	.133D-03	.176D-03	.263D-03
MIN 9 T	.952D-02	.380D-03	.503D-03	.916D-02	.158D-03	.620D-02	.317D-03	.645D-03	.427D-03	.203D-03
MIN10 T	.263D-01	.264D-01	.488D-01	.488D-01	.219D-01	.472D-01	.292D-01	.190	.429D-01	.726D-01
MIN11 T	.154D-01	.641D-03	.101D-02	.753D-03	.487D-03	.873D-03	.794D-03	.245D-02	.651D-03	.103D-02
MIN12 T	.621D-01	.206D-02	.372D-02	.719D-02	.204D-02	.152D-02	.337D-02	.159D-02	.252D-02	.112D-02
MIN13 T	.272D-01	.976D-02	.321D-02	.817D-01	.112D-02	.310D-02	.223D-02	.136D-02	.138D-02	.126D-02
MIN14 T	.810D-01	.254D-02	.168D-02	.203D-02	.124D-02	.265D-02	.206D-02	.179D-02	.175D-02	.164D-02
MIN15 K	1.19	.231D-01	.192D-01	.224D-01	.112	.209D-01	.264D-01	.189D-01	.564D-01	.129D-01
MIN16 T	9.60	5.20	.711	78.9	.332	2.11	.479	.674	2.97	.425
MIN21 T	6.41	.228	.147	.196	.913	.181	.187	.146	.120	.151
MIN23 T	.372D-01	.855D-03	.808D-03	.129D-02	.457D-03	.861D-03	.943D-03	.699D-03	.544D-03	.435D-03
MIN24 T	.569D-01	.916D-02	.276D-02	.533D-02	.101D-02	.346D-02	.108D-01	.354D-02	.132D-02	.137D-02
MIN25 T	.381	.833D-02	.381D-01	.802D-02	.318D-02	.102D-01	.153D-01	.794D-02	.705	.345D-02
MIN26 T	.621D-01	.482D-01	1.15	.224	.200D-01	.621D-01	.149	.501D-01	8.15	.207D-01
MIN27 T	.272	.522D-01	.991D-01	.303D-01	.582D-02	.192D-01	.679D-01	.211D-01	.728D-02	.758D-02
MIN28 T	.167	.983D-01	.806D-02	.398D-01	.135D-01	.204D-01	.210D-01	.109D-01	.503D-02	.441D-02
CHM 1 T	.930	.239	.177D-01	.378D-01	.737D-02	.500D-01	.210D-01	.215D-01	.110D-01	.994D-02
CHM 2 T	.182D-01	.548D-03	.435D-03	.680D-03	.625D-03	.526D-03	.627D-03	.411D-03	.825D-03	.371
SCR 1 T	.230D-01	.124D-01	.196D-01	.418D-01	.752D-01	.734D-02	.156D-01	.759D-02	.623D-02	.883D-02
SCR 2 T	.149D-02	.172D-03	.141D-03	.148D-03	.189D-03	.175D-03	.120D-03	.151D-03	.170D-03	.207D-03
SCR 3 T	.218D-03	.146D-03	.123D-03	.995D-02	.367D-02	.108D-03	.116D-03	.863D-04	.114D-03	.171D-03
SCR 4 T	.103D-01	.226D-03	.227D-03	.228D-02	.881D-04	.158D-02	.234D-03	.262D-03	.152D-03	.103D-03
SCR 5 T	.138D-02	.195D-03	.553D-03	.119D-02	.266D-03	.109D-03	.426D-03	.146D-03	.386D-03	.124D-03
SCR 6 T	.701D-01	.236D-01	.219D-01	.306D-01	.916D-01	.284D-01	.163D-01	.267D-01	.175D-01	.255D-01
SCR 7 T	.465D-01	.360D-01	.425D-01	.722D-01	.853D-01	.387D-01	.291D-01	.397D-01	.296D-01	.324D-01
SCR 8 K	.292D-02	.294D-02	.542D-02	.542D-02	.244D-02	.524D-02	.324D-02	.211D-01	.476D-02	.806D-02
SCR 9 T	.816D-02	.356D-03	.519D-03	.490D-03	.251D-03	.449D-03	.409D-03	.126D-02	.335D-03	.528D-03
SCR10 T	.673D-03	.637D-04	.512D-04	.606D-04	.234D-03	.138D-03	.432D-04	.456D-04	.104D-03	.121D-03
SCR11 T	.811D-02	.291D-02	.957D-03	.243D-01	.335D-03	.925D-03	.663D-03	.404D-03	.412D-03	.374D-03
SCR12 T	.218	.109D-01	.329D-01	.702D-01	.137D-01	.679D-02	.259D-01	.852D-02	.250D-01	.397D-02
SCR13 T	.259D-03	.760D-05	.606D-05	.931D-05	.876D-05	.739D-05	.884D-05	.578D-05	.116D-04	.527D-02
IMM 5 T	13.4	.589	.758	1.46	2.34	.406	.717	.378	.260	.338
IMM 6 T	.783	1.17	.455	1.10	.395	.503	13.0	2.51	.915	2.15
IMM17 T	.688	.344D-01	.104	.222	.434D-01	.214D-01	.817D-01	.269D-01	.790D-01	.125D-01

Table G–1 continued

	MAN24 $	MAN25 $	MET 1 $	MET 2 $	MET 3 $	MET 4 $	MET 5 $	MET 6 $	MET 7 $	MET 8 $
MIN 1 T	.142D-01	.168D-01	.257D-01	.192D-01	.291D-01	.331D-01	.348D-01	.335D-01	.278D-01	.151D-01
MIN 2 T	.787D-02	.121D-01	.194D-01	.109D-01	.510D-01	.229D-01	.257D-01	.181D-01	.318D-01	.195D-01
MIN 3 T	.561D-01	.750D-01	1.10	.990D-01	.323	.624	.527	.647	.399	.992D-01
MIN 4 T	.171D-02	.234D-02	.600D-02	.298D-02	.178D-01	.585D-02	.730D-02	.583D-02	.111D-01	.181D-02
MIN 7 T	.801D-03	.354D-03	.683D-03	.961D-03	.745D-03	.112D-02	.117D-02	.963D-03	.592D-03	.566D-03
MIN 8 T	.179D-03	.247D-03	.209D-03	.206D-03	.177D-03	.192D-03	.164D-03	.161D-03	.207D-03	.337D-03
MIN 9 T	.144D-03	.245D-03	.180D-03	.194D-03	.306D-03	.184D-03	.212D-03	.186D-03	.194D-03	.625D-03
MIN10 T	.231D-01	.478D-01	.414D-01	.341D-01	.236D-01	.378D-01	.468D-01	.419D-01	.288D-01	.139D-01
MIN11 T	.227D-03	.368D-03	.594D-03	.589D-03	.697D-03	.574D-03	.637D-03	.768D-03	.504D-03	.101D-02
MIN12 T	.378	.436D-02	.152D-02	.286D-02	.217D-02	.261D-02	.291D-02	.221D-02	.108D-02	.354D-02
MIN13 T	.781D-03	.140D-02	.130D-02	.709D-03	.149D-02	.214D-02	.272D-02	.614D-03	.867D-03	.261D-02
MIN14 T	.122D-02	.205D-02	.320D-02	.179D-02	.801D-02	.406D-02	.504D-02	.470D-02	.455D-02	.472D-02
MIN15 K	.117D-01	.638D-01	.820D-02	.107D-01	.272D-01	.803D-02	.156D-01	.107D-01	.636D-02	.641D-01
MIN16 T	2.57	2.52	.384	3.34	.468	.370	.362	.325	.314	1.32
MIN21 T	.147	.177	.185	.196	.306	.256	.305	.273	.191	.394
MIN23 T	.182D-02	.956D-03	.118D-01	.211D-02	.374D-02	.578D-02	.515D-02	.141D-02	.398D-02	.329D-02
MIN24 T	.109D-02	.182D-02	.122D-02	.150D-02	.197D-02	.156D-02	.209D-02	.815D-03	.777D-03	.333D-02
MIN25 T	.254D-02	.697D-02	.642D-02	.323D-02	.852D-02	.236D-02	.444D-02	.354D-02	.217D-02	.199D-01
MIN26 T	7.57	.211	.284D-01	.215D-01	.226D-01	.421D-01	.334D-01	.255D-01	.277D-01	.130D-01
MIN27 T	.604D-02	.982D-02	.685D-02	.831D-02	.102D-01	.880D-02	.115D-01	.435D-02	.441D-02	.163D-01
MIN28 T	.378D-02	.645D-02	.113D-01	.131D-01	.144D-01	.141D-01	.131D-01	.910D-02	.110D-01	.142D-01
CHM 1 T	.736D-02	.181D-01	.565D-02	.834D-02	.209D-01	.581D-02	.117D-01	.790D-02	.470D-02	.492D-01
CHM 2 T	.374	.333D-01	.696D-03	.103D-02	.862D-03	.905D-03	.593D-03	.699D-03	.510D-03	.128D-02
SCR 1 T	.614D-02	.724D-02	.111D-01	.829D-02	.126D-01	.143D-01	.150D-01	.144D-01	.120D-01	.651D-02
SCR 2 T	.527D-03	.182D-03	.448D-03	.629D-03	.421D-03	.749D-03	.752D-03	.628D-03	.391D-03	.168D-03
SCR 3 T	.116D-03	.160D-03	.136D-03	.134D-03	.115D-03	.125D-03	.107D-03	.104D-03	.134D-03	.219D-03
SCR 4 T	.726D-04	.184D-03	.766D-04	.976D-04	.238D-03	.763D-04	.137D-03	.101D-03	.711D-04	.562D-03
SCR 5 T	.719D-01	.143D-02	.117D-03	.109D-03	.105D-03	.112D-03	.108D-03	.102D-03	.978D-04	.119D-03
SCR 6 T	.228D-01	.313D-01	.463	.418D-01	.136	.264	.222	.273	.169	.419D-01
SCR 7 T	.279D-01	.362D-01	.436	.106	.153	.285	.248	.207	.189	.826D-01
SCR 8 K	.279D-02	.531D-02	.459D-02	.378D-02	.318D-02	.420D-02	.520D-02	.465D-02	.331D-02	.155D-02
SCR 9 T	.220D-03	.467D-03	.306D-03	.303D-03	.359D-03	.296D-03	.328D-03	.365D-03	.259D-03	.518D-03
SCR10 T	.738D-04	.101D-03	.259D-03	.129D-03	.769D-03	.253D-03	.316D-03	.252D-03	.480D-03	.783D-04
SCR11 T	.232D-03	.417D-03	.386D-03	.211D-03	.442D-03	.636D-03	.809D-03	.183D-03	.258D-03	.776D-03
SCR12 T	.485D-02	.577D-02	.518D-02	.539D-02	.895D-02	.657D-02	.773D-02	.490D-02	.483D-02	.131D-01
SCR13 T	.531D-02	.472D-03	.980D-05	.127D-04	.118D-04	.119D-04	.806D-05	.957D-05	.649D-05	.169D-04
TAM 5 T	.326	.362	.404	.421	.663	.532	.626	.536	.385	.839
TAM 6 T	.224	.435	4.70	.424	1.34	2.63	2.33	1.97	1.59	.472
TAM 7 T	.153D-01	.182D-01	.164D-01	.170D-01	.282D-01	.207D-01	.244D-01	.155D-01	.152D-01	.414D-01

	MET 9 $	MET10 $	MET11 $	MET12 $	MET13 $	MET14 $	MET15 $	MET16 $	MET17 $	MET18 $
MIN 1 T	.351D-01	.173D-01	.199D-01	.227D-01	.208D-01	.180D-01	.164D-01	.179D-01	.212D-01	.264D-01
MIN 2 T	.349D-01	.977D-02	.122D-01	.132D-01	.152D-01	.110D-01	.862D-02	.175D-01	.224D-01	.259D-01
MIN 3 T	.163	.224	.209	.188	.189	.194	.713D-01	.141	.143	.182
MIN 4 T	.113D-01	.360D-02	.348D-02	.345D-02	.299D-02	.346D-02	.194D-02	.592D-02	.646D-02	.837D-02
MIN 7 T	.588D-03	.343D-03	.418D-03	.382D-03	.681D-03	.386D-03	.469D-03	.414D-03	.551D-03	.691D-03
MIN 8 T	.192D-03	.174D-03	.223D-03	.301D-03	.276D-03	.181D-03	.251D-02	.187D-03	.269D-03	.246D-03
MIN 9 T	.176D-03	.243D-03	.188D-03	.165D-03	.169D-03	.186D-03	.173D-03	.226D-03	.231D-03	.319D-03
MIN10 T	.286D-01	.162D-01	.317D-01	.377D-01	.276D-01	.269D-01	.182D-01	.229D-01	.322D-01	.103
MIN11 T	.622D-03	.401D-03	.574D-03	.644D-03	.506D-03	.510D-03	.412D-03	.428D-03	.554D-03	.129D-02
MIN12 T	.176D-02	.994D-03	.113D-02	.106D-02	.191D-02	.104D-02	.123D-02	.872D-03	.138D-02	.115D-02
MIN13 T	.200D-02	.100D-02	.121D-02	.103D-02	.180D-02	.103D-02	.104D-02	.104D-02	.157D-02	.175D-02
MIN14 T	.474D-02	.194D-02	.194D-02	.180D-02	.155D-02	.192D-02	.172D-02	.272D-02	.289D-02	.386D-02
MIN15 K	.799D-02	.111D-01	.103D-01	.119D-01	.948D-02	.945D-02	.853D-02	.766D-02	.845D-02	.113D-01
MIN16 T	.296	.230	.206	.207	.286	.221	.179	.393	.613	.469
MIN21 T	.189	.147	.158	.171	.168	.146	.143	.135	.161	.194
MIN23 T	.920D-02	.962D-03	.212D-02	.141D-02	.630D-01	.108D-01	.211D-02	.151D-02	.284D-01	.565D-02
MIN24 T	.167D-02	.277D-02	.124D-02	.122D-02	.136D-02	.155D-02	.108D-02	.157D-02	.114D-02	.157D-02
MIN25 T	.245D-02	.340D-02	.290D-02	.293D-02	.232D-02	.282D-02	.265D-02	.249D-02	.263D-02	.348D-02
MIN26 T	.270D-01	.209D-01	.143D-01	.146D-01	.167D-01	.157D-01	.124D-01	.211D-01	.228D-01	.265D-01
MIN27 T	.938D-02	.155D-01	.683D-02	.672D-02	.764D-02	.863D-02	.594D-02	.884D-02	.637D-02	.872D-02
MIN28 T	.185D-01	.204	.174D-01	.150D-01	.121D-01	.159	.202D-01	.849D-01	.128D-01	.763D-01
CHM 1 T	.685D-02	.842D-02	.718D-02	.729D-02	.668D-02	.714D-02	.669D-02	.651D-02	.691D-02	.880D-02
CHM 2 T	.547D-02	.820D-03	.753D-03	.203D-02	.221D-02	.416D-02	.740D-03	.614D-03	.124D-02	.208D-02
SCR 1 T	.151D-01	.747D-02	.858D-02	.478D-02	.827D-02	.776D-02	.786D-02	.770D-02	.915D-02	.114D-01
SCR 2 T	.382D-03	.204D-03	.259D-03	.234D-03	.446D-03	.238D-03	.296D-03	.264D-03	.357D-03	.443D-03
SCR 3 T	.124D-03	.118D-03	.145D-03	.125D-03	.178D-03	.117D-03	.163D-02	.121D-03	.174D-03	.168D-03
SCR 4 T	.767D-04	.115D-03	.947D-04	.887D-04	.759D-04	.918D-04	.873D-04	.894D-04	.916D-04	.130D-03
SCR 5 T	.207D-03	.690D-04	.691D-04	.775D-04	.835D-04	.658D-04	.538D-04	.610D-04	.765D-04	.856D-04
SCR 6 T	.688D-01	.948D-01	.883D-01	.795D-01	.800D-01	.818D-01	.300D-01	.594D-01	.603D-01	.768D-01
SCR 7 T	.161	.937D-01	.192	.165	.125	2.17	.201	.114	.125	.995
SCR 8 K	.318D-02	.183D-02	.352D-02	.419D-02	.306D-02	.299D-02	.202D-02	.255D-02	.357D-02	.115D-01
SCR 9 T	.320D-03	.206D-03	.295D-03	.332D-03	.261D-03	.262D-03	.212D-03	.220D-03	.285D-03	.664D-03
SCR10 T	.488D-03	.156D-03	.150D-03	.149D-03	.129D-03	.149D-03	.837D-04	.256D-03	.279D-03	.362D-03
SCR11 T	.596D-03	.299D-03	.361D-03	.306D-03	.536D-03	.305D-03	.309D-03	.310D-03	.468D-03	.521D-03
SCR12 T	.130D-01	.400D-02	.383D-02	.388D-02	.440D-02	.363D-02	.340D-02	.379D-02	.451D-02	.551D-02
SCR13 T	.724D-04	.101D-04	.662D-05	.257D-04	.692D-05	.769D-05	.602D-05	.660D-05	.572D-05	.695D-05
IMM 5 T	.550	.315	.315	.343	.370	.294	.277	.293	.356	.387
IMM 6 T	.705	.356	.564	.490	.496	3.58	.504	.373	.446	1.89
IMM17 T	.412D-01	.126D-01	.121D-01	.123D-01	.139D-01	.115D-01	.107D-01	.120D-01	.142D-01	.174D-01

Table G–1 continued

	METI9 $	METI20 $	METI21 $	METI22 $	METI23 $	MAN26 $	MAN27 $	MAN28 $	MAN29 $	MAN30 $
MIN 1 T	.201D-01	.305D-01	.267D-01	.420D-01	.211D-01	1.35	.226D-01	.287D-01	.210D-01	.369
MIN 2 T	.212D-01	.561D-01	.228D-01	.177	.334D-01	.202D-01	.226D-01	.299D-01	.112	.278D-01
MIN 3 T	.129	.126	4.75	4.77	.232	.427	1.54	.373	1.91	.792
MIN 4 T	.662D-02	.209D-01	.790D-02	.345D-01	.111D-01	.595D-02	.637D-02	.992D-02	.723D-02	.611D-01
MIN 7 T	.562D-03	.603D-03	.514D-03	.725D-03	.425D-03	.435D-03	.680D-03	.573D-03	.605D-03	.118D-02
MIN 8 T	.194D-03	.148D-03	.137D-03	.181D-03	.173D-03	.557D-03	.237D-02	.221D-03	.253D-03	.290D-03
MIN 9 T	.285D-03	.169D-03	.166D-03	.196D-03	.233D-03	.343D-03	.291D-03	.255D-03	.403D-03	.216D-03
MIN10 T	.305D-01	.260D-01	.197D-01	.256D-01	.257D-01	.315D-01	.487D-01	.314D-01	.499D-01	.394D-01
MIN11 T	.570D-03	.548D-03	.423D-03	.515D-03	.429D-03	.630D-03	.740D-03	.550D-03	.767D-03	.100D-02
MIN12 T	.124D-02	.123D-02	.107D-02	.127D-02	.864D-03	.987D-01	.171D-02	.134D-02	.171D-02	.133D-01
MIN13 T	.172D-02	.832D-03	.650D-03	.922D-03	.986D-03	.238D-02	.159D-02	.104D-02	.156D-02	.131D-02
MIN14 T	.317D-02	.788D-02	.461D-02	.121D-01	.419D-02	.289D-02	.383D-02	.413D-02	.363D-02	.513D-02
MIN15 K	.990D-02	.108D-01	.112D-01	.124D-01	.662D-02	.991D-02	.790D-02	.859D-02	.184D-01	.544D-02
MIN16 T	.710	.262	.262	.322	.346	1.88	1.02	.486	.928	.289
MIN21 T	.161	.229	.210	.300	.149	.150	.171	.198	.198	2.26
MIN23 T	.384D-02	.144D-01	.278D-02	.457D-02	.926D-02	.664D-02	.512D-02	.403D-02	.311D-02	.191D-02
MIN24 T	.193D-02	.964D-03	.128D-02	.102D-02	.954D-03	.115D-02	.931D-03	.914D-03	.138D-02	.524D-03
MIN25 T	.480D-02	.340D-02	.372D-02	.365D-02	.204D-02	.367D-02	.771D-02	.368D-02	.618D-02	.186D-02
MIN26 T	.438D-01	.190D-01	.215D-01	.262D-01	.227D-01	.303D-01	.171	.216	.294	.325D-01
MIN27 T	.188D-01	.362D-02	.653D-02	.556D-02	.539D-02	.641D-02	.528D-02	.512D-02	.743D-02	.294D-02
MIN28 T	.484D-01	.864D-02	.446D-01	.129D-01	.211D-01	.789D-02	.753D-02	.713D-02	.977D-02	.515D-02
CMM 1 T	.177D-01	.831D-02	.846D-02	.910D-02	.568D-02	.915D-02	.585D-02	.740D-02	.139D-01	.389D-02
CMM 2 T	.746D-03	.917D-03	.568D-03	.151D-02	.676D-03	.538D-03	.762D-03	.502D-03	.683D-03	.589D-03
SCR 1 T	.867D-02	.132D-01	.115D-01	.181D-01	.911D-02	.583	.974D-02	.124D-01	.907D-02	.159
SCR 2 T	.358D-03	.382D-03	.319D-03	.462D-03	.275D-03	.270D-03	.447D-03	.369D-03	.364D-03	.793D-03
SCR 3 T	.129D-03	.959D-04	.886D-04	.104D-03	.112D-03	.361D-03	.154D-02	.143D-03	.164D-03	.188D-03
SCR 4 T	.116D-03	.975D-04	.101D-03	.109D-03	.845D-04	.132D-03	.101D-03	.103D-03	.188D-03	.713D-04
SCR 5 T	.804D-04	.698D-04	.668D-04	.910D-04	.616D-04	.187D-01	.111D-03	.132D-03	.103D-03	.235D-02
SCR 6 T	.546D-01	.533D-01	2.01	2.02	.980D-01	.180	.653	.158	.806	.335
SCR 7 T	.987D-01	.900D-01	.684D-01	16.4	.168	.696	.711	.726	.693	.555
SCR 8 K	.338D-02	.289D-02	.218D-02	.284D-02	.286D-02	.350D-02	.541D-02	.349D-02	.554D-02	.437D-02
SCR 9 T	.293D-03	.282D-03	.218D-03	.265D-03	.221D-03	.325D-03	.381D-03	.283D-03	.395D-03	.515D-03
SCR10 T	.286D-03	.903D-03	.342D-03	.149D-02	.478D-03	.257D-03	.275D-03	.429D-03	.312D-03	.264D-02
SCR11 T	.511D-03	.248D-03	.193D-03	.275D-03	.294D-03	.710D-03	.474D-03	.308D-03	.463D-03	.389D-03
SCR12 T	.491D-02	.592D-02	.413D-02	.281D-01	.453D-02	1.11	.589D-02	.694D-02	.270D-01	.487D-02
SCR13 T	.821D-05	.758D-05	.111D-03	.109D-03	.485D-05	.635D-05	.104D-04	.701D-05	.945D-05	.797D-05
IMM 5 T	.356	.486	.431	.613	.312	17.7	.356	.391	.402	.544
IMM 6 T	.440	.339	.294	25.7	.836	3.77	3.46	3.51	2.98	2.01
IMM17 T	.155D-01	.187D-01	.130D-01	.887D-01	.143D-01	3.51	.186D-01	.219D-01	.852D-01	.154D-01

	MAN31 $	MAN32 $	MAN33 $	MAN34 $	MAN35 $	MAN36 $	MAN37 $	MAN38 $	MAN39 $	MAN40 $
MIN 1 T	.375	.370	.353	.337	.344	.338	.337	.356	.339	.216D-01
MIN 2 T	.241D-01	.286D-01	.230D-01	1.24	.297D-01	.285D-01	.394D-01	.216D-01	.186D-01	.215D-01
MIN 3 T	.839	.827	.880	.745	.765	.794	.777	.290	.296	.303
MIN 4 T	.111D-01	.185	.850D-02	.507	.100D-01	.100D-01	.134D-01	.436D-02	.546D-02	.701D-02
MIN 7 T	.512D-02	.509D-02	.548D-02	.499D-02	.523D-02	.524D-02	.472D-02	.810D-02	.690D-02	.226D-01
MIN 8 T	.311D-03	.261D-03	.181D-03	.160D-03	.203D-03	.182D-03	.202D-03	.484D-03	.221D-03	.274D-03
MIN 9 T	.449D-03	.320D-03	.261D-03	.138D-03	.210D-03	.178D-03	.157D-03	.638D-02	.140D-01	.274D-03
MIN10 T	.437D-01	.380D-01	.411D-01	.387D-01	.653D-01	.553D-01	.373D-01	1.04	.141	.249
MIN11 T	.978D-03	.627D-03	.670D-03	.511D-03	.716D-03	.584D-03	.752D-03	.540D-02	.350D-01	.169D-02
MIN12 T	.134D-01	.227D-02	.142D-01	.220D-02	.135D-01	.296D-02	.123D-02	.228D-02	.504D-01	.628D-01
MIN13 T	.136D-02	.792D-03	.784D-03	.666D-03	.957D-03	.628D-03	.901D-03	.932D-03	.173D-02	.144D-02
MIN14 T	.651D-02	.543D-01	.564D-02	.157	.518D-01	.513D-01	.523D-02	.186D-02	.304D-02	.288D-02
MIN15 K	.536D-02	.521D-02	.532D-02	.613D-02	.607D-02	.623D-02	.664D-02	.215D-01	.929D-02	.133D-01
MIN16 T	.488	.411	.317	.281	.438	.403	.276	.411	.590	.583
MIN21 T	2.28	2.24	2.17	2.05	2.11	2.08	2.04	3.67	2.19	1.69
MIN23 T	.178D-02	.159D-02	.198D-02	.155D-02	.168D-02	.182D-02	.172D-02	.850D-03	.194D-02	.197D-02
MIN24 T	.679D-03	.584D-03	.617D-03	.580D-03	.813D-03	.634D-03	.595D-03	.837D-03	.798D-03	.102D-02
MIN25 T	.210D-02	.180D-02	.175D-02	.194D-02	.306D-02	.184D-02	.183D-02	.546D-02	.311D-02	.433D-02
MIN26 T	.398D-01	.353D-01	.133D-01	.240D-01	.455D-01	.213	.238D-01	.578D-01	.429D-01	.376D-01
MIN27 T	.383D-02	.328D-02	.347D-02	.323D-02	.480D-02	.357D-02	.333D-02	.469D-02	.446D-02	.561D-02
MIN28 T	.453D-02	.409D-02	.483D-02	.406D-02	.563D-02	.537D-02	.507D-02	.468D-02	.746D-02	.864D-02
CHM 1 T	.511D-02	.418D-02	.439D-02	.452D-02	.435D-02	.457D-02	.459D-02	.710D-02	.648D-02	.933D-02
CHM 2 T	.664D-03	.823D-03	.625D-03	.843D-03	.653D-03	.785D-03	.750D-03	.610D-03	.759D-03	.942D-03
SCR 1 T	.162	.159	.156	.145	.148	.146	.145	.154	.146	.929D-02
SCR 2 T	.348D-02	.347D-02	.373D-02	.340D-02	.356D-02	.356D-02	.321D-02	.552D-02	.469D-02	.154D-01
SCR 3 T	.202D-03	.169D-03	.117D-03	.104D-03	.132D-03	.118D-03	.131D-03	.314D-03	.143D-03	.178D-03
SCR 4 T	.128D-03	.953D-04	.826D-04	.570D-04	.709D-04	.636D-04	.622D-04	.153D-02	.332D-02	.115D-03
SCR 5 T	.236D-02	.186D-03	.221D-02	.666D-04	.219D-02	.101D-03	.718D-04	.530D-02	.833D-02	.108D-03
SCR 6 T	.355	.350	.372	.315	.323	.336	.328	.122	.125	.128
SCR 7 T	.550	.539	.550	.554	.499	.510	.598	.270	.316	.159
SCR 8 K	.485D-02	.422D-02	.456D-02	.430D-02	.726D-02	.614D-02	.414D-02	.115	.157D-01	.277D-01
SCR 9 T	.503D-03	.323D-03	.345D-03	.263D-03	.368D-03	.301D-03	.387D-03	.278D-02	.180D-01	.870D-03
SCR10 T	.482D-03	.800D-02	.367D-03	.219D-01	.434D-03	.432D-03	.581D-03	.188D-03	.236D-03	.303D-03
SCR11 T	.406D-03	.236D-03	.234D-03	.198D-03	.285D-03	.187D-03	.268D-03	.278D-03	.515D-03	.428D-03
SCR12 T	.678D-01	.690D-01	.593D-02	.678D-01	.264D-01	.585D-02	.488D-02	.128D-01	.749D-01	.654D-01
SCR13 T	.948D-05	.117D-04	.994D-05	.123D-04	.894D-05	.116D-04	.111D-04	.861D-05	.108D-04	.130D-04
IAM 5 T	5.09	5.10	5.01	4.80	4.90	4.86	.410	2.99	5.19	3.78
IMM 6 T	2.11	2.04	2.19	1.85	1.86	1.91	1.94	1.68	1.96	.998
IMM19 T	.214	.218	.187D-01	.214	.833D-01	.185D-01	.154D-01	.403D-01	.237	.208

Table G–1 continued

	MAN41 $	MAN42 $	MAN43 $	MAN44 $	MAN45 $	MAN46 $	MAN47 $	MAN48 $	MAN49 $	MAN50 $
MIN 1 T	.163	.155	.160	.163D-01	.211D-01	.386	.367	.347	.189D-01	.111D-01
MIN 2 T	.742D-01	.733D-01	.826D-01	.776D-01	.221D-01	.855D-01	.898D-01	.732D-01	.179D-01	.963D-02
MIN 3 T	.716	.474	.449	.245	.286	.496	.411	.509	.167	.109
MIN 4 T	.648D-02	.893D-01	.483D-02	.412D-02	.693D-02	.165D-01	.172D-01	.153D-01	.482D-02	.146D-02
MIN 7 T	.201D-02	.222D-01	.266D-01	.240D-01	.219D-01	.360D-02	.484D-02	.314D-02	.160D-02	.952D-03
MIN 8 T	.272D-03	.252D-03	.340D-02	.308D-02	.285D-02	.570D-02	.549D-02	.494D-02	.251D-03	.181D-03
MIN 9 T	.602D-02	.334D-03	.591D-02	.403D-03	.364D-03	.540D-02	.500D-02	.489D-02	.302D-03	.257D-03
MIN10 T	.200	.961D-01	1.60	8.49	.141	.625D-01	.417	.932D-01	3.45	.312
MIN11 T	.413D-01	.185D-02	.336D-01	.395D-01	.405D-01	.200D-02	.444D-02	.279D-02	.274D-02	.159
MIN12 T	.354D-01	.244D-02	.120D-02	.141D-02	.126D-02	.136D-01	.119D-01	.139D-01	.218D-02	.952D-03
MIN13 T	.338D-02	.359D-01	.112D-02	.112D-02	.628D-01	.274D-02	.122D-02	.186D-02	.386D-01	.149D-02
MIN14 T	.292D-02	.242D-02	.174D-02	.210D-02	.313D-02	.359D-01	.340D-01	.311D-01	.203D-02	.120D-02
MIN15 K	.153D-01	.106D-01	.141	.135	.766D-02	.105D-01	.169D-01	.920D-02	.141	.130D-01
MIN16 T	1.12	.671	.357	.439	.388	.746	.749	1.02	.464	.436
MIN21 T	2.46	.206	.722	.701	2.16	2.74	2.56	2.44	.167	.112
MIN23 T	.211D-02	.179D-02	.854D-03	.112D-02	.136D-02	.171D-02	.108D-02	.234D-02	.116D-02	.837D-03
MIN24 T	.114D-02	.102D-02	.857D-03	.115D-02	.106D-02	.919D-03	.757D-03	.165D-02	.968D-03	.122D-02
MIN25 T	.723D-02	.238D-01	.787D-02	.212D-01	.328D-02	.132D-01	.258D-02	.692D-02	.739D-02	.727D-02
MIN26 T	.771D-01	.268	.864D-01	.224	.393D-01	.165	.328D-01	.906D-01	.845D-01	.610D-01
MIN27 T	.624D-02	.565D-02	.483D-02	.622D-02	.599D-02	.507D-02	.438D-02	.938D-02	.563D-02	.675D-02
MIN28 T	.793D-02	.816D-02	.592D-02	.736D-02	.964D-02	.583D-02	.413D-02	.730D-02	.822D-02	.669D-02
CHM 1 T	.122D-01	.101D-01	.702D-02	.119D-01	.829D-02	.898D-02	.437D-02	.762D-02	.872D-02	.103D-01
CHM 2 T	.918D-03	.738D-03	.632D-03	.852D-03	.731D-03	.104D-02	.793D-03	.108D-02	.631D-03	.570D-03
SCR 1 T	.703D-01	.666D-01	.691D-01	.704D-02	.908D-02	.166	.158	.150	.813D-02	.477D-02
SCR 2 T	.134D-02	.151D-01	.181D-01	.164D-01	.150D-01	.243D-02	.329D-02	.213D-02	.107D-02	.624D-03
SCR 3 T	.177D-03	.164D-03	.221D-02	.200D-02	.185D-02	.370D-02	.356D-02	.320D-02	.163D-03	.117D-03
SCR 4 T	.147D-02	.121D-03	.142D-02	.163D-03	.113D-03	.131D-02	.119D-02	.118D-02	.102D-03	.111D-03
SCR 5 T	.600D-02	.851D-04	.495D-02	.145D-03	.822D-04	.248D-02	.248D-02	.242D-02	.110D-03	.635D-04
SCR 6 T	.303	.201	.190	.104	.121	.210	.174	.215	.706D-01	.458D-01
SCR 7 T	.264	.168	.149	.145	.172	.292	.320	.257	.200	.215
SCR 8 K	.223D-01	.107D-01	.178	.943	.157D-01	.695D-02	.464D-01	.104D-01	.383	.346D-01
SCR 9 T	.213D-01	.953D-03	.173D-01	.203D-01	.208D-01	.103D-02	.228D-02	.144D-02	.141D-02	.817D-01
SCR10 T	.280D-03	.386D-02	.209D-03	.178D-03	.300D-03	.711D-03	.744D-03	.660D-03	.208D-03	.629D-04
SCR11 T	.100D-02	.107D-01	.334D-03	.333D-03	.187D-01	.817D-03	.365D-03	.554D-03	.115D-01	.444D-03
SCR12 T	.661D-01	.635D-01	.758D-01	.698D-01	.572D-02	.355D-01	.368D-01	.334D-01	.855D-02	.530D-02
SCR13 T	.132D-04	.101D-04	.844D-05	.115D-04	.959D-05	.145D-04	.115D-04	.148D-04	.860D-05	.776D-05
IMM 5 T	11.6	.487	.230	.306	6.97	5.14	4.63	4.65	.338	.214
IMM 6 T	1.33	1.03	.815	.840	1.02	1.44	1.05	1.34	.819	.836
IMM17 T	.209	.200	.239	.220	.181D-01	.112	.116	.105	.270D-01	.167D-01

	MAN51 $	SRV 1 $	SRV 2 $	SRV 3 $	SRV 4 $	SRV 5 $	SRV 6 $	SRV 7 $	SRV 8 $	SRV 9 $
MIN 1 T	.157	.230D-01	.592D-02	.256D-02	.380D-01	.587D-02	.467D-02	.348D-02	.203D-01	.995D-02
MIN 2 T	.125D-01	.121D-01	.298D-02	.135D-02	.260D-01	.321D-02	.229D-02	.193D-02	.587D-02	.502D-02
MIN 3 T	.533	.453D-01	.170D-01	.684D-02	.659D-01	.117D-01	.106D-01	.115D-01	.341D-01	.285D-01
MIN 4 T	.217D-02	.137D-02	.342D-03	.165D-03	.170D-02	.354D-03	.260D-03	.302D-03	.676D-03	.716D-03
MIN 7 T	.265D-02	.342D-03	.634D-03	.797D-04	.260D-03	.107D-03	.166D-03	.917D-04	.303D-03	.345D-03
MIN 8 T	.266D-03	.465D-03	.128D-03	.453D-04	.303D-03	.110D-03	.120D-03	.844D-04	.150D-03	.173D-03
MIN 9 T	.512D-03	.278D-03	.165D-03	.448D-04	.134D-03	.850D-04	.707D-04	.747D-04	.197D-03	.167D-03
MIN10 T	11.6	.332D-01	.566D-01	.202D-01	.211D-01	.176D-01	.350D-01	.102D-01	.147	.832D-01
MIN11 T	.122	.633D-03	.102D-02	.675D-03	.494D-03	.416D-03	.820D-03	.253D-03	.212D-02	.208D-02
MIN12 T	.124D-02	.102D-02	.288D-03	.209D-03	.888D-03	.300D-03	.288D-03	.419D-03	.144D-02	.814D-03
MIN13 T	.163D-02	.552D-03	.211D-03	.122D-03	.616D-03	.166D-03	.173D-03	.285D-03	.498D-03	.412D-03
MIN14 T	.159D-02	.144D-02	.184D-03	.167D-03	.808D-03	.314D-03	.268D-03	.163D-03	.648D-03	.797D-03
MIN15 K	.705D-01	.792D-02	.414D-02	.175D-02	.921D-02	.244D-02	.262D-02	.179D-02	.720D-02	.740D-02
MIN16 T	1.11	.258	.129	.827D-01	.287	.133	.162	.197	.303	.393
MIN21 T	.130	.157	.339D-01	.196D-01	.229	.391D-01	.333D-01	.237D-01	.818D-01	.888D-01
MIN23 T	.149D-02	.212D-03	.687D-04	.577D-04	.246D-03	.662D-04	.813D-04	.670D-04	.268D-03	.261D-03
MIN24 T	.213D-02	.595D-03	.299D-03	.152D-02	.907D-03	.322D-03	.430D-03	.798D-03	.582D-02	.794D-03
MIN25 T	.512D-02	.175D-02	.668D-03	.640D-03	.205D-02	.103D-02	.109D-02	.546D-03	.543D-02	.315D-02
MIN26 T	.375D-01	.175D-01	.791D-02	.778D-02	.150D-01	.919D-02	.760D-02	.974D-02	.688D-01	.169D-01
MIN27 T	.120D-01	.340D-02	.172D-02	.874D-02	.507D-02	.191D-02	.252D-02	.456D-02	.353D-01	.454D-02
MIN28 T	.150D-01	.228D-02	.908D-03	.290D-02	.268D-02	.115D-02	.142D-02	.170D-02	.113D-01	.315D-02
CHM 1 T	.191D-01	.327D-02	.119D-02	.143D-02	.152D-01	.207D-02	.268D-02	.114D-02	.583D-02	.767D-02
CHM 2 T	.113D-02	.309D-03	.179D-03	.852D-04	.486D-03	.942D-04	.102D-03	.314D-03	.255D-03	.208D-03
SCR 1 T	.678D-01	.989D-02	.255D-02	.111D-02	.164D-01	.253D-02	.201D-02	.150D-02	.874D-02	.429D-02
SCR 2 T	.178D-02	.224D-03	.430D-03	.503D-04	.161D-03	.695D-04	.108D-03	.594D-04	.192D-03	.216D-03
SCR 3 T	.172D-03	.302D-03	.834D-04	.294D-04	.196D-03	.711D-04	.778D-04	.548D-04	.974D-04	.113D-03
SCR 4 T	.184D-03	.832D-04	.443D-04	.184D-04	.635D-04	.273D-04	.271D-04	.239D-04	.749D-04	.757D-04
SCR 5 T	.990D-04	.146D-03	.136D-03	.281D-04	.101D-03	.441D-04	.421D-04	.648D-04	.227D-03	.805D-04
SCR 6 T	.225	.192D-01	.720D-02	.288D-02	.278D-01	.495D-02	.446D-02	.487D-02	.143D-01	.120D-01
SCR 7 T	.491	.239D-01	.758D-02	.452D-02	.348D-01	.600D-02	.641D-02	.587D-02	.243D-01	.181D-01
SCR 8 K	1.28	.369D-02	.629D-02	.224D-02	.235D-02	.195D-02	.389D-02	.113D-02	.164D-01	.924D-02
SCR 9 T	.629D-01	.366D-03	.526D-03	.348D-03	.254D-03	.213D-03	.422D-03	.130D-03	.189D-02	.197D-02
SCR10 T	.938D-04	.598D-04	.148D-04	.715D-05	.736D-04	.153D-04	.112D-04	.131D-04	.292D-04	.316D-04
SCR11 T	.464D-03	.164D-03	.627D-04	.362D-04	.144D-03	.183D-04	.514D-04	.848D-04	.148D-03	.123D-03
SCR12 T	.536D-02	.389D-02	.225D-02	.753D-03	.273D-02	.107D-02	.119D-02	.850D-03	.109D-01	.289D-02
SCR13 T	.150D-04	.437D-05	.252D-05	.120D-05	.688D-05	.133D-05	.143D-05	.445D-05	.358D-05	.293D-05
IMM 5 T	.281	.332	.488D-01	.434D-01	.534	.868D-01	.727D-01	.584D-01	.309	.178
IMM 6 T	1.79	.124	.460D-01	.254D-01	.149	.369D-01	.367D-01	.342D-01	.147	.101
IMM17 T	.169D-01	.975D-02	.709D-02	.238D-02	.861D-02	.336D-02	.377D-02	.269D-02	.344D-01	.911D-02

Table G–1 continued

	SRV10 $	SRV11 $	SRV12 $	SRV13 $	SRV14 $	SCR27 $
MIN 1 T	.111	.683D-02	.744D-02	.541D-02	.178D-01	.0
MIN 2 T	.259D-01	.385D-02	.363D-02	.279D-02	.875D-02	.0
MIN 3 T	.168	.167D-01	.178D-01	.187D-01	.528D-01	.0
MIN 4 T	.514D-02	.394D-03	.443D-03	.382D-03	.168D-02	.0
MIN 7 T	.149D-02	.156D-03	.159D-03	.211D-02	.377D-03	.0
MIN 8 T	.153D-02	.111D-03	.127D-03	.110D-03	.333D-03	.0
MIN 9 T	.196D-02	.110D-03	.120D-03	.776D-04	.280D-03	.0
MIN10 T	.407D-01	.568D-01	.823D-01	.133D-01	.394D-01	.0
MIN11 T	.221D-02	.226D-02	.152D-02	.644D-02	.891D-03	.0
MIN12 T	.532D-02	.535D-03	.490D-03	.276D-03	.183D-02	.0
MIN13 T	.157D-02	.311D-03	.699D-03	.158D-03	.179D-02	.0
MIN14 T	.888D-02	.438D-03	.337D-03	.319D-03	.863D-03	.0
MIN15 K	.686D-02	.462D-02	.475D-02	.217D-02	.879D-02	.0
MIN16 T	.689	.175	.204	.922D-01	.826	.0
MIN21 T	.778	.501D-01	.414D-01	.377D-01	.111	.0
MIN23 T	.643D-03	.152D-03	.128D-03	.646D-04	.332D-03	.0
MIN24 T	.536D-03	.654D-02	.112D-02	.211D-03	.135D-02	.0
MIN25 T	.752D-02	.168D-02	.271D-02	.769D-03	.286D-02	.0
MIN26 T	.939D-01	.265D-01	.406D-01	.851D-02	.399D-01	.0
MIN27 T	.300D-02	.375D-01	.787D-02	.140D-02	.786D-02	.0
MIN28 T	.307D-02	.118D-01	.275D-02	.792D-03	.377D-02	.0
CHM 1 T	.474D-02	.360D-02	.320D-02	.152D-02	.591D-02	.0
CHM 2 T	.624D-03	.204D-03	.184D-03	.881D-04	.136D-02	.0
SCR 1 T	.480D-01	.294D-02	.320D-02	.233D-02	.681D-02	.0
SCR 2 T	.101D-02	.950D-04	.101D-03	.143D-02	.239D-03	.0
SCR 3 T	.990D-03	.722D-04	.821D-04	.711D-04	.216D-03	.0
SCR 4 T	.481D-03	.476D-04	.423D-04	.250D-04	.102D-03	.0
SCR 5 T	.946D-03	.675D-04	.685D-04	.408D-04	.271D-03	.0
SCR 6 T	.710D-01	.699D-02	.752D-02	.451D-02	.222D-01	.0
SCR 7 T	.106	.112D-01	.123D-01	.618D-02	.265D-01	.0
SCR 8 K	.452D-02	.631D-02	.914D-02	.148D-02	.438D-02	.0
SCR 9 T	.114D-02	.116D-02	.781D-03	.332D-02	.459D-03	.0
SCR10 T	.222D-03	.170D-04	.192D-04	.131D-04	.728D-04	.0
SCR11 T	.467D-03	.927D-04	.208D-03	.471D-04	.532D-03	.0
SCR12 T	.131D-01	.202D-02	.256D-02	.914D-03	.406D-02	.0
SCR13 T	.880D-05	.288D-05	.259D-05	.125D-05	.193D-04	.0
IMM 5 T	1.48	.118	.112	.833D-01	.276	.0
IMM 6 T	.545	.689D-01	.827D-01	.353D-01	.232	.0
IMM17 T	.414D-01	.637D-02	.807D-02	.288D-02	.128D-01	.0

Table G–2
Inverse Matrix (''Minerals Portion''):2000

	AGR 1 $	AGR 2 $	AGR 3 $	AGR 4 $	AGR 5 $	EXT 1 $	EXT 2 $	EXT 3 $	EXT 4 $	EXT 5 $
MIN 1 T	.224D-01	.150D-01	.173D-01	.212D-01	.202D-01	.287D-01	.191D-01	.318D-01	.374D-01	.290D-01
MIN 2 T	.642D-02	.963D-02	.108D-01	.958D-02	.857D-02	.777D-02	.574D-02	.722D-02	.817D-02	.645D-02
MIN 3 T	.350D-01	.324D-01	.373D-01	.399D-01	.476D-01	.925D-01	.677D-01	.972D-01	.112	.907
MIN 4 T	.632D-03	.781D-03	.733D-03	.762D-03	.102D-02	.398D-02	.366D-02	.566D-02	.637D-02	.458D-02
MIN 7 T	.264D-03	.413D-03	.333D-03	.291D-03	.400D-03	.508D-03	.346D-03	.575D-03	.542D-03	.460D-03
MIN 8 T	.188D-03	.198D-03	.246D-03	.214D-03	.261D-03	.255D-03	.141D-03	.166D-03	.283D-03	.204D-03
MIN 9 T	.179D-03	.303D-03	.224D-03	.208D-03	.247D-03	.293D-03	.246D-03	.180D-03	.206D-03	.142D-03
MIN10 T	.837D-02	.670D-02	.667D-02	.121D-01	.192D-01	.221D-01	.769D-02	.479D-01	.161D-01	.425D-01
MIN11 T	.150D-03	.224D-03	.175D-03	.175D-03	.303D-03	.238D-03	.120D-03	.391D-03	.190D-03	.331D-03
MIN12 T	.181D-02	.241D-02	.181D-02	.192D-02	.193D-02	.179D-02	.109D-02	.156D-02	.157D-02	.130D-02
MIN13 T	.835D-03	.157D-02	.139D-02	.119D-02	.107D-02	.939D-03	.124D-02	.124D-02	.679D-03	.866D-03
MIN14 T	.448D-03	.961D-03	.678D-03	.652D-03	.744D-03	.115D-02	.812D-03	.109D-02	.128D-02	.927D-03
MIN15 K	.557D-02	.136D-01	.955D-02	.758D-02	.871D-02	.607D-02	.467D-02	.590D-02	.598D-02	.440D-02
MIN16 T	.320	.436	.343	.377	.471	.364	.268	.293	.299	.262
MIN21 T	.149	.275	.216	.188	.228	.258	.171	.264	.292	.219
MIN23 T	.408D-03	.922D-03	.631D-03	.550D-03	.615D-03	.640D-03	.454D-03	.645D-03	.615D-03	.531D-03
MIN24 T	.366D-01	.536D-01	.986D-01	.105	.391D-01	.298D-02	.433D-02	.191D-02	.371D-02	.101D-02
MIN25 T	.257D-02	.441D-02	.295D-02	.245D-02	.307D-02	.209D-02	.158D-02	.199D-02	.178D-02	.135D-02
MIN26 T	.130	.141	.257	.276	.118	.272D-01	.267D-01	.221D-01	.233D-01	.163D-01
MIN27 T	.210	.306	.564	.603	.224	.164D-01	.243D-01	.104D-01	.207D-01	.539D-02
MIN28 T	.438D-01	.645D-01	.116	.124	.475D-01	.510D-02	.645D-02	.362D-02	.593D-02	.246D-02
CHM 1 T	.124D-01	.301D-01	.202D-01	.167D-01	.203D-01	.140D-01	.109D-01	.128D-01	.123D-01	.974D-02
CHM 2 T	.372D-03	.737D-03	.548D-03	.430D-03	.599D-03	.726D-03	.908D-03	.794D-03	.356D-02	.423D-03
SCR 1 T	.967D-02	.648D-02	.747D-02	.916D-02	.872D-02	.124D-01	.822D-02	.137D-01	.161D-01	.125D-01
SCR 2 T	.130D-03	.148D-03	.140D-03	.128D-03	.192D-03	.289D-03	.194D-03	.337D-03	.319D-03	.277D-03
SCR 3 T	.135D-03	.142D-03	.176D-03	.154D-03	.187D-03	.183D-03	.101D-03	.119D-03	.203D-03	.147D-03
SCR 4 T	.691D-04	.162D-03	.109D-03	.922D-04	.107D-03	.917D-04	.719D-04	.724D-04	.725D-04	.516D-04
SCR 5 T	.215D-03	.122D-03	.120D-03	.183D-03	.160D-03	.152D-03	.782D-04	.112D-03	.133D-03	.105D-03
SCR 6 T	.652D-02	.597D-02	.673D-02	.731D-02	.896D-02	.176D-01	.129D-01	.185D-01	.213D-01	.982D-01
SCR 7 T	.206D-01	.188D-01	.220D-01	.239D-01	.307D-01	.511D-01	.378D-01	.607D-01	.781D-01	.432
SCR 8 K	.177D-02	.141D-02	.141D-02	.255D-02	.404D-02	.467D-02	.162D-02	.101D-01	.340D-02	.898D-02
SCR 9 T	.269D-03	.400D-03	.314D-03	.313D-03	.543D-03	.427D-03	.216D-03	.700D-03	.341D-03	.593D-03
SCR10 T	.372D-04	.459D-04	.431D-04	.448D-04	.599D-04	.234D-03	.215D-03	.333D-03	.374D-03	.269D-03
SCR11 T	.249D-03	.467D-03	.413D-03	.353D-03	.319D-03	.280D-03	.368D-03	.369D-03	.202D-03	.258D-03
SCR12 T	.123D-02	.958D-03	.783D-03	.111D-02	.810D-03	.685D-03	.533D-03	.672D-03	.720D-03	.542D-03
SCR13 T	.240D-05	.477D-05	.354D-05	.277D-05	.387D-05	.469D-05	.585D-05	.513D-05	.231D-04	.274D-05
IMM 5 T	.443	.603	.494	.482	.497	.537	.380	.486	.596	.455
IMM 6 T	.204	.372	.285	.261	.308	.405	.298	.420	.454	1.82
IMM17 T	.369D-01	.288D-01	.235D-01	.332D-01	.243D-01	.206D-01	.160D-01	.202D-01	.216D-01	.163D-01

Table G–2 continued

	EXT 6 $	EXT 7 $	CON 1 $	CON 2 $	MAN 1 $	MAN 2 $	MAN 3 $	MAN 4 $	MAN 5 $	MAN 6 $
MIN 1 T	.304D-01	.313	.221	.286D-01	.379D-01	.603D-01	.773D-02	.182D-01	.181D-01	.137D-01
MIN 2 T	.586D-02	.125	.813D-02	.784D-02	.144D-01	.520D-02	.335D-02	.576D-02	.580D-02	.340D-02
MIN 3 T	.164	.361	.290	.118	.316	.437D-01	.214D-01	.535D-01	.551D-01	.754D-01
MIN 4 T	.821D-02	.155D-02	.258D-02	.245D-02	.458D-02	.739D-03	.370D-03	.710D-03	.773D-03	.450D-03
MIN 7 T	.524D-03	.359D-03	.326D-02	.893D-03	.101D-01	.218D-03	.129D-03	.320D-03	.319D-03	.234D-03
MIN 8 T	.153D-03	.925D-04	.774D-03	.693D-03	.712D-02	.169D-03	.103D-03	.138D-03	.153D-03	.101D-03
MIN 9 T	.159D-03	.723D-04	.303D-02	.574D-03	.424D-03	.190D-03	.139D-03	.207D-03	.320D-03	.134D-03
MIN10 T	.118D-01	.962D-02	.294D-01	.415D-01	.141	.987D-02	.786D-02	.154D-01	.174D-01	.135
MIN11 T	.141D-03	.978D-04	.343D-03	.540D-03	.645D-03	.149D-03	.110D-03	.223D-03	.268D-03	.871D-03
MIN12 T	.108D-02	.104D-02	.124D-01	.522D-02	.431D-01	.454D-02	.619D-03	.145D-02	.143D-02	.776D-03
MIN13 T	.611D-03	.491D-03	.167D-02	.362D-02	.815D-03	.763D-03	.413D-03	.330D-02	.321D-02	.151D-02
MIN14 T	.135D-02	.353D-03	.464D-02	.412D-03	.151D-02	.407D-03	.234D-03	.693D-03	.712D-03	.345D-03
MIN15 K	.466D-02	.163D-02	.461D-02	.432D-02	.553D-02	.435D-02	.234D-02	.601D-02	.605D-02	.323D-02
MIN16 T	.261	.225	.858	2.61	.311	.502	.287	1.31	1.38	.706
MIN21 T	.235	1.71	.888	.188	.299	.114	.653D-01	.173	.172	.916D-01
MIN23 T	.509D-03	.212D-03	.811D-03	.491D-03	.108D-02	.472D-03	.183D-03	.451D-03	.462D-03	.233D-03
MIN24 T	.609D-02	.493D-03	.201D-02	.108D-02	.940D-03	.199D-01	.230D-01	.757D-02	.496D-02	.332D-02
MIN25 T	.145D-02	.579D-03	.170D-02	.155D-02	.145D-02	.595D-02	.921D-03	.338D-02	.353D-02	.148D-02
MIN26 T	.284D-01	.134D-01	.731D-01	.100	.267D-01	.210	.672D-01	.769D-01	.725D-01	.344D-01
MIN27 T	.344D-01	.270D-02	.113D-01	.596D-02	.514D-02	.121	.132	.433D-01	.282D-01	.192D-01
MIN28 T	.837D-02	.108D-02	.478D-02	.331D-02	.359D-02	.248D-01	.279D-01	.211D-01	.174D-01	.950D-02
CHM 1 T	.102D-01	.339D-02	.791D-02	.731D-02	.848D-02	.113D-01	.708D-02	.571D-01	.546D-01	.256D-01
CHM 2 T	.510D-03	.380D-03	.456D-02	.439D-02	.414D-03	.316D-03	.166D-03	.416D-03	.417D-03	.241D-03
SCR 1 T	.131D-01	.135	.953D-01	.123D-01	.163D-01	.260D-01	.333D-02	.785D-02	.780D-02	.593D-02
SCR 2 T	.315D-03	.230D-03	.218D-02	.581D-03	.679D-02	.113D-03	.670D-04	.162D-03	.161D-03	.134D-03
SCR 3 T	.110D-03	.664D-04	.556D-03	.497D-03	.512D-02	.121D-03	.738D-04	.989D-04	.110D-03	.726D-04
SCR 4 T	.583D-04	.219D-04	.463D-03	.104D-03	.692D-04	.591D-04	.384D-04	.789D-04	.958D-04	.409D-04
SCR 5 T	.776D-04	.725D-04	.219D-02	.845D-03	.822D-02	.767D-03	.610D-04	.126D-03	.120D-03	.780D-04
SCR 6 T	.312D-01	.689D-01	.553D-01	.226D-01	.603D-01	.825D-02	.402D-02	.101D-01	.104D-01	.143D-01
SCR 7 T	.124	.158	.191	.752D-01	.257	.191D-01	.144D-01	.246D-01	.241D-01	.248D-01
SCR 8 K	.248D-02	.203D-02	.619D-02	.877D-02	.297D-01	.208D-02	.166D-02	.324D-02	.368D-02	.285D-01
SCR 9 T	.253D-03	.175D-03	.614D-03	.967D-03	.116D-02	.266D-03	.197D-03	.400D-03	.480D-03	.156D-02
SCR10 T	.483D-03	.908D-04	.152D-03	.144D-03	.269D-03	.434D-04	.217D-04	.417D-04	.454D-04	.265D-04
SCR11 T	.182D-03	.146D-03	.497D-03	.108D-02	.243D-03	.227D-03	.123D-03	.984D-03	.956D-03	.449D-03
SCR12 T	.730D-03	.269D-03	.258D-02	.868D-03	.920D-03	.469D-02	.340D-03	.645D-03	.643D-03	.348D-03
SCR13 T	.330D-05	.246D-05	.295D-04	.285D-04	.265D-05	.202D-05	.107D-05	.267D-05	.268D-05	.154D-05
IMM 5 T	.517	4.07	2.98	.496	.527	.893	.159	.401	.399	.218
IMM 6 T	.625	.668	.910	.402	1.11	.210	.107	.280	.287	.298
IMM17 T	.219D-01	.809D-02	.774D-01	.261D-01	.276D-01	.141	.102D-01	.194D-01	.193D-01	.104D-01

	MAN 7 $	MAN 8 $	MAN 9 $	MAN10 $	MAN11 $	MAN12 $	MAN13 $	MAN14 $	CHM 1 $	CHM 2 $
MIN 1 T	.144D-01	.177D-01	.137D-01	.132D-01	.139D-01	.156D-01	.146D-01	.936D-02	.451D-01	.374D-01
MIN 2 T	.443D-02	.960D-02	.629D-02	.829D-02	.867D-02	.735D-02	.882D-02	.459D-02	.132D-01	.643D-01
MIN 3 T	.733D-01	.188	.120	.265	.314	.578D-01	.481D-01	.304D-01	.824D-01	.938D-01
MIN 4 T	.633D-03	.131D-02	.107D-02	.115D-02	.195D-02	.949D-03	.172D-02	.608D-03	.109D-02	.992D-03
MIN 7 T	.254D-03	.332D-03	.242D-03	.272D-03	.320D-03	.307D-03	.377D-03	.308D-03	.349D-03	.256D-03
MIN 8 T	.120D-03	.300D-03	.200D-03	.280D-03	.243D-03	.202D-03	.173D-03	.214D-02	.213D-03	.167D-03
MIN 9 T	.257D-03	.257D-03	.147D-03	.441D-03	.413D-03	.297D-03	.288D-03	.229D-03	.182D-03	.128D-03
MIN10 T	.473D-01	.167D-01	.191D-01	.348D-01	.288D-01	.148D-01	.177D-01	.357D-01	.152D-01	.126D-01
MIN11 T	.396D-03	.200D-03	.174D-03	.287D-03	.234D-03	.201D-03	.260D-03	.711D-03	.175D-03	.142D-03
MIN12 T	.111D-02	.151D-02	.888D-03	.100D-02	.101D-02	.156D-02	.200D-02	.139D-02	.183D-02	.205D-02
MIN13 T	.191D-02	.137D-02	.753D-03	.183D-02	.148D-02	.174D-02	.146D-02	.939D-03	.222D-01	.274D-02
MIN14 T	.497D-03	.690D-03	.457D-03	.463D-03	.543D-03	.704D-03	.102D-02	.622D-03	.714D-03	.504D-03
MIN15 K	.450D-02	.645D-02	.344D-02	.376D-02	.286D-02	.770D-02	.108D-01	.729D-02	.446D-02	.294D-02
MIN16 T	2.72	.855	.444	1.40	1.13	7.60	3.36	1.49	.466	.361
MIN21 T	.127	.175	.114	.108	.108	.183	.231	.152	.257	.148
MIN23 T	.365D-03	.572D-03	.532D-03	.696D-03	.136D-02	.590D-03	.895D-03	.583D-03	.505D-03	.379D-03
MIN24 T	.422D-02	.140D-01	.608D-02	.336D-02	.193D-02	.427D-02	.303D-02	.186D-02	.115D-01	1.78
MIN25 T	.241D-02	.276D-02	.145D-02	.303D-02	.309D-02	.853D-02	.600D-02	.345D-02	.154D-02	.110D-02
MIN26 T	.490D-01	.870D-01	.456D-01	.995D-01	.106	.313D-01	.198D-01	.143D-01	.484D-01	4.52
MIN27 T	.241D-01	.797D-01	.346D-01	.191D-01	.109D-01	.239D-01	.162D-01	.988D-02	.657D-01	10.2
MIN28 T	.118D-01	.182D-01	.834D-02	.701D-02	.495D-02	.273D-01	.142D-01	.691D-02	.156D-01	2.07
CHM 1 T	.334D-01	.142D-01	.724D-02	.144D-01	.982D-02	.102	.606D-01	.307D-01	.111D-01	.835D-02
CHM 2 T	.324D-03	.792D-03	.435D-03	.552D-03	.455D-03	.620D-03	.676D-03	.467D-03	.476D-03	.356D-03
SCR 1 T	.620D-02	.761D-02	.589D-02	.568D-02	.601D-02	.672D-02	.630D-02	.403D-02	.194D-01	.161D-01
SCR 2 T	.133D-03	.171D-03	.138D-03	.155D-03	.195D-03	.138D-03	.152D-03	.138D-03	.201D-03	.152D-03
SCR 3 T	.864D-04	.215D-03	.144D-03	.201D-03	.175D-03	.145D-03	.124D-03	.153D-02	.153D-03	.120D-03
SCR 4 T	.727D-04	.858D-04	.446D-04	.905D-04	.789D-04	.105D-03	.135D-03	.965D-04	.576D-04	.378D-04
SCR 5 T	.100D-03	.129D-03	.862D-04	.986D-04	.880D-04	.109D-03	.107D-03	.789D-04	.224D-03	.315D-03
SCR 6 T	.139D-01	.359D-01	.229D-01	.506D-01	.599D-01	.110D-01	.914D-02	.577D-02	.149D-01	.138D-01
SCR 7 T	.216D-01	.737D-01	.434D-01	.199	.192	.326D-01	.260D-01	.183D-01	.431D-01	.477D-01
SCR 8 K	.998D-02	.352D-02	.404D-02	.735D-02	.609D-02	.311D-02	.373D-02	.753D-02	.321D-02	.266D-02
SCR 9 T	.710D-03	.358D-03	.311D-03	.515D-03	.419D-03	.360D-03	.465D-03	.127D-02	.313D-03	.254D-03
SCR10 T	.372D-04	.768D-04	.630D-04	.693D-04	.114D-03	.558D-04	.101D-03	.358D-04	.643D-04	.583D-04
SCR11 T	.569D-03	.407D-03	.224D-03	.546D-03	.440D-03	.518D-03	.433D-03	.280D-03	.660D-02	.816D-03
SCR12 T	.496D-03	.694D-03	.390D-03	.606D-03	.529D-03	.655D-03	.785D-03	.536D-03	.111D-02	.175D-02
SCR13 T	.208D-05	.512D-05	.280D-05	.352D-05	.284D-05	.400D-05	.435D-05	.300D-05	.295D-05	.225D-05
IMM 5 T	.299	.387	.245	.247	.242	.409	.499	.327	.695	.552
IMM 6 T	.321	.728	.447	.943	1.10	.324	.369	.244	.288	.267
IMM17 T	.149D-01	.208D-01	.117D-01	.182D-01	.159D-01	.197D-01	.236D-01	.161D-01	.335D-01	.526D-01

Table G–2 continued

	CHM 3 $	MAN15 $	MAN16 $	MAN17 $	MAN18 $	MAN19 $	MAN20 $	MAN21 $	MAN22 $	MAN23 $
MIN 1 T	.533D-01	.288D-01	.454D-01	.970D-01	.175	.170D-01	.361D-01	.176D-01	.145D-01	.205D-01
MIN 2 T	.133	.921D-02	.587D-02	.869D-02	.979D-01	.910D-02	.646D-02	.537D-02	.606D-02	.661D-02
MIN 3 T	.155	.529D-01	.472D-01	.661D-01	.201	.604D-01	3.77	.713	.156	.690
MIN 4 T	.115D-01	.108D-02	.871D-03	.103D-02	.399D-02	.235D-02	.734D-03	.776D-03	.176D-02	.205D-02
MIN 7 T	.830D-02	.354D-03	.288D-03	.311D-03	.313D-03	.350D-03	.297D-03	.303D-03	.285D-03	.341D-03
MIN 8 T	.295D-03	.198D-03	.166D-03	.135D-01	.523D-02	.145D-03	.156D-03	.117D-03	.154D-03	.231D-03
MIN 9 T	.710D-02	.284D-03	.375D-03	.684D-02	.118D-03	.463D-02	.236D-03	.481D-03	.319D-03	.152D-03
MIN10 T	.139D-01	.139D-01	.257D-01	.257D-01	.115D-01	.248D-01	.154D-01	.100	.226D-01	.382D-01
MIN11 T	.455D-02	.199D-03	.290D-03	.274D-03	.140D-03	.251D-03	.228D-03	.705D-03	.187D-03	.295D-03
MIN12 T	.621D-01	.206D-02	.372D-02	.719D-02	.204D-02	.152D-02	.337D-02	.158D-02	.252D-02	.112D-02
MIN13 T	.272D-01	.976D-02	.321D-02	.817D-01	.112D-02	.310D-02	.223D-02	.136D-02	.138D-02	.126D-02
MIN14 T	.270D-01	.845D-03	.559D-03	.675D-03	.415D-03	.884D-03	.687D-03	.597D-03	.583D-03	.547D-03
MIN15 K	.396	.771D-02	.640D-02	.748D-02	.372D-01	.697D-02	.881D-02	.630D-02	.188D-01	.428D-02
MIN16 T	9.60	5.20	.711	78.9	.332	2.11	.479	.674	2.97	.425
MIN21 T	6.41	.228	.147	.196	.913	.181	.187	.146	.120	.151
MIN23 T	.275D-01	.634D-03	.599D-03	.953D-03	.338D-03	.637D-03	.699D-03	.518D-03	.403D-03	.323D-03
MIN24 T	.569D-01	.916D-02	.276D-02	.533D-02	.101D-02	.346D-02	.108D-01	.354D-02	.132D-02	.137D-02
MIN25 T	.135	.295D-02	.135D-01	.284D-02	.113D-02	.363D-02	.544D-02	.282D-02	.250	.122D-02
MIN26 T	.621D-01	.482D-01	1.15	.224	.200D-01	.621D-01	.149	.501D-01	8.15	.207D-01
MIN27 T	.272	.522D-01	.991D-01	.303D-01	.582D-02	.192D-01	.679D-01	.211D-01	.728D-02	.758D-02
MIN28 T	.115	.678D-01	.556D-02	.275D-01	.932D-02	.140D-01	.145D-01	.751D-02	.347D-02	.304D-02
CHM 1 T	.920	.237	.175D-01	.374D-01	.730D-02	.495D-01	.208D-01	.213D-01	.109D-01	.984D-02
CHM 2 T	.181D-01	.543D-03	.430D-03	.673D-03	.619D-03	.520D-03	.621D-03	.407D-03	.817D-03	.367
SCR 1 T	.230D-01	.124D-01	.196D-01	.418D-01	.752D-01	.734D-02	.156D-01	.759D-02	.623D-02	.883D-02
SCR 2 T	.146D-02	.169D-03	.138D-03	.145D-03	.185D-03	.172D-03	.118D-03	.148D-03	.166D-03	.203D-03
SCR 3 T	.212D-03	.142D-03	.120D-03	.966D-02	.376D-02	.104D-03	.112D-03	.838D-04	.111D-03	.166D-03
SCR 4 T	.477D-02	.105D-03	.105D-03	.105D-02	.408D-04	.732D-03	.109D-03	.121D-03	.702D-04	.475D-04
SCR 5 T	.138D-02	.195D-03	.553D-03	.119D-02	.266D-03	.109D-03	.426D-03	.146D-03	.386D-03	.124D-03
SCR 6 T	.293D-01	.974D-02	.893D-02	.124D-01	.383D-01	.114D-01	.720	.136	.298D-01	.132
SCR 7 T	.403D-01	.295D-01	.236D-01	.272D-01	.895D-01	.351D-01	.173D-01	.380D-01	.249D-01	.346D-01
SCR 8 K	.292D-02	.294D-02	.542D-02	.542D-02	.244D-02	.524D-02	.324D-02	.211D-01	.476D-02	.806D-02
SCR 9 T	.816D-02	.356D-03	.519D-03	.490D-03	.251D-03	.449D-03	.409D-03	.126D-02	.335D-03	.528D-03
SCR10 T	.673D-03	.637D-04	.512D-04	.606D-04	.234D-03	.138D-03	.432D-04	.456D-04	.104D-03	.121D-03
SCR11 T	.811D-02	.291D-02	.957D-03	.243D-01	.335D-03	.925D-03	.663D-03	.404D-03	.412D-03	.374D-03
SCR12 T	.229D-01	.114D-02	.346D-02	.738D-02	.144D-02	.714D-03	.272D-02	.896D-03	.263D-02	.417D-03
SCR13 T	.117D-03	.344D-05	.274D-05	.421D-05	.396D-05	.334D-05	.400D-05	.261D-05	.525D-05	.238D-02
IMM 5 T	13.4	.589	.758	1.46	2.34	.406	.717	.378	.260	.338
IMM 6 T	9.40	.297	.265	.336	.438	.323	13.2	2.57	.579	2.42
IMM17 T	.688	.344D-01	.104	.222	.434D-01	.214D-01	.817D-01	.269D-01	.790D-01	.125D-01

	MAN24 $	MAN25 $	MET 1 $	MET 2 $	MET 3 $	MET 4 $	MET 5 $	MET 6 $	MET 7 $	MET 8 $
MIN 1 T	.142D-01	.168D-01	.257D-01	.192D-01	.291D-01	.331D-01	.348D-01	.335D-01	.278D-01	.151D-01
MIN 2 T	.455D-02	.702D-02	.112D-01	.633D-02	.295D-01	.133D-01	.149D-01	.105D-01	.184D-01	.113D-01
MIN 3 T	.378	.120	1.01	.907D-01	.297	.560	.479	.603	.362	.906D-01
MIN 4 T	.126D-02	.172D-02	.441D-0	.219D-02	.131D-01	.430D-02	.537D-02	.429D-02	.817D-02	.133D-02
MIN 7 T	.793D-03	.351D-03	.676D-03	.952D-03	.738D-03	.111D-02	.116D-02	.954D-03	.587D-03	.560D-03
MIN 8 T	.157D-03	.217D-03	.183D-03	.181D-03	.155D-03	.169D-03	.144D-03	.141D-03	.182D-03	.296D-03
MIN 9 T	.107D-03	.220D-03	.134D-03	.145D-03	.224D-03	.137D-03	.159D-03	.139D-03	.145D-03	.466D-03
MIN10 T	.132D-01	.252D-01	.218D-01	.179D-01	.151D-01	.199D-01	.246D-01	.220D-01	.157D-01	.733D-02
MIN11 T	.123D-03	.261D-03	.171D-03	.169D-03	.200D-03	.165D-03	.183D-03	.204D-03	.145D-03	.289D-03
MIN12 T	.378	.836D-02	.152D-02	.286D-02	.217D-02	.261D-02	.291D-02	.221D-02	.108D-02	.354D-02
MIN13 T	.781D-03	.140D-02	.130D-02	.709D-03	.149D-02	.214D-02	.272D-02	.614D-03	.867D-03	.261D-02
MIN14 T	.407D-03	.683D-03	.107D-02	.598D-03	.267D-02	.135D-02	.168D-02	.157D-02	.152D-02	.157D-02
MIN15 K	.390D-02	.213D-01	.273D-02	.355D-02	.908D-02	.268D-02	.521D-02	.355D-02	.212D-02	.214D-01
MIN16 T	2.57	2.52	.384	3.34	.468	.370	.362	.325	.314	1.32
MIN21 T	.147	.177	.185	.196	.306	.256	.305	.273	.191	.394
MIN23 T	.135D-02	.708D-03	.877D-02	.156D-02	.277D-02	.428D-02	.382D-02	.105D-02	.295D-02	.244D-02
MIN24 T	.109D-02	.182D-02	.122D-02	.150D-02	.197D-02	.156D-02	.209D-02	.815D-03	.777D-03	.333D-02
MIN25 T	.899D-03	.247D-02	.859D-03	.114D-02	.302D-02	.837D-03	.171D-02	.126D-02	.770D-03	.705D-02
MIN26 T	7.57	.211	.284D-01	.215D-01	.226D-01	.421D-01	.334D-01	.255D-01	.277D-01	.130D-01
MIN27 T	.604D-02	.982D-02	.685D-02	.831D-02	.102D-01	.880D-02	.115D-01	.435D-02	.441D-02	.163D-01
MIN28 T	.261D-02	.445D-02	.781D-02	.903D-02	.990D-02	.972D-02	.900D-02	.627D-02	.757D-02	.978D-02
CHM 1 T	.728D-02	.179D-01	.559D-02	.825D-02	.207D-01	.575D-02	.116D-01	.782D-02	.465D-02	.487D-01
CHM 2 T	.370	.329D-01	.690D-03	.102D-02	.853D-03	.896D-03	.588D-03	.692D-03	.505D-03	.127D-02
SCR 1 T	.614D-02	.724D-02	.111D-01	.829D-02	.126D-01	.143D-01	.150D-01	.144D-01	.120D-01	.651D-02
SCR 2 T	.516D-03	.178D-03	.440D-03	.617D-03	.413D-03	.734D-03	.737D-03	.615D-03	.384D-03	.165D-03
SCR 3 T	.113D-03	.156D-03	.132D-03	.130D-03	.111D-03	.121D-03	.104D-03	.101D-03	.130D-03	.212D-03
SCR 4 T	.336D-04	.851D-04	.355D-04	.452D-04	.110D-03	.353D-04	.634D-04	.467D-04	.329D-04	.260D-03
SCR 5 T	.719D-01	.143D-02	.117D-03	.109D-03	.105D-03	.112D-03	.108D-03	.102D-03	.978D-04	.119D-03
SCR 6 T	.718D-01	.227D-01	.194	.173D-01	.566D-01	.107	.914D-01	.115	.691D-01	.173D-01
SCR 7 T	.305D-01	.400D-01	.665D-01	.416D-01	.623D-01	.114	.937D-01	.690D-01	.714D-01	.200D-01
SCR 8 K	.279D-02	.531D-02	.459D-02	.378D-02	.318D-02	.420D-02	.520D-02	.465D-02	.331D-02	.155D-02
SCR 9 T	.220D-03	.467D-03	.306D-03	.303D-03	.359D-03	.296D-03	.328D-03	.365D-03	.259D-03	.518D-03
SCR10 T	.738D-04	.101D-03	.259D-03	.129D-03	.769D-03	.253D-03	.316D-03	.252D-03	.480D-03	.783D-04
SCR11 T	.232D-03	.417D-03	.386D-03	.211D-03	.442D-03	.636D-03	.809D-03	.183D-03	.258D-03	.776D-03
SCR12 T	.510D-03	.607D-03	.545D-03	.567D-03	.941D-03	.691D-03	.813D-03	.516D-03	.508D-03	.138D-02
SCR13 T	.240D-02	.214D-03	.411D-05	.573D-05	.533D-05	.538D-05	.365D-05	.433D-05	.294D-05	.764D-05
IMM 5 T	.326	.382	.404	.421	.663	.532	.626	.536	.385	.839
IMM 6 T	1.33	.513	3.51	.346	1.18	1.93	1.71	2.12	1.25	.753
IMM17 T	.153D-01	.182D-01	.164D-01	.170D-01	.282D-01	.207D-01	.244D-01	.155D-01	.152D-01	.414D-01

Table G–2 continued

	MET 9 $	MET10 $	MET11 $	MET12 $	MET13 $	MET14 $	MET15 $	MET16 $	MET17 $	MET18 $
MIN 1 T	.351D-01	.173D-01	.199D-01	.227D-01	.208D-01	.180D-01	.164D-01	.179D-01	.212D-01	.264D-01
MIN 2 T	.202D-01	.565D-02	.702D-02	.766D-02	.880D-02	.636D-02	.498D-02	.101D-01	.130D-01	.150D-01
MIN 3 T	.139	.206	.193	.174	.174	.178	.644D-01	.126	.127	.159
MIN 4 T	.830D-02	.265D-02	.256D-02	.254D-02	.220D-02	.254D-02	.142D-02	.435D-02	.475D-02	.615D-02
MIN 7 T	.582D-03	.339D-03	.414D-03	.378D-03	.674D-03	.382D-03	.464D-03	.410D-03	.546D-03	.684D-03
MIN 8 T	.168D-03	.157D-03	.196D-03	.264D-03	.242D-03	.159D-03	.220D-02	.164D-03	.236D-03	.216D-03
MIN 9 T	.127D-03	.182D-03	.141D-03	.123D-03	.126D-03	.139D-03	.129D-03	.169D-03	.172D-03	.238D-03
MIN10 T	.150D-01	.853D-02	.167D-01	.199D-01	.145D-01	.142D-01	.957D-02	.121D-01	.169D-01	.545D-01
MIN11 T	.179D-03	.115D-03	.165D-03	.185D-03	.145D-03	.146D-03	.118D-03	.123D-03	.159D-03	.371D-03
MIN12 T	.176D-02	.994D-03	.113D-02	.106D-02	.191D-02	.104D-02	.123D-02	.872D-03	.138D-02	.115D-02
MIN13 T	.200D-02	.100D-02	.121D-02	.103D-02	.180D-02	.103D-02	.104D-02	.104D-02	.157D-02	.175D-02
MIN14 T	.158D-02	.648D-03	.663D-03	.601D-03	.517D-03	.639D-03	.572D-03	.907D-03	.963D-03	.129D-02
MIN15 K	.266D-02	.370D-02	.343D-02	.396D-02	.316D-02	.315D-02	.284D-02	.255D-02	.282D-02	.375D-02
MIN16 T	.296	.230	.206	.207	.286	.221	.179	.393	.613	.469
MIN21 T	.189	.147	.158	.171	.168	.146	.143	.135	.161	.194
MIN23 T	.681D-02	.727D-03	.157D-02	.105D-02	.467D-01	.801D-02	.156D-02	.112D-02	.210D-01	.419D-02
MIN24 T	.167D-02	.277D-02	.124D-02	.122D-02	.136D-02	.155D-02	.108D-02	.157D-02	.114D-02	.157D-02
MIN25 T	.870D-03	.121D-02	.103D-02	.104D-02	.821D-03	.999D-03	.940D-03	.883D-03	.933D-03	.123D-02
MIN26 T	.270D-01	.209D-01	.143D-01	.146D-01	.167D-01	.157D-01	.124D-01	.211D-01	.228D-01	.265D-01
MIN27 T	.938D-02	.155D-01	.683D-02	.672D-02	.764D-02	.863D-02	.594D-02	.884D-02	.637D-02	.872D-02
MIN28 T	.128D-01	.141	.120D-01	.103D-01	.833D-02	.109	.139D-01	.585D-01	.882D-02	.527D-01
CHM 1 T	.678D-02	.834D-02	.711D-02	.722D-02	.661D-02	.707D-02	.663D-02	.644D-02	.684D-02	.872D-02
CHM 2 T	.542D-02	.812D-03	.745D-03	.201D-02	.219D-02	.412D-02	.733D-03	.608D-03	.122D-02	.206D-02
SCR 1 T	.151D-01	.747D-02	.856D-02	.978D-02	.897D-02	.776D-02	.706D-02	.770D-02	.915D-02	.114D-01
SCR 2 T	.375D-03	.200D-03	.254D-03	.230D-03	.437D-03	.233D-03	.290D-03	.258D-03	.350D-03	.434D-03
SCR 3 T	.121D-03	.112D-03	.140D-03	.190D-03	.174D-03	.114D-03	.158D-02	.118D-03	.169D-03	.155D-03
SCR 4 T	.355D-04	.531D-04	.435D-04	.411D-04	.351D-04	.425D-04	.404D-04	.416D-04	.424D-04	.603D-04
SCR 5 T	.207D-03	.690D-04	.691D-04	.775D-04	.835D-04	.658D-04	.538D-04	.610D-04	.765D-04	.856D-04
SCR 6 T	.265D-01	.394D-01	.368D-01	.331D-01	.333D-01	.340D-01	.123D-01	.241D-01	.242D-01	.303D-01
SCR 7 T	.832D-01	.327D-01	.344D-01	.359D-01	.355D-01	.340D-01	.311D-01	.407D-01	.484D-01	.678D-01
SCR 8 K	.318D-02	.180D-02	.352D-02	.419D-02	.306D-02	.299D-02	.202D-02	.255D-02	.357D-02	.115D-01
SCR 9 T	.320D-03	.206D-03	.295D-03	.332D-03	.261D-03	.262D-03	.212D-03	.220D-03	.285D-03	.664D-03
SCR10 T	.488D-03	.156D-03	.150D-03	.149D-03	.129D-03	.149D-03	.837D-04	.256D-03	.279D-03	.362D-03
SCR11 T	.596D-03	.299D-03	.361D-03	.306D-03	.536D-03	.305D-03	.309D-03	.310D-03	.468D-03	.521D-03
SCR12 T	.137D-02	.421D-03	.402D-03	.408D-03	.463D-03	.382D-03	.357D-03	.399D-03	.474D-03	.580D-03
SCR13 T	.328D-04	.456D-05	.300D-05	.116D-04	.313D-05	.348D-05	.272D-05	.298D-05	.259D-05	.315D-05
IMM 5 T	.550	.315	.315	.343	.370	.294	.277	.293	.356	.387
IMM 6 T	.486	.756	.701	.629	.613	.649	.258	.455	.452	.577
IMM17 T	.412D-01	.126D-01	.121D-01	.123D-01	.139D-01	.115D-01	.107D-01	.120D-01	.142D-01	.174D-01

	MET19 $	MET20 $	MET21 $	MET22 $	MET23 $	MAN26 $	MAN27 $	MAN28 $	MAN29 $	MAN30 $
MIN 1 T	.201D-01	.305D-01	.267D-01	.420D-01	.211D-01	1.35	.226D-01	.287D-01	.210D-01	.369
MIN 2 T	.123D-01	.325D-01	.132D-01	.102	.193D-01	.117D-01	.130D-01	.173D-01	.650D-01	.161D-01
MIN 3 T	.114	.114	4.43	4.45	.202	.390	1.42	.338	1.30	.731
MIN 4 T	.487D-02	.154D-01	.581D-02	.254D-01	.813D-02	.437D-02	.469D-02	.729D-02	.531D-02	.449D-01
MIN 7 T	.556D-03	.597D-03	.509D-03	.718D-03	.421D-03	.431D-03	.673D-03	.567D-03	.599D-03	.117D-02
MIN 8 T	.175D-03	.130D-03	.120D-03	.141D-03	.152D-03	.488D-03	.208D-02	.194D-03	.222D-03	.254D-03
MIN 9 T	.212D-03	.126D-03	.124D-03	.146D-03	.174D-03	.256D-03	.217D-03	.190D-03	.301D-03	.161D-03
MIN10 T	.160D-01	.137D-01	.103D-01	.135D-01	.135D-01	.166D-01	.256D-01	.166D-01	.263D-01	.207D-01
MIN11 T	.164D-03	.158D-03	.122D-03	.148D-03	.123D-03	.181D-03	.213D-03	.158D-03	.220D-03	.288D-03
MIN12 T	.124D-02	.120D-02	.107D-02	.127D-02	.864D-03	.987D-01	.171D-02	.134D-02	.171D-02	.133D-01
MIN13 T	.172D-02	.832D-03	.650D-03	.922D-03	.986D-03	.238D-02	.159D-02	.104D-02	.156D-02	.131D-02
MIN14 T	.106D-02	.262D-02	.154D-02	.403D-02	.137D-02	.965D-03	.101D-02	.138D-02	.121D-02	.171D-02
MIN15 K	.330D-02	.360D-02	.375D-02	.414D-02	.221D-02	.330D-02	.263D-02	.286D-02	.545D-02	.181D-02
MIN16 T	.710	.262	.262	.322	.346	1.88	1.02	.486	.928	.289
MIN21 T	.161	.229	.210	.300	.149	.150	.171	.198	.198	2.26
MIN23 T	.284D-02	.103D-01	.205D-02	.338D-02	.708D-02	.492D-02	.380D-02	.298D-02	.230D-02	.141D-02
MIN24 T	.193D-02	.964D-03	.120D-02	.102D-02	.954D-03	.115D-02	.931D-03	.919D-03	.138D-02	.524D-03
MIN25 T	.170D-02	.121D-02	.132D-02	.130D-02	.724D-03	.130D-02	.273D-02	.131D-02	.219D-02	.660D-03
MIN26 T	.438D-01	.190D-01	.215D-01	.262D-01	.227D-01	.303D-01	.171	.216	.294	.325D-01
MIN27 T	.108D-01	.522D-02	.653D-02	.552D-02	.539D-02	.641D-02	.528D-02	.512D-02	.743D-02	.294D-02
MIN28 T	.334D-01	.596D-02	.308D-01	.888D-02	.146D-01	.544D-02	.519D-02	.492D-02	.674D-02	.355D-02
CHM 1 T	.175D-01	.823D-02	.838D-02	.901D-02	.563D-02	.906D-02	.579D-02	.733D-02	.138D-01	.385D-02
CHM 2 T	.738D-03	.908D-03	.562D-03	.149D-02	.670D-03	.533D-03	.754D-03	.497D-03	.676D-03	.583D-03
SCR 1 T	.867D-02	.132D-01	.115D-01	.181D-01	.911D-02	.583	.974D-02	.124D-01	.907D-02	.159
SCR 2 T	.351D-03	.374D-03	.312D-03	.453D-03	.269D-03	.265D-03	.439D-03	.362D-03	.357D-03	.777D-03
SCR 3 T	.126D-03	.931D-04	.860D-04	.101D-03	.109D-03	.351D-03	.149D-02	.139D-03	.160D-03	.183D-03
SCR 4 T	.538D-04	.451D-04	.468D-04	.505D-04	.391D-04	.610D-04	.467D-04	.475D-04	.873D-04	.330D-04
SCR 5 T	.804D-04	.698D-04	.668D-04	.910D-04	.616D-04	.187D-01	.111D-03	.132D-03	.103D-03	.235D-02
SCR 6 T	.217D-01	.217D-01	.846	.849	.384D-01	.744D-01	.272	.645D-01	.248	.140
SCR 7 T	.517D-01	.680D-01	13.8	13.8	.672D-01	.517D-01	1.11	.127	1.08	.626
SCR 8 K	.338D-02	.289D-02	.218D-02	.284D-02	.286D-02	.350D-02	.541D-02	.349D-02	.554D-02	.437D-02
SCR 9 T	.293D-03	.282D-03	.218D-03	.265D-03	.221D-03	.325D-03	.381D-03	.283D-03	.395D-03	.515D-03
SCR10 T	.286D-03	.903D-03	.342D-03	.149D-02	.478D-03	.257D-03	.275D-03	.429D-03	.312D-03	.264D-02
SCR11 T	.511D-03	.248D-03	.193D-03	.275D-03	.294D-03	.710D-03	.474D-03	.308D-03	.463D-03	.389D-03
SCR12 T	.517D-03	.622D-03	.434D-03	.296D-02	.476D-03	.117	.620D-03	.730D-03	.284D-02	.513D-03
SCR13 T	.371D-05	.343D-05	.500D-04	.492D-04	.219D-05	.287D-05	.470D-05	.317D-05	.428D-05	.361D-05
IMM 5 T	.356	.486	.431	.613	.312	17.7	.356	.391	.402	.544
IMM 6 T	.422	.431	15.4	15.4	.699	1.38	4.58	1.19	4.22	2.51
IMM17 T	.155D-01	.187D-01	.130D-01	.887D-01	.143D-01	3.51	.186D-01	.219D-01	.852D-01	.154D-01

Table G–2 continued

	MAN31 $	MAN32 $	MAN33 $	MAN34 $	MAN35 $	MAN36 $	MAN37 $	MAN38 $	MAN39 $	MAN40 $
MIN 1 T	.375	.370	.353	.337	.344	.338	.337	.356	.339	.216D-01
MIN 2 T	.139D-01	.165D-01	.133D-01	.717	.166D-01	.165D-01	.227D-01	.125D-01	.107D-01	.124D-01
MIN 3 T	.775	.765	.812	.691	.705	.735	.715	.261	.261	.276
MIN 4 T	.819D-02	.136	.625D-02	.373	.738D-02	.736D-02	.989D-02	.320D-02	.401D-02	.516D-02
MIN 7 T	.506D-02	.504D-02	.542D-02	.494D-02	.518D-02	.518D-02	.467D-02	.802D-02	.683D-02	.223D-01
MIN 8 T	.273D-03	.229D-03	.159D-03	.141D-03	.178D-03	.160D-03	.177D-03	.424D-03	.194D-03	.240D-03
MIN 9 T	.335D-03	.239D-03	.195D-03	.103D-03	.156D-03	.133D-03	.117D-03	.476D-02	.105D-01	.205D-03
MIN10 T	.230D-01	.200D-01	.216D-01	.204D-01	.344D-01	.291D-01	.196D-01	.546	.743D-01	.131
MIN11 T	.281D-03	.180D-03	.193D-03	.147D-03	.206D-03	.168D-03	.216D-03	.155D-02	.101D-01	.486D-03
MIN12 T	.134D-01	.227D-02	.142D-01	.220D-02	.135D-01	.296D-02	.123D-02	.228D-02	.504D-01	.628D-01
MIN13 T	.136D-02	.792D-03	.784D-03	.666D-03	.957D-03	.628D-03	.901D-03	.932D-03	.173D-02	.144D-02
MIN14 T	.217D-02	.181D-01	.188D-02	.522D-01	.173D-01	.171D-01	.174D-02	.618D-03	.101D-02	.958D-03
MIN15 K	.179D-02	.174D-02	.177D-02	.204D-02	.202D-02	.208D-02	.221D-02	.718D-02	.310D-02	.442D-02
MIN16 T	.488	.411	.317	.281	.438	.403	.276	.411	.590	.583
MIN21 T	2.28	2.24	2.17	2.05	2.11	2.08	2.04	3.67	2.19	1.69
MIN23 T	.132D-02	.118D-02	.146D-02	.115D-02	.124D-02	.135D-02	.127D-02	.629D-03	.144D-02	.146D-02
MIN24 T	.679D-03	.584D-03	.617D-03	.580D-03	.813D-03	.634D-03	.595D-03	.837D-03	.798D-03	.102D-02
MIN25 T	.744D-03	.640D-03	.619D-03	.690D-03	.109D-02	.653D-03	.647D-03	.194D-02	.110D-02	.154D-02
MIN26 T	.399D-01	.353D-01	.333D-01	.240D-01	.455D-01	.213	.238D-01	.576D-01	.429D-01	.376D-01
MIN27 T	.383D-02	.328D-02	.347D-02	.323D-02	.460D-02	.357D-02	.333D-02	.469D-02	.446D-02	.561D-02
MIN28 T	.316D-02	.282D-02	.333D-02	.280D-02	.389D-02	.370D-02	.350D-02	.323D-02	.514D-02	.596D-02
CHM 1 T	.506D-02	.414D-02	.435D-02	.447D-02	.431D-02	.453D-02	.455D-02	.703D-02	.641D-02	.924D-02
CHM 2 T	.657D-03	.815D-03	.619D-03	.835D-03	.647D-03	.778D-03	.742D-03	.604D-03	.752D-03	.932D-03
SCR 1 T	.162	.159	.152	.145	.148	.146	.145	.154	.146	.929D-02
SCR 2 T	.341D-02	.340D-02	.366D-02	.333D-02	.349D-02	.349D-02	.315D-02	.541D-02	.460D-02	.151D-01
SCR 3 T	.196D-03	.164D-03	.114D-03	.101D-03	.128D-03	.115D-03	.127D-03	.305D-03	.139D-03	.172D-03
SCR 4 T	.592D-04	.441D-04	.383D-04	.264D-04	.328D-04	.294D-04	.288D-04	.709D-03	.154D-02	.534D-04
SCR 5 T	.236D-02	.186D-03	.221D-02	.666D-04	.219D-02	.101D-03	.718D-04	.530D-02	.833D-02	.108D-03
SCR 6 T	.148	.146	.155	.132	.135	.140	.136	.499D-01	.498D-01	.528D-01
SCR 7 T	.613	.602	.744	.621	.576	.638	.662	.204	.187	.316
SCR 8 K	.485D-02	.422D-02	.456D-02	.4[illegible]0D-02	.726D-02	.614D-02	.414D-02	.115	.157D-01	.277D-01
SCR 9 T	.503D-03	.323D-03	.345D-03	.263D-03	.368D-03	.301D-03	.387D-03	.278D-02	.180D-01	.870D-03
SCR10 T	.482D-03	.800D-02	.367D-03	.219D-01	.434D-03	.432D-03	.581D-03	.188D-03	.236D-03	.303D-03
SCR11 T	.406D-03	.236D-03	.234D-03	.198D-03	.285D-03	.187D-03	.268D-03	.278D-03	.515D-03	.428D-03
SCR12 T	.713D-02	.726D-02	.624D-03	.713D-02	.277D-02	.615D-03	.513D-03	.134D-02	.788D-02	.687D-02
SCR13 T	.429D-05	.529D-05	.450D-05	.556D-05	.404D-05	.524D-05	.502D-05	.390D-05	.490D-05	.590D-05
IMM 5 T	5.09	5.10	5.01	4.80	4.90	4.86	.410	2.99	5.19	3.78
IMM 6 T	2.53	2.49	2.66	2.26	2.29	2.40	2.33	.768	.765	.982
IMM17 T	.214	.218	.187D-01	.214	.833D-01	.185D-01	.154D-01	.403D-01	.237	.206

	MAN41 $	MAN42 $	MAN43 $	MAN44 $	MAN45 $	MAN46 $	MAN47 $	MAN48 $	MAN49 $	MAN50 $
MIN 1 T	.163	.155	.160	.163D-01	.211D-01	.386	.367	.347	.189D-01	.111D-01
MIN 2 T	.429D-01	.423D-01	.477D-01	.449D-01	.128D-01	.494D-01	.519D-01	.423D-01	.104D-01	.557D-02
MIN 3 T	.653	.438	.411	.220	.255	.446	.375	.468	.147	.979D-01
MIN 4 T	.477D-02	.656D-01	.355D-02	.303D-02	.510D-02	.121D-01	.127D-01	.112D-01	.354D-02	.107D-02
MIN 7 T	.199D-02	.219D-01	.263D-01	.238D-01	.217D-01	.356D-02	.479D-02	.311D-02	.158D-02	.943D-03
MIN 8 T	.239D-03	.221D-03	.298D-02	.271D-02	.250D-02	.500D-02	.482D-02	.433D-02	.221D-03	.158D-03
MIN 9 T	.449D-02	.249D-03	.441D-02	.301D-03	.272D-03	.403D-02	.373D-02	.365D-02	.225D-03	.192D-03
MIN10 T	.106	.506D-01	.842	4.47	.743D-01	.329D-01	.220	.491D-01	1.82	.164
MIN11 T	.119D-01	.532D-03	.965D-02	.114D-01	.116D-01	.575D-03	.127D-02	.803D-03	.787D-03	.456D-01
MIN12 T	.354D-01	.244D-02	.120D-02	.141D-02	.126D-02	.136D-01	.119D-01	.139D-01	.218D-02	.952D-03
MIN13 T	.338D-02	.359D-01	.112D-02	.112D-02	.628D-01	.274D-02	.122D-02	.186D-02	.386D-01	.149D-02
MIN14 T	.974D-03	.808D-03	.582D-03	.700D-03	.104D-02	.120D-01	.113D-01	.104D-01	.676D-03	.401D-03
MIN15 K	.511D-02	.354D-02	.471D-01	.451D-01	.255D-02	.350D-02	.565D-02	.307D-02	.471D-01	.432D-02
MIN16 T	1.12	.671	.357	.439	.388	.746	.749	1.02	.464	.436
MIN21 T	2.46	.206	.722	.701	2.16	2.74	2.56	2.44	.167	.112
MIN23 T	.156D-02	.132D-02	.633D-03	.831D-03	.100D-02	.127D-02	.803D-03	.173D-02	.858D-03	.620D-03
MIN24 T	.114D-02	.102D-02	.857D-03	.115D-02	.106D-02	.919D-03	.757D-03	.165D-02	.968D-03	.122D-02
MIN25 T	.256D-02	.845D-02	.279D-02	.751D-02	.116D-02	.469D-02	.916D-03	.245D-02	.262D-02	.258D-02
MIN26 T	.771D-01	.268	.864D-01	.224	.393D-01	.165	.328D-01	.906D-01	.845D-01	.610D-01
MIN27 T	.624D-02	.565D-02	.483D-02	.622D-02	.599D-02	.507D-02	.438D-02	.938D-02	.563D-02	.675D-02
MIN28 T	.547D-02	.563D-02	.408D-02	.507D-02	.665D-02	.402D-02	.285D-02	.503D-02	.567D-02	.461D-02
CHM 1 T	.121D-01	.100D-01	.695D-02	.117D-01	.821D-02	.889D-02	.432D-02	.755D-02	.864D-02	.102D-01
CHM 2 T	.909D-03	.731D-03	.626D-03	.844D-03	.724D-03	.102D-02	.785D-03	.107D-02	.625D-03	.564D-03
SCR 1 T	.703D-01	.666D-01	.691D-01	.704D-02	.908D-02	.166	.158	.150	.813D-02	.477D-02
SCR 2 T	.132D-02	.148D-01	.178D-01	.160D-01	.147D-01	.239D-02	.323D-02	.209D-02	.105D-02	.612D-03
SCR 3 T	.172D-03	.159D-03	.214D-02	.194D-02	.179D-02	.359D-02	.346D-02	.311D-02	.158D-03	.114D-03
SCR 4 T	.680D-03	.562D-04	.658D-03	.753D-04	.522D-04	.608D-03	.552D-03	.547D-03	.471D-04	.512D-04
SCR 5 T	.600D-02	.851D-04	.495D-02	.145D-03	.822D-04	.248D-02	.248D-02	.242D-02	.110D-03	.635D-04
SCR 6 T	.125	.836D-01	.785D-01	.420D-01	.487D-01	.851D-01	.716D-01	.893D-01	.280D-01	.187D-01
SCR 7 T	.280	.209	.198	.189	.230	.253	.318	.269	.148	.108
SCR 8 K	.223D-01	.107D-01	.178	.943	.157D-01	.695D-02	.464D-01	.104D-01	.383	.346D-01
SCR 9 T	.213D-01	.953D-03	.173D-01	.203D-01	.208D-01	.103D-02	.228D-02	.144D-02	.141D-02	.817D-01
SCR10 T	.280D-03	.386D-02	.209D-03	.178D-03	.300D-03	.711D-03	.744D-03	.660D-03	.208D-03	.629D-04
SCR11 T	.100D-02	.107D-01	.334D-03	.333D-03	.187D-01	.817D-03	.365D-03	.554D-03	.115D-01	.444D-03
SCR12 T	.695D-02	.667D-02	.797D-02	.734D-02	.602D-03	.374D-02	.387D-02	.351D-02	.899D-03	.558D-03
SCR13 T	.596D-05	.459D-05	.382D-05	.521D-05	.434D-05	.658D-05	.522D-05	.672D-05	.389D-05	.351D-05
IMM 5 T	11.6	.487	.230	.306	6.97	5.14	4.63	4.65	.338	.214
IMM 6 T	1.16	.920	.783	.809	.881	1.25	.974	1.32	.514	.378
IMM17 T	.209	.200	.239	.220	.181D-01	.112	.116	.105	.270D-01	.167D-01

Table G–2 continued

	MAN51 $	SRV 1 $	SRV 2 $	SRV 3 $	SRV 4 $	SRV 5 $	SRV 6 $	SRV 7 $	SRV 8 $	SRV 9 $
MIN 1 T	.157	.230D-01	.592D-02	.256D-02	.380D-01	.587D-02	.467D-02	.348D-02	.203D-01	.995D-02
MIN 2 T	.721D-02	.697D-02	.172D-02	.780D-03	.150D-01	.186D-02	.132D-02	.111D-02	.339D-02	.290D-02
MIN 3 T	.501	.406D-01	.157D-01	.667D-02	.610D-01	.107D-01	.978D-02	.107D-01	.321D-01	.254D-01
MIN 4 T	.160D-02	.100D-02	.251D-03	.122D-03	.125D-02	.260D-03	.191D-03	.222D-03	.497D-03	.527D-03
MIN 7 T	.262D-02	.338D-03	.628D-03	.789D-04	.257D-03	.106D-03	.165D-03	.908D-04	.300D-03	.341D-03
MIN 8 T	.233D-03	.408D-03	.113D-03	.398D-04	.266D-03	.961D-04	.105D-03	.740D-04	.132D-03	.152D-03
MIN 9 T	.382D-03	.207D-03	.123D-03	.334D-04	.100D-03	.634D-04	.528D-04	.557D-04	.147D-03	.125D-03
MIN10 T	6.08	.175D-01	.298D-01	.106D-01	.111D-01	.925D-02	.184D-01	.535D-02	.776D-01	.438D-01
MIN11 T	.351D-01	.182D-03	.294D-03	.194D-03	.142D-03	.119D-03	.236D-03	.727D-04	.610D-03	.597D-03
MIN12 T	.124D-02	.102D-02	.288D-03	.209D-03	.888D-03	.300D-03	.288D-03	.419D-03	.144D-02	.814D-03
MIN13 T	.163D-02	.552D-03	.211D-03	.122D-03	.616D-03	.166D-03	.173D-03	.285D-03	.498D-03	.412D-03
MIN14 T	.531D-03	.480D-03	.629D-04	.557D-04	.269D-03	.105D-03	.894D-04	.544D-04	.216D-03	.266D-03
MIN15 K	.235D-01	.264D-02	.138D-02	.582D-03	.307D-02	.812D-03	.873D-03	.597D-03	.240D-02	.247D-02
MIN16 T	1.11	.258	.129	.827D-01	.287	.133	.162	.197	.303	.393
MIN21 T	.130	.157	.339D-01	.196D-01	.229	.391D-01	.333D-01	.237D-01	.818D-01	.888D-01
MIN23 T	.110D-02	.157D-03	.509D-04	.427D-04	.182D-03	.490D-04	.602D-04	.496D-04	.198D-03	.193D-03
MIN24 T	.213D-02	.595D-03	.299D-03	.152D-02	.907D-03	.322D-03	.430D-03	.798D-03	.582D-02	.794D-03
MIN25 T	.182D-02	.621D-03	.237D-03	.227D-03	.727D-03	.364D-03	.387D-03	.194D-03	.192D-02	.112D-02
MIN26 T	.375D-01	.175D-01	.791D-02	.778D-02	.150D-01	.919D-02	.760D-02	.974D-02	.688D-01	.169D-01
MIN27 T	.120D-01	.340D-02	.172D-02	.874D-02	.507D-02	.191D-02	.252D-02	.456D-02	.353D-01	.454D-02
MIN28 T	.104D-01	.157D-02	.626D-03	.200D-02	.185D-02	.793D-03	.980D-03	.117D-02	.778D-02	.217D-02
CHM 1 T	.189D-01	.324D-02	.118D-02	.141D-02	.150D-01	.205D-02	.265D-02	.113D-02	.578D-02	.759D-02
CHM 2 T	.11[illegible]D-02	.306D-03	.177D-03	.844D-04	.481D-03	.932D-04	.101D-03	.311D-03	.252D-03	.206D-03
SCR 1 T	.678D-01	.989D-02	.255D-02	.111D-02	.164D-01	.253D-02	.201D-02	.150D-02	.874D-02	.429D-02
SCR 2 T	.174D-02	.220D-03	.422D-03	.493D-04	.157D-03	.681D-04	.106D-03	.582D-04	.188D-03	.212D-03
SCR 3 T	.167D-03	.293D-03	.809D-04	.285D-04	.191D-03	.690D-04	.755D-04	.532D-04	.945D-04	.109D-03
SCR 4 T	.850D-04	.385D-04	.205D-04	.850D-05	.294D-04	.126D-04	.126D-04	.110D-04	.347D-04	.350D-04
SCR 5 T	.990D-04	.146D-03	.136D-03	.281D-04	.101D-03	.441D-04	.421D-04	.648D-04	.227D-03	.805D-04
SCR 6 T	.957D-01	.774D-02	.299D-02	.127D-02	.116D-01	.204D-02	.186D-02	.204D-02	.609D-02	.483D-02
SCR 7 T	.498	.252D-01	.879D-02	.419D-02	.331D-01	.654D-02	.626D-02	.670D-02	.196D-01	.179D-01
SCR 8 K	1.28	.369D-02	.629D-02	.224D-02	.235D-02	.195D-02	.389D-02	.113D-02	.164D-01	.924D-02
SCR 9 T	.629D-01	.326D-03	.526D-03	.348D-03	.254D-03	.214D-03	.422D-03	.130D-03	.109D-02	.107D-02
SCR10 T	.938D-04	.590D-04	.148D-04	.715D-05	.736D-04	.153D-04	.112D-04	.131D-04	.292D-04	.310D-04
SCR11 T	.484D-03	.164D-03	.627D-04	.362D-04	.184D-03	.493D-04	.514D-04	.848D-04	.148D-03	.123D-03
SCR12 T	.564D-03	.325D-03	.236D-03	.792D-04	.287D-03	.112D-03	.125D-03	.894D-04	.115D-02	.303D-03
SCR13 T	.677D-05	.198D-05	.114D-05	.545D-06	.311D-05	.602D-06	.649D-06	.201D-05	.162D-05	.133D-05
IMM 5 T	.281	.332	.488D-01	.434D-01	.534	.868D-01	.727D-01	.584D-01	.309	.178
IMM 6 T	1.78	.131	.431D-01	.283D-01	.178	.373D-01	.396D-01	.379D-01	.128	.117
IMM17 T	.169D-01	.975D-02	.709D-02	.238D-02	.861D-02	.336D-02	.377D-02	.268D-02	.344D-01	.911D-02

Table G–2 continued

	SRV10 $	SRV11 $	SRV12 $	SRV13 $	SRV14 $	SCR27 $
MIN 1 T	.111	.683D-02	.744D-02	.541D-02	.158D-01	.0
MIN 2 T	.150D-01	.211D-02	.210D-02	.161D-02	.506D-02	.0
MIN 3 T	.149	.174D-01	.165D-01	.997D-02	.485D-01	.0
MIN 4 T	.378D-02	.290D-03	.326D-03	.222D-03	.124D-02	.0
MIN 7 T	.148D-02	.154D-03	.158D-03	.209D-02	.374D-03	.0
MIN 8 T	.134D-02	.976D-04	.111D-03	.962D-04	.292D-03	.0
MIN 9 T	.146D-02	.822D-04	.893D-04	.579D-04	.209D-03	.0
MIN10 T	.214D-01	.299D-01	.433D-01	.702D-02	.208D-01	.0
MIN11 T	.636D-03	.650D-03	.436D-03	.185D-02	.256D-03	.0
MIN12 T	.532D-02	.535D-03	.490D-03	.276D-03	.183D-02	.0
MIN13 T	.157D-02	.311D-03	.699D-03	.158D-03	.179D-02	.0
MIN14 T	.296D-02	.146D-03	.112D-03	.106D-03	.288D-03	.0
MIN15 K	.229D-02	.154D-02	.158D-02	.722D-03	.293D-02	.0
MIN16 T	.689	.175	.204	.922D-01	.826	.0
MIN21 T	.778	.501D-01	.414D-01	.377D-01	.111	.0
MIN23 T	.476D-03	.113D-03	.946D-04	.479D-04	.246D-03	.0
MIN24 T	.536D-03	.654D-02	.112D-02	.211D-03	.135D-02	.0
MIN25 T	.267D-02	.597D-03	.962D-03	.273D-03	.101D-02	.0
MIN26 T	.939D-01	.265D-01	.406D-01	.851D-02	.399D-01	.0
MIN27 T	.300D-02	.375D-01	.787D-02	.140D-02	.786D-02	.0
MIN28 T	.212D-02	.813D-02	.189D-02	.546D-03	.260D-02	.0
CHM 1 T	.470D-02	.356D-02	.317D-02	.150D-02	.585D-02	.0
CHM 2 T	.617D-03	.202D-03	.182D-03	.873D-04	.135D-02	.0
SCR 1 T	.480D-01	.294D-02	.320D-02	.233D-02	.681D-02	.0
SCR 2 T	.989D-03	.932D-04	.993D-04	.141D-02	.234D-03	.0
SCR 3 T	.961D-03	.701D-04	.797D-04	.691D-04	.210D-03	.0
SCR 4 T	.223D-03	.220D-04	.196D-04	.116D-04	.470D-04	.0
SCR 5 T	.946D-03	.675D-04	.685D-04	.408D-04	.271D-03	.0
SCR 6 T	.284D-01	.329D-02	.314D-02	.190D-02	.922D-02	.0
SCR 7 T	.893D-01	.104D-01	.109D-01	.594D-02	.304D-01	.0
SCR 8 K	.452D-02	.631D-02	.914D-02	.148D-02	.438D-02	.0
SCR 9 T	.114D-02	.116D-02	.781D-03	.332D-02	.459D-03	.0
SCR10 T	.222D-03	.170D-04	.192D-04	.131D-04	.728D-04	.0
SCR11 T	.467D-03	.927D-04	.208D-03	.471D-04	.532D-03	.0
SCR12 T	.138D-02	.212D-03	.269D-03	.961D-04	.427D-03	.0
SCR13 T	.398D-05	.130D-05	.117D-05	.564D-06	.875D-05	.0
IMM 5 T	1.48	.118	.112	.833D-01	.276	.0
IMM 6 T	.434	.756D-01	.639D-01	.[illegible]48D-01	.180	.0
IMM17 T	.414D-01	.637D-02	.807D-02	.288D-02	.128D-01	.0

Note: Column names are defined in table A–1.

Appendix H Regional and Sectoral Classifications in the World Model

Table H–1
Regional Classification

Code	*Region*	*Country or Territory*
AAF	Africa, arid	1. Chad
		2. Comoro Islands
		3. Egypt
		4. Ethiopia
		5. Djibouti
		6. Israel
		7. Jordan
		8. Lebanon
		9. Mali
		10. Mauritania
		11. Morocco
		12. Niger
		13. Somalia
		14. Sudan
		15. Syrian Arid Republic
		16. Tunisia
		17. Upper Volta
		18. Western Sahara
ASC	Asia, centrally planned	1. China
		2. Democratic Kampuchea
		3. Democratic People's Republic of Korea
		4. Democratic Republic of Viet-Nam
		5. Mongolia
		6. Republic of South Viet-Nam
ASL	Asia, low-income	1. Afghanistan
		2. Bangladesh
		3. British Solomon Islands
		4. Brunei
		5. Bhutan

Table H-1 continued

Code	*Region*	*Country or Territory*
		6. Burma
		7. Fiji Islands
		8. Hong Kong
		9. India
		10. Indonesia
		11. Republic of Korea
		12. Laos
		13. Malaysia
		14. Maldive Islands
		15. Macao
		16. Nepal
		17. New Hebrides
		18. Pacific Territories and islands, not elsewhere classified
		19. Pakistan
		20. Papua New Guinea
		21. Philippines
		22. Sikkim
		23. Singapore
		24. Sri Lanka
		25. Taiwan
		26. Thailand
CAN	Canada	1. Canada
EEM	Eastern Europe	1. Albania
		2. Bulgaria
		3. Czechoslovakia
		4. German Democratic Republic
		5. Hungary
		6. Poland
		7. Romania
JAP	Asia, high-income	1. Japan
		2. Ryukyu Islands
LAL	Latin America, resource-rich	1. Barbados
		2. Bolivia
		3. British Honduras
		4. Colombia
		5. Costa Rica
		6. Dominican Republic
		7. Ecuador

		8. El Salvador
		9. French Guiana
		10. Guadeloupe
		11. Guatemala
		12. Guyana
		13. Haiti
		14. Honduras
		15. Jamaica
		16. Martinique
		17. Nicaragua
		18. Panama
		19. Paraguay
		20. Peru
		21. Surinam
		22. Trinidad and Tobago
		23. Venezuela
LAM	Latin America, medium-income	1. Argentina
		2. Bahamas
		3. Bermuda
		4. Brazil
		5. Chile
		6. Cuba
		7. Mexico
		8. St. Lucia/Grenada/St. Vincent/Dominica/ St. Kitts/Nevis/Anguilla/Netherlands/Antilles/ Turks and Caicos Islands/Montserrat
		9. Uruguay
OCH	Oceania	1. Australia
		2. New Zealand
OIL	Middle East and Africa (oil producers)	1. Algeria
		2. Bahrain
		3. Democratic Yemen
		4. Gabon
		5. Iran
		6. Iraq
		7. Kuwait
		8. Libyan Arab Republic
		9. Muscat/Trucial/Oman
		10. Nigeria
		11. Qatar
		12. Saudi Arabia
		13. United Arab Emirates

Table H-1 continued

Code	*Region*	*Country or Territory*
		14. Yemen
RUH	Soviet Union	1. Union of Soviet Socialist Republics
SAF	Southern Africa	1. South Africa, including Namibia
TAF	Africa, tropical	1. Angola
		2. Benin
		3. Botswana
		4. Burundi
		5. Cameroon
		6. Cape Verdi
		7. Central African Republic
		8. Congo
		9. Equatorial Guinea
		10. Gambia
		11. Ghana
		12. Guinea
		13. Guinea-Bissau
		14. Ivory Coast
		15. Kenya
		16. Lesotho
		17. Liberia
		18. Madagascar
		19. Malawi
		20. Mauritius
		21. Rwanda
		22. Sao Tome and Principe
		24. Senegal
		25. Seychelles Islands
		26. Sierra Leone
		27. Swaziland
		28. Togo
		29. United Republic of Tanzania
		30. Uganda
		31. Zaire
		32. Zambia
		33. Zimbabwe
USA	United States	1. United States of America
WEH	Western Europe, high-income	1. Andora
		2. Austria
		3. Belgium

		4. Denmark, including Greenland
		5. Faeroe Islands
		6. Finland
		7. France
		8. Germany, Federal Republic of
		9. Iceland
		10. Italy
		11. Luxembourg
		12. Netherlands
		13. Norway
		14. Sweden
		15. Switzerland
		16. United Kingdom of Great Britain and Northern Ireland (including the Channel Islands and Isle of Man)
WEM	Western Europe, medium-income	1. Cyprus
		2. Gibraltar
		3. Greece
		4. Malta
		5. Portugal
		6. Spain
		7. Turkey
		8. Yugoslavia

Source: Leontief, Carter and Petri, *The Future of the World Economy*. New York: Oxford University Press, 1977. Reprinted with permission.

Table H-2
Sectoral Classification

Type of variable and variable symbol			*Full name of variable*	*Unit of measurement*
Level of abatement activities				
AB	1 ABATE	23.1	Air pollution control	T(M)
AB	2 ABATE	23.2	Primary water treatment	T(M)
AB	3 ABATE	23.3	Secondary water treatment	T(M)
AB	4 ABATE	23.4	Tertiary water treatment	T(M)
AB	5 ABATE	23.5	Solid waste disposal	T(M)
Abatement equations slacks				
AB	6 ABSLK	93.1	Air pollution control	T(M)
AB	7 ABSLK	93.2	Primary water treatment	T(M)
AB	8 ABSLK	93.3	Secondary water treatment	T(M)
AB	9 ABSLK	93.4	Tertiary water treatment	T(M)
AB	10 ABSLK	93.5	Solid waste disposal	T(M)

Table H-2 continued

Type of variable and variable symbol			*Full name of variable*	*Unit of measurement*
Calibration slack for resources (shortage)				
CA	1 RSSLK	86.1	Copper	T(M)
CA	2 RSSLK	86.2	Bauxite	T(M)
CA	3 RSSLK	86.3	Nickel	T(000s)
CA	4 RSSLK	86.4	Zinc	T(M)
CA	5 RSSLK	86.5	Lead	T(M)
CA	6 RSSLK	86.6	Iron	T(M)
CA	7 RSSLK	86.7	Petroleum	T(M coal equivalent)
CA	8 RSSLK	86.8	Natural gas	T(M coal equivalent)
CA	9 RSSLK	86.9	Coal	T(M coal equivalent)
Historical selected resource output				
CU	1 HRSS	90.1	Copper	T(M)
CU	2 HRSS	90.2	Bauxite	T(M)
CU	3 HRSS	90.3	Nickel	T(000s)
CU	4 HRSS	90.4	Zinc	T(M)
CU	5 HRSS	90.5	Lead	T(M)
CU	6 HRSS	90.6	Iron	T(M)
CU	7 HRSS	90.7	Petroleum	T(M coal equivalent)
CU	8 HRSS	90.8	Natural gas	T(M coal equivalent)
CU	9 HRSS	90.9	Coal	T(M coal equivalent)
Cumulative resource output at end of period				
CU	*10 ECUMR*	*74.1*	*Copper*	*T(M)*
CU	*11 ECUMR*	*74.2*	*Bauxite*	*T(M)*
CU	*12 ECUMR*	*74.3*	*Nickel*	*T(000s)*
CU	*13 ECUMR*	*74.4*	*Zinc*	*T(M)*
CU	*14 ECUMR*	*74.5*	*Lead*	*T(M)*
CU	*15 ECUMR*	*74.6*	*Iron*	*T(M)*
CU	*16 ECUMR*	*74.7*	*Petroleum*	*T(M coal equivalent)*
CU	*17 ECUMR*	*74.8*	*Natural gas*	*T(M coal equivalent)*
CU	*18 ECUMR*	*74.9*	*Coal*	*T(M coal equivalent)*
Cumulative resource output at start of period				
CU	19 SCUMR	75.1	Copper	T(M)
CU	20 SCUMR	75.2	Bauxite	T(M)
CU	21 SCUMR	75.3	Nickel	T(000s)
CU	22 SCUMR	75.4	Zinc	T(M)
CU	23 SCUMR	75.5	Lead	T(M)
CU	24 SCUMR	75.6	Iron	T(M)
CU	25 SCUMR	75.7	Petroleum	T(M coal equivalent)
CU	26 SCUMR	75.8	Natural gas	T(M coal equivalent)

CU	27 SCUMR	75.9	Coal	T(M coal equivalent)
EM	12 EMTOT	13.4	Nitrogen water pollution	T(M)
EM	13 EMTOT	13.5	Phosphates	T(M)
EM	14 EMTOT	13.6	Suspended solids	T(M)
EM	15 EMTOT	13.7	Dissolved solids	T(M)
EM	16 EMTOT	13.8	Solid waste	T(M)
Fishing variables				
FS	1 XFISH	76.1	Fish catch	T(M)
FS	2 NFISH	94.1	Non-human consumption of fish	T(M)
FS	3 MFISH	77.1	Fish imports	T(M)
FS	4 EFISH	78.1	Fish exports	T(M)
Investment and capital variables				
IN	1 IEQP	26.1	Equipment investment	$(B)
IN	2 IPLT	25.1	Plant investment	$(B)
IN	3 INVCH	96.1	Inventory change investment	$(B)
IN	4 IIRR	27.1	Irrigation investment	H(M)
IN	5 ILAND	28.1	Land development	H(M)
IN	6 SEQP	30.1	Equipment capital stock	$(B)
IN	7 SPLT	29.1	Plant capital stock	$(B)
IN	8 SINVY	95.1	Inventory stocks	$(B)
IN	9 SFAS	31.1	Stock of foreign assets	$(B)
IN	10 SLAND	33.1	Cultivated land area	H(M)
IN	11 HEQP	35.1	Historical equipment capital stock	$(B)
IN	12 HPLT	34.1	Historical plant capital stock	$(B)
IN	13 HINVY	97.1	Historical inventory stock	$(B)
IN	14 HFAS	36.1	Historical stock of foreign assets	$(B)
IN	15 HRFAS	91.1	Historical net inward capital flow	$(B)
IN	16 INICF	92.1	1970 net foreign investment income	$(B)
Macro-economic variables				
MA	1 GDP	1.1	Gross domestic product	$(B)
MA	2 CONS	2.1	Consumption level	$(B)
MA	3 DSAVE	8.1	Excess savings potential	$(B)
MA	4 INV	3.1	Investment level	$(B)
MA	5 GOV	4.1	Government expenditures	$(B)
MA	6 BAL	5.1	Balance of payments	$(B)
MA	7 IMPRT	6.1	Total imports	$(B)
MA	8 EXPRT	7.1	Total exports	$(B)
MA	9 POP	8.1	Population	(M)
MA	10 URBAN	9.1	Urban population	(M)
MA	11 LABOR	10.1	Employment	MY(M)
Consumption of food nutrition units				
MA	12 NUTRN	11.1	Calories	(KCAL/DAY)(B)

Table H-2 continued

Type of variable and variable symbol			*Full name of variable*	*Unit of measurement*
MA	13 NUTRN	11.2	Proteins	(GRAMS/DAY)
Imports of selected agriculture				
MM	1 MAGS	39.1	Livestock	T(M)
MM	2 MAGS	39.2	Oil-crops	T(M)
MM	3 MAGS	39.3	Grains	T(M)
MM	4 MAGS	39.4	Roots	T(M)
MM	5 MAGR	40.1	Imports of residual agriculture	$(B)
Imports of selected resources				
MM	6 MRSS	41.1	Copper	T(M)
MM	7 MRSS	41.2	Bauxite	T(M)
MM	8 MRSS	41.3	Nickel	T(000s)
MM	9 MRSS	41.4	Zinc	T(M)
MM	10 MRSS	41.5	Lead	T(M)
MM	11 MRSS	41.6	Iron	T(M)
MM	12 MRSS	41.7	Petroleum	T(M coal equivalent)
MM	13 MRSS	41.8	Natural gas	T(M coal equivalent)
Imports of selected resources				
MM	14 MRSS	41.9	Coal	T(M coal equivalent)
MM	15 MRSR	42.1	Imports of residual resources	$(B)
MM	16 MAGM	43.1	Imports of agricultural margins	$(B)
Imports of resource margins				
MM	17 MRSM	44.1	Petroleum refining	$(B)
MM	18 MRSM	44.2	Primary metal processing	$(B)
Imports of traded goods				
MM	19 MXT	45.1	Textiles, apparel	$(B)
MM	20 MXT	45.2	Wood and cork	$(B)
MM	21 MXT	45.3	Furniture, fixtures	$(B)
MM	22 MXT	45.4	Paper	$(B)
MM	23 MXT	45.5	Printing	$(B)
MM	24 MXT	45.6	Rubber	$(B)
MM	25 MXT	45.7	Industrial chemicals	$(B)
MM	26 MXT	45.8	Fertilizer	T(M)
MM	27 MXT	45.9	Miscellaneous chemical products	$(B)
MM	28 MXT	45.10	Cement	$(B)
MM	29 MXT	45.11	Glass	$(B)
MM	30 MXT	45.12	Motor vehicles	$(B)
MM	31 MXT	45.13	Shipbuilding	$(B)
MM	32 MXT	45.14	Aircraft	$(B)
MM	33 MXT	45.15	Metal products	$(B)

MM	34 MXT	45.16	Machinery	$(B)
MM	35 MXT	45.17	Electric machinery	$(B)
MM	36 MXT	45.18	Professional instruments	$(B)
MM	37 MXT	45.19	Watches, clocks	$(B)
MM	38 MSER	46.1	Imports of service	$(B)
MM	39 MTR	79.1	Imports of transport	$(B)
MM	40 MAID	48.1	Aid (inflow)	$(B)
MM	41 MSLK	49.1	Import slack	$(B)
MM	42 MCAP	47.1	Foreign capital (outflow)	$(B)
Resource import slacks				
MS	1 MRSLK	99.1	Copper	T(M)
MS	2 MRSLK	99.2	Bauxite	T(M)
MS	3 MRSLK	99.3	Nickel	T(000s)
MS	4 MRSLK	99.4	Zinc	T(M)
MS	5 MRSLK	99.5	Lead	T(M)
MS	6 MRSLK	99.6	Iron	T(M)
MS	7 MRSLK	99.7	Petroleum	T(M coal equivalent)
MS	8 MRSLK	99.8	Natural gas	T(M coal equivalent)
MS	9 MRSLK	99.9	Coal	T(M coal equivalent)
Selected agricultural activities				
SE	1 AGS	14.1	Livestock	T(M)
SE	2 AGS	14.2	Oil-crops	T(M)
SE	3 AGS	14.3	Grains	T(M)
SE	4 AGS	14.4	Roots	T(M)
Selected resource activities				
SE	5 RSS	16.1	Copper	T(M)
SE	6 RSS	16.2	Bauxite	T(M)
SE	7 RSS	16.3	Nickel	T(000s)
SE	8 RSS	16.4	Zinc	T(M)
SE	9 RSS	16.5	Lead	T(M)
SE	10 RSS	16.6	Iron	T(M)
SE	11 RSS	16.7	Petroleum	T(M coal equivalent)
SE	12 RSS	16.8	Natural gas	T(M coal equivalent)
SE	13 RSS	16.9	Coal	T(M coal equivalent)
Export pool of selected agriculture				
WW	1 PAGS	63.1	Livestock	T(M)
WW	2 PAGS	63.2	Oil-crops	T(M)
WW	3 PAGS	63.3	Grains	T(M)
WW	4 PAGS	63.4	Roots	T(M)
WW	5 PAGR	64.1	Export pool of residual agriculture	$(B)
Export pool of selected resources				
WW	6 PRSS	65.1	Copper	T(M)

Table H-2 continued

Type of variable and variable symbol			*Full name of variable*	*Unit of measurement*
WW	7 PRSS	65.2	Bauxite	T(M)
WW	8 PRSS	65.3	Nickel	T(000s)
WW	9 PRSS	65.4	Zinc	T(M)
WW	10 PRSS	65.5	Lead	T(M)
WW	11 PRSS	65.6	Iron	T(M)
WW	12 PRSS	65.7	Petroleum	T(M coal equivalent)
WW	13 PRSS	65.8	Natural gas	T(M coal equivalent)
WW	14 PRSS	65.9	Coal	T(M coal equivalent)
WW	15 PRSR	66.1	Export pool of residual resources	$(B)
WW	16 PAGM	67.1	Export pools of agricultural margins	$(B)
Export pool of resource margins				
WW	17 PRSM	68.1	Petroleum refining	$(B)
WW	18 PRSM	68.2	Primary metal processing	$(B)
Export pool of traded goods				
WW	19 PXT	69.1	Textiles, apparel	$(B)
WW	20 PXT	69.2	Wood and cork	$(B)
WW	21 PXT	69.3	Furniture, fixtures	$(B)
WW	22 PXT	69.4	Paper	$(B)
WW	23 PXT	69.5	Printing	$(B)
WW	24 PXT	69.6	Rubber	$(B)
WW	25 PXT	69.7	Industrial chemicals	$(B)
WW	26 PXT	69.8	Fertilizer	T(M)
WW	27 PXT	69.9	Miscellaneous chemical products	$(B)
WW	28 PXT	69.10	Cement	$(B)
WW	29 PXT	69.11	Glass	$(B)
WW	30 PXT	69.12	Motor vehicles	$(B)
WW	31 PXT	69.13	Shipbuilding	$(B)
WW	32 PXT	69.14	Aircraft	$(B)
WW	33 PXT	69.15	Metal products	$(B)
WW	34 PXT	69.16	Machinery	$(B)
WW	35 PXT	69.17	Electric machinery	$(B)
WW	36 PXT	69.18	Professional instruments	$(B)
WW	37 PXT	69.19	Watches, clocks	$(B)
WW	38 PSER	70.1	Export pool of services	$(B)
WW	39 PTR	81.1	Export pool of transport	$(B)
WW	40 PAID	72.1	Pool of aid (inflow)	$(B)
WW	41 PCAP	71.1	Pool of foreign capital (inflow)	$(B)

Output slacks				
XS	1 XSLK	98.1	Primary metal	$(B)
XS	2 XSLK	98.2	Rubber	$(B)
XS	3 XSLK	98.3	Fertilizer	T(M)
XS	4 XSLK	98.4	Cement	$(B)
XS	5 XSLK	98.5	Motor vehicles	$(B)
XS	6 XSLK	98.6	Aircraft	$(B)
XS	7 XSLK	98.7	Machinery	$(B)
XS	8 XSLK	98.8	Electric machinery	$(B)
Output variables				
XX	1 AGR	17.1	Residual agriculture	$(B)
XX	2 RSR	18.1	Residual resource activities	$(B)
XX	3 AGM	19.1	Agricultural margins (food)	$(B)
Resource margins (refining)				
XX	4 RSM	20.1	Petroleum refining	$(B)
XX	5 RSM	20.2	Primary metal processing	$(B)
Output of traded commodities				
XX	6 XT	21.1	Textiles, apparel	$(B)
XX	7 XT	21.2	Wood and cork	$(B)
XX	8 XT	21.3	Furniture, fixtures	$(B)
XX	9 XT	21.4	Paper	$(B)
XX	10 XT	21.5	Printing	$(B)
XX	11 XT	21.6	Rubber	$(B)
XX	12 XT	21.7	Industrial chemicals	$(B)
XX	13 XT	21.8	Fertilizer	T(M)
XX	14 XT	21.9	Miscellaneous chemical products	$(B)
XX	15 XT	21.10	Cement	$(B)
XX	16 XT	21.11	Glass	$(B)
XX	17 XT	21.12	Motor vehicles	$(B)
XX	18 XT	21.13	Shipbuilding	$(B)
XX	19 XT	21.14	Aircraft	$(B)
XX	20 XT	21.15	Metal products	$(B)
XX	21 XT	21.16	Machinery	$(B)
XX	22 XT	21.17	Electric machinery	$(B)
XX	23 XT	21.18	Professional instruments	$(B)
XX	24 XT	21.19	Watches, clocks	$(B)
Output of non-traded commodities				
XX	25 XNT	22.1	Electricity, water	$(B)
XX	26 XNT	22.2	Construction	$(B)
XX	27 XNT	22.3	Trade	$(B)
XX	28 XNT	22.4	Transport	$(B)
XX	29 XNT	22.5	Communication	$(B)

Table H-2 continued

Type of variable and variable symbol			*Full name of variable*	*Unit of measurement*
XX	30 XNT	22.6	Services	$(B)
Dummy and constant variables				
DU	1 GMSUR	82.1	Grain substituted for meat	T(M)
DU	2 GMSUR	61.1	Constant vector	
Dummy variables				
DU	3 DUMA	62.1	Dummy	
DU	4 DUMA	62.2	Dummy	
DU	5 DUMMY	73.1	Dummy	
DU	6 DUMMY	73.2	Dummy	
Exports of selected agriculture				
EE	1 EAGS	50.1	Livestock	T(M)
EE	2 EAGS	50.2	Oil-crops	T(M)
EE	3 EAGS	50.3	Grains	T(M)
EE	4 EAGS	50.4	Roots	T(M)
EE	5 EAGR	51.1	Exports of residual agriculture	$(B)
Exports of selected resources				
EE	6 ERSS	52.1	Copper	T(M)
EE	7 ERSS	52.2	Bauxite	T(M)
EE	8 ERSS	52.3	Nickel	T(000s)
EE	9 ERSS	52.4	Zinc	T(M)
EE	10 ERSS	52.5	Lead	T(M)
EE	11 ERSS	52.6	Iron	T(M)
EE	12 ERSS	52.7	Petroleum	T(M coal equivalent)
EE	13 ERSS	52.8	Natural gas	T(M coal equivalent)
EE	14 ERSS	52.9	Coal	T(M coal equivalent)
EE	15 ERSR	53.1	Exports of residual resources	$(B)
EE	16 EAGM	54.1	Exports of agricultural margins	$(B)
Exports of resource margins				
EE	17 ERSM	55.1	Petroleum refining	$(B)
EE	18 ERSM	55.2	Primary metal processing	$(B)
Exports of traded goods				
EE	19 EXT	56.1	Textiles, apparel	$(B)
EE	20 EXT	56.2	Wood and cork	$(B)
EE	21 EXT	56.3	Furniture, fixtures	$(B)
EE	22 EXT	56.4	Paper	$(B)
EE	23 EXT	56.5	Printing	$(B)
EE	24 EXT	56.6	Rubber	$(B)
EE	25 EXT	56.7	Industrial chemicals	$(B)
EE	26 EXT	56.8	Fertilizer	T(M)

EE	27 EXT	56.9	Miscellaneous chemical products	$(B)
EE	28 EXT	56.10	Cement	$(B)
EE	29 EXT	56.11	Glass	$(B)
EE	30 EXT	56.12	Motor vehicles	$(B)
EE	31 EXT	56.13	Shipbuilding	$(B)
EE	32 EXT	56.14	Aircraft	$(B)
EE	33 EXT	56.15	Metal products	$(B)
EE	34 EXT	56.16	Machinery	$(B)
EE	35 EXT	56.17	Electric machinery	$(B)
EE	36 EXT	56.18	Professional instruments	$(B)
EE	37 EXT	56.19	Watches, clocks	$(B)
EE	38 ESER	57.1	Exports of services	$(B)
EE	39 ETR	80.1	Exports of transport	$(B)
EE	40 EAID	59.1	Aid (inflow)	$(B)
EE	41 ESLK	60.1	Export slack	$(B)
EE	42 ECAP	58.1	Foreign capital (inflow)	$(B)
New emissions of abatable pollutants				
EM	1 EMA	12.1	Pesticides	T(M)
EM	2 EMA	12.2	Particulates	T(M)
EM	3 EMA	12.3	Biological oxygen demand	T(M)
EM	4 EMA	12.4	Nitrogen water pollution	T(M)
EM	5 EMA	12.5	Phosphates	T(M)
EM	6 EMA	12.6	Suspended solids	T(M)
EM	7 EMA	12.7	Dissolved solids	T(M)
EM	8 EMA	12.8	Solid waste	T(M)
Net total emissions				
EM	9 EMTOT	13.1	Pesticides	T(M)
EM	10 EMTOT	13.2	Particulates	T(M)
EM	11 EMTOT	13.3	Biological oxygen demand	T(M)

Source: Leontief, Carter and Petri, *The Future of the World Economy*. New York: Oxford University Press, 1977, pp. 75–82. Reprinted with permission.

References

1. "A World Coal Market for the U.S. to Mine." *Business Week*, November 9, 1981.

2. "BP Statistical Review of the World Oil Industry 1979," British Petroleum Company Ltd., London, 1980.

3. Barnett, H.J., and Morse, C. *Scarcity and Growth: The Economics of Natural Resource Availability*. Baltimore: Johns Hopkins, 1963.

4. Battelle Memorial Institute, Columbus Laboratories. *A Study to Identify Opportunities for Increased Solid Waste Utilization,* vols. 1–7. U.S. Environmental Protection Agency, 1972.

5. Congress of the United States, Office of Technology Assessment. *Technical Options for Conservation of Metals,* September 1979.

6. Council on Environmental Quality and U.S. Department of State. *The Global Report to the President,* vol. 2, 1980.

7. Exxon Background Series. *World Energy Outlook*. Exxon Corporation, December 1980.

8. Fischman, L.L. *World Mineral Trends and U.S. Supply Problems.* Washington, D.C.: Resources for the Future, 1980.

9. Forrester, J.W. *World Dynamics*. Cambridge, Mass.: Wright-Allen, 1971.

10. Friedman, T.L. "Hawking Arms Overseas." *The New York Times,* February 14, 1982, p. 22.

11. Gallagher, J.M., and Bechtel Group, Inc. "Coal: A Global Energy Option for the 1980s." Paper presented at the Third International Conference on Energy Use Management, Global Energy Supply Assessment session, Berlin, October 28, 1981.

12. Gianessi, L.P. "Appendix B1: Estimates of National Water Pollutant Discharges by Polluting Sector: 1972," in H.M. Peskin, *The Distributional Implications of Federal Water Pollution Control Policy*. Washington, D.C.: Resources for the Future, October 1978.

13. Gianessi, L.P., Peskin, N.M., and Wolff, E. "The Distributional Implications of National Air Pollution Damage Estimates," data appendix in F.T. Juster, *Distribution of Economic Wellbeing*. New York: National Bureau of Economic Research, 1977.

14. Goeller, H.E., and Weinberg, A.M. "The Age of Substitutability." *Science,* February 20, 1976. Material reprinted with permission.

15. Haloran, R. "Pentagon Urges 50 Cargo Planes For Rapid Forces." *The New York Times,* January 21, 1982.

16. Herrera, A.O., and others. *Catastrophe or New Society? A Latin American World Model.* Ottawa: International Development Research Center, 1976.

17. Holder, C. "Getting Serious About Strategic Minerals." *Science,* April 17, 1981, vol. 212, pp. 305–307. Material reprinted with permission.

18. Huddle, F.P. "The Evolving National Policy for Materials," in *Materials: Renewable and Nonrenewable Resources,* ed. P. Abelson and A.L. Hammond. Washington, D.C.: American Association for the Advancement of Science, 1976.

19. Landsberg, H.H. "Materials: Some Recent Trends and Issues," in *Materials: Renewable and Nonrenewable Resources,* ed. P. Abelson and A.L. Hammond. Washington, D.C.: American Association for the Advancement of Science, 1976.

20. Landsberg, H.H., Fischman, L.L., and Fisher, J.L. *Resources in America's Future.* Baltimore, Md.: Johns Hopkins, 1963.

21. Leontief, W. *Input-Output Economics.* New York: Oxford University Press, 1966.

22. Leontief, W., Carter, A.P., and Petri, P.A. *The Future of the World Economy.* New York: Oxford University Press, 1977.

23. Leontief, W., and Duchin, F. *Military Spending: Facts and Figures, Worldwide Implications, and Future Outlook.* New York: Oxford University Press, 1983.

24. Leontief, W., in collaboration with F. Duchin and I. Sohn, and with the assistance of G.D. Gorbenko. "Population Growth and the Future of the World Economy," in *Economic and Demographic Change: Issues for the 1980's.* Proceedings of the 1978 Helsinki Conference, International Union for the Scientific Study of Population, Belgium, 1979.

25. Leontief, W., in collaboration with C. Gray and R. Kleinberg. "The Future of World Ports," in *Ports and Harbors,* September 1979. Reprinted as "The Growth of Maritime Traffic and the Future of World Ports," in *International Journal of Transport Economics,* vol. 6, no. 3, Rome, December 1979.

25a. Leontief, W., Koo, J., Nasar, S., and Sohn, I. *The Production and Consumption of Non-fuel Minerals to the Year 2030 Analyzed within an Input-Output Framework of the U.S. and World Economy. (Techniques for Consistent Forecasting of Future Demand for Major Minerals Using an Input-Output Framework).* Prepared by The Institute for Economic Analysis, New York University for the National Science Foundation and the U.S. Bureau of Mines, 122564, Washington, D.C., June 1982, and published by the National Technical Information Service, NTIS No. PB83–122564.

26. Leontief, W., and Sohn, I. "Economic Growth" in *Population and the World Economy in the 21st Century,* edited by Just Faaland, Oxford: Basil Blackwell, 1982.

27. Linneman, H. *MOIRA: A Model of International Relations in Agriculture.* Amsterdam: North Holland, 1980.

28. Malenbaum, W. *World Demand for Raw Materials in 1985 and 2000.* New York: McGraw-Hill, 1978.

29. Meadows, D.H., and others. *The Limits to Growth: A Report for the Club of Rome's Project on the Predicament of Mankind.* New York: Potomac Associates-Universe Books, 1972.

30. Mesarovic, M.D., and Pestel, E. *Mankind at the Turning Point.* New York: Dutton, 1974.

31. *Metal Statistics 1970–1980,* 68th ed., ed. Willy Bauer. Frankfurt: Metallgesellschaft AG., 1981.

32. National Commission on Supplies and Shortages, *Government and the Nation's Resources,* Government Printing Office, 1977.

33. Peccei, A. *The Human Quality.* Oxford: Pergamon, 1977.

34. Penner, P.S. *Direct Energy Transaction Matrix for 1971,* technical memorandum 98. Urbana–Champaign: Center for Advanced Computation, 1978.

35. Postner, H.H. *Canada and the Future of the International Economy: A Global Modeling Analysis.* Economic Council of Canada, March 1979.

36. The President's Materials Policy Commission. *Resources for Freedom,* (Government Printing Office, 1952) vols. 1 to 5.

37. *Reshaping the International Order: A Report to the Club of Rome.* Ian Tinberger, coordinator. New York: Dutton, 1976.

38. Ridker, R.G., and Watson, W.D. *To Choose a Future.* Baltimore: John Hopkins, 1980.

39. Smith, V.K. *Scarcity and Growth Reconsidered.* Washington, D.C.: Resources for the Future, 1979.

40. Statistics Canada. *The Input-Output Structure of the Canadian Economy, 1961–1971,* catalogue 15-506. Ottawa, March 1977.

41. Statistics Canada. *The Input-Output Structure of the Canadian Economy in Constant 1961 Prices, 1961–1971,* catalogue 15-507. Ottawa, December 1977.

42. Tilton, J.E. *The Future of Nonfuel Minerals.* Washington, D.C.: The Brookings Institution, 1977.

43. U.S. Department of Commerce, Bureau of Economic Analysis. *The Detailed Input-Output Structure of the U.S. Economy: 1972,* vol. 2: *Total Requirements for Commodities and Industries,* 1979.

44. U.S. Department of Commerce, Bureau of The Census. *Statistical Abstract of the U.S.,* 100th ed., 1979.

45. U.S. Department of the Interior, Bureau of Mines. *Mineral Facts and Problems,* bulletin 650, 1970.

46. ———. *Mineral Facts and Problems,* bulletin 667, bicentennial ed., 1976.

47. ———. *Mineral Facts and Problems,* bulletin 671, 1980 ed., 1981.

48. ———. *Mineral Facts and Problems,* bulletin 671, 1980 ed., 1981. Reprinted in *Mining Engineering,* September 1980, p. 1344.

49. ———. *Minerals in the U.S. Economy: Ten-Year Supply-Demand Profiles for Mineral and Fuel Commodities (1966–75),* 1977.

50. ———. *Minerals Yearbook,* vol. 1: *Metals, Minerals, and Fuels,* 1972.

51. ———. *Minerals Yearbook,* vol. 1: *Metals, Minerals, and Fuels,* 1973.

52. ———. *Minerals Yearbook,* vol. 1: *Metals, Minerals, and Fuels,* 1974.

53. U.S. Department of Labor, Bureau of Labor Statistics. *Employment Projections for the 1980's,* bulletin 2030, 1979.

54. ———. *The Structure of the United States Economy in 1980 and 1985,* bulletin 1831, 1975.

55. ———. *The U.S. Economy in 1985,* bulletin 1809, 1974.

56. Vogely, W.A. "Non-Fuel Resources," *A Report to Resources for the Future.* Washington, D.C., 1976.

57. The Minerals Availability System Staff. *The Bureau of Mines Availability System,* draft manuscript. U.S. Bureau of Mines, 1982.

Index

About the Authors

Wassily Leontief graduated from the University of Leningrad in 1925 and received the Ph.D. in economics from the University of Berlin in 1928. After his arrival in the United States in 1931 Dr. Leontief joined Harvard University as professor of economics and was later director of the Harvard Economic Research Project. After joining the faculty of New York University in 1975, Dr. Leontief established the Institute for Economic Analysis. He now serves as its director.

In 1973, Dr. Leontief was awarded the Nobel Prize in economic science for the development of the input-output method and for its application to economic problems. The author of seven books and over 190 scholarly articles, Dr. Leontief was president of the American Economic Association in 1970 and is a member of the National Academy of Sciences. He holds honorary degrees from eleven universities in the United States and abroad.

James C.M. Koo received the B.S. in finance from Long Island University and the M.A. in economics from New York University, where he is now a candidate for the Ph.D. Mr. Koo is a member of the research staff of the Institute for Economic Analysis, New York University, where he models fuel and nonfuel minerals and studies the impact of population growth on the future of the world economy. His publications include a recently completed study for the U.S. Bureau of Mines, *Evaluating the Impact of Prospective Changes in Materials Use: An Input-Output Approach* (1983). He has also contributed an article to *The Journal of Resource Management and Technology.*

Sylvia Nasar is a senior economist at Control Data Corporation. She has also been a director of energy programs at the Scientists' Institute for Public Information and, from 1977 to 1981, she was a member of the research staff of the Institute for Economic Analysis. Ms. Nasar received the M.A. in economics from New York University in 1976 and has since contributed to professional journals and meetings on such topics as input-output modeling, metals recycling, and materials productivity.

Ira Sohn received the M.A. in 1972 and the Ph.D in 1976 from New York University. Since then he has carried out research projects at the Institute for Economic Analysis, New York University, in the fields of fuel- and nonfuel-minerals modeling, global economic modeling, and fisheries modeling, in addition to a recently completed study on the Soviet economy. Currently Dr. Sohn is engaged in research on the Italian economy and the global implications of growing protectionism in Western Europe and the United States.

Dr. Sohn has contributed to *The McGraw-Hill Dictionary of Economics* (1981) and *Population and the World Economy in the 21st Century* (1982). His articles have been published in *Materials and Society, The Journal of Policy Modeling* and *The Journal of Resource Management and Technology.*